Process Analytical Technology

Process Analytical Technology

Spectroscopic Tools and Implementation
Strategies for the Chemical and
Pharmaceutical Industries

Edited by

Katherine A. Bakeev

Blackwell
Publishing

© 2005 by Blackwell Publishing Ltd

Editorial offices:
Blackwell Publishing Ltd, 9600 Garsington Road, Oxford OX4 2DQ, UK
Tel: +44 (0) 1865 776868
Blackwell Publishing Professional, 2121 State Avenue, Ames, Iowa, 50014-8300, USA
Tel: +1 515 292 0140
Blackwell Publishing Asia Pty Ltd, 550 Swanston Street, Carlton, Victoria 3053, Australia
Tel: +61 (0)3 8359 1011

First published 2005

2   2006

ISBN-10: 1-4051-2103-3
ISBN-13: 978-1-4051-2103-3

Library of Congress Cataloging-in-Publication Data
Process analytical technology / edited by Katherine A. Bakeev.
    p.   cm.
Includes bibliographical references and index.
ISBN-10: 1-4051-2103-3 (acid-free paper)
ISBN-13: 978-1-4051-2103-3 (acid-free paper)
1. Chemical process control—Industrial applications.   2. Chemistry, Technical.   3.
Chemistry, Analytic—Technological innovations.   4. Chemistry, Analytic—
Technique.   5. Spectrum analysis.   6. Pharmaceutical chemistry.   I. Bakeev,
Katherine A.

TP155.75.P737 2005
660'.2—dc22

2004065962

A catalogue record for this title is available from the British Library

Set in 10 / 12pt Minion
by Integra Software Services Pvt. Ltd, Pondicherry, India
Printed and bound by Replika Press Pvt. Ltd., India

The publisher's policy is to use permanent paper from mills that operate a sustainable forestry
policy, and which has been manufactured from pulp processed using acid-free and elementary
chlorine-free practices. Furthermore, the publisher ensures that the text paper and cover board
used have met acceptable environmental accreditation standards.

For further information on
Blackwell Publishing, visit our website:
www.blackwellpublishing.com

Contents

The Colour Plate Section appears after page 204

Contributors

Mr Manel Alcalá	Department of Chemistry, Faculty of Sciences, Autonomous University of Barcelona, E-08193 Bellaterra, Spain
Dr Katherine A. Bakeev	Glaxo SmithKline, 709 Swedeland Rd, King of Prussia, PA 19406
Dr Ernest Baughman	PO Box 1171, Rancho Cucamonga, CA 91729-1171
Dr Lewis C. Baylor	Equitech International Corporation, 903 Main St. South, New Ellenton, SC 29809
Prof Marcelo Blanco	Department of Chemistry, Faculty of Sciences, Autonomous University of Barcelona, E-08193 Bellaterra, Spain
Dr Ann M. Brearley	2435 Shadyview Lane N, Plymouth MN 55447
Dr John P. Coates	Coates Consulting, 12 N. Branch Rd, Newtown, CT 06470-1858
Dr Steven J. Doherty	Eli Lilly and Company, Lilly Corporate Center, Indianapolis, IN 46285
Dr Robert Guenard	Merck and Co, Inc, PO Box 4, West Point, PA 19486
Dr Nancy L. Jestel	General Electric, Advanced Materials, 1 Noryl Ave, Selkirk, NY 12158
Dr Charles N. Kettler	Eli Lilly and Company, Lilly Corporate Center, Indianapolis, IN 46285
Dr Linda H. Kidder	Spectral Dimensions, 3416 Olandwood Ct, #210, Olney, MD 20832
Dr Eunah Lee	Spectral Dimensions, 3416 Olandwood Ct, #210, Olney, MD 20832
Dr E. Neil Lewis	Spectral Dimensions, 3416 Olandwood Ct, #210, Olney, MD 20832
Dr Charles E. Miller	Dupont Engineering Technologies, 140 Cypress Station Drive, Ste 135, Houston, TX 77090

Dr Patrick E. O'Rourke   Equitech International Corporation, 903 Main St. South, New Ellenton, SC 29809

Dr Joseph W. Schoppelrei   Spectral Dimensions, 3416 Olandwood Ct, #210, Olney, MD 20832

Dr Michael B. Simpson   ABB Analytical and Advanced Solutions, 585 Charest Boulevard East, Ste 300, Quebec GIK 9H4, Canada

Dr Gert Thurau   Merck and Co, Inc, PO Box 4, West Point, PA 19486

Preface

A subject as broad as Process Analytical Technology (PAT) is difficult to capture in one volume. It can be covered from so many different angles, covering engineering, analytical chemistry, chemometrics, and plant operations, that one needs to set a perspective and starting point. This book is presented from the perspective of the spectroscopist who is interested in implementing PAT tools for any number of processes. Not all spectroscopic tools are covered. Included in this book are UV-vis spectroscopy, mid-infrared and near-infrared, Raman, and near-infrared chemical imaging. Hopefully the treatment of these subjects and discussion of applications will provide information useful to newcomers to the field and also expand the knowledge of existing practitioners. The chapter on implementation is written in a general way making it applicable to those who have worked in the field, and to those who are new to it. It discusses some of the issues that need to be addressed when undertaking any type of process analytical project. In the applications of NIR chapters, many diverse applications are presented. Implementation in terms of business value and teamwork is discussed, along with applications in the chemical industry in Chapter 11. Chemometrics is a critical tool in PAT and is discussed thoroughly in Chapter 8, with salient points relevant to process analytical data analysis.

Since the field of Process Analytical Technology is currently a very active field in various industries, the contents of such a book will rapidly become historical, requiring revision to reflect the current state of the work. It is hoped that the discussions presented here will serve as a basis for those seeking to understand the area better, and provide information on many practical aspects of work done thus far.

I would like to thank all those who contributed to this book. Writing a book chapter is an activity that is taken on in addition to one's work, and I appreciate the commitment of all of the authors who took on this task. I would also like to thank my friends who helped me by reading through items for me including John Hellgeth, Bogdan Kurtyka, and Glyn Short.

List of Abbreviations

2D	second derivative
ALS	alternating least squares
ANN	artificial neural network
ANOVA	analysis of variance
AOTF	acousto-optic tunable filter
APC	all possible combinations
APCI	atmospheric pressure chemical ionization
API	active pharmaceutical ingredient
ASTM	American Society for Testing of Materials
ATR	attenuated total reflectance
BEST	Bootstrap Error-adjusted Single-sample Technique
CCD	charge-coupled device
CDER	Center for Drug Evaluation and Research
CENELEC	Comité Européen de Normalisation Electrotechnique
CLS	classical least squares
CMC	chemical manufacturing and controls
CPAC	Center for Process Analytical Chemistry
CPACT	Center for Process Analytics and Control
CVF	circular variable filter
DA	discriminant analysis
DCS	distributed control system
DOE	design of experiments
DS	direct standardization
DSC	differential scanning calorimetry
DTGS	deuterated triglycine sulfate
EMEA	European Medicines Agency
EPA	Environmental Protection Agency
EP-IR	encoded photometric IR
ESI	electrospray ionization
FDA	Food and Drug Administration
FDPM	frequency domain photon migration
FIR	finite impulse response
FM	factory mutual
FOV	field of view
FPA	focal-plane array

FPE	Fabry-Perot étalon
FTIR	Fourier transform infrared
FTNIR	Fourier transform near-infrared
GA	genetic algorithm
GC	gas chromatography
GLP	good laboratory practice
GMP	good manufacturing practice
HCA	hierarchical cluster analysis
HDPE	high-density polyethylene
InGaAs	Indium-Gallium-Arsenide
IQ	installation qualification
ISE	ion-selective electrode
ITTFA	iterative target transformation factor analysis
IVS	interactive variable selection
KNN	K-nearest neighbor
LCTF	liquid crystal tunable filter
LDA	linear discriminant analysis
LDPE	low-density polyethylene
LED	light emitting diode
LEL	lower explosive limit
LIF	light-induced fluorescence
LVF	linear variable filter
LVQ	learning vector quantization
MCT	mercury cadmium telluride
MEMS	micro-electrical mechanical systems
MIR	mid-infrared
MLR	multiple linear regression
MSC	multiplicative scatter correction or multiplicative signal correction
MST	minimal spanning tree
MSZW	metastable zone width
MWS	multivariate wavelength standardization
NDIR	non-dispersive infrared
NIR	near-infrared
NIR-CI	near-infrared chemical imaging
NIST	National Institute of Standards and Technology
NMR	nuclear magnetic resonance
OQ	operational qualification
PAC	process analytical chemistry
PAT	process analytical technology
PbS	lead sulfide
PbSe	lead selenide
PC	principal component or personal computer
PCR	principal component regression
PDA	photodiode array
PDS	piecewise direct standardization

PFM	potential function method
PLS	projection to latent structures or partial least squares
PLSR	partial least squares regression
PQ	performance qualification or personnel qualification
QA	quality assurance
QC	quality control
QFD	quality function deployment
RMSEE	root mean square error of estimate
RMSEP	root mean square error of prediction
ROI	return on investment
SEC	standard error of calibration
SEP	standard error of prediction
SIMCA	soft independent modeling of class analogies
SMCR	self-modeling curve resolution
SMLR	stepwise multiple linear regression
SMV	spectral match value
SMZ	sulfamethoxazole
SNR	signal to noise ratio (also abbreviated S/N)
SOM	self-organizing maps
SPC	statistical process control
SNV	standard normal variate
SVM	support vector machines
SWS	simple wavelength standardization
TE	thermo-electric
UL	Underwriters Laboratory
USDA	United States Department of Agriculture
USP	United States Pharmacopeia
UV-vis	ultraviolet-visible

Chapter 1
Process Analytical Chemistry: Introduction and Historical Perspective

Ernest Baughman

Process analytical chemistry (PAC) is a field that has existed for several decades in many industries. It is now gaining renewed popularity as the pharmaceutical industry begins to embrace it as well. PAC encompasses a combination of analytical chemistry, process engineering, process chemistry, and multivariate data analysis. It is a multidisciplinary field that works best when supported by a cross-functional team including members from manufacturing, analytical chemistry, and plant maintenance.

The use of PAC enables one to gain a deeper understanding of the process. This in turn can lead to more consistent product, reduced waste, improved manufacturing efficiencies, overall improvement in the use of resources, improved safety, and the reduced costs that can be garnered from each of these. In the most rudimentary application of PAC, one can gain a descriptive knowledge of a process. The process signature that is measured permits a determination of trending in process parameters. Measurements can be made to give a direct indication of reaction progress or the composition of a mixture at a given time. This information about the process can be used to make changes to keep a process running within some set limits. With additional process information, and knowledge, there is the possibility of understanding the parameters that impact a process and the product quality. Full process understanding includes the identification of key parameters that can be used to control a process.

In this chapter I mainly discuss my experience over the time period from 1978 to 1994 during which I worked at Amoco Oil and Amoco Chemical companies, now part of British Petroleum (BP). The last 10 years of that period were spent in a group charged with finding, modifying or inventing the on-line analytical tools to fit Amoco's needs. Among the instruments used were on-line gas chromatographs (GCs), filter spectrometers, pH meters, density meters, flash point analyzers, and both infrared and near-infrared (NIR) spectrometers (both filter photometers and full wavelength ones). Later in this chapter, some of the tools that were developed will be discussed.

This chapter deals with two main issues: Why PAC is done and how it is done. The history of PAC is discussed from the perspective of its early origins in petrochemicals. Further, aspects on early instrument development are presented. Why PAC is done is easy to answer: to increase the bottom line via improved production efficiency. The benefits of PAC have already been mentioned – all of which impact a company's bottom line. How

PAC is done is much more complex, and is difficult to generalize. Many of those who work in this field continually discover new ways to glean information that allows for optimization of processing parameters and better control of the process. Better control is the prime goal as it will improve product quality, result in less waste, increase the safety of operations, and thus increase profitability.

This chapter is written in first person because I am trying to communicate to YOU, the person working in this world of PAC who is involved in selecting the projects as well as the people to work on these process analytical projects. Let me introduce myself as my view of PAC is colored by what I have seen and done. My PhD is in chemical kinetics, the measurement of speed of reactions. Since concentration of the reactants plays a role in governing the speed of the reaction, one needs to know accurately what is present and at what amounts. In the laboratory setting, this is reasonably easy to do. In the real world of manufacturing it can be much more involved. Sometimes the label may not match what is in the barrel. In a batch process, one may not know when a process has reached completion. In continuous processes, one never knows exactly what is where in the process without a means to measure it directly. Add in the uncertainty of the composition of something like crude oil and it is amazing that the refinery, with its wide variety of feedstocks, operates at all. (The refinery feed depends on which crude oil can be delivered and at what price, normally the cheaper the better: a parameter which frequently changes.)

1.1   Historical perspective

Process analytical chemistry is a field that has developed over many years. Over the past decade there has been a 5% annual growth in the use of process analytical instruments, and this growth is expected to continue.[1] More recently, the term process analytical technology (PAT) has been used to describe this area that is the application of analytical chemistry and process chemistry with multivariate tools for process understanding. Much of the industrial PAC work is considered proprietary and does not appear in published literature until after it has become standard. Some work is found in the patent literature, especially that dealing with analyzer development. A classic book on process analyzers by Clevett[2] does a good job of comparing laboratory and process analyzers that use the same name and is very useful as a general overview. An article[3] by Callis *et al.* in 1987 defined five different eras of PAC. These eras discuss an evolution from off-line and at-line analysis, to intermittent on-line analysis, continuous on-line analysis, then in-line analysis, and non-invasive analysis. The terminology defining the different eras or applications of PAC is still in use today, though the generic term of on-line analysis is often used without distinction from the specific type of interface used. A review[4] of the applications and instrumentation for PAC with 507 references covers advances until 1993. This was the first in what have come to be biannual reviews on PAC published in Analytical Chemistry. In this series the authors cover the important developments in the area that occurred in the previous 2 years – a demanding task. The current review[5] has 119 references to the literature and covers topics including biosensors, chromatography, optical spectroscopy, mass spectrometry, chemometrics, and flow injection analysis (FIA). To really help a company to be first-rate in PAC, it is best to take a leadership role in developing on-line analyzers that will solve the problems specific to your company. It is important to focus attention on PAC projects that have tangible benefits to the company. Another approach is to work closely with an instrument manufacturer to find optimal measurement solutions.

Process analytical chemistry has been performed in the petrochemical industry for several decades. Some of the earliest univariate tools such as pH meters, oxygen sensors, and flow meters are still in use today. Along with these, there are many more tools available including on-line chromatography, spectroscopy (NIR, mid-infrared, mass spectroscopy, nuclear magnetic resonance spectroscopy, and others), viscosity measurements, and X-ray analysis. An early work[2] on process analyzers has a thorough discussion of the state-of-the-art of process analyzer technology in 1986, and is still a valuable reference for those new to the field. It covers analyzers, as well as sampling systems, data systems, analyzer installation and maintenance, and analyzer calibration – all relevant topics today when undertaking a PAC project.

Many of the process analytical instruments currently in use were developed by the oil and petrochemical industries and then adopted by other industries. The very high throughput in many of these industries requires quick information attainable only by direct on-line measurements. To define 'very high throughput,' let me give an example from one of the refineries where I have had the pleasure of working. This typical refinery makes about one million gallons of gasoline per hour, that is about 15 000 gallons per minute. If something goes wrong and no corrective action is taken for a few hours, thousands of dollars in product can be lost. In the specialty chemical or pharmaceutical industries where raw material costs may be orders of magnitude higher, the cost of lost product is proportionally larger. The typical laboratory response time can be in the range of 4–6 hours. Why so long? The operator must manually extract a sample from the process and take it to the laboratory, which may be far from the actual sampling point. The laboratory will then need to calibrate the instruments because one does not want to give misleading information to the process operators. After instrument calibration or validation, the sample analysis is started, which may require more than an hour to prepare and run for each sample. Under such conditions it is not unusual for the results to be available only 6 hours after the sample was taken. In situations where an unexpected result is reported, the operator may not take the result to be accurate and another process sample is taken to the laboratory for analysis before any action is taken on any process adjustments.

This type of case was observed when we were starting up a new distillate desulfurization unit at one refinery, and a new type of analyzer was installed. The analyzer reported a sharp change in the sulfur level of the product. As this was an unexpected result, even though the laboratory had confirmed it, it required about 15 hours for the operations people to believe the result. Their argument was that 'big units couldn't change that fast.' If they had accepted the analyzer result, several thousand gallons of diesel fuel would not have needed to be reprocessed. The change in sulfur level was later traced to a change in the crude oil feedstock and had been accurately detected by the process analyzer.

So, it takes some time. What does this cost in terms of productivity, utilities or work force? The only way to be sure the petroleum product stays on specification is to make the product that is over specification, so the 'bumps' in the process never put the product under specification. This amounts to product giveaway. To show how expensive this can be, one refinery implemented faster on-line tools and saved about US$14 000 000/year in the first year, new spectroscopic tools were installed, merely by reducing product giveaway. The product must be manufactured to, at least, the minimum specification value. In the oil world no one worries about getting more for their money than was anticipated. In industries where products have defined specification ranges, material that is not manufactured within the range for a premium-grade product may be sold at a lower price as a lower grade. In many cases, materials that are not manufactured to the specification may be harmful to the consumer and must be discarded or, perhaps, can be reprocessed to be within acceptable limits.

How is the incentive calculated for installation of process analytical instrumentation? My approach was to determine the current costs of the operation, then find the costs of the operation when everything was at the optimum and were normally significantly less. I would take about half the difference to determine my potential savings. Calculating the economic benefits is an important step in choosing projects to pursue. In the project planning, it is necessary to do this to secure funding for a project with costs that will include the time for choosing an analyzer, the analyzer, sampling system, calibration development, and installation with post-installation training. The business justification of process analytical projects is critical to the success of such projects and this is discussed further in Chapters 2 and 11 of this book. No operation that I ever had the pleasure of working with was perfect. However, the closer one ran to perfection, the greater the economic payout of the operation. In Clevett's book,[2] he notes, 'Success that can be attributed to the use of process analyzers includes savings in production, product give-away, operating work force, and energy conservation.' Other authors[1] note that in the chemical industry one plant went from 2 to 80 process analyzers and was able to achieve a 35% capacity increase without making any expansions to the plant.

Savings on implementation of real-time analysis can come from the better use of raw material, less energy consumption, higher throughput or any combination of the above. Reduced raw material usage results in reduced waste. A better-controlled process yields more products within the specification limits. In batch operations, one frequently 'holds' the batch in the reactor awaiting laboratory analysis. With on-line equipment, hold time is reduced or eliminated. If operating at capacity, eliminating that hold time can contribute to increased manufacturing capacity, which can have a large economic impact.

1.2 Early instrument development

The differences between laboratory and on-line instruments are huge to the point where often one does not see that they are related. What are some of the commonalities and differences between an on-line analyzer and a laboratory piece of equipment? Usually the name is the same. The laboratory is a temperature-controlled, reasonably safe environment with instrumentation operated by trained chemists. In contrast, the process is frequently outdoors so temperature control is what Mother Nature gives you. Many processes also have the possibility of danger from fire, high temperatures and pressures, and hazardous materials. But the biggest difference relative to the laboratory instrument is that a trained operator runs it. In the process, the instrument must run itself. There is no one to check to see if the proper reagents are present, if the sample is clean, and in the physical state the instrument is expecting. A process operator, who has no understanding of how the analyzer calculates values, takes the readout as correct. Therefore, false information must be identified by the instrument and should not be given to the operator. This requires that some intelligence be built into the instrument. The instrument should be capable of determining when it requires recalibration. If there is something wrong with the sample, the instrument should signal that to the process operator. These considerations can be difficult to build into the analyzer; they require a tremendous amount of process knowledge in addition to the electronics to actually do the job. It is often handled by using software tools such as pattern recognition where data from the analyzer are checked to see if they match what is expected. Again, these may appear to be rigorous requirements. They are necessary if the instrument is expected to operate unattended and control the process. The information provided by the process analyzer must be reliable and available without interruptions.

How are on-line analytical tools developed? The trite answer is in any way possible. In the early days, it was generally the user of the equipment doing the majority of the development, as he understood what his measurement requirements were. He had the incentive of saving money on his process and the process knowledge to do the job. What he frequently lacked were the skills necessary to develop an analyzer. How can this be overcome? The approach taken at Amoco was to have a small group that had or could get the detailed process knowledge and had basic skills in instrumentation development. Process analyzer teams at various large companies have contributed to the early development of process instruments as they worked to optimize their particular operations.[4,6] Now there are more commercial instruments available and instrument companies can provide a turnkey solution that includes the analyzer, software, and sample interface. One must work closely with the system developer, providing a tremendous amount of process information to ensure that the best solution is provided. Confidentiality agreements with suppliers are very important when discussing process details.

The success of user-developed analyzers is seen in the on-line GCs developed by Phillips, which later became the starting point for the company known as Applied Automation (AA) (now part of Siemens), and Union Carbide, which started Greenbrier Instruments that was later taken over by ABB. (Both AA and ABB went far beyond the GCs, that is where both companies started.) The goal of the developers in these process analytical teams was to solve problems for their company. In doing this they provided process analytical solutions for the entire industry. Where is the economic incentive in developing analyzers? Amoco engineers said that, if they were in on the analyzer development, setting specifications and learning to use the analyzer to understand and optimize their process, this would give them at least a year's head start in the process development, using the instrument. With that type of head start, no one could ever catch up to their process gain from using that analyzer. Similarly, DuPont needed spectroscopic process measurements and played a major role in getting dual wavelength spectrometers into field locations, as well as developing on-line mass spectrometry measurements.[6]

The process GCs developed by Phillips and Union Carbide were rugged, stable boxes used in applications where the major requirement for the analyzer was stability. The thinking was that if the analyzer was stable and sensitive to process changes, then the process engineer could use the output to optimize the process. Real-time data from the process GC provided data that would have taken hours to get when using off-line measurements. Having rapid results is a very different goal than that of the laboratory GC where the goal is the best separation or the latest technology. Many times in developing the needed stability into the process GC, additional developments needed to take place. Consideration had to be given to controlling gas flows and temperatures, and to detecting column fouling. Guard columns to prevent fouling had to be developed, as well as a system to allow these to be easily changed. In the process world, an 'easily changed part' does not really exist. Someone has to take down the analyzer, cool it close to room temperature, change the part, regardless of time of the day, and bring the analyzer back to working condition. This may not be easy, even if changing the part requires only a few minutes. Following field repair, the analyzer may not be producing believable data for several hours. Once the process has been 'turned over' to the analyzer, the analyzer has to be reliably operating continuously. Normally one considers that, unless the analyzer is operating at least 95%, some say 98%, of the time, it is useless. The down time includes any calibration or validation that must be done to verify the results of the instrument. These are high standards, but necessary in manufacturing. Such reliable operation has been shown to be possible using on-line spectroscopic analyzers and GCs among others. Two things need to happen for the analyzer to meet the 'up time' requirements: solid hardware and solid software to turn the data into information.

How can high-reliability and high-value analyzers be developed? The first requirement is knowledge of the process and the parameters that need to be measured and controlled. Further, one must have some ideas about tools that are available and knowledge about how they can be applied. One good source of information is the academic–industrial consortia that deal with process analysis. The three that have been the longest in existence are: Center for Process Analytical Chemistry (CPAC), Measurement and Control Engineering Center (MCEC), and the Center for Process Analytics and Control (CPACT). CPAC, headquartered at the University of Washington in Seattle has strengths in chemometrics, chromatography, particularly in extracting 'fast' signals, FIA, and optical spectroscopy. MCEC is headquartered at the University of Tennessee in Knoxville and is strong in Raman spectroscopy and mass spectrometry, and in applying analytical technology to the chemical engineering of controlling the process. MCEC has members from the chemistry and chemical engineering departments of several universities working together. CPACT, founded in 1997, is based at the University of Strathclyde in Scotland and comprises the University of Strathclyde, University of Newcastle, and Hull University with many industrial sponsors. Their research areas include chemometrics, acoustic spectroscopy, NMR, and NIR spectroscopy. The research of CPACT closely follows the needs of its member companies, as well as conducting efforts in process analytics for flexible manufacturing.

While one cannot expect these organizations to give specific answers, they can be very useful in finding the answer you need. If the problem can be exposed to the public, both graduate students and faculty are interested in working on your problem. Proposed solutions may sound very strange and may not have industrial merit; however, these solutions can offer very valuable insights as will be seen in the examples given below. Many times, for confidentiality reasons a problem cannot be openly discussed. This can be solved by temporarily hiring a graduate student to work on the problem; this may also be an excellent way to interview a prospective employee. Another benefit of such consortia is the interaction between academic and industrial members from whom one can gain knowledge based on their experiences in addressing PAC issues.

1.3   Sampling systems

One of the ways that process analyzers achieve the required 'up time' is through the use of sampling systems. These take the sample from the process line, filter it, temperature-control it, pressure-control it, deliver it to the analyzer and return the unused portion to the process stream. In addition, the sampling system should allow for calibration of the analyzer with minimal disturbance of the normal process operation. Many have spent their careers in trying to design the proper sampling system. It depends on the state of the sample when things are going right and when things are going bad. Can the system be shut off to avoid getting any cleaning material into it during 'turnaround' between products? This is one area that requires a tremendous amount of process knowledge, analyzer tolerance, and common sense. It is difficult to know which of these three is most critical as all three are required. Many analyzer projects have failed because not enough effort was put into the sampling system. This topic of sampling systems for process analyzers has been the topic of some entire book chapters and books.[7,8]

Some analyzers do not use a sampling system; the analyzer probe is put directly into the process line. Many spectroscopic analyzers can be operated with a probe directly interfaced with the process (called in-line analysis). Some concerns with taking this approach are calibration, and how to introduce a real standard to the system for instrument verification. These are questions one must be asking.

I once started up a new analyzer and the operators said, 'It worked but it's too slow to be useful.' The analyzer updated itself about 1700 times/minute so the analysis time could not be the problem. It was determined that the sample system had about a 45-minute delay time built into it. Once that was removed the analyzer was 'fast.'

One program that had its origins in CPAC is NeSSI, the New Sampling Sensor Initiative.[9] The basic concept behind this industrial–academic consortium is to reduce the sampling system to a small panel that contains flow controllers, filters, pressure controllers, points to introduce standards, etc. The objective is to reduce cost and time in developing the sampling system by having the basic sampling systems uniform between sites. Having modular systems can reduce down time and allows for interchanging of parts.

Optical fibers are now used extensively to bring the analyzing light to and from the sample for optical spectrometers. What was not known about optical fibers in early implementations was the problems they can bring to spectroscopy. In early applications of fiber optics, they were found to have limited lifetime, cause signal drift and attenuation, and cause process contamination.[4] Fiber optics were developed for the on/off application of data communication and applied in PAC as a means to locate analyzers remotely, and to use a single analyzer with a multiplexer to make measurements at several points in a process. In the spectroscopic world, we are worried about intensity changes (and the intensity of light through most fibers is a function of the temperature), fiber bending, and outside light shining on them. All of these effects can be eliminated through proper fiber design and requires that the user specify the proper fiber. Advances in fiber optics have now made them an integral part of process analysis by UV, NIR, infrared, and Raman spectroscopy. Also, fibers that are resistant to temperature change have lower total throughput than fibers that are not. Further, the light that does get through carries the desired information rather than noise.

1.4   Examples

An example of NIR technology taken from CPAC is discussed here. This is one example in which the university was working on a problem of great interest to the supporters of its organization and came up with a novel solution to the problem. Dr Jim Callis stated in the spring meeting of CPAC in 1986 that the octane number of gasoline could be measured by NIR. There were several oil company representatives at the meeting and the questioning was intense. If these results were real, there was a huge payout to implement the technology. There was some confusion with this as the 'oil people' knew there were three octane numbers: the motor octane number (MON), the research octane number (RON), and the pump octane number (PON) which is the numerical average of the other two. Dr Callis was talking only of a reference octane value and was not specific as to which that was. If his preliminary results were just a statistical 'fluke' because of a limited sample size, no one wanted to waste any time on it. Dr Callis came back to the fall meeting of CPAC and showed how various NIR bands correlated to certain functional groups and could be used to measure the RON.[10] NIR is now also used to predict other quality parameters of gasoline including Reid vapor pressure, API gravity, aromatic, olefinic, and saturate contents.

A project at Amoco was undertaken to verify his NIR results and turn the instrument and measurement into something that could be used on-line. At first, there was total failure at duplicating his results with the precision that was needed to be useful. Then, the samples were split by gasoline grade and the results improved dramatically. When the sample sets were split by both grade and refinery it became clear that a very useful correlation between all three octane numbers and the NIR spectra could be constructed for specific grades and refineries.

While there will be much more extensive discussions in later chapters on the topic of analyzer calibration, I will try to give you a quick snapshot here on how NIR analysis is done. The first step is to have a valid sample set, normally 30–50 samples covering the range of interest. The spectra are collected and a correlation is built between the wavelengths and their corresponding absorbencies and the reference analysis of these samples. The chemometric tools used in developing calibrations are discussed in Chapter 8. This is just the first step in developing a calibration but many people make the mistake of stopping here. One must then take several more known samples and apply the NIR calibration equation just built to see if the correct values are predicted. If the calibration model can correctly predict the values of several unknown (validation) samples, then one can have confidence that the unit is working satisfactorily.

Given the good results from the laboratory NIR work for octane measurements at Amoco, an NIR analyzer was installed in a refinery laboratory. At first the results were satisfactory but then failed after some time. The failure was traced to wear on parts in the instrument. As we had recognized the large potential savings of using NIR in place of the knock engines, we made the effort to make the technology work. Dr George Vickers led the effort to build a diode-array NIR spectrometer for the analysis thus eliminating the moving grating. This proved successful but required that the diode-array chamber be temperature-stabilized to prevent wavelength drift similar to the problem seen with the moving grating instruments. These early instruments were basically modified laboratory instruments, and not designed for the process environment.

Amoco selected Perkin Elmer to be the company that would produce the analyzer that Dr Vickers' team had developed. The first item was to go over the specifications for the analyzer. I had the pleasure of presenting those based on laboratory work done with other NIR analyzers which showed that the natural line-width in liquid gasoline was about 3 nm, dramatically less than the 10 nm optical resolution of the NIR analyzers on the market at that time. When the specifications were presented in relationship to resolution and stability, one of the gentlemen from Perkin Elmer stated, 'it can't be done,' at which time Dr Vickers said, 'come on down to the lab and I'll show you how we did it.' This instrument, with a 3 nm resolution over the spectral range of 800–1100 nm, became known as the PIONIR 1024, where PIONIR stands for Process Instrument On-line NIR and 1024 denotes the number of silicon diodes in the diode-array readout.

For more detailed information on NIR applications, please see Chapter 11 by Dr Ann M. Brearley for applications in general, Chapter 10 by Prof Marcelo Blanco and Manel Alcalá where laboratory applications related to the pharmaceutical industry are discussed, and Chapter 9 by Dr Charles Kettler and Dr Steven Doherty regarding on-line pharmaceutical applications.

Another approach to spectrometer design for on-line applications that has worked well is the use of full spectral range instruments, either Fourier transform NIRs or scanning grating instruments. Process NIR instruments are now built with the stability to survive in the process world. The big advantage of a full spectral range instrument is in having the wavelength range that is not available in filter instruments. With the silicon diode array used in the PIONIR 1024, the longest wavelength is about 1100 nm; with full range instruments the whole NIR and IR regions, from 700 nm to 15 000 nm, are available. Even though this broad range is available, for practical reasons, often only a small section of the range is used at a time. For further information on NIR instrumentation, see Chapter 3 by Dr Michael Simpson.

Today, NIR analyzers are being used in the refineries to determine many different properties in addition to the octane numbers of gasoline. One refinery is measuring 25 properties of their gasoline every 45 seconds. Once the spectrum has been acquired, it just takes some calculations on the attached computer to determine additional analytical properties, such as boiling points, olefin content of the gasoline, aromatic content, methyl tertiary butyl ether (MTBE) content, etc. The NIR is also being used on several other liquid streams within the refinery in addition to just gasoline blending. There are reports of the determination of the British Thermal Units (BTU), a measure of the heating value of natural gas by NIR. NIR is also used for process analysis in the food industry,[11] pulp and paper industries,[12] chemical and pharmaceutical industries as will be discussed in later chapters.

Another laboratory tool that has met great success in process analysis is FIA. The laboratory version has a peristaltic pump that moves reagent and sample through the system. For the process version, this pump had to be eliminated due to high wear and hence a constantly changing system. A common replacement is the use of pressurized reservoirs to hold the reagents and solvent, and use the pressure difference between the reservoir and the detector as the 'pump' to keep liquid flowing. One unrealized benefit of the peristaltic pump was that the pulsing flow of reagents and sample introduced some mixing as the sample moved through the system. Using the reservoir concept, which gives a very steady flow, eliminated this mixing. With process FIA, a different flow pattern is required to achieve this mixing.

Flow injection analysis can be used on-line but care must be taken in developing both the hardware and the application. One example of a system that I developed measured hydrogen sulfide (H_2S) and carbon dioxide (CO_2) in an aqueous amine stream. (The amine system is used in many natural gas plants and refineries to remove H_2S and/or CO_2 from vapor streams. For economic reasons, after the amine stream extracts the H_2S and the CO_2 from the vapor, it is taken to a recovery tower where the pressure is reduced and the temperature raised. This frees the H_2S and CO_2 and the amine goes back to the contactor for another load.) Knowing the concentration of H_2S is important for control of this process. One gas plant in Canada recovers about 6000 tons of sulfur/day from the H_2S and one refinery is recovering about 1500 tons/day from the sulfur in the crude oil. The FIA system works by injecting a small amount of the amine stream into a strong acid bath (6N H_2SO_4). The acid bath is constantly mixed and purged with nitrogen. This nitrogen stream goes past a UV detector and an IR detector for analyses of H_2S and CO_2, respectively.

This analyzer was installed at several plants and produced savings from the reduced energy required to strip the amine stream and from the reduced the amount of good gas lost to the process. At one gas plant, this 'lost gas' was estimated to be worth over US\$1 000 000/year and was an unexpected benefit. The installation driving force was to reduce energy usage in the system; a benefit shown to be about US\$360 000/year.[13] Such large savings are realized due to the large volume of material that flows through this type of plant. The numbers also make another point about the installation of process analyzers. One can calculate the benefit achievable by process analyzers but many times there are unanticipated benefits that outweigh the anticipated ones.

I have discussed some of the history of PAC, and some early approaches to its implementation in the petrochemical industry. So, what does it take to 'Do Process Analysis?' It takes freedom to try new things with an eye on how it can improve process control in order to make the plant more economical. The dollars drive the technology. Someone, preferably a group, needs to have the freedom to fail because not everything is going to work. This freedom means one tries various approaches looking for the best, most viable solution. Sometimes that solution is the one with which the plant is most comfortable. The people in the PAC group must understand the process and know, or have access to some analyzer technology. The latter can be learned at conferences such as IFPAC, International Forum of Process Analytical Chemistry, or the university consortiums, CPAC, MCEC, or CPACT. There is a developing educational program for PAC at the University of Texas at Dallas, which may aid in the development of future practitioners.[1]

Why the emphasis on freedom? An example will help. I was asked to measure the sulfuric acid content in a gasoline component. My first idea was to separate the acid from the stream and measure the amount of acid and relate that back to the original concentration. My separator worked well at high levels but when I got down to the level of interest, below 100 ppm, the separator did not work at all. However, when we got to that level, the solution became turbid. The solution was to install an on-line turbidity detector instead of building a complex separator and analysis device, to measure acid in the 10–100-ppm level. This worked, but certainly was not the first approach.

As you go into PAC, it is a fun and frustrating field. Fun because progress can be made and it makes the company money and that is good for your career. Frustrating because there are many wrong approaches and you will try most of them. The key to success is being willing to try things, fail, and try other things. This book will provide further information on techniques that have been applied to various problems in the chemical and pharmaceutical industries, and steps taken for successful PAC implementations. Get to know people in your field and others working on Process Analysis at IFPAC, CPAC, MCEC, or CPACT. They will not solve your problem but they are a major source of ideas and suggestions that have worked on other problems and that may lead you to the solution of your problem.

From one who has done it, have fun. It can be very rewarding.

References

1. Chauvel, J.P.; Henslee, W.W. & Melton, L.A., Teaching Process Analytical Chemistry; *Anal. Chem.* 2002, 74, 381A–384A.
2. Clevett, K.J., *Process Analyzer Technology*; Wiley-Interscience; New York, 1986.
3. Callis, J.B.; Illman, D.L. & Kowalski, B.R., Process Analytical Chemistry; *Anal. Chem.* 1987, 59, 624A–631A.
4. Beebe, K.R.; Blaser, W.W.; Bredeweg, R.A.; Chauvel, J.P. Jr; Harner, R.S.; LaPack, M.; Leugers, A.; Martin, D.P.; Wright, L.G. & Yalvac, E.D., Process Analytical Chemistry; *Anal. Chem.* 1993, 65, 199R–216R.
5. Workman, J.; Koch, M. & Veltkamp, D.J., Process Analytical Chemistry; *Anal. Chem.* 2003, 75, 2859–2876.
6. Hassell, D.C. & Bowman, E.M., Process Analytical Chemistry for Spectroscopists; *Appl. Spectrosc.* 1998, 52, 18A–29A.
7. Maestro, J., Sampling and Engineering Considerations. In Chalmers, J.M. (ed.); *Spectroscopy in Process Analysis*; Sheffield Academic Press; Sheffield, 2000, pp. 284–335.
8. Sherman, R.E., *Process Analyzer Sample-Conditioning System Technology*; John Wiley & Sons; New York, 2002.
9. http://www.cpac.washington.edu/NeSSI/NeSSI.html.

10. Kelly, J.J.; Barlow, C.H.; Jinguji, T.M. & Callis, J.B., Prediction of Gasoline Octane Numbers from Near-Infrared Spectral Features in the Range 660–1215 nm; *Anal. Chem.* 1989, 61, 313–320.

11. Osborne, B.G.; Fearn, T. & Hindle, P.H., *Practical NIR Spectroscopy with Applications in Food and Beverage Analysis*, 2nd Edition; Longman Scientific and Technical; Essex, 1993.

12. Antti, H.; Sjöström, M. & Wallbäcks, L., Multivariate Calibration Models using NIR Spectroscopy on Pulp and Paper Industrial Applications; *J. Chemom.* 1996, 10, 591–603.

13. Skinner, F.D.; Carlson, R.L.; Ellis, P.F. & Fisher, K.S., Evaluation of the CS-220 Analyzer for the Optimization of Amine Unit Operations; GRI-95/0341, December 1995; Gas Research Institute; Chicago, IL.

Chapter 2

Implementation of Process Analytical Technologies

Robert Guenard and Gert Thurau

2.1 Introduction to implementation of process analytical technologies (PATs) in the industrial setting

The field of process analytics (PA) has grown by leaps and bounds over the last 10 years or so. This growth is poised to increase dramatically with the drive for bringing about productivity improvement across all industries as well as encouragement by the US Food and Drug Administration (US FDA) to implement PAT in the Pharmaceutical Industry through 'GMPs for the 21st Century'[1] and the guidance 'PAT – A Framework for Innovative Pharmaceutical Manufacture and Quality Assurance.'[2] The purpose of this chapter is to provide a generalized work process that will serve as a foundation for the fundamentals of implementing PATs in an industrial setting. A collective body of experience from the process analytics community across industries reveals that there are certain tasks that must be completed in a certain order to ensure the success of the implemented PATs. It is our hope to convey a work process general enough to be tailored to fit diverse industries of various sizes. This chapter will be somewhat unorthodox since it will focus on the approach to implementation of process analyzers rather than discuss any one technology or application. We feel that this topic is under-represented in the literature and would like to add our perspective based upon our own experience as well as that gleaned from the broader PA community. We intend to serve two purposes here:

(1) To provide our experience and knowledge base of implementation of PAT to new and experienced practitioners in the field as well as to observers and stakeholders of PA.
(2) To set the stage for further discussions amongst the PA community related to the very important business issues and opportunities surrounding the use of PATs.

Our premise in this chapter is to give a general description of important considerations and also a general work process for implementing PATs. We recognize that our viewpoint is a contemporary snapshot and highly dependent upon the organization in which the technologies are applied. It is not our contention that this chapter will be all-encompassing or provide a one-size-fits-all template for strategic implementation of PA.

2.1.1 Definition of process analytics

A broadly accepted definition of 'PA' is difficult to capture as the scope of the methodology has increased significantly over the course of its development. What was once a sub-category of analytical chemistry or measurement science has developed into a much broader system for process understanding and control. Historically, a general definition of 'PA' could have been:

> Chemical or physical analysis of materials in the process stream through the use of an in-line or on-line analyzer.

This definition can be described as 'analysis *in* the process' and is closely related to the traditional role of analytical chemistry in process control. The classical scope of a process analytical method is to supplement the control scheme of a chemical, pharmaceutical or agricultural manufacturing process with data from a process analyzer that directly measures chemical or physical attributes of the sample.

More recently however, the definition has broadened to encompass all factors influencing the quality and efficiency of a chemical or pharmaceutical manufacturing process. Driven by developments in operational excellence programs, an extended definition includes such items as:

- Development of the process, namely the identification of critical-to-quality attributes and their relationship to the quality of the product.
- Design of a robust process to control the critical-to-quality attributes.
- Simple sensors and more complex process analyzers.
- A systems approach to use and correlate all significant process information.
- Data mining approaches to detect long-term trends and interactions.
- Potent data management systems to process the large amounts of data generated.

This very wide definition can be summarized as the 'analysis *of* the process' and has most recently been proposed in the pharmaceutical industry[1] to encourage better use of the information content of classical process analytical methods for the improvement of process development and control. Particularly in the pharmaceutical industry, the acronym PAT for 'process analytical technology' is often used to describe this newer definition of PA.

2.1.2 Differences between process analyzers and laboratory analysis

Several attributes distinguish process analytical methods from classic laboratory analysis. Most prominent are those differences that refer to the 'directness' of the analysis in the process versus that of one in a controlled laboratory environment:

- The speed of the analysis, which opens the opportunity for live feedback.
- The elimination of manual sample handling with the inherent gain in safety and, to some extent, the elimination of operator error (and being able to maintain sample integrity).
- The general ability to overcome the traditional separation of analytical chemistry and the manufacturing process for even more complex analytical problems.

Maybe less obvious, but equally significant, is the fact that the integrity of the sample is more likely retained when it is not removed from the process which is to be characterized. In contrast to the laboratory analysis, this approach can offer true 'process' analysis vs. merely sample analysis. This inherent characteristic can be a double-edged sword, however. On one hand, the signal from a well-designed and operated analyzer contains a variety of information that can be effectively used for process understanding and control. On the other hand, the 'sample selection and preparation' is done largely 'by the process,' which can lead to unique challenges for the robust performance of a process analytical method even if the fundamental science for the analysis is straightforward.

Additional advantages of process analyzers when compared to laboratory analyses are the possibility to automate all or most parts of the analysis and thereby to offer the opportunity of a 24/7 unattended operation: a convenience of use and low level of operator training that most laboratory methods do not afford. On the downside, the level of complexity and up-front effort during the design, implementation and even maintenance stages can be high. Also, there can be the need for specialized expertise that can add significant costs to the implementation of process analyzer technology.

2.1.3 General industrial drivers for process analytics

The growing use of more complex PAT (vs. the historically used simple univariate sensors such as pressure, temperature, pH, etc.) within manufacturing industries is driven by the increased capabilities of these systems to provide scientific and engineering controls. Increasingly complex chemical and physical analyses can be performed in, on, or immediately at the process stream. Drivers to implement PA include the opportunity for live feedback and process control, cycle time reduction, laboratory test replacement as well as safety mitigation. All of these drivers can potentially have a very immediate impact on the bottom line, since product quality and yield may be increased and labor cost reduced. If these benefits can be exploited by a process analytical application then the business case is typically readily available.

In addition to these immediate benefits, the element of increased process information with the opportunity for continuous improvement is an important driver. With the availability of process data of higher density, especially compared to off-line laboratory tests, the analysis of the existing process scenario, which is one of the first steps of process improvement, is greatly enhanced. This driver might not appear to be as readily attractive as those with immediate financial benefits such as that gained through yield improvement. However, the ability of a manufacturing organization to continuously improve its processes is not only attractive from a technological point of view, but can also ultimately generate monetary value. This aspect will be discussed in greater detail in Section 2.2.

For regulated industries, an additional driver is the use of process-monitoring data to support changes in the manufacturing process to show the equivalency of an enhanced process to a previously registered version. Regulatory agencies worldwide encourage this use of process analytical information. In particular, the US FDA has promoted the use of PAT through recent guidance and publications.[1,2]

2.1.4   Types of applications (R&D vs. Manufacturing)

There are several types of PA applications that can accompany the lifespan of a product or process. The drivers, challenges and benefits associated with these types of methods can be very different from process research and development through to full-scale production.

Some methods are only applied to the initial phases of the development with the goal of defining and understanding critical process parameters or product quality attributes. In this early phase of process development, a broad range of PA methods can be used to gain process understanding, with no intention of retaining all possible methods during the transfer to full-scale production manufacturing. This phase is often called process optimization – finding the most efficient and robust operating conditions that allow the process to produce quality product. The requirements for hardware and the procedures are typically aligned along those of other R&D technologies. In these types of applications, only a subset of stakeholders are involved in the PA implementation compared with the number involved in implementation for a full-scale production plant. For example, maintenance, automation and control, and quality organizations may not be involved in multi-purpose systems used in the research and development environment. The applicable project phases might be reduced to 'project identification' and 'application development,' with some modified aspects of a short-term 'implementation phase,' but without the need for a routine operations phase. Details of these project stages will be explained in Section 2.2.1.

The next stage of process development is typically the scale-up phase. The goal of PA in this phase is to continue to gain process knowledge and/or to verify scalability of the process by comparing process data obtained during initial development with pilot-plant or full-scale data. Depending on the scale, more stringent requirements for safety, installation and procedures are typical, especially when actual manufacturing equipment is utilized.

Finally, PA methods can be used in commercial manufacturing, either as temporary methods for gaining process information or troubleshooting, or as permanent installations for process monitoring and control. The scope of these applications is often more narrowly defined than those in development scenarios. It will be most relevant for manufacturing operations to maintain process robustness and/or reduce variability. Whereas the scientific scope is typically much more limited in permanent installations in production, the practical implementation aspects are typically much more complex than in an R&D environment. The elements of safety, convenience and reliability,

Table 2.1 Differences in PA applications in R&D and manufacturing

Item	R&D	Manufacturing
Installation type	Temporary, often mobile setup	Permanent, hard-wired and integrated
Flexibility	Very flexible to changing needs, continuous method improvement	Typically less flexible to change as part of integrated system solution
Analytical support	Readily available specialized analytical support	Site analytical support less specialized and with more limited resources
Method validation	Basic method validation (short-term use, little robustness), data will see 'expert eye' prior to release	Extended method validation, robustness and ruggedness tests important for unsupervised operation
Training needs	Expert user, only initial training – vendor introductory training + manuals sufficient	Continued training on different levels (administrator, operator) – dedicated training program
System qualification	Minor qualification needs as long as the method is reliable	Qualification as part of larger system, and with more focus on long-term operation
Maintenance effort	High maintenance of equipment acceptable	Complexity and cost of maintenance is significant factor
Data format	Various (proprietary) formats acceptable, conversion to information manually	Need for standardized data format, validated system for conversion of data
Data volume	Limited amount of data	Large amount of data, archiving and retention issues
Owner of policies	Process analytics developer – R&D policies	Manufacturing policies (including GMP if regulated industry)

validation and maintenance are of equal importance for the success of the application. Some typical attributes of PA applications and how they are applied differently in R&D and manufacturing are listed in Table 2.1.

2.1.5 Organizational considerations

2.1.5.1 Organizational support

One of the most important factors in ensuring the success of process analytical methods is the strategic organizational support that is afforded to design, implement and maintain a system. Since most process analytical methods have significant up-front costs and effort, they require managerial support at least on the site operations level, but ideally through a company-wide strategy.

Most expertise/resources for all phases of process analytical methodology development are typically available in the different functional areas of a manufacturing plant that has some laboratory capacity, but they need to be effectively organized. Thus, there is not necessarily the need for additional personnel to implement PA. However, the coordination of this existing expertise in the different areas, and the approval of the area management to redirect some resources to PA and not merely consider this an 'add-on' project are critical for long-term success. Alternatively, most expertise can be contracted to third-party organizations such as instrument vendors or implementation specialists (though care should be taken in this approach since there will be some loss of competencies related to the individual organization). PA development can be undertaken either on the individual site/production operations level or through the initiative or the support of a central corporate group. Both approaches have their distinct advantages and challenges.

2.1.5.1.1   Site/production operations development

Process analytics development as initiated by the site will typically allow the highest level of interaction with the plant personnel and management and more direct access to process experience and knowledge. The business case can originate from needs identified by site personnel or management, and evolve in a process that is managed with local organizational structures and work processes in mind. Most often the aspects of ownership and roles and responsibilities are handled very effectively on the site/production operations level as long as there is continuity in management support of the approach. The downside of a pure site development of PA methods is the potential initial lack of specialized PA expertise until know-how is obtained by virtue of developing the first system. Also, it is important that site-based projects leverage technology and expertise available in other parts of the organization.

2.1.5.1.2   Central group development

The development of PA applications by central groups and subsequent transfer of ownership to the plant can have certain advantages: specialized know-how and training is immediately available and will allow synergies with other implementations done previously. A dedicated central group with a significant portfolio of successful implementations can afford better leverage of internal and external resources, such as vendor or other third-party support. The development of a PA strategy, previously identified as one of the key elements of the successful use of process analytical methodology in any organization, is often much more effectively done by a group that has some overview and oversight of several different areas and implementations, rather than a site group that works with a more limited scope. An easier alignment of process analytical efforts with central research and development objectives is another advantage of central group development.

The most often cited disadvantages of central group development of PA are associated with the interaction and technology transfer to the site: the identification of a local technical or business justification, the relaying of ownership of the PA system to the site and the long-term support by site operations groups. It is critical to success that the interests of central organizations and sites are aligned to effectively develop and support the process analytical systems, particularly those with relatively high complexity.

2.1.5.2   Necessary roles

Necessary roles in PA implementation can be filled by personnel from different departments, depending on the organizational setup of the company and the complexity of the analyzer installation. A list of required expertise/necessary roles includes:

- *Management* – provides overall project management, business support and expertise as well as strategic oversight.
- *Procurement* – works with vendors to facilitate and coordinate purchases of equipment.
- *Site/plant operations* – contributes technical know-how as well as process/manufacturing logistics. This is typically the recipient and final owner of the technology.
- *Central engineering* – includes project, process, mechanical and civil engineering services.
- *Process analytics* – contributes project management, analytical knowledge, analyzer technology development, analyzer engineering, implementation support and training.
- *Automation and control* – handles efforts in the area of process control and automation of process equipment. This group or person may also include IT expertise to the project.
- *Maintenance* – will be ultimately responsible for care and feeding of process analytical equipment.
- *Vendor* – will provide products and services for PA implementation.
- *Quality* – potential oversight of data and product quality.
- *Regulatory expertise* – necessary for regulatory compliance and strategic partnerships with regulatory bodies.

As has been stated before, not all of these roles need to be occupied based on the scope of the project. Nevertheless, a subset of these skills is required for most implementations, and the coordination of efforts is a key element of success.

2.1.5.3   Profile of a process analytical scientist/engineer

Based upon our collective experience, successful process analytical personnel have some combination of the following competencies:

(1) *Technical*: Good measurement scientist who understands manufacturing processes.
(2) *Interpersonal effectiveness*: Project management, interacts well cross-functionally, team player, creates strong partnerships and can resolve conflicts.
(3) *Initiative*: Fills roles as needed, responsive and not easily discouraged.
(4) *Business focus*: Makes value-based decisions, evaluates on an opportunity basis and not purely technical.
(5) *Innovative*: Looks for new ways to solve problems; monitors pulse of technology.
(6) *Learning*: Continues to learn new skills and grows in expertise throughout the career.
(7) *Overall leadership*: Provides central role in PA implementations; guides and influences to keep often diverse implementation teams motivated and on mission.

Currently, there are few colleges or graduate programs that address at least the specific technical qualifications for process analytical chemists. The University of Texas, Dallas, and the Center for Process Analytical Chemistry (CPAC) at the University of Washington are the two graduate programs in the US with specific training in process analytical chemistry. The above-listed competencies that fall into the 'soft skills' category are, however, just as essential as the technical abilities and most often provide the key to the success of PA in an organization.

2.2   Generalized process analytics work process

There are many ways to execute capital projects, particularly those which ultimately lead to the implementation of PATs. Differences in project execution arise from the type of industry, the corporate organizational structure as well as the technology that is implemented. Furthermore, a distinction can be made between implementations in brand new facilities and those in existing facilities. It must also be defined whether PA is done in R&D vs. a full-scale production setting: rules can be very different in each case. Though no two companies will apply PA exactly alike, there are some commonalities surrounding the stakeholders that need to be involved and the fundamental project stages that must be completed for successful implementation. In this section a generalized work process is presented which illustrates a rudimentary project structure. Figure 2.1 shows a swim lane flowchart giving a high-level overview of a PA project from concept through to its release-to-operations (RTO) and continuous operation. Though not a perfect fit for all organizations and projects, this chart gives a basic road map and was spawned through knowledge and experience across several companies and industries. Sub-steps and roles will change depending on the industry, but the basic structure should be broadly applicable. A fundamental premise of the work process suggested here is that the implementation will be into a production-scale facility for a permanent installation. We chose this type of installation as a basic premise because it typically requires the most comprehensive project management and work process. Other types of implementations will most likely be some subset of this work process.

The basic idea of this flowchart is to graphically depict the order and roles necessary in the various stages of a project. As can be seen in Figure 2.1, the stakeholders (or roles) are listed in the first column of the chart and occupy a unique row or 'swim lane.' Project steps are sequentially orientated from left to right, with the start located at the far left. Those stakeholders participating in a given step are denoted by inclusion in the project step block. If the swim lane is not included in the project block or the block has a dotted line in a lane, this stakeholder does not have primary responsibility in that project step. Flowchart arrow location indicates the primary stakeholder facilitating each step, gives temporal order to project flow, and allows for the use of decision blocks. Swim lane charts are very useful for depicting complex workflows in a logical and concise manner. Laying out a work process this way maximizes project alignment and serves as a blueprint to success. Given this foundation, each of the steps in the generalized work process will be surveyed in the following sections.

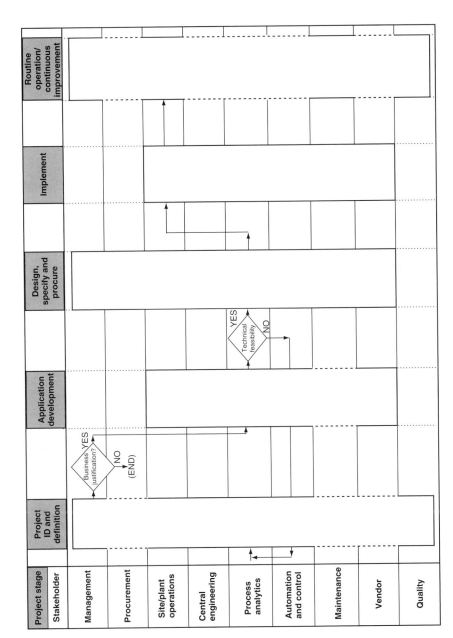

Figure 2.1 Swim lane flowchart of generalized PA work process.

2.2.1 *Project identification and definition*

In the initial stage of a process analyzer project, which we call project identification and definition (see Figure 2.1), an assessment is performed to evaluate the business opportunity for application of PA, as well as to develop a project plan to ensure a successful process analytics implementation. Specifically, identification (ID) refers to the opportunity assessment and definition to codifying the scope and drafting a plan. It is in this project phase that success should be defined and agreed upon by all stakeholders. Ideas for PA projects can spawn from many places and people including process control engineers, site production operations and technical support organizations, business leadership, business system improvement personnel, strategic manufacturing initiatives, process analytical chemists and the results of benchmarking within or across industries, to name a few.

A small but diverse team of personnel is typically involved in the activities of the project identification and definition stage. Representatives from groups such as PA, process control, project management, central engineering and production operations should populate the team in the first stage. The size and diversity of the team should grow from the ID step to the definition step as resources are allocated and plans developed.

Regardless of the source of the project idea, each idea should be passed through a PA opportunity assessment screen of some sort. It has been our experience that a formalized screening process, developed under the premise of consensus, with input from all key stakeholders, is the most effective method of ensuring an implementation that is business-prudent, technically feasible and sustainable. An opportunity screen should asses the value added with the PA implementation, look at alternatives to implementing PA, consider the strategic fit to the production plant and company goals, and assess any associated risks (impact and probability). One of the major contributors to the failure of PA implementations is a missing or poorly executed opportunity assessment. Projects that proceed based upon an attractive technology alone, without establishing criteria for success, will have a severely diminished likelihood of producing business value. In all cases, the business opportunity must be clearly articulated. Though it is preferable to have implementations that add financial value, some might be considered necessary to conduct business from a safety, regulatory or technical basis (cannot operate the plant without it). Therefore, overall, it is recommended that PA projects be treated such that they are screened like every other business proposal.

When calculating the cost-benefit ratio of PA, care should be exercised to include all costs and value added over the lifetime of the implementation. Long-term cost of ownership (LTCO) is often used to describe the total cost of an analyzer over the lifetime of the installation: usually 10–15 years. Line items in this calculation include costs to engineer, implement, run, maintain and disconnect the analyzer. Costs as ostensibly small as utilities are included in LTCO because kilowatt hours and purge gas volume, for example, can become significant given an analyzer is in continuous operation around-the-clock, 365 days a year. Costs are typically much simpler to estimate than the value gained by the implementation, particularly if even only fairly accurate estimates are desired. Cost-accounting techniques such as return on investment (ROI), return on assets

(ROA), index of profitability (IP), initial rate of return (IRR) and net present value (NPV) are often used to estimate the benefit of capital projects including those of PA nature. NPV is represented by the equation

$$NPV = CF \times HR - II \qquad (2.1)$$

where CF is the cash flow from operation, HR is the hurdle rate or required minimum return and II is the initial investment.[3] Advantages of NPV are that it accounts for the time-value of money, provides a value estimate for the lifetime of the project (10 years), and can always be calculated (unlike IRR). HRs are set by central finance groups or, in some cases, the project team. If a positive NPV is obtained at a given HR a project is considered profitable and should be executed.

Project definition can be executed once the PA opportunity has been deemed attractive through cost analysis, risk assessment and strategic fit evaluation. In this step, the majority of the planning is done and considers resources, technology, process control, engineering and project execution. Logical first steps in project definition are chartering the project team and developing an overall project plan that includes roles, detailed tasks and timelines. Various software tools are available to aid in this planning and tracking. There are essentially three major portions of the plan: the analyzer technology itself, the implementation plan and a process control plan (if necessary).

Essential elements of the technology plan are: completing some type of analyzer questionnaire (AQ), selecting an appropriate technology and surveying potential vendors for the technology. Probably the most critical activity to achieving long-term success of a PA implementation from a technical standpoint is defining the application through an AQ. An AQ is a survey that aids in determining the requirements of the application (e.g. desired precision, accuracy, range, components, etc.), asks for the physico-chemical conditions of the process at the sample point and inquires into how the data will be used. In other words, this information defines technical success for the application. Armed with the AQ information, developers of the PA solution will have a target and can optimize the chosen technology and method for the best fit. Process developers/ engineers and control engineers are usually a good source for the information necessary in the AQ. Clevett describes basic information requirements and the need to have the necessary data early in the project.[4] Once the application target is developed, technology selection can take place. Key operating performance and functional attributes should be addressed in choosing the right technology for the application.[5] Simply stated, the technology should comply with safety and regulatory bodies, give the necessary analytical figures of merit for the application, have acceptable LTCO, be reliable and easily maintained and be user-friendly to those using the data generated. If any of these attributes are sub-standard, the likelihood of success for the application is severely diminished. Since there are typically several vendors for a given process analyzer, a selection process must be completed. In the definition step, one has only to list potential vendors, with the formal vendor-selection process to take place in the design, specify and procure stage.

During the design, specify and procure stage, preliminary process control and implementation plans are developed. These plans lay out the basic timing/milestones for each effort, as well as define the roles necessary to complete all the given tasks. It is expected that these plans will change over the course of the project with cross-functional coordination and increase in technical knowledge. Key elements of a process control plan include automation considerations and the potential role of PA in production recipes. Typically, representatives from automation and control, PA and central engineering are key players in the development of a process control plan. An implementation plan lays out the specific tasks and timing for receiving, installing, checking and commissioning of the technology. This plan often includes training schedules as well. All of the stakeholders in the project play some role in the implementation plan because it involves the coordination of so many resources and timing with production.

2.2.2 Analytical application development

Using the technical information obtained in the AQ (from the definition stage), which states the desired analytical figures of merit required from the analyzer and details the physico-chemical properties of the process, technology selection and feasibility studies are done. This stage of the project usually begins as a thought experiment and ends with data showing the technical proof of concept for a PA solution. It does not include all of the details necessary to have a successful implementation; these important details are addressed in the design and specify the stage of the project. Development of the analytical application is primarily the role of the PA scientist, but involves iterative feedback with technical partners in central engineering, plant operations, and automation and control to assure that the application is on course to fulfill the needs of the process. An experienced and efficient practitioner of developing process analytical applications will have a keen awareness of the available technologies and current science, use designed experiments to gather feasibility data, and perform the experiments necessary to fulfill the needs of the analysis. If available, it is useful to develop and evaluate the technology using a pilot or development-scale production facility. In lieu of these, simulated process conditions in the lab may have to suffice. Closure of this step typically comes with a technical report showing the proof of concept with recommendations for the analytical application that fulfills the needs of the analysis.

2.2.3 Design, specify and procure

Properly designing a PA system is crucial to achieving a successful implementation. It is this stage of the project that is arguably the most technically and logistically challenging. Armed with the AQ, business justification and feasibility data, the project team must design, specify and facilitate the purchase and delivery of equipment to the production site. Equally important are the logistics of assuring that lead times are codified and coordinated with site operations milestones, particularly if the process analytical system is being installed in a plant under construction. Key stakeholders in this effort include PA, engineering, operations, and automation and control on the technical side with management, procurement, project engineering and PA groups working on the logistics and resource needs. It is most critical at this stage of the project that all of the team's expertise is heard from and communication between stakeholders is free-flowing. Even the smallest exclusion in this stage of the project can cause large difficulties during implementation and routine operation. A comprehensive description of all the necessary steps to complete the design specifications of a process analyzer system could on its own provide enough material for a book. We have decided here to provide a survey of some key points and considerations in this phase of the project. The basic sub-steps of this stage of the project are designing the process analysis system (drawings included), estimating costs, developing specifications with detailed criteria, selection of suppliers, procurement of equipment, and logistics and resource plans for implementation.

Process analytics systems consist of the following components: the sampling systems, the analyzer(s), automation, and a data management system. These components must be carefully selected and tested during the design phase because each part profoundly affects all others. When designing an analyzer system, the following considerations should be addressed:

- the physical layout (must be designed to fit the physical space)
- utility needs (electricity, gases, etc.)
- equipment identification (tagging)
- environmental influence (not affected by outside environment)
- user interface (easily learned and operated)
- signal input/output (I/O) (compatible with control system, fault detection, etc.)
- diagnostics (automated to communicate problems)
- regulatory code compliance (EPA, FDA, CENELEC, etc.)
- reliability (e.g. long mean time between failures – MTBF)
- safety (to the process and to the operators)
- maintainability (e.g. mean time to repair – MTTR)
- forward/backward compatibility
- long-term cost of ownership (cost over life of installation).

The impact of certain items on this list (reliability, maintainability, compatibility and long-term cost of ownership) will be felt more profoundly after some time of operation but should be addressed in the design phase as well.

A well-designed sampling system: should provide a sample that is representative of the process, does not introduce any significant time delays, maintains the physico-chemical state of the sample, is highly reliable (long MTBF), is easily maintained (short MTTR), is cost-effective and allows safe operation. For general chemical or petrochemical implementations, upwards of 80% of all maintenance problems occur in sampling.[6] A general philosophy of a useful sampling design includes 'the lowest total cost of ownership' over the expected lifetime of the system.[7] Automation and I/O design depend upon the ultimate use of the PA and the process control equipment that will be communicating with the analyzer(s). A complex automation scenario would be an analyzer that has an automated sampling system (complete with automated calibration), is used in closed-loop control of the process and is controlled by the distributed control system (DCS) of the production plant. A simple case would be simplex communication from the analyzer to the DCS used in a process-monitoring mode. A general rule of thumb when designing process analysis systems is to minimize the complexity while fulfilling the needs of the analysis. Complex implementations can lead to reliability and maintenance problems. The foregoing considerations will take the project team a significant amount of collaboration and time to address. Unfortunately, all too often, operations and maintenance groups are left out of design efforts. Because they are the ultimate owners and caretakers of the systems, they have unique perspective on the long-term considerations that central groups may overlook.

Once the process analysis system is designed, the analyzer specification process can take place. Essentially this involves setting the detailed criteria necessary for the design so that vendors can be selected and equipment procured. Specifications for the analyzer system must be set such that the desired performance of the analysis can be met or exceeded. In particular, this requires a strong operational understanding of the analyzer, which should have been acquired during application development, and how it relates to analytical figures of merit required by the analysis. With this information, a detailed specification for the operating performance of the analyzer and sampling system can be written.

With the specifications for the analyzer, sampling system and automation equipment in hand, a supplier selection process can be initiated. Based upon experience, choosing suppliers is often an emotionally and politically charged event amongst project team members. A nice way of avoiding this, as well as making the best business decision, is to set up a data-driven decision matrix that incorporates the needs of the implementation and assesses the ability of the supplier to fulfill them. It is here that procurement groups become highly valuable to the project. They can assist in setting up a decision matrix, sending out RFPs (request for proposals) to potential suppliers and aiding in the selection decision. An RFP from each potential supplier basically provides much of the data needed to make the decision. Once vendor(s) have been selected, quotes can be issued and cost estimates made. These cost estimates and recommendations can then be presented to management for approval. Once approved, procurement can take over to purchase and coordinate delivery of the equipment. An often-requested service of the supplier is a factory acceptance test (FAT). FATs are usually co-designed between the customer and supplier. These tests provide data proving that the supplied equipment can meet the specifications and also serve as a baseline of performance.

2.2.4 *Implementation in production*

Site/plant operations and maintenance groups take center stage in the implementation stage of the project: they will become the eventual owners of the process analysis system. Operations personnel will do most of the work, usually with support from the central PA group and engineering (especially during a new plant startup). A detailed implementation plan should be followed that contains these basic elements chronologically: train personnel, receive and install equipment, safety check, demonstrate operating performance, test automation, document and release to operations. The implementation stage really begins prior to the arrival of any equipment with training of operations and maintenance personnel. If the equipment or technology is new to the site, training is given on the scientific basics of the technique and the details of the hardware. Furthermore, training on the methodology, calibration and routine operation is also recommended.

In collaboration with central groups, and quite often the vendor, operations will install and test the equipment. After receiving the equipment on-site and prior to installation, a thorough visual inspection should take place to look for any obvious damage to the equipment. Timeline and liability problems can often be mitigated through early detection of shipping damage. The actual installation involves physically fixing the equipment in its permanent location, connecting utilities and signal I/O. Sampling systems and interfaces are installed and connected with the analyzer. Once the analyzer is installed, safety and installation inspections are performed, the analyzer is energized, internal diagnostics are performed and FATs are repeated to assure the performance of the instrument remains within specification limits. Sampling systems are inspected, and pressure and flow tested to assure the safety and operation of the system. Loop checks of IO connections are performed to test continuity with the control system. In parallel to the analyzer system installation and individual component testing, process automation is implemented by operations, control engineering and automation personnel.

After the checkout of all of the components of the total analyzer system, the instrument(s) can be calibrated or standardized as necessary. For many types of analyzers (e.g. gas chromatographs, oxygen analyzers, filter photometers) standards are analyzed and calibrations developed on the installed equipment. On some types of sophisticated spectroscopic analyzers, calibration methods can be transferred electronically by standardizing the instrument.[8] With the calibration/standardization complete, commissioning can take place. Commissioning is essentially the performance check of the total process analysis system. In commissioning, process monitoring is simulated with some type of sample while the analyzer and automation are running. This tests the system as a whole and allows any minor glitches to be addressed before actual production runs. As part of this, there may be some method validation work that occurs by using known samples or with grab sample and lab analysis comparisons to process analyzer measurements. This process may be referred to as performance qualification, particularly in the pharmaceutical industry (see Section 2.3). After this work has been successfully completed and documented, the ownership of the analyzer is transferred to operations, also known as RTO. Support from PA does not stop suddenly after RTO; there is usually a transition period when operations gains experience with the system. In addition, often further validation data is needed to assure quality of the measured results after RTO.

2.2.5　*Routine operation*

There are several key factors to the success of the long-term reliability and viability of a PA system. Based on experience, it is necessary to have an on-site owner central to facilitating performance monitoring, maintenance and updating of the system to obtain long-term value out of a PA system. Overall, production operations are responsible for a large number of systems and equipment. By assigning one or two people as owners of the PA system, the issues and intricacies of the implementation will be focused in a small effort. To keep process analysis systems running reliably, the performance should be monitored, calibrations updated or method verified, and preventative maintenance (PM) performed. It is optimal if performance monitoring of both the instrument method and the performance can be automated. Most modern process analyzers have the capability of performing diagnostics on-the-fly and can communicate these results to the DCS. In this manner, a fault detection alarm is triggered when there is an instrument problem, allowing plant operators to address problems as necessary. Example situations that can be automated as fault alarms are fouled probes, clogged filters, burned-out light sources and low purge air pressure to name a few.

Regardless of the type of analyzer system or application, verification or calibration of methods must be performed periodically or be triggered by some type of event. There are two main reasons for a method to need updating: a change in the instrument response or a change in the sample. All instruments have some inherent drift, but the amount of that drift and its effect on prediction are what is really important. Therefore, a periodic check of the analyzer predictions through either check standards or grab sampling and comparison with lab analysis should be done. Furthermore, when adjustments or replacement of analyzer hardware takes place, the method predictions should be verified. Changes in the sample such as raw material supplier, composition, or physical state could affect the calibration method. Thus, anytime there is a change in the process, it must be communicated to the analyzer owner so that the effect of the change on the method can be evaluated. This is usually taken care of procedurally in a change control process (see Section 2.3).

Analyzer systems that contain consumable or limited-lifetime parts must undergo periodic PM to keep performing at a high level. Typical consumable parts include spectrometer light sources, chromatography columns, sampling filters and pump seals to name just a few. Maintenance organizations are commonly charged with keeping maintenance schedules, logging performance, performing the actual maintenance and shelving spare parts. Central groups and production operations often work closely with the maintenance group to establish PM schedules and inventories on spare parts. Both of these will be optimized as experience is gained with a particular analyzer system.

2.2.6 Continuous improvement

With the increasingly competitive environment in the manufacturing sector, organizations are always searching for new ways to reduce costs, and improve efficiency and product quality. To remain competitive, many companies have devised and implemented corporate-based continuous improvement initiatives. In many cases these programs are based upon some variation of Six Sigma (trademark, Motorola, Inc.). Six Sigma was developed by Bill Smith in the 1980s at Motorola and has proliferated throughout the corporate world such that most manufacturing-based organizations utilize it in some fashion. There are many publications on Six Sigma methodology. We have found that books written by Barney and McCarty[9] as well as by Pande et al.[10] are particularly useful. This methodology uses detailed steps to measure the capability of any process (Cpk) and provides a road map and tools to improve that Cpk. PA can play a major role in continuous improvement because it can provide data to improve the fundamental understanding of a process to manufacture materials. It can also be used for process control to decrease process variability and increase efficiency. Continuous improvement can be applied to PAT itself, which should be continually improved to enable further production process improvements. As part of continuous improvement, PA implementations should be periodically evaluated to determine if they are still necessary and adding value to the production process.

For implementations of PA in the pharmaceutical industry, the continuous improvement aspect has received increased attention in recent years. PA or PAT (see previous definition) has been identified as a key factor in continuous improvement initiatives and is seen as a central element of emerging regulatory strategies to enable these efforts. This relationship between PA and continuous improvement has been discussed at several recent technical meetings.[11]

2.3   Differences between implementation in chemical and pharmaceutical industries

2.3.1   Introduction

The purpose of this section is to describe the differences between PA implementation in what are broadly called the chemical and the pharmaceutical industries. In the present context, petrochemicals, polymers, specialty and basic chemicals are examples from the chemical industry. In contrast, this section reflects on PA in the pharmaceutical industry. In this industry, the manufacturing process can be divided into two distinct phases: the chemical synthesis of the active pharmaceutical ingredient (API), and subsequently the formulation of the API into a drug product, often a solid dosage form like a tablet or a capsule. While the first phase, the synthesis of the API, is technically very similar to manufacturing processes in the chemical industries, there are still unique regulatory and economic drivers that set the pharmaceutical industry apart. For the second phase, the formulation of the drug product, the distinction from the chemical industry extends also into the technical sector, with an even larger element of regulatory influence. To some extent the technical similarities to PA in the formulation of the drug product can be found within the food and, to a lesser extent, the agricultural industry. The class of biologicals, including vaccines, comprises another distinct application area within pharmaceuticals that is outside of the scope of this contribution.

The best practices for the successful implementation of PA applications are common to all industries. There are, however, differences in the business model (driver), technical and particularly regulatory areas that are unique to the pharmaceutical industry. The following section seeks to briefly outline these distinctions.

2.3.2   Business model

The business model of the innovative pharmaceutical industry is to transform the results of basic medical research into products that provide health benefits to the patient. This process is characterized by a high risk of failure in the development of new products, coupled with the benefit of market exclusivity for a number of years if a product can be successfully developed and is approved by the regulatory agencies. Only very few chemical entities complete the development process from drug discovery to a commercial product. But those products that do so reach sufficient financial rewards to assure that the overall medical research program is productive. It is fair to say that the manufacturing costs in the pharmaceutical industry account for a much smaller portion of the overall cost during the lifetime of a product, i.e. from discovery to administration to the patient. Of greater influence to the profit over product lifetime is the speed-to-market, by which a product can be moved from discovery through development to a commercialized product. The gains for getting individual products to market fast and the benefits of standardization in technology must be constantly balanced against the potential cost savings that could be achieved through frequent innovation in manufacturing. Delays in gaining manufacturing approval for new drugs are costly because they negatively impact the period of marketing exclusivity. Therefore, new pharmaceutical process applications are usually modeled after existing ones. Not only does a high degree of standardization of the processes lead to the commonly known economic benefits, but the existing processes have previously passed regulatory scrutiny, an important factor in shortening approval times. Since PA is seen not only as a promising tool for manufacturing innovation and continuous improvement on one hand, but also as a development timeline risk and an evolving technology on the other hand, the proper balance between the risks and the benefits is currently being identified in the pharmaceutical industry. Very closely related to the regulatory framework of the industry is also the approach to principles of Six Sigma and 'continuous improvement' of processes after the initial product launch. These principles are widely used in the chemical industry due to their large potential for economic benefits. They have been adopted rather hesitantly in the pharmaceutical industry, partly because of the significant historic regulatory hurdles for implementing change as stated in manufacturing validation guidelines.

2.3.3 Technical differences

For the two pronounced phases of drug manufacturing, the technical differences for the use of PA between the chemical and the pharmaceutical industries are more pronounced in the manufacturing of the formulated drug product than in the chemical synthesis and purification of the API.

2.3.3.1 Active pharmaceutical ingredient (API) manufacturing

The manufacturing of the API is in principle a synthesis of fine chemicals with very high quality requirements for the final product and a tight external regulatory framework. In API manufacturing, the process steps, physical and chemical characteristics of the sample and equipment are very similar to the chemical industries:

- a series of synthetic, separation and purification steps;
- samples are often liquid or slurries;

- solid samples, if present, are processed as batch units;
- common unit operations equipment like reactors, separation columns, large-scale dryers.

Many technical solutions developed for PA in the chemical industry consequently also apply to the pharmaceutical industry. Analyzer systems and implementation principles are essentially the same in both industries. Nevertheless, there are two distinct technical differences even in API manufacturing that make PA in the pharmaceutical industry standout: the absence of continuous manufacturing processes, and the well-established work process for testing including a relative abundance of quality control laboratory resources.

Most API processes are run, in all steps, as batch processes for a number of reasons, among them the fact that the typical development of an API process with small quantities is done as a batch process (development history). Also it is straightforward to define a subset of material as a 'batch' for regulatory and release purposes.

The relatively extensive availability of laboratory resources for quality control is mostly a reaction to the regulatory framework of product release criteria testing. The existence of quality control laboratories in most API manufacturing sites then leads to the development of in-process tests with an off-line laboratory method.

Both of these differences to the chemical industry can present an obstacle to the implementation of PA, which can often show direct economic benefit in a continuous 24/7 operation by minimizing the amount of necessary laboratory quality control resources.

2.3.3.2   Drug product formulation

The techniques used in the manufacturing processes for formulating final drug products are often very different from the chemical industries, with some similarities to food or agricultural manufacturing processes. Although significant research has been invested and progress made in academia and industry to develop a better first-principle-based understanding of the chemical and physical aspects of individual process steps (often referred to as 'unit operations'), there is still an opportunity for improved mechanistic understanding of the influence of process parameters on critical-to-quality attributes in pharmaceutical manufacturing. This can be attributed to many different factors, for example:

- Other than the predominance of small molecules, there are very few chemical commonalities between different APIs within a company's portfolio.
- Even though crystalline forms are preferred for APIs, different API crystalline forms or polymorphs can exhibit vastly different processing performance with the same set of excipients (pharmacologically inert raw materials).
- As solid dosage forms such as tablets or capsules are a desired final product, the reaction systems tend to be blends of solids and are therefore often heterogeneous.

With respect to the use of PA, the heterogeneity of the reaction systems often poses the largest technical challenge. The aspect of sampling, particularly with the elements of a representative sample size, sample presentation to analyzers and probe fouling deserve special attention in method development. Some similarities to the food or agricultural industry are evident in this aspect: both industries often deal with heterogeneous samples, with the added complexity that they often use natural products and not raw materials obtained via a synthetic route or through extensive purification processes, as is the case in the pharmaceutical industry.

Another industry-specific aspect of quality control is both a technical challenge for the implementation of PA in the pharmaceutical manufacturing process, and one of the great opportunities: the typical reliance on finished product testing to assess the quality of the product. This 'quality filter' at the end of the manufacturing process traditionally allows putting less scrutiny on minor variations of the process on the way to the final product. Process analyzers typically show sensitivity to process variability or drift even at magnitudes where the quality of the final product, as tested by regulatory tests, is not affected. On the downside, the process analytical methods have to be made rugged enough to deliver reliable results for decision-making even in the presence of otherwise insignificant process variations. On the upside, this additional information about the variability of the process can be used then for identifying inefficiencies (through unwanted variability) in the manufacturing process. The potential of quantifying and reducing process variability in pharmaceutical manufacturing is just recently being explored as one of the most promising drivers for the use of PA.

A short summary of conceptual technical differences with respect to PA between the chemical and the pharmaceutical industries are listed in Table 2.2.

2.3.4 Regulatory differences

The chemical and pharmaceutical industries are governed by distinctly different regulations. In the US, for example, the Environmental Protection Agency (EPA) regulates business practices in the chemical industry whereas the pharmaceutical industry is primarily regulated by the FDA (in regions outside of the US other environmental and regulator agencies are responsible, such as the 'European Medicines Agency' (EMEA) in Europe or the 'Pharmaceuticals and Medical Devices Agency' in Japan). While the EPA is

Table 2.2 Comparison between chemical and pharmaceutical processes with respect to PA

Industry	Chemical	Pharmaceutical
Process type	Continuous	Batch
Process inputs	Crude feed stocks with variable properties	Pure, high-quality raw materials
Sample type	Homogeneous or slurry	Heterogeneous, often blends of solids
Control	Approaching control limit that gives the highest yield	On target at the center of the specification range
	In-process controls with limited final product testing	Reliance on final product testing (release testing)
Product specifications	Broad range of products tailored for distinct applications (potentially very tight tolerances)	Narrow range of product specifications

primarily concerned with the control of waste and emissions, the FDA is concerned with all aspects of pharmaceutical research and manufacturing.

The current Good Manufacturing Practices (cGMPs) – established in the Code of Federal Regulations (CFR)[12] for the US and in other regulatory documents in other countries – were created to ensure that the drugs produced are 'safe, pure and effective.' While the detailed interpretation of these regulations is the responsibility of individual companies, the overall guidance of the cGMP has ensured the manufacture of high-quality products but has not specifically encouraged innovation in manufacturing. Recent efforts by the US FDA outlined in 'cGMPs for the 21st Century' provide a risk-based approach to regulatory inspections that is designed to remove the barriers to innovation. Further, a new regulatory framework for PAT[2] developed through cooperation with industry and academia has been issued by the FDA's Center for Drug Evaluation and Research (CDER). A draft version of the guidance was published for industry input in August 2003 and the final guidance was issued in September 2004. The document recommends a path forward for the pharmaceutical industry to implement PAT systems to achieve the cost-benefits that have been realized in the chemical industry while maintaining the quality of the product as delivered currently with conventional controls. Other regulatory bodies worldwide such as the EMEA have recently indicated similar interest in PAT but until the time of writing did not issue any written guidance.

Of the many regulatory aspects that influence the extent of the use of PAT in the pharmaceutical industry in comparison to the chemical industry, two items with particular significance should be discussed in detail in this section: the concept of process validation and the need to comply with cGMPs in development manufacturing.

2.3.4.1   Concept of pharmaceutical process validation

The historical way of demonstrating to regulatory agencies that a pharmaceutical manufacturing process can deliver a safe, pure and effective drug product is to 'validate' the process through 'demonstration batches.' The 'validation' of the process assumes that the process is well understood as a result of process development and can be controlled with a limited number of critical controls during commercial manufacturing. In order to assure that the variability can be reduced, is limited to the control of critical variables in routine manufacturing, and delivers the required quality in a reliable and reproducible manner, the fully developed process (at full scale) is tested extensively using a much larger number of physico-chemical tests for a few selected 'demonstration batches.' These are typically performed on three demonstration batches prior to product approval. The design of this extensive 'validation testing' has to be described in a process validation protocol and the results summarized in the process validation report. The time and resource requirements for the preparation and execution of this process validation are significant, particularly given the consequences of a failed process validation which will most likely result in a delay of approval, delay of market launch and therefore loss of time of market exclusivity, previously identified as the main economic benefit of innovative companies in the drug marketplace.

Once the process is validated, only very limited changes can be made to the process without leaving the framework of the 'filed and approved' process. Effectively, this has historically served as an inhibitor to the concept of continuous improvement – instead of using the economic driver of increasing efficiency, the benefits of changes have to be balanced with the extensive cost, and risk, of a subsequent validation and regulatory filing of the changed process. Also, process validation resources originate not only in manufacturing organizations but also in the technology transfer and R&D organizations of the company. The use of these personnel to change processes for existing products is highly unattractive in an industry that thrives on the development, and the speed to market, of new products. With all of this, 'change' has been viewed as a 'bad' thing in the pharmaceutical industry.

Why does all of this matter for PA? While PA can be very beneficial in process development, and even in creating efficiencies in the initially validated versions of the manufacturing processes, the value of PA is often in the identification, and subsequently elimination of inefficiencies in commercial manufacturing (see Section 2.2.6). This consequently amounts to 'change' with the above-mentioned challenges in resource and risk analysis.

More recently, however, the FDA has moved away from the concept of 'process validation' through demonstration batches, and removed the language requiring at least three batches to validate a process.[13] Although the details of the FDA initiative in this area are still to be outlined by the agency, it seems evident that a goal for pharmaceutical companies should be to develop a thorough process understanding in process development and then continue to use this understanding to control the key quality attributes through ongoing process monitoring, including the use of PA. One commonly used phrase is to 'not validate the process but to validate the controls and to control the process.'[14] In this scenario of an 'ongoing process validation' the need for extensive testing of demonstration batches after process changes would be minimized or eliminated without sacrificing the quality and safety of the pharmaceutical product. This approach would be much more compatible for the concept of 'continuous improvement.'

2.3.4.2 Compliance with cGMP in the pharmaceutical industry

As established previously, the cGMPs form the regulatory framework for R&D and manufacturing in the pharmaceutical industry. The specific impact of operating under cGMP rules on process analytical work will be discussed in the following section with special focus on current regulatory guidance and the impact of cGMP on new technology introduction in the field of PA.

2.3.4.2.1 Regulatory guidance on PA

The specifics of regulatory guidance in the pharmaceutical industry are defined in a number of publications such as:

- Country- or region-specific government documents such as the CFR in the US.
- Pharmaceutical compendia issued in countries or regions that detail procedures to test the quality of pharmaceuticals (US Pharmacopia, Pharmeuropa, Japanese Pharmacopia).

- The International Conference on Harmonization (ICH), guidances.
- Other standard organizations – the American Society for Testing of Materials (ASTM) who has been active in the area of PAT since late 2003 in the development of standards for the use of PAT in the pharmaceutical industry internationally in their standard committee E55.

Most guidance documents on testing in the pharmaceutical industry do not presently address the relevant differences of PA systems to conventional laboratory systems. For example, the very narrow definition of required performance characteristics for a laboratory analyzer, combined with the often-mandated use of certified standards for instrument suitability tests in pharmaceutical compendia, can be challenging to the implementation of a PA system using the same technology in the factory. Along the same line, method validation elements that are scientifically sound and essential in the laboratory setting might be less relevant and technically challenging to satisfy in the PA domain.

2.3.4.3 *Innovative technology introduction in a regulated environment*

In order to comply with regulations, pharmaceutical companies need to establish quality systems that define in detail all aspects of the process that are used to manufacture product. The main element of such a quality system is the thorough documentation of the manufacturing process and the related product testing. The documentation requirements are in fact a major part of the work process when developing or implementing PA in the pharmaceutical industry. This section does not intend to describe these requirements in detail or to serve as a manual for proper documentation. Instead, two elements that are unique to the pharmaceutical industry with particular relevance for the introduction of modern analyzer technology, as it is often used in process analytics, will be discussed: system qualification and change control.

2.3.4.3.1 *System qualification*
System qualification is a process that ensures that an analyzer system is installed and operated according to requirements that are aligned with the intended use of the system. The commonly used approach in the pharmaceutical industry is the 'System Life Cycle' or SLC process. In the SLC approach, the definition of intended use, design, configuration, installation and operation is linked and documented over the lifetime of a system.

The SLC document typically consists of technical documents with system descriptions. It also has references to standard operating procedures (SOPs) and administrative and maintenance systems. A completed SLC documentation set is typically required prior to using analyzer systems in the pharmaceutical laboratory or factory. It is also required prior to use of the data obtained by the system in GMP.

Some of the concepts of instrument qualification using the SLC concept can potentially delay the introduction of PA. With evolving technologies as they are used in PA, information for some of the elements of a traditionally designed SLC set is likely to be generated only when the system is put into use. The development of a process analytics method might require some feasibility studies with access to real process equipment, as the requirement specification of the system or method might be related to the process capabilities, which can be quantified only by actually measuring them with the PA system. In contrast, for laboratory analytical systems that use external standards instead of real process samples to verify system suitability, the SLC concept historically mandates that these specifications are set prior to the start of experiments. It will be important for the application of SLC requirements for PA analyzers to not put the 'cart before the horse' and instead to conduct a thorough quality and business risk analysis for the intended use of the instrument rather than prescriptively following the historic practices set forth in the application of the SLC concept to laboratory analyzer systems.[15,16]

2.3.4.3.2 Change control

Change control is an important principle in both the pharmaceutical and the chemical industries. There are several types of change control: for example change control of systems (process analyzer), change control of analytical methods or change control for processes. Change control takes on a regulatory meaning in the pharmaceutical industry. Common to these different types of change control is the need to monitor the status of the system/method/process with respect to the initial qualification/validation and to put measures in place to verify that the changes do not affect the GMP status of the system and the validity of the assumptions under which the system/method/process was operated prior to the change. One rationale behind change control has obvious practical benefits for system integrity as it allows the operation of systems independent of a particular owner, and minimizes the risk of unwarranted modifications to the system by non-expert or unauthorized parties. In the field of PA, where the performance of systems/methods/process can be influenced by a large number of factors, the formal documentation helps to identify and manage the risk involved in a modification. Typical quality systems in pharmaceutical companies handle modification through equipment, analytical method and process change requests. However, similar to the concept of process validation discussed previously, it can also have a negative impact on the use of continuous improvement principles and puts a high resource cost even on necessary changes to the system. One practical example of the consequences of change control is that in the regime of system/instrument change control, hardware, software and operational aspects such as routine operation, maintenance and data integrity are locked down in fixed configurations. Any modification that has not been anticipated in the initial qualification requires testing and documentation.

Analytical change control is the monitoring of any changes to analytical methodology, which has to be justified and evaluated for its performance and impact on the quality of the product. For process analytical methods, the change control approach requires significant documentation and can impede the efficient optimization of the method in the early phases of implementation if parameters have to be locked down based on an insufficient dataset.

The goal of this section is to outline current differences between the implementation of PA in the chemical and the pharmaceutical industries, specifically in the areas of business model, technology and regulations.[1,16] The reader of this chapter should be aware that this status quo is changing rapidly in the pharmaceutical industry due to a changing business climate, combined with a more technology-aware regulatory environment in the major pharmaceutical markets. The authors of this chapter would expect that future editions of this book would see major revisions to this comparison of the industries.

2.4 Conclusions

The intent of the chapter is to present a generalized work process for implementing PATs in the industrial setting. Based upon a work flow captured in a swim lane chart, each of the major stages in a PA project have been described, with some detail as to the sub-steps, necessary organization input and expertise, as well as an introduction to some of the project tools and implementation considerations. As stated previously, it is not claimed that this is a comprehensive, one-size-fits-all approach, but merely a thumbnail sketch of what must occur to successfully implement PA. In describing the work process, we spoke of the differences and considerations when implementing in the chemical industry and the pharmaceutical industry, where regulations in manufacturing are significant. We hope that this presentation of a work process will open discussion and spawn further work to improve how companies can optimize the benefit gained through PA.

References

1. US FDA http://www.fda.gov/oc/guidance/gmp.html.
2. US FDA http://www.fda.gov/cder/guidance/6419fnl.pdf.
3. Polimeni, R.S.; Handy, S.A. & Cashin, J.A., *Cost Accounting*, 3rd Edition; McGraw-Hill; New York, 1994, pp. 158–160.
4. Clevett, K.J., *Process Analyzer Technology*; Wiley-Interscience; New York, 1986, pp. 839–842.
5. Tate, J.D.; Chauvel, P.; Guenard, R.D. & Harner, R., Process Monitoring by Mid- and Near-Infrared Fourier Transform Spectroscopy. In Chalmers, J.M. and Griffiths, P.R. (eds); *Handbook of Vibrational Spectroscopy*; Volume 4, Wiley; New York, 2002, pp. 2738–2750.
6. Clevett, K.J., *J. Process Contr. Qual.* 1994, 6, 81–90.
7. Crandall, J., *NIR News* 1994, 5(5), 6–8.
8. Feudale, R.N.; Woody, N.A.; Tan, H.; Miles, A.J.; Brown, S.D. & Ferre, J., *Chemom. Intell. Lab. Syst.* 2002, 64, 181–192.
9. Barney, M. & McCarty, T., *The New Six Sigma: A Leader's Guide to Achieving Rapid Business Improvement and Sustainable Results*; Prentice-Hall, PTR; Indianapolis, IN, 2003.
10. Pande, P.S.; Cavanagh, R.R. & Newman, R.P., *The Six Sigma Way: How GE, Motorola, and Other Top Companies Are Honing Their Performance*; McGraw-Hill; New York, NY, 2000.
11. See for example proceedings of International Forum for Process Analytical Chemistry (IFPAC), January 2004, Arlington, VA; AAPS 39th annual Arden House Conference US, Harriman, NY, January 2004; 9th Arden House European Conference, London, March 2004.
12. Code of Federal Regulations 21 CFR part 210 'Current good manufacturing practice in manufacturing, processing, packing, or holding of drugs; general' and part 211 'Current good manufacturing practice for finished pharmaceuticals', Food and Drug Administration, Department of Health and Human Services.

13. See announcement on FDA website of 17 March 2004 at: http://www.fda.gov/cder/gmp/processvalidation.htm.

14. Presentation of Ajaz Hussein, PAT team of CDER, FDA at International Forum for Process Analytical Chemistry (IFPAC) in Arlington, VA, January 2004.

15. Schadt, R., Process Analytical Technology: Changing the Validation Paradigm; *Pharmaceutical Review* 2004, 7(1), 58–61.

16. Watts, D.C.; Afnan, A.M. & Hussain, A.S., Process Analytical Technology and ASTM Committee E55; *ASTM Standardization News* 2004, 32(5), 25–27.

Chapter 3

Near-Infrared Spectroscopy for Process Analytical Chemistry: Theory, Technology and Implementation

Michael B. Simpson

3.1 Introduction

Process analytical chemistry (PAC) applications of near-infrared spectroscopy (NIRS) range in difficulty from the tolerably simple (e.g. moisture content) to the hugely ambitious (e.g. full characterisation of hydrocarbon composition and properties in refinery process streams). The various demands of these applications, in order to yield useful process control information, may be quite different. The key criteria may be in some cases speed of response, in others sensitivity, and still in others long-term analyser response and wavelength stability. It is therefore not surprising that implementations of NIRS for chemical and pharmaceutical process monitoring cover a very wide range of NIR technologies, sample interfaces and calibration methodologies.

However, what unite all applications of NIRS for PAC are the unique features of the NIR spectrum. The NIR is in effect the chemical spectroscopy of the hydrogen atom in its various molecular manifestations. The frequency range of the NIR from about $4000 \, \text{cm}^{-1}$ up to $12\,500 \, \text{cm}^{-1}$ (800–2500 nm) covers mainly overtones and combinations of the lower-energy fundamental molecular vibrations that include at least one X—H bond vibration. These are characteristically significantly weaker in absorption cross-section, compared with the fundamental vibrational bands from which they originate. They are faint echoes of these mid-IR absorptions. Thus, for example, NIR absorption bands formed as combinations of mid-IR fundamental frequencies (for example $\nu 1 + \nu 2$), typically have intensities ten times weaker than the weaker of the two original mid-IR bands. For NIR overtone absorptions (for example $2\nu 1, 2\nu 2$) the decrease in intensity can be 20–100 times that of the original band.

It follows that all mid-IR fundamental absorptions in the fingerprint region (roughly below $1700 \, \text{cm}^{-1}$) arising from the molecular skeleton (C—C, C=C, C=O, etc.) or from functional groups containing heavier atoms (C—Cl, C—N, etc.) are already into their second or third overtone above $4000 \, \text{cm}^{-1}$, and hence too weak to contribute significantly in the NIR. The single practically useful exception to this is the appearance of weak C=O absorption bands in the second overtone just below $5000\,\text{cm}^{-1}$, where they can be observed among the normal hydrocarbon combination bands.

However, what saves the day, in terms of the analytical specificity of NIRS, is the (sometimes extreme) sensitivity of the frequency and intensity of these X—H NIR absorption bands to near neighbours in the molecular structure. The local electronic environment has a particularly strong influence on the X—H bond force constants, and from this derives a remarkably high information content in NIR spectra. Furthermore some particular functional groups (for example O—H, N—H) are very strongly affected by both intermolecular and intramolecular H-bonding effects, with sometimes dramatic influences on the intensity and band-shapes in the NIR. NMR functions analogously where much molecular structure is deduced from neighbouring influences on the nuclear magnetic resonance frequency for a particular atom type except that in NIR one can view a number of marker frequencies such as O—H, C—H and N—H simultaneously.

The functional groups almost exclusively involved in NIRS are those involving the hydrogen atom: C—H, N—H, O—H (Figure 3.1). These groups are the overtones and combinations of their fundamental frequencies in the mid-IR and produce absorption bands of useful intensity in the NIR. Because the absorptivities of vibrational overtone and combination bands are so much weaker, in NIRS the spectra of condensed phase, physically thick samples, can be measured without sample dilution or the need to resort to difficult short-pathlength sampling techniques. Thus conventional sample preparation is redundant, and fortunately so, because most PAC applications require direct measurement of the sample either in situ, or after extraction of the sample from the process in a fast-loop or bypass.

A further benefit of the low absorptivity of most samples in the NIR is that measurements involving scattering effects (both diffuse transmission and diffuse reflectance) are possible. The penetration depth of the sampling beam in diffuse reflectance measurements of powders in the NIR can be on the scale of millimetres, despite the numerous refractive

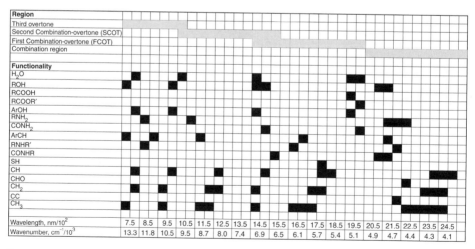

Figure 3.1 NIR correlation chart.

index interfaces and extended optical path involved. Thus relatively large volumes of material can be interrogated, avoiding problems of surface contamination and sample non-homogeneity.

Whilst it is true that the majority of NIR analysers installed in the field for quality monitoring purposes are based on measurements of discrete wavebands using multiple interference bandpass filters, these analysers are generally in agricultural, food or bulk commodity material QA applications. This article is focused on NIRS applications in PAC and will concentrate on NIR technologies capable of yielding full-spectral information – the starting point for the complex data reduction and calibration methods used for PAC quantitative applications. Even with this restriction, there are a surprising number of distinct NIR technologies available, which will be discussed in detail but which are briefly described here to set the scene.

The most conventional analyser format used to achieve full-spectrum information is the scanning grating monochromator (Figure 3.2). In its current modern format this is a robust piece of equipment, normally based on a concave holographic grating, controlled via an industrial motor drive and optical encoder for precise positioning. Under some circumstances this format, using a standard tungsten-halogen source at 2800 K and suitable detector can achieve a near-perfect match of the monochromator throughput with detector sensitivity, and hence very high signal-to-noise ratio (SNR). However, this system may have limitations; wavelength registration is mechanically derived, and under real-life circumstances (for example in multiplexed fibre-optic sampling) the geometric etendue limitations of a slit arrangement (even when the input and output fibres are used to define the slit) can make it interesting to look at higher optical throughput alternatives.

An obvious extension of the scanning monochromator is the grating polychromator (Figure 3.3), fitted with a photodiode array (PDA). Whilst this system still requires an entrance slit to the concave grating, the exit slit is replaced with a PDA, allowing measurement in a number of detection channels simultaneously. This allows for a multiplex measuring advantage (all frequencies are measured at the same time, rather than scanned sequentially as in a scanning monochromator), so response time can be short and SNR kept high. However, as with all diffraction optics systems, there are issues of stray

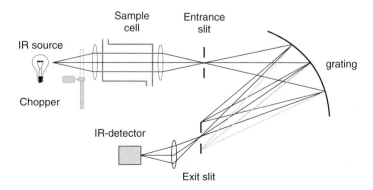

Figure 3.2 Scanning grating monochromator.

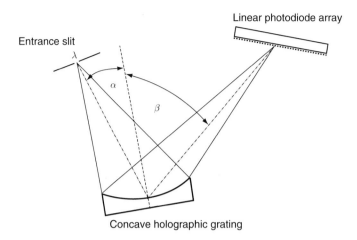

Figure 3.3 PDA grating polychromator.

light and diffraction order sorting to be addressed, which in practice restrict the spectral free range of the diode array. A significant advantage of the PDA grating polychromator is that it continuously views the full spectrum. This makes measurements immune to spectral distortions due to intensity variations resulting from inhomogeneous samples transiting through the optical beam.

A further example of a high-throughput full-spectrum device is the tunable filter instrument (Figure 3.4), normally based on an acousto-optical tunable filter (AOTF). In this case no entrance slit is required. As much of the source radiant energy as can be collected and collimated, and which the thermal properties of the AOTF filter can withstand, can be used. The AOTF filter is based on the use of a carefully oriented birefringent crystal, optically transparent across the NIR (TeO_2 is a typical example). The crystal is driven by an RF-powered piezoelectric transducer, which sets up an acoustic wave that propagates through the crystal and interacts with the incident broadband NIR radiation. Most of the NIR light passes through unaffected, but one specific narrow waveband is diffracted – with a change of orientation due to the birefringence of the crystal. This allows the broadband light (which passes straight through) to be physically blocked (or partly used as a reference signal), and the diffracted narrow bandpass to be channelled into the sample. In principle, the broadband and diffracted narrow bandpass elements can

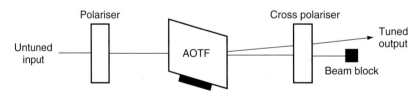

Figure 3.4 AOTF monochromator.

be separated using crossed polarisers on the input and exit of the crystal, but this requires extremely high rejection efficiency to avoid straylight issues. In practice, commercial devices use physical beam blocks. AOTF devices can be used in scanning or random frequency access mode, depending on the way in which the RF power input is driven.

Finally, in the field of full-spectrum NIRS methods, Fourier transform near-infrared (FTNIR) analysers are included (Figure 3.5). FTIR techniques are predominant in mid-IR spectroscopy because there are clear and absolute advantages for the FTIR analyser in the mid-IR compared with any other available technology. This arises because of the low power output of mid-IR sources and the low specific detectivity of mid-IR detectors, which mean that the classical Jacquinot (optical throughput) and Fellgett (multiplex sampling) advantages can be fully realised. The argument is a little more complex as applied to NIRS, and will be discussed in more detail later. An FTIR device is in effect an optical modulator which allows wavelength encoding. The input beam is unmodulated broadband NIR, and the exit beam from the interferometer is still broadband NIR, but with each optical frequency uniquely amplitude-modulated in the acoustic frequency range. This allows the detector signal (which is a combination of signals from all the incident broadband NIR frequencies) to be decomposed using a Fourier transform, and the individual amplitudes of each optical frequency in the broadband input signal to be obtained. The system thus has no input or output restrictions other than a resolution defining Jacquinot stop. Thus in the conventional arrangement using one exit beam of a two-port interferometer (the other beam returns to the source), 50% of the source power is available at all frequencies all the time. The amplitude modulation of the NIR beam is normally achieved using a beam division/retardation/recombination sequence, but there are a variety of physical methods to implement this. The uniquely defining feature of FTIR methods in NIRS is that they all contain some inherent form of direct wavelength registration. Most practical devices employ a visible frequency laser to define the optical retardation sampling interval. This translates directly into the frequency axis of the recorded spectrum, linked intimately with the known HeNe laser frequency. This final

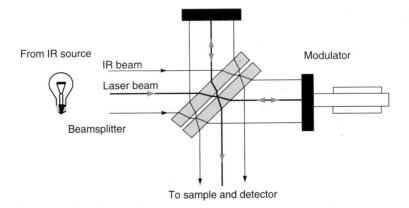

Figure 3.5 FTNIR modulator.

FTIR advantage (the Connes advantage) turns out perhaps to be the major practical contribution of FTIR to NIRS PAC applications.

These four very different NIR technologies represent the mainstream analyser types used in many PAC applications. They also cover the main types of established commercially available analysers. Given their various characteristics and physical principles upon which they are based, it is obvious that they will have different advantages and disadvantages depending on the detailed circumstances of the application. These issues will be covered in more detail below.

3.2 Theory of near-infrared spectroscopy

3.2.1 *Molecular vibrations*

The atoms in a molecule are never stationary, even close to the absolute zero temperature. However, the physical scale of the vibrational movement of atoms in molecules is rather small – of the order of 10^{-9}–10^{-10} cm. The movement of the atoms in a molecule is confined within this narrow range by a potential energy well, formed between the binding potential of the bonding electrons, and the repulsive (mainly electrostatic) force between the atomic nuclei. Whenever atomic-scale particles are confined within a potential well, one can expect a quantum distribution of energy levels (Figure 3.6).

The solution to the Schrödinger equation for a particle confined within a simple harmonic potential well is a set of discrete allowed energy levels with equal intervals of energy between them. It is related to the familiar simple solution for a particle in an infinite square well, with the exception that in the case of the simple harmonic potential, the particle has a non-zero potential energy within the well. The restoring force in a simple harmonic potential well is kx, and thus the potential energy $V(x)$ is $\frac{1}{2} kx^2$ at

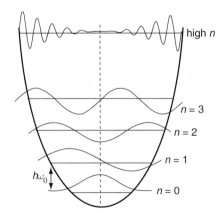

Figure 3.6 Simple harmonic potential well.

displacement x, and the Schrödinger equation to be solved for the quantised energy levels in the well becomes

$$\left(\frac{-\hbar^2}{2m}\right)\frac{\mathrm{d}^2\psi}{\mathrm{d}x^2} + \frac{1}{2}kx^2\psi(x) = E\psi(x) \tag{3.1}$$

where, here and throughout, $\hbar = h/2\pi$. Note that the potential energy $V(x)$ rises to infinite values at sufficiently large displacement. One should expect this boundary condition to mean that the vibrational wavefunction will fall to zero amplitude at large displacement (as in the square well case, but less abruptly). One should also expect that the confining potential well would lead to quantised solutions, as is indeed the case:

$$E_n = \left(n + \frac{1}{2}\right)\hbar\omega, \tag{3.2}$$

where $n = 0$, 1, 2 and $\omega = (k/m)^{1/2}$ and $\hbar = h/2\pi$.

The first important conclusion from this solution of a simple harmonic potential well is that the gap between adjacent energy levels is exactly the same:

$$\Delta E = E_{(n+1)} - E_n = \hbar\omega \tag{3.3}$$

The only changes in energy level for the particle, which are allowed, involve a unit change in quantum number.

$$\Delta n = \pm 1 \tag{3.4}$$

At room temperature the value of Boltzmann factor (kT) is about $200\,\mathrm{cm}^{-1}$, so the vast majority of molecules are in the their vibrational ground state $(n = 0)$. Thus in the case of the simple harmonic model described above, only transitions to $(n = 1)$ are allowed, and NIR spectroscopy would not exist.

3.2.2   Anharmonicity of the potential well

This is clearly not the case, and the answer lies in the non-conformity of the potential energy well in real molecules with the simple harmonic model. In the simple harmonic (parabolic well) model, it can be seen that the potential well is entirely symmetrical. The potential energy rises equally with displacement in both positive and negative directions from the equilibrium position. This is counter-intuitive because the forces responsible for the rise in potential energy are different in the two cases. In the bond-stretching case the dominant factor is the shifting of the molecular orbital away from its minimum energy configuration. In the bond-compression case, there is the additional factor of electrostatic repulsive energy as the positively charged atomic nuclei approach each other. Thus one would expect the potential energy curve to rise more steeply on the compression cycle, and (due to the weakening of the bond with displacement) to flatten off at large displacements on the decompression cycle.

As with the Schrödinger equation itself, the most useful estimate of the form of the potential well in real molecules was an inspired hypothesis, rather than a function derived rigorously from first principles. The Morse Curve is a very realistic approximation to the potential energy well in real molecular bonds.

The expression for potential energy within the well becomes:

$$V(r) = D_e\{1 - \exp[-a(r - r_e)]\}^2 \tag{3.5}$$

Where D_e is the depth of the well, and $a = \frac{1}{2}\omega\,(2m/D_e)^{1/2}$ and m is the reduced mass.

When the function $V(r)$ is used in the vibrational Hamiltonian instead of the simple harmonic $V(x)$, the quantised vibrational energy levels are

$$E_n = \left(n + \frac{1}{2}\right)h\omega - \left(n + \frac{1}{2}\right)^2 X_e h\omega \tag{3.6}$$

and the new factor X_e is known as the anharmonicity constant $X_e = ha^2/2m\omega$.

There are two effects of the anharmonicity of the quantised energy levels described above, which have significance for NIRS. First, the gap between adjacent energy levels is no longer constant, as it was in the simple harmonic case. The energy levels converge as n increases. Secondly, the rigorous selection rule that $\Delta n = \pm 1$ is relaxed, so that weak absorptions can occur with $\Delta n = \pm 2$ (first overtone band), or ± 3 (second overtone band), etc.

For the transition from $n = 0$ to $n = 1$ (the fundamental vibration) one sees the absorption at

$$h\nu = h\omega - 2X_e h\omega \tag{3.7}$$

However, an overtone transition in an anharmonic well involving $\Delta n = 2$ has an energy of

$$h\nu = \Delta E = 2h\omega - 2(2n + 3)X_e h\omega \tag{3.8}$$

where $n =$ the initial state quantum number.

Thus for the transition from $n = 0$ to $n = 2$ the first overtone absorption is seen at

$$h\nu = \Delta E = 2h\omega - 6X_e h\omega \tag{3.9}$$

which is not twice the fundamental frequency, but is lower than this, due to the increased effect of the anharmonicity constant in transitions to higher values of n.

Note that the value of the anharmonicity constant can thus be derived from a comparison between the fundamental $n = 0$ to $n = 1$ absorption band frequency and the first and subsequent overtone band positions. Given the form of the Morse Curve, X_e can be calculated from D_e, and hence an estimate of the bond dissociation energy

$$D_0 = D_e - \frac{1}{2}h\omega \tag{3.10}$$

and it is a very useful method of estimating chemical bond strengths.

3.2.3   Combination and overtone absorptions in the near-infrared

So far, the effect of anharmonicity in determining the frequency of overtone $(\Delta n > 1)$ absorptions, and its effect in relaxing quantum selection rules to allow these transitions to have some absorption intensity have been considered.

However, in polyatomic molecules transitions to excited states involving two vibrational modes at once (combination bands) are also weakly allowed, and are also affected by the anharmonicity of the potential. The role of combination bands in the NIR can be significant. As has been noted, the only functional groups likely to impact the NIR spectrum directly as overtone absorptions are those containing C—H, N—H, O—H or similar functionalities. However, in combination with these hydride-bond overtone vibrations, contributions from other, lower-frequency fundamental bands such as C=O and C=C can be involved as overtone-combination bands. The effect may not be dramatic in the rather broad and overcrowded NIR absorption spectrum, but it can still be evident and useful in quantitative analysis.

Taking the example of the water molecule, a non-linear triatomic, one expects three fundamental vibrations. If a simple harmonic potential is applied, one could write

$$E = \left(n_1 + \frac{1}{2}\right)h\omega_1 + \left(n_2 + \frac{1}{2}\right)h\omega_2 + \left(n_3 + \frac{1}{2}\right)h\omega_3 \qquad (3.11)$$

And to include anharmonicity one might suppose

$$En_1 = \left(n_1 + \frac{1}{2}\right)h\omega_1 - \left(n_1 + \frac{1}{2}\right)2X_1h\omega_1$$

$$En_2 = \left(n_2 + \frac{1}{2}\right)h\omega_2 - \left(n_2 + \frac{1}{2}\right)2X_2h\omega_2$$

$$En_3 = \left(n_3 + \frac{1}{2}\right)h\omega_3 - \left(n_3 + \frac{1}{2}\right)2X_3h\omega_3$$

For a simple combination band such as the transition from $(0, 0, 0)$ to $(0, 1, 1)$ the ΔE would then be calculated as

$$\Delta E\ (0, 0, 0)\ \text{to}\ (0, 1, 1) = (h\omega_2 - 2X_2h\omega_2) + (h\omega_3 - 2X_3h\omega_3) \qquad (3.12)$$

Note that to first order this is simply the sum of the fundamental frequencies, after allowing for anharmonicity. This is an oversimplification, because, in fact, combination bands consist of transitions involving simultaneous excitation of two or more normal modes of a polyatomic molecule, and therefore mixing of vibrational states occurs and additional anharmonicity cross-terms need to be included, such as X_{12}, X_{13} and X_{23}. These appear in expressions such as

$$E(n_1, n_2, n_3) = \left(n_1 + \frac{1}{2}\right)h\omega_1 - \left(n_1 + \frac{1}{2}\right)2X_1 h\omega_1 \cdots - \left(n_1 + \frac{1}{2}\right)\left(n_2 + \frac{1}{2}\right)X_{12}h\omega_{12} \cdots$$

$$- \left(n_1 + \frac{1}{2}\right)\left(n_3 + \frac{1}{2}\right)X_{13}h\omega_{13} \ldots \text{ and higher terms.} \tag{3.13}$$

These cross-terms have the effect of reducing the actual frequency of combination absorption bands in exactly the same way as overtone bands, even though only excitation to $n = 1$ levels are involved.

3.2.4   Examples of useful near-infrared absorption bands

As previously mentioned, the principal functional groups contributing to NIRS are C—H, N—H and O—H, with additional contributions from C=O and C=C groups influencing particularly the shape and location of combination bands especially at lower frequencies ($4000–5500 \, cm^{-1}$). However, though one is dealing with only three major functional groups, all containing the H atom, it does not follow that the information content is limited. One might suppose that the vibrational frequencies of the X—H fundamentals would be similar, since the reduced mass of an X—H diatomic unit is dominated only by the light H atom. For example, the reduced masses of C—H, N—H and O—H all lay within the range of 0.85–0.90 grams. However, in real polyatomic molecules the normal modes of vibration can involve multiple X—H units in different environments, and the individual X—H bond strengths are significantly affected by the local bonding environment. Following is a specific example of the application of NIRS to speciation in hydrocarbon mixtures.

3.2.4.1   Hydrocarbons

The principal saturated hydrocarbon functional groups of concern are methyl, methylene and methyne ($-CH_3$, $-CH_2-$, $= CH-$). The spectra of typical hydrocarbon mixtures (for example as in gas oil and gasoline) are dominated by two pairs of strong bands in the first overtone and combination regions ($5900–5500 \, cm^{-1}$ and $4350–4250 \, cm^{-1}$). These are predominantly methylene ($-CH_2-$). The methyl end groups typically show up as a weaker higher-frequency shoulder to these methylene doublets.

Olefinic, unsaturated hydrocarbons have CH groups adjacent to C=C double bonds. The C=C vibrations do not show up directly except as weaker contributions in the combination region, but the nucleophilic effect of the C=C bond tends to shift nearby CH groups to significantly higher wavenumber, and this is indeed seen in hydrocarbon spectra.

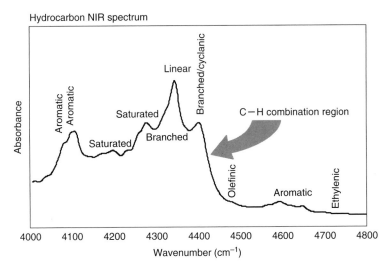

Figure 3.7 Typical hydrocarbon spectrum with assignments.

As can be seen in the Figure 3.7, not only are there clearly distinguishable saturated hydrocarbon functionalities, but there is also a set of absorption band groups between 4760–$4450\,cm^{-1}$ and 4090–$4000\,cm^{-1}$. These features are due to aromatic C—H groups appearing as combination bands. Similar, though less clearly separated and defined aromatic C—H features are also observable in the first and the second overtone C—H stretch regions (approx. 6250–$5550\,cm^{-1}$ and 9100–$8000\,cm^{-1}$, respectively). The band group in the combination region at approximately $4600\,cm^{-1}$ is particularly informative because it arises from aromatic C—H and ring C—C modes in combination.

3.2.4.2   Hydrocarbon speciation

Figure 3.8 shows an example dataset of mixed hydrocarbons used as a petrochemical feedstock. These are straight-run naphthas which consist of a wide range of alkane, alkene, aromatic and naphthenic hydrocarbons, mainly in the range of C_4–C_9. The conventional analytical method for naphtha analysis is temperature-programmed gas chromatography (GC), which can provide a full analysis including C-number breakdown, but which is rather slow for process optimisation purposes.

If NIRS can be demonstrated to have sufficient hydrocarbon speciation capability to reproduce the analysis with the same precision as the GC method, but in a fraction of the time, then a useful process analytical goal will be achieved.

3.2.4.3   Esters, alcohols and organic acids

The ester group C=O functionality makes a fleeting and weak appearance in the NIR spectrum only as the second overtone at around $4650\,cm^{-1}$, where it may well be hidden by overlapping aromatic C—H features if the sample mixture also includes aromatics.

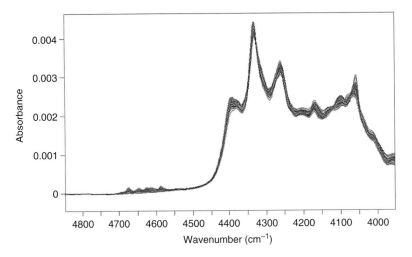

Figure 3.8 Naphtha dataset.

However, as in the case of olefinic compounds, the effect of the carbonyl group on the absorption of nearby C—H groups is quite dramatic, and the bands of both methylene and methyl CH groups are shifted to significantly higher wavenumber.

Alcohols and organic acids both contain the O—H functionality, which is the next most important NIR signature after C—H. However, it behaves very differently. The dominant feature of the —OH group is its capacity for both intra- and inter-molecular hydrogen bonding. The essential —OH group contributions in the NIR spectrum are combination and overtone stretching modes that occur around 5250–4550 cm^{-1} and 7200–6000 cm^{-1} (Figure 3.9).

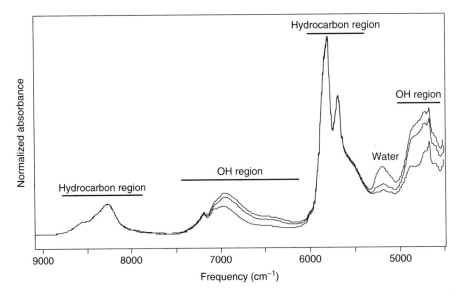

Figure 3.9 NIR spectra of polyols.

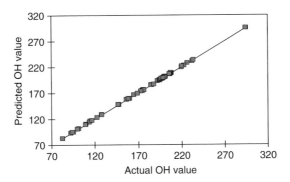

Figure 3.10 Calibration curve for hydroxyl number.

However, the position, shape and intensity of the —OH absorptions which appear in these regions are very strongly influenced by hydrogen bonding effects, which generally tend to randomise the electronic environment of the —OH group, spreading out the range of absorption frequencies, and reducing absorbance intensity. Moreover, where these are intermolecular effects, they are strongly influenced by sample temperature, changes of state and sample dilution.

Nonetheless, there is a very wide range of industrially significant —OH-containing materials, often used as building blocks in polymer and oleochemical processes. Amongst these materials, such as glycols, polyester polyols, polyurethane prepolymers, etc., applications such as hydroxyl number or acid value determination are highly dependent upon the NIRs of the —OH group. A typical scatter plot for a NIR calibration for hydroxyl number is shown in Figure 3.10.

3.3 Analyser technologies in the near-infrared

In this section the mainstream NIRS technologies in current widespread practical use for PAC applications will be examined. These are scanning grating monochromators, grating polychromator PDA spectrometer, AOTF analysers and FTNIR analysers.

3.3.1 The scanning grating monochromator

What follows is a detailed discussion of the principles of operation and practical limitations of analysers based on scanning monochromators. Included in this discussion will also be the background information on NIR source and detector technologies, which are in fact common to all of the NIR spectroscopic technologies listed above.

3.3.1.1 Principles of operation: The diffraction grating

The diffraction grating monochromator is a specific example of multiple-beam interference effects. Interference between multiple beams can be generated by both division of amplitude (as in the Fabry-Perot interferometer) or by division of wavefront (as in the diffraction grating) (Figures 3.11 and 3.12).

The two examples shown demonstrate how an interference effect can be produced either by amplitude division of an incident beam, followed by retardation (achieved in this case by multiple reflection between the partially reflective parallel plates of a Fabry-Perot resonator) and recombination, or by division of the wavefront at the multiple equally spaced slits of a diffraction grating, again followed by recombination.

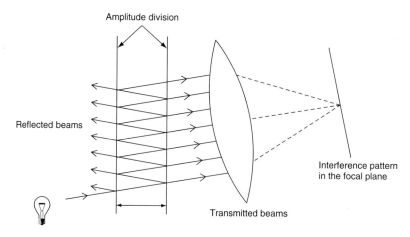

Figure 3.11 Multiple-beam interference by division of amplitude. Reprinted with kind permission of Prentice-Hall International.

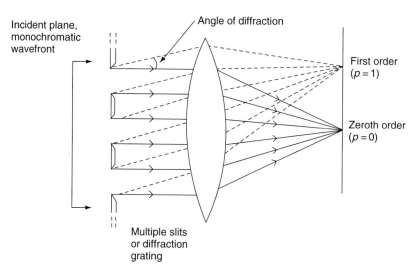

Figure 3.12 Multiple-beam interference by division of wavefront. Reprinted with kind permission of Prentice-Hall International.

The condition for the formation of an interference maximum at the point of recombination is that the contributions to the total irradiance at that point from all the slits of the grating should be in phase. This condition is met when the classical diffraction equation for normal incidence is satisfied:

$$p\lambda = (a + b)\sin\theta = d\sin\theta \qquad (3.14)$$

where in this expression a and b are the width and spacing between the slits of the grating, and hence $d\,(=a+b)$ is defined as the grating constant. Here d and λ are measured in the same units of length.

Note that for a specific wavelength, λ, interference maxima will occur at a series of integer values of p, the order of diffraction, including $p = 0$. As λ varies, so the position, defined by θ, of the interference maximum varies. So for an incident polychromatic beam, a series of interference maxima for each wavelength is formed at varying θ. If the range of wavelengths in the incident polychromatic beam is large enough, then for certain ranges of θ there will be contributions from different diffraction orders superimposed. From this it follows that separation of the diffraction orders will be necessary to avoid excessive straylight contributions to the detected signal. Note also that all wavelengths are coincident at zeroth order, so this signal must also be isolated in some way to avoid a scattered light contribution to the detected signal.

The current practical realisation of the diffraction grating monochromator for PAC using NIRS takes the form of a concave holographic grating. In this the diffraction effect of a grating, and the image-forming effect of a concave reflector are combined. Thus compared with a conventional plane-grating arrangement (for example a Czerny-Turner type monochromator) there is no requirement that the incident and the diffracted beams are collimated. Holographic gratings are formed from a grating blank with a photoresist coating. The photoresist is subjected to a holographic image of interference fringes formed from a rigorously monochromatic source (a laser). The combination of holographic imaging and the curved surface of the photoresist layer on the grating blank are capable of exactly reproducing the hyperbolic grating pattern first described by Rowland in the late nineteenth century. Holographic gratings are effectively capable of producing a diffracted image free from optical aberration and very low in scattered (stray) light. Also the optical (non-mechanical) means of production lends itself very well to repeatable manufacturing techniques, so that the performance and characteristics of the gratings installed in a set of analysers can be made reasonably reproducible.

The concave grating format (Figure 3.13) also allows the analyser design more easily to be optimised for focal length and optical throughput. The longer focal length gives easier, more stable alignment, but raises f-number; however, the larger grating diameter possible for holographic gratings decreases f-number for the same resolution and slit area and recovers throughput. Also, and significantly from the point of view of manufacturing design, the use of a concave grating reduces the number of optical components involved, and therefore the number of optical surfaces generating stray light and imaging errors, as well as manufacturing complexity and the risk of misalignment.

One further point to mention about the design and use of concave holographic gratings concerns the exit focal plane. For a scanning monochromator it is of no great concern that

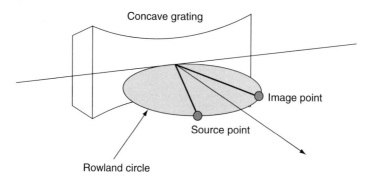

Figure 3.13 The concave grating monochromator.

the focal plane is not necessarily flat since the focal point can be maintained in position by the rotation of the grating. However, for the polychromator PDA-based analyser this is an issue, since the linear PDA is, of necessity normally fabricated as a flat device, and all the exit wavelengths are intended to be imaged simultaneously across the detector plane. Therefore the polychromator should produce a more-or-less flat image plane at the detector surface. In the case of holographically produced concave gratings the shape of the focal curve can be modified and can be made to approach a flat focal plane.

In an analyser based on scanning monochromator technology, one should be aware of a number of critical design issues:

- There is a direct compromise between spectroscopic resolution and optical throughput. In order to achieve wavelength separation at the detector, both input and output beams to and from the grating need to be imaged at entrance and exit slits, and the image formed at the exit slit refocused onto the NIR detector. In order to achieve an appropriate level of optical throughput the slits need to be sufficiently wide. In order to achieve adequate optical resolution (a narrow instrument line shape) the slits need to be sufficiently narrow. In practice the entrance and exit slits are normally set to the same size, which is selected depending on whether optical resolution or SNR is the more important parameter for a specific application.

- Another point of practical significance is the issue of rotation of the grating to achieve reproducible wavelength scanning and accurate wavelength registration. This has been the Achilles' heel of scanning monochromator design. Originally it depended upon purely mechanical scanning mechanisms. It has been greatly facilitated by the use of miniature industrial electric-motor drives and optical encoding (position sensing) of the grating. This is particularly important in terms of the development of complex full-spectrum Partial Least Squares (PLS) or similar calibration models for more difficult applications, which require sub-nanometer wavelength stability over an extended period. For example, the calibration stability of applications involving refinery hydrocarbons, particularly middle distillates or heavier products, can be seen to require a frequency reproducibility between analysers of at least $0.1\,\text{cm}^{-1}$ at $5000\,\text{cm}^{-1}$. This is equivalent to about $0.05\,\text{nm}$ at $2000\,\text{nm}$ – a demanding objective for a scanning device.

3.3.1.2   Order sorting

The function of the diffraction grating monochromator is to physically separate wave-lengths along the focal plane of the concave grating. As seen above, for a broadband source different wavelengths diffracted at different diffraction orders might be coincident at the detector. For example, if the scanning monochromator is required to scan an NIR spectrum between 2000 and 2500 nm, in the first order, then there will be coincident radiation from the source hitting the detector at the same time from the range of 1000–1250 nm in the second order, which is certainly not negligible in terms of typical source output.

The effect of multiple-order diffraction, if not corrected, is to introduce a straylight signal at the detector, which badly affects the measured linearity of the required absorb-ance signal. In the example above, if the required measurement is at 2000 nm, and only 0.1% of stray light at 1000 nm reaches the detector, then appreciable absorbance non-linearity is seen at absorbances above 2.0 AU.

In general the diffraction at first order is the more intense, so is normally used for NIRS; occasionally the second order is used also to extend the available spectral range for a given physical grating rotation. This situation requires that order-sorting bandpass optical filters are introduced into the beam, normally just before the detector, to eliminate unwanted diffraction orders. A stepper motor or similar arrangement will introduce or remove specific bandpass filters for certain wavelength scan ranges.

Multiple diffraction orders are not the only potential source of stray light in a scanning monochromator. Internal reflections and scattering from the surfaces of optical compon-ents, including the grating, any additional focusing optics, order-sorting or blocking filters, detector windows and any zeroth-order diffraction may all contribute.

3.3.2   Light sources and detectors for near-infrared analysers

3.3.2.1   The tungsten-halogen source and the halogen cycle

The tungsten-halogen source is almost universally used for NIR spectroscopy. It has a broadband, pseudo-blackbody emission spectrum with no significant structure. It is inexpensive and remarkably long-lived if operated at appropriate filament temperature and lamp wattage. The peak spectral radiance of a tungsten lamp operating at a filament temperature of 2800 K is located at approximately $10\,000\,\text{cm}^{-1}$ (1000 nm) – very slightly higher frequency than that of a blackbody at the same temperature, but at less than half the peak intensity. Increasing the filament temperature above 3000 K might boost the peak intensity of the spectral radiance to two-thirds of the blackbody at 2800 K, but with a dramatic reduction in source lifetime. For practical process NIR analysers (of all tech-nologies, for which the tungsten-halogen lamp is the standard broadband NIR source, even though we discuss it here in relation to scanning monochromators) the required source lifetime between changes needs to be of the order of 12 months' continuous operation. Source lifetime of 10 000 hours is not an unreasonable expectation when operated at 2800 K and at or below the rated voltage for the filament. For a hot filament one would expect the room-temperature resistivity to be much lower than when at operating temperature. This is the case, and unless the source power supply is designed to be current-limited, there will be a significant initial current surge, with potential for excessive filament vapourisation and redeposition.

The tungsten-halogen lamp operates in a very specific mode, dependent on the detailed chemistry of the atmosphere within the lamp enclosure. The lamp envelope is normally a thin quartz bulb, filled with an inert gas such as krypton or xenon and a trace of a halogen, often bromine or iodine. Under operating conditions, even with the increased pressure of the inert gas filling, tungsten is vapourised from the filament. Without taking care to deal with this, the evaporated tungsten metal would simply condense on the coolest part of the lamp available – which would normally be the inside of the glass envelope, leading to blackening, loss of emission intensity and premature filament failure. The trace of halogen in the lamp reacts with the deposited elemental tungsten on the inside of the envelope (provided that the lamp is run at a high-enough temperature – the envelope needs to be above about 200°C for this process to work efficiently). This halogen reaction not only cleans the envelope, but also allows for transport of the tungsten halide back into the vapour phase, where it eventually reaches the surface of the filament, where it is destroyed, and the elemental tungsten is redeposited back at the filament, whilst the halogen is released back into the vapour phase – hence the halogen cycle. Incidentally, this is the reason why tungsten-halogen lamps should never be touched with unprotected fingers during lamp changes. The diffusion of sodium (as the salt in sweat) back through the thin glass envelope eventually causes the loss of the trace halogen from the lamp atmosphere, and the interruption of the halogen cycle.

3.3.2.2   Photodetectors for the near-infrared

The earliest detectors used commonly in NIR process analysers were lead sulphide and lead selenide (PbS and PbSe) photoconductive devices, which are both highly sensitive and inexpensive. They were ideally suited for use in scanning grating monochromators. They do however have limitations in linearity, saturation and speed of response. Consequently the adoption of NIR analyser technologies requiring some or all of these characteristics saw the rise in use of photodiode devices such as the Indium-Gallium-Arsenide (InGaAs) detector. These are both very linear and very fast and sensitive, making them the ideal all-round detector for NIR applications.

There are two basic types of photodetectors available for use in NIRS: thermal detectors and photon devices. Photon devices can be divided into photoconductive and photodiode detectors. The latter can be operated in either photoconductive or photovoltaic modes (with or without an applied bias voltage respectively). Thermal detectors are not common in most NIR analysers, but they have a unique and useful place, as pyroelectric detectors in FTNIR instruments. The higher overall optical throughput of an FT instrument allows the far less intrinsically sensitive pyroelectric device to be used successfully, with the added advantage that depending on the choice of optical materials and source technology, extended range instruments can be built spanning simultaneously both NIR and mid-IR applications.

3.3.2.2.1   The pyroelectric detector

Pyroelectric detectors depend on the use of a thin slice of ferroelectric material (deuterated triglycine sulphate (DTGS), Figure 3.14, is the standard example) – in which the molecules of the organic crystal are naturally aligned with a permanent electric dipole. The thin slab is cut and arranged such that the direction of spontaneous polarisation is normal to the large faces. Typically one surface of the crystal is blackened to enhance thermal absorption, and the entire assembly has very low thermal mass.

Changes in the temperature of the DTGS crystal, induced by incident broadband infrared radiation, cause the degree of polarisation to change, and hence changes the amount of charge stored at the crystal surface. Electrodes arranged as in Figure 3.14 are driven by this difference in surface charge between the back and the front faces of the crystal, and a current will flow generating a signal voltage across the load resistor (R_L).

Note: The temperature change arises from the incident (photon) energy per unit time absorbed by the low thermal mass but subsequently dissipated by thermal conductivity. Thermal conductivity in the device is created by the electrodes, and is tailored depending on the incident power levels and speed of response desired. For high-throughput FTIR, the required thermal conductivity is high. There is some improvement in SNR for low-throughput measurements with lower thermal conductivity. The detector is not fast; the signal response reduces with electrical frequency. But the noise reduces likewise until beyond several kHz. And so the SNR remains consistent up to several kHz. This makes it suitable for the fast modulation requirements of FTIR.

Important features of the DTGS detector are:

- Since it is a thermal and not a photon detector, there is no band-gap and no specific cut-off frequency. The device is a broadband detector capable of operation from the NIR to the long-wavelength mid-IR.
- Because there is no band-gap, and electron-hole pair generation is not involved in the detector response, there is no issue of bleaching (carrier depletion) under high levels of irradiance, so the detector can achieve good linearity performance.
- To generate an output signal the DTGS detector requires a modulated NIR signal. A constant irradiance will generate no change in voltage across the load. Even with a low thermal mass detector element, and appropriate values of R_L, the response time of a

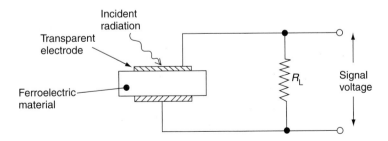

Figure 3.14 The DTGS pyroelectric detector. Reprinted with kind permission of Prentice-Hall International.

DTGS device will be slow, but can extend into the kHz range, which fits well with the amplitude modulation audio frequencies typical of an FT instrument.

- Because the response time of the detector depends on the thermal time constant of the detector element/electrode assembly, coupled with the electrical time constant of the device capacitance and load resistor, the response vs. modulation frequency (f_m) shows a typical $1/f_m$ form.

The pyroelectric DTGS detector is a very useful, low-cost, general purpose, wideband NIR detector well suited for use in FT-based analysers. It is not normally used in scanning monochromators where higher sensitivity detectors are needed to match the lower optical throughput and discrete wavelength scanning requirements.

3.3.2.2.2 Photoconductive and photodiode detectors

Both photoconductive and photodiode detectors are semiconductor devices. The difference is that in the photoconductive detector there is simply a slab of semiconductor material, normally intrinsic to minimise the detector dark current, though impurity-doped materials (such as B-doped Ge) can be used for special applications. By contrast, the photodiode detector uses a doped semiconductor fabricated into a *p–n* junction.

In an intrinsic semiconductor material there is an energy gap, known as the band-gap, between the valence band (populated with electrons) and the higher-energy, nominally empty conduction band. An incident photon of sufficient energy can interact with a valence band electron, and push it to a higher energy state in the conduction band, where it is free to migrate, under the influence of an applied bias voltage, to the collection electrodes deposited on the lateral surfaces of the detector element. Any change in detector element conductivity will result in a flow of current around the circuit, driven by the bias voltage, and detected across a load resistor.

Typical materials used in NIR photoconductive detectors are PbS, PbSe, InSb and InAs (lead sulphide, lead selenide, indium antimonide and indium arsenide).

Since photon energies decrease with increasing wavelength, there will be a maximum wavelength beyond which the photons lack sufficient energy to cause electron transitions into the conduction band.

$$\lambda_g = \frac{hc}{\Delta E_g} \tag{3.15}$$

There are a few complications to this simple picture. First it has been assumed the conduction band is normally empty of electrons. This is never entirely true, and for many semiconductor materials used in photoconductive devices with reasonably large values of λ_g, it is not even approximately true. Where kT is an appreciable fraction of ΔE_g there will be significant thermal promotion of electrons into the conduction band, and a consequently high detector dark current. For this reason, many photoconductive detectors are cooled, often using thermo-electric devices (for easy packaging) in order to reduce kT, and hence the dark current.

Secondly, not all electrons generated by incident photons are swept out into the detection circuit by the bias voltage. In fact there is a continuous dynamic process of

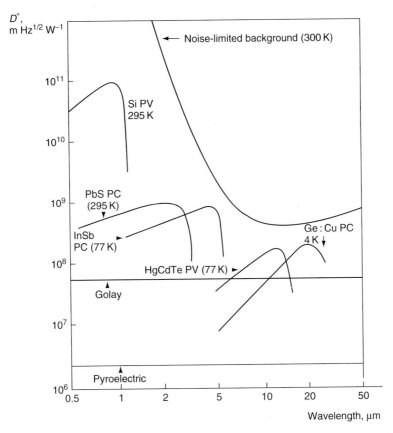

Figure 3.15 Response curves of various detector materials. Reprinted with kind permission of Prentice-Hall International.

electron-hole pair generation by incident photons, and electron-hole pair recombination via a variety of lattice relaxation mechanisms. This balance between generation and recombination is known as the photoconductive gain, and in general it increases with applied bias voltage, and decreases with detector element size. The detector noise associated with this dynamic generation–recombination process (even assuming the detector is operated at a temperature where $kT \ll \Delta E_g$ so that thermally generated processes can be neglected) is known as generation–recombination noise, and is the dominant noise contribution in photoconductive devices.

The difference between a photoconductive detector and a photodiode detector lies in the presence of a thin p-doped layer at the surface of the detector element, above the bulk n-type semiconductor. Holes accumulate in the p-layer, and electrons in the n-type bulk, so between the two there is a region with a reduced number density of carriers, known as the depletion layer. The important effect of this is that electron-hole pairs, generated by photon absorption within this depletion layer, are subjected to an internal electric field (without the application of an external bias voltage) and are automatically swept to the p

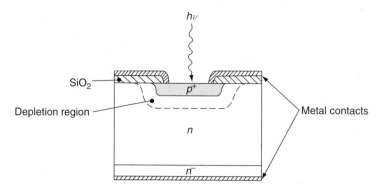

Figure 3.16   Schematic of a photodiode *p–n* junction. Reprinted with kind permission of Prentice-Hall International.

and *n* regions, and a measurable external voltage will appear at the electrodes (Figure 3.16). The signal is measured as the voltage across an external load resistor in parallel. This situation represents a photodiode detector operated in photovoltaic mode.

In the photovoltaic mode of operation, the generated external voltage is a logarithmic function of the incident irradiance.

$$V_{ext} = \left(\frac{kT}{e}\right) \ln(\text{constant} \times I_0) \tag{3.16}$$

Of course it is possible to apply an external bias voltage to a photodiode, under which circumstance it operates in photoconductive mode, and the signal is measured as the voltage across a series load resistor through which the photocurrent flows. This has the advantage that the external signal (now a photocurrent) is linearised with respect to the incident irradiance. The disadvantage is that the bias voltage now drives a detector dark current noise.

The most common photodiode materials applied to NIRS are Si, Ge and InGaAs. The latter alloy-semiconductor has the significant advantage that the band-gap can be tuned for different wavelength cut-offs, and hence for different applications, by varying the alloy composition. For example, the basic InGaAs photodiode produced by epitaxial deposition on an InP substrate is constrained to an alloy composition of In (53%)/Ga (47%) in order to retain lattice-matching with the substrate. This has a cut-off around 1700 nm, and very high sensitivity even at ambient temperature. In order to achieve different alloy compositions, graded lattice structures must be deposited, absorbing the lattice stresses until the main active diode layer can be formed at, say, In (82%)/Ga (18%), giving a much longer wavelength cut-off at about 2600 nm. The narrower band-gap requires this photodiode to be thermo-electric (TE)-cooled for optimum performance.

There are two significant variations on the simple photodiode arrangement described above. First, in the basic arrangement the depletion layer (between the *p*-doped surface and the *n*-type bulk) can be rather thin. In this case for photons absorbed below the depletion layer, although still generating electron-hole pairs, their contribution to the photosignal will be dependent on their diffusion into the depletion layer (where they will be swept to the detector electrodes). In order to maximise the size and sweep volume of the depletion layer a *p*–I–*n* structure can be developed, where an intrinsic zone separates the *p*-doped surface and the *n*-type bulk. Upon application of a bias voltage the depletion layer can be extended right through the intrinsic region, so that photogenerated carriers throughout this zone are swept without the need for carrier diffusion.

The second significant variant to the photodiode is the Si avalanche photodiode. An avalanche photodiode is designed for use under conditions of very low incident irradiance, where high levels of external amplification of the photovoltage would be required. The structure is a basic *p*–*n* diode, but operated at very high reverse bias, close to the breakdown voltage. This is one reason why they are relatively easily realised in silicon. The effect of the large bias voltage is rapidly to accelerate carriers across the depletion region, to the extent that they are able to generate additional electron-hole pairs simply by relaxation of their kinetic energy. This process can be repeated many times during the transit across the depletion region, and signal gains of 100 times or more are quite realisable. However, the design is tricky and a number of features such as extremely stable bias voltage power supplies and precise temperature control requirements make this a device for specialised applications.

3.3.2.2.3   Detector performance characteristics

The fundamental performance parameter of any detector is its noise equivalent power (NEP). This is simply the input irradiance power necessary to achieve a detector output just equal to the noise. This NEP is dependent on a number of detector and signal variables such as modulation frequency and wavelength (the input signal is defined as sine wave–modulated monochromatic light), and detector area, bandwidth and temperature.

In order to eliminate some of these variables in comparing detector performance characteristics, it is usual to use a parameter described as specific detectivity (D^*) and defined as:

$$D^* = \frac{(A\Delta f_{\mathrm{m}})^{1/2}}{\mathrm{NEP}} \qquad (3.17)$$

The parameter D^* depends on the test wavelength and the modulation frequency, and should ideally be quoted in some form such as $1 \times 10^{10}\,\mathrm{mHz}^{1/2}\,\mathrm{W}^{-1}$ (1000 nm, 1 kHz).

Fortunately, for most photon device photodetectors, the dependence on modulation frequency is fairly flat across most of the region likely to be encountered with conventional NIR analyser technology. Provided the D^* is satisfactory at the required measurement wavelengths, it is one useful parameter for detector selection (Figure 3.15).

Photodiodes provide an electric current in response to incident light by the photovoltaic effect. In the NIR, the conversion factor, or responsivity, is generally in the range of 0.5–1.2 amps/Watt. Multiplying the NEP by the responsivity gives the noise current of the detector. It occurs frequently that the current noise generated by the amplifier exceeds the current noise given by the NEP of the detector. Even with the best amplifier there remains the thermal noise current generated by the feedback resistor, which is proportional to the reciprocal of the square root of its resistance value. In order to reduce the thermal noise current the feedback resistance needs to increase in resistance value. Often the desired resistance value to match the NEP noise current will be so large that there will be far too much amplification and the amplifier will saturate from amplifying the optical signal. For this reason dynamic gain switching is frequently employed so that for low light levels the sensitivity can be higher than for high light levels.

3.3.3 The polychromator photodiode-array analyser

There is a great deal in common, in terms of fundamental optical principles, between a scanning monochromator and a PDA polychromator. For example, both commonly use a concave grating as the dispersive element. However operationally they are quite distinct. The PDA polychromator exploits the semiconductor fabrication capabilities of photodiode manufacturing. In the above mentioned discussion, the basics of an NIR photodiode detector were described. By fabricating a linear array of photodiodes, an assembly can be produced, which sits at the focal plane of the concave grating exit beam and for which each individual photodiode element records a specific waveband in the spectrum. No exit slit is required, and no physical scanning mechanism is involved. Just the electronic accumulation and data logging from each detector element are required. There does, however, need to be a good match between the diode-array size, the linear dispersion of the grating, the bandwidth of the detector and the correct order-sorting filter arrangement. Since the whole PDA is subjected to a dispersed image of the entrance slit (or input fibre-optic), it is even more important than in the scanning monochromator that a correct order-sorting filter, working in conjunction with the spectral response range of the detector, is used to block stray light from irrelevant diffraction orders.

Given the high D^* of Si or TE-cooled InGaAs photodiodes, and the multiplex advantage of simultaneous multichannel detection, it is clear one can expect such devices to have high SNR with fast data acquisition times. One can anticipate a potential SNR enhancement of $n^{1/2}$, where n is the number of resolution elements measured by the PDA compared with a scanning monochromator over the same data acquisition time. However, the linear array detector elements are often much smaller than a typical single detector, and there may be a significant throughput limitation.

There are in general two classes of noise associated with NIR analyser systems: additive and multiplicative noise. Additive noise is mostly wavelength-independent, and does not scale with the signal strength. It is in effect the white noise from the detector and associated read-out electronics, which is added to the pure signal. In PDAs the major contribution is the generation/recombination noise associated with the photoproduction of carriers. Multiplicative noise on the other hand arises from almost anywhere else in the optical train and is mostly associated with instabilities of various sorts – source instability, mechanical vibrational effects on gratings, slits, apertures, fibre-optics, mirrors, etc. This type of noise scales with the signal, and is often not white noise, but may have significant wavelength dependency.

In a polychromator PDA spectrometer it was previously mentioned that there is no exit slit – all diffracted wavelengths (after order sorting) fall onto the diode-array detector. In this case what defines the analyser spectral bandwidth? Each individual detector element integrates the signal falling onto it, and allocates that signal to a specific wavelength. In reality, of course, it is seeing a bandpass of wavelengths, depending on the angular dispersion of the grating and the flatness of the focal plane. For standard longer-wavelength NIR applications – which tend to predominate in real-life PAC situations – the PDA will be a TE-cooled InGaAs assembly. These are costly, especially if a large number of detector elements are required. Consequently most PDA analysers working in this region will tend somewhat to compromise resolution and match the detector element size to a fairly wide entrance slit image. Depending on the exact layout of the detector pixels this can produce a variety of instrument line shape functions, and for real spectral data recording can also result in line shape asymmetry. This brings to mind one of the key operational difficulties of a PDA spectrometer. Wavelength calibration is absolutely necessary, and the variation in grating linear dispersion across the whole range of measurement (and hence a changing overlap between slit image and pixel size) leads to some difficulty in this case. In principle, the PDA device can be built in a reasonably monolithic form so that the influence of the external environment, mechanical and thermal, can be reduced. But for effective transfer of applications and calibrations between analysers, an absolute wavelength accuracy better than 0.05 nm (at 2000 nm) may be required, and that is currently a challenging target for PDA devices.

The read-out signal from a PDA is formed by accumulation of each pixel's photo-current within either the photodiode capacitance itself, or a separate linked capacitance circuit. Often this is a completely integrated circuit so photodiode pixel and read-out circuit are on the same chip. Subsequent digitisation of the accumulated signal from each pixel will be performed by an ADC, typically ranged at 16 bits. This means the maximum SNR at any one wavelength will be approximately 64 000: 1, which matches quite well, for example, with the dynamic range of an InGaAs PDA. Due to the small size and applied bias voltage (when used in photoconductive mode) the response time of each diode element in the PDA can be very fast indeed, so total integration times can also be short – in principle, data rates of 200 spectra per second are achievable but in reality spectral co-adding is a more sensible objective, to improve overall SNR.

3.3.4 The acousto-optic tunable (AOTF) analyser

3.3.4.1 Principle of operation

Both the diffraction grating devices described above make use of a physical diffraction grating, where the diffraction takes place at or from a simple plane or concave surface grating. Thus all incident wavelengths are diffracted at some angle or other, and either the exit slit or the discrete elements of the PDA sort out which wavelength is which. There are two physical models which are useful to describe the principle of operation of an AOTF device. Both will show that in contrast with a surface grating monochromator, for an AOTF device operating in the Bragg regime only a single wavelength (or at least very narrow bandpass of frequencies) can be diffracted for a given tuning condition. This is therefore in principle a full aperture device – input and exit slits are not required. However, there are throughput limitations of a different sort, which will be discussed.

The two physical models that can be used to describe an AOTF device depend as usual on optics, and on wave and particle descriptions of the interactions involved. The basic physical layout of the device will be described, and then its principle of operation will be discussed.

The heart of an AOTF analyser is an NIR-transmitting, optically anisotropic (birefringent) crystal lattice. A piezo-transducer, normally LiNbO$_3$ (lithium niobate) is cold-metal bonded to one surface of this optically polished crystal. An RF driver, typically a PLL (Phase Locked Loop) synthesiser is used to generate a high-frequency electric field that causes (via the piezo-electric effect) a coupled high-frequency vibration of the transducer. Since this is closely bonded to the AOTF crystal, this in turn is subjected to an acoustic wave, which propagates through the crystal, and is normally absorbed at the far surface of the crystal to avoid undue acoustic reflections. Another way to describe the impact of the RF-driven piezo-transducer on the AOTF crystal is as a strong source of lattice phonons which propagate through the crystal with a more or less fixed wavelength (Λ) determined by the RF frequency of the oscillator, and a flux dependent on the RF power.

The two physical models for the operation of an AOTF depend on how one views the interaction of an incident NIR light beam with this acoustically excited crystal lattice.

On one level it is a quantum effect, and can be described in terms of photon–phonon scattering. The incident NIR beam is a source of photons, and the energy from the piezo-transducer provides a source of lattice phonons that propagate through the crystal. As in all collision processes, the twin principles of conservation of momentum and conservation of energy apply. The momentum of a quantum particle is linked to its wavevector by hk. The energy is linked to its frequency by $h\omega$.

Consider first the simpler case of acousto-optic interaction in an isotropic medium (i.e. a standard acousto-optic modulator). If the wavevector of the incident light is k_0, that of the scattered light k_+ or k_- and that of the phonon K, we have for conservation of momentum

$$hk_+ = hk_0 + hK \text{ and } hk_- = hk_0 - hK \tag{3.18}$$

And for conservation of energy

$$\omega_+ = \omega_0 + \Omega \text{ and } \omega_- = \omega_0 - \Omega \tag{3.19}$$

(The alternatives are for photon energies being increased or decreased by the scattering.) Also some interesting things can be noticed (Figure 3.17). First, the acoustically scattered photons are Doppler-shifted up or down in frequency. This may be by a small amount, but it can form the basis for frequency modulation of the output beam in acousto-optic devices (varying the RF driver frequency – and therefore the phonon frequency not only changes the output wavevector, i.e. direction, but also the frequency).

Secondly, this seemingly innocent relationship between input and scattered wavevectors has a significant effect on the scattered (diffracted) beam. Because in all realistic cases the scalar value of k_0 will greatly exceed that of K, if one plots the wavevector diagrams one will see that k_+ and k_0 form the two sides of a very nearly flattened triangle, and for

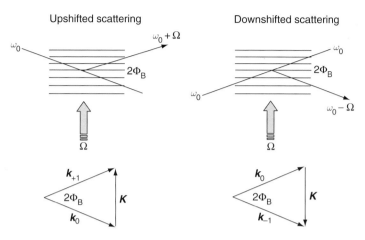

Figure 3.17 Photon–phonon scattering in an AO modulator.

a given input wavelength, and given K, there will be just one scattering angle, the Bragg angle, given by

$$\Phi_B = \sin^{-1}\left(\frac{K}{2k_0}\right) = \sin^{-1}\left(\frac{\lambda}{2\Lambda}\right) \tag{3.20}$$

where K and k_0 are the wavevector magnitudes, λ is the wavelength of incident light, and Λ is the wavelength of the sound.

The physical description of the Bragg diffraction regime is similar to the case of X-ray diffraction in a crystal lattice (Figure 3.18). The diffracted beam is extensively re-diffracted inside the crystal due to its thick optical pathlength, and the diffraction takes place at a

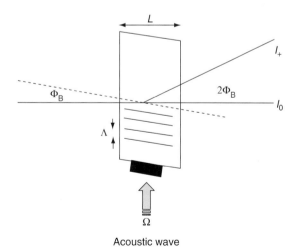

Figure 3.18 Bragg-type AO diffraction.

series of planes rather than at the lines of a surface grating. Simple geometric optical arguments concerning the need for light scattered from a given plane, and light from successive planes, to arrive at the diffracted wavefront in-phase, if significant optical energy is to appear in the diffracted beam, lead to the same Bragg angle equation as that derived above from photon–phonon scattering arguments.

$$\sin \Phi_B = \left(\frac{\lambda}{2\Lambda}\right) \tag{3.21}$$

where $\theta_1 = \theta_d = \Phi_B$ for phase matching. Note that this condition is extremely wavelength-selecting. For a given angle of incidence between the input optical beam and the acoustic wavefront, and a given value of the acoustic wavelength, only one value of λ is permitted.

The Bragg angle can be calculated for reasonable conditions. For TeO_2 the acoustic velocity is around $650\,ms^{-1}$, so at an RF frequency of 50 MHz the phonon wavelength is approximately $1.3 \times 10^{-5}\,m$, and the Bragg angle of deflection for an incident beam of 1000 nm light is 40 milliradians (2°). Note that an increase in RF driver frequency to 100 MHz would halve the acoustic wavelength, and the Bragg angle for the same incident wavelength would increase to approximately 4.5°.

So far, a modulator device has been described, not a filter. This can be used on a mono-chromatic (for example laser) input beam to vary the Doppler-shifted frequency or direction or intensity of an output beam as a function of input RF frequency and power. In order to understand how an AOTF device goes further than this, it is necessary to look at the effect of Bragg regime diffraction and the impact of the anisotropy of the AOTF crystal medium.

In an AOTF analyser the input optical beam to the crystal is polychromatic from a broadband source. For a given acoustic frequency and angle of incidence, only a very narrow bandpass of wavelengths from the polychromatic input beam will satisfy the condition for anisotropic Bragg diffraction. That bandpass is angularly deflected *and* its polarisation is rotated through 90°. As the frequency of the RF driver is changed, the bandpass of optical frequencies satisfying the phase-matching requirement will also change. Note that Bragg regime diffraction can often achieve >90% efficiency, i.e. up to 90% of the input beam at that wavelength can be diffracted into the first order – but that is only for the selected diffractable wavelength. The remainder of the polychromatic input beam carries straight on through the crystal, and needs to be separated from the monochromatic diffracted and deflected beam. Above, one saw that the values of the angular deflection in Bragg-regime AO diffraction are not particularly large. Consequently the polarisation rotation of the diffracted beam is a very useful feature in AOTF devices because it allows for a greater chance to separate the monochromatic diffracted beam from the polychromatic non-diffracted beam using a combination of physical beam blockers and crossed polarisers.

The raw bandpass of an AOTF has a sinc squared function line shape with sidebands, which if ignored may amount to 10% of the pass optical energy in off-centre wavelengths. This apodisation issue is normally addressed by careful control of the transducer coupling to the crystal.

What happens to the diffracted beam in an anisotropic medium (e.g. as in an AOTF device)? As given above, the birefringence induced by the acoustic wave causes the dif-fracted beam polarisation to change. So if, for example, the input beam is *P*-polarised, then

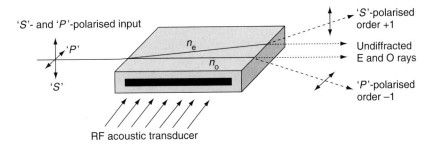

Figure 3.19 AOTF with polarisation rotation.

the diffracted exit beam will be S-polarised for the upshifted ray. If the crystal medium is anisotropic then, in addition to the Bragg diffraction, the polarisation-rotated exit beam will experience the anisotropic refractive index, n_e, compared with n_o for the input beam (Figure 3.19). It will propagate as an extraordinary ray which allows for additional beam separation. Clearly, an important factor in the set-up of an AOTF device is a proper analysis of the crystal symmetry, and a proper cleaving of the crystal to sit with the correct orientation with respect to the input beam polarisation and acoustic wavefront.

3.3.4.2 Practical aspects of operation

The very specific nature of the AO effect requires one to consider some particular issues regarding device performance. These include:

- Thermal conductivity of the AOTF crystal
- The maximum input aperture width to avoid bandpass compromise
- The full optical acceptance angle
- Diffraction efficiency
- Wavelength stability and the temperature-tuning coefficient
- Blocking efficiency
- RF driver tuning range.

Quite a large amount of input optical power is being channelled through the AOTF device, as well as the acoustic wave power being coupled in via the piezo-transducer. The AOTF crystal must have a high thermal conductivity to allow this power to dissipate without heating up the crystal. This is the main reason that TeO_2 has become the standard material for NIR AOTF, since it has a wide transparency from approximately 370 nm up to 3500 nm. However, the useable wavelength range for any particular device set-up is normally more limited, both by the detector range (typically 900–2600 nm for a TE-cooled InGaAs detector), and more importantly by the frequency drive range of the RF oscillator. At the level of stability required this can rarely exceed one octave. So a typical AOTF device might have a tuning bandwidth of 900–1800 nm, 1000–2000 nm, or 1200–2400 nm.

Although there are no slits associated with an AOTF device, there are some necessary compromises regarding throughput. First, there are physical restrictions on the size of available crystals, and in the case of the normal non-collinear arrangement where the input beam direction is traversed by the acoustic wavefront the open beam aperture needs to be limited so that the transit time of the acoustic wavefront across the crystal is limited, in order that a narrow optical bandpass can be maintained. Typically this is of the order of a few millimetres.

It has also been shown that the diffraction condition requires control of the incident beam angle. Normally a collimated beam is incident on the AOTF crystal, but there will be some beam divergence from a real non-point source. The full acceptance angle is often in the range of a few degrees only, so input beam divergence must be limited. It is for these reasons that AOTF analysers are very often designed as fibre-optic-coupled devices. The small input beam diameter is very well matched with a fibre-optic input, although the relatively high numerical aperture (NA) of a fibre-optic makes beam collimation within the full acceptance angle of the crystal difficult.

It has been noted that the undiffracted beam, which retains all the other wavelengths which have not been coupled into the narrow bandpass of the diffracted beam, is still present and separated from the emerging diffracted beam only by a small angle and its retained polarisation, which will now be at 90° to that of the diffracted beam. An important practical consideration is how well a combination of physical beam blocking and crossed polarisers will eliminate the untuned, out-of-band wavelengths which represent stray light. Again the use of fibre-optic coupling is one way to help separate these small-diameter exit beams. There are some design complications here because multimode optical fibres do not normally preserve input polarisation, so any polarisation selection must be done after the input fibre output, and before the output fibre input. The typical spectral resolution for an AOTF device is >5 nm at 2000 nm (roughly $12\,\text{cm}^{-1}$), so comparable to a scanning grating analyser.

Finally, and possibly, the most important practical issues for an AOTF device are those of temperature control and wavelength stability. As in the case of the grating polychromator, there is no inherent calibration for wavelength in the device. For any particular device, for a known angle of incidence and RF frequency the Bragg wavelength is determined in theory, but is dependent in fact on manufacturing tolerances and operational environment. We saw earlier that for demanding NIR PAC applications, a wavelength accuracy of at worst 0.05 nm at 2000 nm is typically required.

The temperature tuning coefficient of TeO_2 is 0.025 nm/°C, so it is clear that in order to combat the effects of thermal heating of the crystal due to incident optical and acoustic power, as well as environmental considerations, a good level of thermal control is needed. This can be in the form of a TE-cooled enclosure. This will deal to a certain extent with the need to achieve analyser wavelength repeatability. However, it does not address the issue of wavelength reproducibility between AOTF modules, which impacts the transportability of calibrations and datasets between analysers.

The key operational advantages of an AOTF device lie in the very rapid and random access wavelength selection possible by programming the RF driver frequency. For example, once an application is defined, an isophotonic scanning regime can be defined so that data acquisition is extended in spectral regions with high sample absorptivity, helping to give a uniform SNR across the spectrum. Similarly, combinations of discrete wavelengths can be accessed with a switching time below 10 μs. This, along with the low-voltage power requirements of the RF oscillator and the tungsten lamp source, lend the unit to miniaturisation and remote location. Thus AOTF systems have been used very effectively in 100% whole-tablet analysis applications and oil seed content discrimination applications on production lines.

3.3.5 Fourier transform near-infrared analysers

3.3.5.1 Principle of operation

So far, devices for NIRS which depend on some form of wavelength selection have been considered. This ranges from the physical selection by slits in a scanning grating analyser, through the selection by discrete detector elements in a PDA polychromator, to the electronic tuning selection of an AOTF device. For all of these devices, wavelength identification, and therefore spectral repeatability and reproducibility, is a design issue, albeit one for which many elegant and effective solutions have been found.

Fourier transform NIR analysers do not depend on any form of wavelength selection, but instead provide a method of wavelength-encoding so that all transmitted wavelengths may be measured simultaneously at a single detector. Moreover the mechanism by which the wavelength-dependent modulation is introduced into the d.c. NIR source output is itself controlled by a very precise wavelength-registered mechanism (a reference laser channel) – so that inherent frequency calibration of the NIR spectrum is achieved. The situation is not, of course, completely plain-sailing. To achieve this objective, quite demanding issues of optical alignment and integrity need to be maintained, but again elegant and effective solutions have been found and will be described below.

Two-beam interference requires the bringing together of two at least partially coherent wave-trains. In practice, the most efficient way to do this is to start with a single light source, generate two beams from this by division of amplitude, introduce an optical retardation into one of the beams, relative to the other, and then recombine the beams. This means that the two interfering beams are generated from the same light source, and the optical retardation introduced is small enough to be within the coherence length of the source.

For a wavelength λ and an optical retardation δ, one expects that the amplitude of the resultant beam after recombination (the interference signal) will be

$$I(\delta) = A_0 \left[1 + m \cos\left(\frac{2\pi\delta}{\lambda}\right)\right] \tag{3.22}$$

where A_0 is effectively the total intensity of the input beam, factored by the efficiency of the division–recombination process, and m is the modulation efficiency of the recombining beams. In the ideal case this is 1, but one will see the effect of changes in this parameter later.

Thus the a.c. part of this signal is

$$I'(\delta) = A_0 \; m \cos\left(\frac{2\pi\delta}{\lambda}\right) \tag{3.23}$$

Remembering that the wavenumber $\nu = 1/\lambda$, one has

$$I'(\delta) = A_0 \; m \cos(2\pi\delta\nu) \tag{3.24}$$

Therefore for fixed ν and a linearly changing δ, the output of the ideal two-beam interferometer is a cosine function. In other words, it is a cosine modulator of the original d.c. light source. The modulation frequency observed in the a.c. output of the FTIR spectrometer detector is dependent on the rate at which δ is increased or decreased, and is given by

$$f_m = \delta'\nu \tag{3.25}$$

where δ' is the rate of change of the optical retardation, in units of cm s^{-1}.

For a typical broadband NIR source (for example the tungsten-halogen lamp described previously) ν will cover a range from around 3500–12 000 cm^{-1}, and δ' might be 1 cm s^{-1} for a typical fast-scanning interferometer, so the observed range of frequencies in the a.c. detector output could be 3.5–12 kHz. In other words, signals typically in the audio-frequency range.

Assuming that the rate of change of the optical retardation introduced by the interferometer is the same for all the input radiation frequencies (which is normally the case), each individual value of ν in the broadband source output contributes a different value of f_m in the a.c. component of the detector output (see Figure 3.20). These contributions are, of course, summed by the detector. The combined detector output (see Figure 3.21) arising from the simultaneous measurement of all modulated input signals, is a sum of cosine functions. Such a sum is a Fourier series, and the amplitudes of the individual

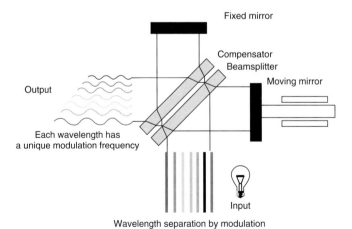

Figure 3.20 The generation of a Fourier-modulated NIR beam. Reprinted with kind permission of ABB Analytical and Advanced Solutions.

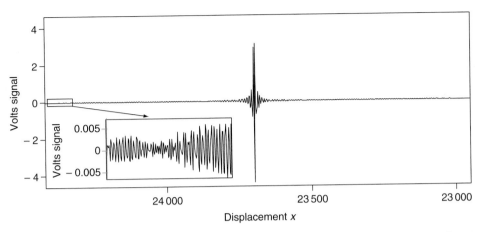

Figure 3.21 The interferogram of a broadband NIR source. Reprinted with kind permission of ABB Analytical and Advanced Solutions.

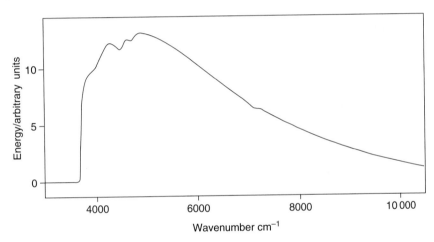

Figure 3.22 The Fourier transform of the interferogram. Reprinted with kind permission of ABB Analytical and Advanced Solutions.

input modulations at each Fourier frequency can be recovered by a Fourier decomposition. These amplitudes are in proportion to the optical intensity throughput of the spectrometer at each wavelength. This combination of modulation, summation and Fourier decomposition constitutes the essential and basic principle of FTIR, and explain how it allows to measure the intensity of all source wavelengths simultaneously. The combined detector signal for all incident radiation frequencies is thus

$$I_{TOT}(\delta) = \Sigma A_{\nu 0} m_{\nu} \cos(2\pi\delta\nu) \tag{3.26}$$

where the summation applies in principle over all wavelengths from zero to infinity, but in practice covers the throughput range of the overall source–interferometer–detector combination (see Figure 3.22).

There are a number of ways in which the necessary optical retardation can be introduced within the FTIR modulator. Historically, when FTIR spectrometers were developed initially for mid-IR operation, the classical Michelson interferometer design predominated, with one fixed plane mirror and a linearly scanned moving plane mirror (to provide the variation in δ). Although functional in the mid-IR, and capable of generating very high spectral resolution, such designs are not at all suitable for industrial and PAC applications in the NIR. The majority of FTIR analysers used for industrial process analytical applications use interferometer designs based on retroreflector mirrors (cube-corner mirrors or variants thereof, Figure 3.23) and rotational motion to achieve the variation in δ. There are also designs based on the use of a linearly mechanically scanned refractive optical wedge driven into one beam of the interferometer to achieve the necessary optical retardation.

The typical NIR source employed is the tungsten-halogen lamp, as described previously, although high-power silicon carbide (SiC) sources can also be used in combination with suitable beamsplitter materials, for example zinc selenide (ZnSe) for combined mid-IR and NIR operation. Typical detectors are pyroelectric DTGS for general-purpose applications, and cooled photoconductive detectors such as InAs and InGaAs for low throughput or fibre-optic-based applications.

3.3.5.2 Practical aspects of operation

3.3.5.2.1 The advantages of FTIR-based analysers
It is commonplace that FTIR-based analysers are the predominant technology for mid-IR applications. This arises from a unique tie-in between the inherent advantages of the FTIR method and serious limitations in the mid-IR range. The most serious problem for mid-IR spectroscopy is the very low emissivity of mid-IR sources combined with the low detectivity of mid-IR thermal detectors. This causes a direct and severe conflict between the desire for

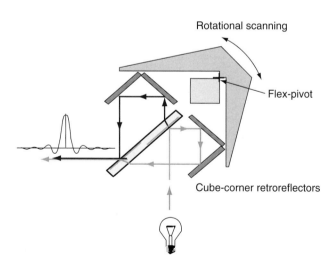

Figure 3.23 The retroreflector rotational modulator.

good SNR and optical resolution, when any form of wavelength-selective analyser design is employed.

The classical FTIR advantages are based on (1) geometry (the Jacquinot or throughput advantage), (2) statistics (the Fellgett or multiplex advantage), and (3) physics (the Connes or wavelength advantage).

(1) The Jacquinot advantage arises because of the lack of significant optical throughput restrictions in the optical layout of an FTIR analyser. There are no slits or other wavelength-selecting devices, and an essentially cylindrical collimated beam from the source travels unhindered through the beam paths to the detector. There is one proviso to this, in that a circular aperture Jacquinot-stop (J-stop) must be included at an intermediate focal plane to allow proper definition of achievable optical resolution, but the effect on throughput is minor. In fact, for many FTNIR analysers the source, the sample device itself or the detector element act as the J-stop and no additional throughput restriction occurs.

(2) The Fellgett advantage is based on the simultaneous modulation of all incident wavelengths, which allows the FTIR analyser to measure all wavelengths all of the time. For a total data acquisition time of T, a wavelength-selective device measures each wavelength for only T/m, where m is the number of resolution elements (determined by the slit width), whereas in the FTIR analyser each wavelength is measured for the whole of time T. If the detector noise is the dominant noise contribution, this allows an SNR advantage of $m^{1/2}$.

(3) The Connes advantage arises from the fact that the a.c. modulation frequency introduced into each wavelength of the source output is in effect measured using a very high wavelength-precision and wavelength-accuracy device (a HeNe laser). The single-frequency interferogram signal from a HeNe laser is used as an internal clock both to control the rate of change of δ, and to trigger the data acquisition of the resulting modulated NIR detector signal.

(4) There is in addition a fourth, unnamed FTIR advantage, in that FTIR analysers are immune to straylight effects. The detector is acting in effect as a lock-in amplifier, looking only at modulated signal in the audio-frequency range (those signals having been generated by the FTIR modulator). Thus incident external d.c. light components have no impact.

It is easy to see that for mid-IR operation the advantages (1) and (2) have an absolutely overriding significance. They produce by themselves a decisive advantage for the FTIR technique in mid-IR applications. The situation is not so clear at first sight in the NIR, but careful consideration also shows here that FTNIR-based analysers can demonstrate some significant advantages over alternative technologies.

The argument for the applicability (or lack of it) of the above factors in FTIR-based NIR applications centres on the issues of optical throughput and detector responsivity. NIR sources are comparatively very bright, and detector responsivities may be many orders of magnitude better than in the mid-IR. Hence in order to avoid detector saturation when operating in open-beam mode with no additional attenuation due to sample or sampling devices, an FTIR-based analyser using a photoconductive detector will need the NIR source to be significantly attenuated. The issue of SNR then becomes limited by the digital resolution of the fast analog-to-digital circuit (ADC) used to capture the detector signal, with the open-beam optical throughput and detector gain adjusted so that the detector's dark noise is just providing noise to the least significant ADC bits for co-adding. However, this situation can be manipulated to advantage. First, the availability of high optical throughput in an FTNIR analyser means that lower-cost, more robust and also more linear DTGS detectors (see Section 3.3.2.2.1) can be used routinely. These in turn allow (through the choice of suitable beamsplitter materials and source characteristics) a wider range of operation than the conventional pure NIR, including, for example, combined NIR and mid-IR operation in the same analyser in the same scan.

Secondly, many real-life situations in PAC applications are in fact severely throughput-limited. This might include multiplexed applications using fibre-optic interfaces, highly scattering samples (tablets, powders, slurries), highly absorbing samples (crude oil, bitumen), etc. In these circumstances the application can quickly become detector noise–limited, and the full power of the Jacquinot and Fellgett advantages apply.

Finally, however, it is the advantages (3) and (4) which really have a marked effect on NIR applications. The wavelength precision and reproducibility achievable with a good FTIR analyser design translates into long-term spectroscopic stability both within one analyser and across multiple analysers, so that measurement errors between analysers are in the region of a single milliabsorbance unit. For the untroubled development of NIR calibration models, and their maintainability over time, this of itself is critically important.

3.3.5.2.2 The implementation of FTIR technology for NIR process analytical applications

The achievement in practice of the theoretical advantages of FTIR technology, especially for NIR process analytical applications, is a demanding task. Although essentially a simple device, the FTIR modulator is very unforgiving in certain areas. For example, failure exactly to control the relative alignment of beamsplitters and mirrors during the scanning of optical retardation, and the divergence of the collimated input beam from the NIR source both lead to degradation in achievable optical resolution and error in absorbance bandshapes and wavelengths. These are important errors, especially in quantitative applications, and their presence would severely limit the usefulness of FTNIR, were no ways available to control them.

An approximate calculation for the differences in optical retardation for the central and the extreme ray of a beam-path which is diverging as it travels through the FTIR modulator, separated by a half-angle (in radians) of α, shows that the maximum allowable value of α to achieve a required resolution $\Delta\nu$ at a wavenumber of ν_{MAX} is

$$\alpha_{\text{MAX}} = \left(\frac{\Delta\nu}{\nu_{\text{MAX}}}\right)^{1/2} \tag{3.27}$$

More seriously, the effect of the above beam divergence also impacts wavenumber accuracy, and the deviation in measured wavenumber for the extreme ray vs. the central ray is approximated by

$$\nu' = \nu\left(1 - \frac{1}{2}\alpha^2\right) \tag{3.28}$$

Taking an example of a measurement at $5000\,\text{cm}^{-1}$ with a required resolution of $2\,\text{cm}^{-1}$ but a wavenumber accuracy requirement of $0.04\,\text{cm}^{-1}$ (which are entirely typical of the requirements for a demanding NIR quantitative application), the limit on α_{MAX} is about $0.8°$ for the resolution, but around $0.2°$ for the wavenumber accuracy.

However, the most significant practical issue for the implementation of an FTNIR device suitable for quantitative process analytical applications arises from the need to maintain a very high degree of alignment at the point of recombination of the divided interferometer beam paths – in other words at the exit of the interferometer.

In Equation 3.24 the output signal from an FTIR modulator was described as

$$I'(\delta) = A_0\, m\cos(2\pi\delta\nu)$$

where m was described as the modulation efficiency, and should ideally have a value of 1. Figure 3.24 shows how the modulation efficiency of the FTIR device is affected when the recombining exit beams are no longer parallel and coincident. The effect is that the theoretically uniform irradiance across the circular exit aperture, at any given value of

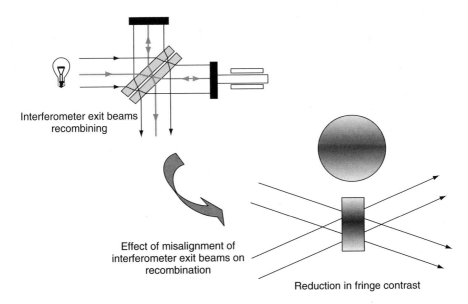

Interferometer exit beams recombining

Effect of misalignment of interferometer exit beams on recombination

Reduction in fringe contrast

Figure 3.24 Recombination at the exit of the interferometer.

optical retardation, is broken up into a series of fringes, and the detector signal is then an average over these fringes. This amounts to a loss of contrast in the interferogram signal, introducing the effects described above for beam divergence. This misalignment of the exit beams is precisely what will happen in case there is any tilt misalignment of plane mirrors used in a standard Michelson arrangement. For mid-IR FTIR devices, especially those required to operate at very high optical resolution (and therefore requiring large amounts of optical retardation), successful solutions have been found using dynamic alignment systems in which the scanning plane mirror is dynamically tracked and realigned during the scanning process. This type of arrangement is however totally unsuitable for a process analytical environment, and for that type of FTNIR application the universally adopted and successful strategy has been to use optical layouts with inherent tilt-compensation. The most successful designs typically use a combination of cube-corner retroreflectors, and a rotational motion for the mirror-scanning mechanism around a fixed balance point (Figure 3.25). This avoids the need for a linear mechanical scanning mechanism, and allows for minimal individual mirror movement, since both mirrors are moving in opposition, and the effective optical retardation is doubled.

There are of course other causes for exit beam misalignment other than simply the process of optical retardation by mirror scanning. The overall alignment integrity of the cube-corner/beamsplitter/cube-corner module as well as its optical relationship to the collimated input beam from the NIR source, and the collinear HeNe laser are all critical. The thermal and mechanical instability of this whole assembly is roughly a function of its volume. Small compact FTIR modulators will perform better in process analytical

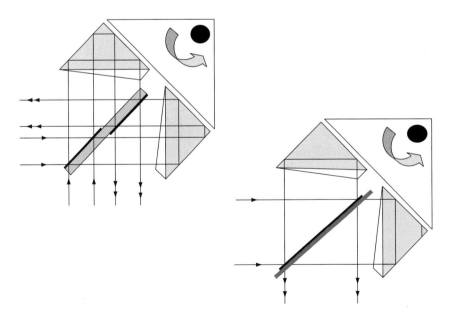

Figure 3.25 Cube-corner retroreflector interferometers with rotational scanning.

applications than those less compactly designed. Fortunately the cube-corner modulator layout in Figure 3.25 allows for both a compact layout and a tilt compensation. The cube-corner retroreflector works perfectly only when the three plane mirror sections of which it is composed are set exactly orthogonal at 90° to each other. Under that condition it also displays near-perfect optical stability with respect to temperature changes. For NIR applications the accuracy of orthogonality should be around one second of arc, which is now an achievable manufacturing specification on a routine basis.

The optimum FTIR modulator design, based on the above principles, will depend on factors such as the wavelength range of operation, the required optical resolution and also issues such as the packaging requirements and the need to deal with field-mounting or hostile environments. This is an area of active current development with emerging possibilities and variations around the central theme of rotating retroreflector modulator designs.

What is quite clear from current experience is that the optical and mechanical simplicity of FTNIR analysers, coupled with their compact implementation and appropriate packaging, allows the exploitation of the advantages of FTIR devices in NIR process analytical applications. The current generation of industrial FTIR devices has a robustness and longevity entirely in keeping with the needs of quantitative industrial process monitoring applications, despite the very real engineering challenges described above. The other NIR technologies discussed in this chapter can also be successfully used in process analytical applications.

3.4   The sampling interface

3.4.1   Introduction

Section 3.3 described a number of alternative design and implementation strategies for NIR analysers, suitable for operation in a process analytical environment. However, none of these analysers can operate without a robust, maintainable and repeatable sampling interface with the process sample under consideration. In terms of practical issues this is certainly the most demanding aspect of all. It requires a number of issues to be addressed and clearly understood for each specific application to be attempted. These include:

- The physical nature of the process stream. Is it single phase or two phase? Is it liquid, solid, vapour or slurry? What is its temperature and pressure at the sampling point, and how far can these be allowed to change during sampling? What is its viscosity at the appropriate sample measurement temperature?
- The chemical nature of the process stream. Is it at equilibrium (a final product) or is it to be measured mid-reaction? Is sample transport possible, or must the sample be measured in situ? Is it corrosive, and what material and metallurgical constraints exist?
- The optical nature of the process stream. Is it a clear fluid or scattering, highly absorbing, diffusely transmitting or reflecting?

- What issues concerning sample system or probe fouling and process stream filtration or coalescing need to be addressed?
- Where is it possible or desirable to locate the NIR analyser module? What hazardous area certification requirements apply to both analyser and sampling system?
- Are there multiple process streams to be sampled? If so, can they be physically stream-switched, or does each stream require a dedicated sampling interface?
- If fibre-optics are to be used, what physical distances are involved, and what civil engineering costs are implied (for gantry, cable-tray, conduit, tunnelling, etc.)?
- What are the quantitative spectroscopic demands of the calibration? How much impact will vary in the sampling interface (as opposed to analyser instability) has on the likely calibration success and maintainability? What issues concerning transport of laboratory-developed calibration models to the process analyser need to be addressed?
- What NIR spectral region will be targeted for the calibration development; for example, combination region, first combination/overtone (FCOT), second combination/overtone (SCOT)? Each region will imply a different transmission pathlength, and hence cell or probe design.

Depending on the answers to these questions, various sampling interface strategies may be developed which will include:

- *Extractive sample conditioning systems.* Based on a sample fast-loop with a sample take-off probe in the main process line, and a return to either a pump-suction or an ambient pressure sample recovery system (Figure 3.26). The fast-loop provides a filtered slip-stream to the analyser, along with flow control and monitoring (low-flow alarm), a sample autograb facility for sample capture, and allows for a very high degree of sample temperature control (typically 0.1°C) within the analyser (see Figure 3.27). This option is mainly applicable to single-phase liquid equilibrium process streams – and is the *de facto* standard for refinery hydrocarbon applications, where the equilibrium (non time-dependant) sample condition is met and where the quantitative stability and reproducibility aspects of the application demand the highest level of sampling control. Where physical stream-switching is acceptable, this arrangement allows for multiple-sampling fast-loops to feed a single NIR analyser, with stream switching controlled by the analyser. The analyser will most likely need to be hazardous area compliant, but may be either located in a dedicated analyser shelter, or field-mounted.

- *Local extractive/fibre-optic-based systems.* This is a development of the above where a fibre-optic-linked liquid sample transmission cell is integrated with the sample fast-loop cabinet (Figures 3.28 and 3.29). There can be multiple sample streams, take-offs and fast-loops, each with its own separate fibre-optic transmission cell. The analyser can be either local with short fibre-optic runs to the sampling cabinet(s), or remote, where a safe area location for the analyser module may be feasible, but at the cost of longer, potentially less stable fibre-optic runs. This system allows to avoid physical stream switching.

- *Remote fibre-optic-based systems.* These can be in a variety of formats depending on the nature of the sample, and examples of these are shown in Figures 3.30–3.33. In the case of an equilibrium sample, but at high process temperatures, this can allow for the

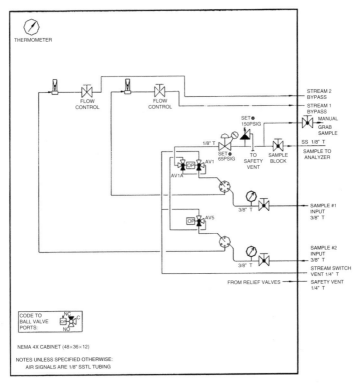

Figure 3.26 Extractive sample conditioning fast-loop. Reprinted with kind permission of ABB Analytical and Advanced Solutions.

location of a dedicated high-temperature fibre-optic sample cell cabinet very close to the sample take-off point, minimising the need for heated sample transport lines. A number of such cabinets can sample multiple process streams and feedback to a centrally located hazardous area–certified fibre-optic multiplexed analyser module, or at the cost and risk of longer fibre runs, out to a safe area. Alternatively, for non-equilibrium process streams, where a pumped reactor sample recycle line is available, in-line fibre-optic transmission cells or probes can be used to minimise sample transport. It is highly desirable that some form of pumped sample bypass loop is available for installation of the cell or probe, so that isolation and cleaning can take place periodically for background reference measurement. In the worst case, where either sample temperature, or pressure or reactor integrity issues make it impossible to do otherwise, it may be necessary to consider a direct in-reactor fibre-optic transmission or diffuse reflectance probe. However, this should be considered the position of last resort. Probe retraction devices are expensive, and an in-reactor probe is both vulnerable to fouling and allows for no effective sample temperature control. Having said that, the process chemical applications that normally require this configuration often have rather simple chemometric modelling development requirements, and the configuration can be made to work.

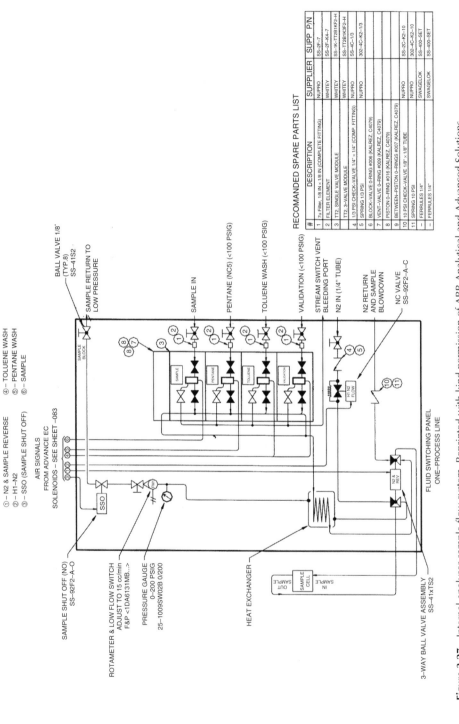

Figure 3.27 Internal analyser sample flow system. Reprinted with kind permission of ABB Analytical and Advanced Solutions.

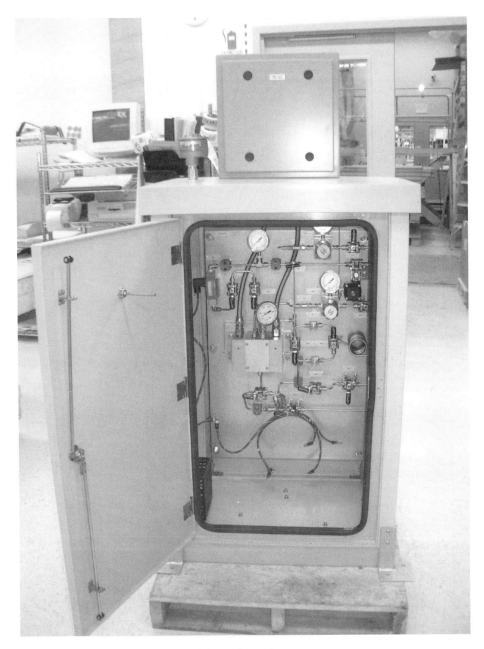

Figure 3.28   Example fibre-optic-coupled flow cell sample system.

Figure 3.29   Example fibre-optic-coupled flow cell detail.

Figure 3.30   Example fibre-optic-coupled in-line transmission cell. (Courtesy of Specac Limited, www.specac.co.uk)

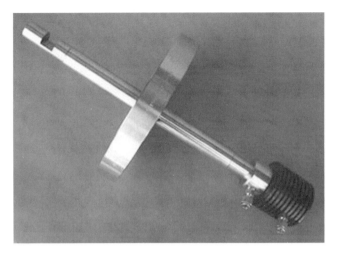

Figure 3.31   Fibre-optic-coupled insertion transmittance and diffuse reflectance probes. (Courtesy of Axiom Analytical Inc., www.goaxiom.com)

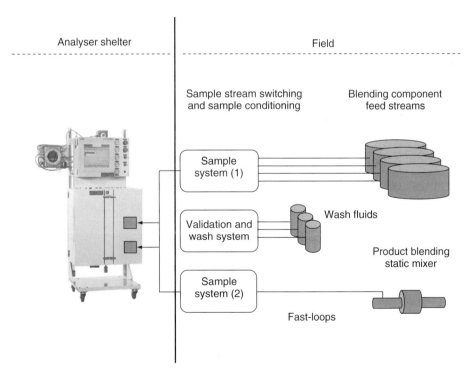

Figure 3.32   Typical field configuration for extractive on-line process analyser.

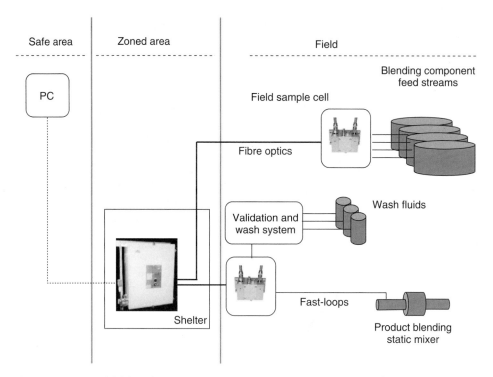

Figure 3.33 Typical field configuration for fibre-optic on-line process analyser.

3.4.2 *Further discussion of sampling issues*

- Liquids vs. Solids vs. Slurries

These three main classes of process sample streams are in increasing order of difficulty for NIR process analysis. In general, liquid streams are best measured in a transmission sampling mode, solids (powders) in diffuse reflectance mode, and slurries in either diffuse reflectance or diffuse transmission mode according to whether the liquid phase or the suspended phase is of greater analytical significance. If the latter and diffuse transmission is used, a specialised set-up using an NIR analyser, a fibre-optic bundle for illumination and a local dedicated high-sensitivity large area detector (for example, a silicon avalanche photodiode detector, as described earlier) in close proximity to the sample (in order to maximise light collection) will very likely be needed.

- When considering liquid sampling, issues of sample pathlength, sample temperature and pressure, and sample viscosity, along with the overall process engineering environment, will influence the final decision concerning analyser and sample interface format. Many of these issues are interconnected. Determination of the optimum spectral region for calibration model development may suggest use of the combination region

(4000–5000 cm^{-1}), which will imply the use of relatively short pathlengths (around 0.5 mm or less). This option is both feasible and desirable for light, clean hydrocarbon streams. This set-up is often the chosen standard for refinery and petrochemical applications, where the convenience and additional sample control provided by a filtered bypass loop, sample back-pressure and precise sample temperature control all pay large dividends relative to the rather trivial extra effort of sample system maintenance. However, for both higher-temperature and higher-viscosity sample streams this degree of sample transport and manipulation may not be possible, and selection of a wavelength region requiring relatively short pathlengths may not be viable. In those cases, in increasing order of potential difficulty, remote fibre-optic-coupled sample transmission cell cabinets, in-line fibre-optic transmission cells installed in a pumped bypass loop, in-line fibre-optic transmission probes and finally in-reactor fibre-optic transmission probes may be used.

• Although the use of a sample fast-loop sample conditioning system is not always possible, it is important to realise the benefits, which are sacrificed, if the decision is made to use alternative interface formats. First and foremost, the sample conditioning system allows for easy sample cell cleaning, maintenance and the use of reference or calibration fluids. Also it allows to incorporate a local process sample take-off or autograb facility, to ensure that recorded spectra and captured samples are correctly associated for calibration development or maintenance. In addition many process streams operate at elevated pressure, and although this has a negligible effect on the NIR spectrum, uncontrolled pressure drops can cause cavitation and bubbling. A sample system allows for either sample back-pressure and flow control, or a sample shut-off valve to stop flow against a back-pressure during measurement to eliminate this risk. Some of these benefits are also realisable in a pumped bypass loop for in-line cell or probe installation. Here it is normally possible to arrange isolation valves to allow for cell or probe removal between process shutdowns. Also wash or calibration fluids can be introduced, and to some extent sample flow and pressure can be controlled. The control of sample temperature is however somewhat compromised, and calibration modelling strategies will have to be adopted to deal with that. It is for these reasons, rather than any fundamental issues relating to NIR technology or spectroscopy, that those applications where a reliable bypass or fast-loop cannot realistically be envisaged (for example hot polymerisation or polycondensation processes) have proved the most demanding for NIR process analysis. Where (as in these cases) direct in-reactor insertion transmission probe installation is the only physically practical option, issues of probe fouling, reference measurement, sample capture and probe resistance to process conditions must be addressed.

• As mentioned above, the sampling of solids (powders) is also a challenging objective, but one for which quite satisfactory solutions can be found. The typical NIR process analysis applications for solids are for powder drying (either moisture or solvents) and for blend content uniformity determinations. Though the energy throughput of a fibre-optic-coupled diffuse reflectance probe in contact with a suitably scattering powder sample may be of orders of magnitude less than the throughput of a simple clear liquid transmission measurement, this is an area where the available power output of an FTNIR analyser or use of more sensitive detectors can be beneficial, and allow the use of smaller fibre-optic bundles, multiplexed, multichannel sample outputs, and hence longer fibre-optic runs at less cost. The main difficulty in developing quantitative calibration models for the measurement of solid process streams is that of sample presentation and homogeneity. Unlike liquid process streams, process powders are often transported or manipulated in ways which make them difficult to sample reproducibly. Powders flowing down pipes may not offer a uniform density to a sample probe head. Powders in dryers or blenders may either cake the probe head or simply sit around it, unrefreshed by new sample. Thus sample probe location and environment are critical considerations when setting up solids applications. To make this decision appropriately, knowledge of the process dynamics is needed.

3.4.3   The use of fibre-optics

3.4.3.1   Principles of operation

Simply stated, fibre-optic cable allows for the transfer of light from one point to another without going in a straight line, and as such their use in process analytical applications is highly seductive. In a likely NIR process application there may be multiple sample stream take-offs, each with different sample characteristics, located in positions spread around a process unit, with the only apparently convenient analyser location some distance away.

Fibre-optic cables work due to the principle of total internal reflection, when an incident light beam falls on an interface between two media of differing refractive indices, going in the direction of higher to lower refractive index. In a fibre-optic cable the higher refractive index is provided by the fibre core, and the lower by a cladding (see Figure 3.34). This can be either a totally different material (for example, a polymeric coating) in close optical contact with the core, or a layer formed around the core made from the same material but with added doping to change the refractive index. In this latter case, the change in refractive index between the core and the cladding can be either abrupt (a step-index fibre) or gradual (a graded-index fibre).

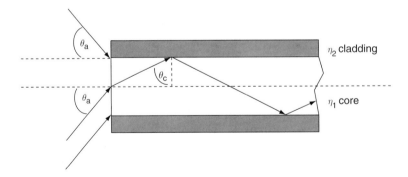

Figure 3.34   Internal reflection and acceptance angle in a fibre-optic cable.

The maximum input acceptance angle for a fibre-optic cable is determined by the critical angle for total internal reflection, which is given by Snell's law.

$$\theta_c = \sin^{-1}\left(\frac{\eta_1}{\eta_2}\right) \tag{3.29}$$

From which the acceptance angle is

$$\theta_a = \sin^{-1}(\eta_1{}^2 - \eta_2{}^2)^{1/2} \tag{3.30}$$

and the fibre-optic NA

$$\mathrm{NA} = \sin\theta_a = (\eta_1{}^2 - \eta_2{}^2)^{1/2} \tag{3.31}$$

Typical values for low-hydroxyl silica fibres are in the range of 0.22–0.3.

Note that the propagating wave through the fibre-optic core can potentially do so in a number of geometric forms (depending on the electric field amplitude distribution across the plane of the core) referred to as modes. Even for a monochromatic input, modes with a more direct path through the core will travel down the fibre with a larger linear velocity than those with a more transverse component. This leads to the effect of mode dispersion, where a short input light pulse is stretched as it travels down the fibre. Hence for digital and telecommunications purposes very narrow step-index fibres, which will typically transmit only a single-mode in a limited bandpass of wavelengths, are more suitable. By contrast, for process analytical spectroscopy, where continuous broadband sources are used and the dominating requirement is the overall transmission efficiency of the fibre, larger-diameter multimode fibres are completely acceptable.

3.4.3.2 Losses in fibres

The throughput of an optical fibre depends on a number of factors including:

- The input fibre diameter
- The input NA
- The length of the fibre
- The number of impurity centres causing scattering
- The concentration of impurities causing absorption
- The absorption edge of the core material itself
- The installed configuration of the fibre, in particular the minimum bending radius to which it is subjected.

All of these factors need to be considered against the cost, the robustness, and the amount of light available to couple into the fibre from the NIR process analyser.

In general the physical dimensions of the source image, as it is focused onto the fibre input, even when the NAs are matched will be larger than the fibre diameter, and there will be consequent injection losses for smaller diameter fibres. However, these may be compensated for by the fact that the spread of input angles, when the source image is stopped down in this way, is reduced, and the overall fibre transmission, once the light is inside, is improved compared with a larger-diameter fibre.

Also, for parallel multichannel devices, the source image can be divided, and injected simultaneously into a number (for example up to eight) of output fibres. Alternatively for more demanding diffuse reflectance or diffuse transmission applications with much lower optical throughput the whole source image dimension can be used, but injected into micro- or macro-bundle fibre optics.

A typical low hydroxyl silica fibre used for NIR liquid sampling applications should be a single-fibre, approximately 250–300 µm in diameter. This provides a good compromise between injection losses, transmission losses, cost and robustness. In general, larger-diameter single fibres are more fragile and more expensive (typically in the ratio of the fibre cross-sectional area). Alternatively for lower-throughput applications micro-bundles of 7×200-µm fibres (effectively a 600-µm fibre overall) can be used, or mini- macro-bundles up to a total of 2-mm diameter. These increase in cost proportionally, but, being made up of multiple small fibres, are far more robust than a large single fibre.

For NIR applications the overwhelmingly most common fibre-optic material is low-hydroxyl silica. For realistic run lengths this allows for use down to about 2000–2200 nm before the silica absorption band edge cuts in too strongly, depending on the exact fibre length. Their successful use is made possible by careful control of the number of −OH groups left as impurities in the silica core. There are significant absorption contributions from −OH centres, and these must be kept to the minimum level possible. This is because the other fibre loss mechanisms such as scattering or bending tend in general to produce fairly benign and easily handled distortions in the resultant sample absorbance spectrum (for example, baseline offsets, tilts or curvature), which can be dealt with by standard mathematical correction procedures (baseline correction or multiplicative scatter correction). However, the −OH absorption bands are chemical in origin, and they are typically temperature-sensitive as regards intensity, position and band shape. Unless these features are minimised, the risk of non-reproducible absorption-like band distortions appearing in the sample spectrum is high, which will be difficult to deal with in the calibration model development.

Alternative fibre-optic materials exist, but the only serious candidate for PAC applications is ZrF_4, which has the property of transmitting down to about 5500 nm, in other words well across the 4000–5000 cm^{-1} combination band region. However, it is fragile and extremely expensive compared with silica fibre. This makes it a good possibility for specialised refinery hydrocarbon applications which require not only the combination band region, but also the physical separation of streams. A typical analyser format would see a hazardous area–compliant multichannel analyser located close-coupled with ZrF_4 fibres to a multi-cell, multi-stream sampling cabinet.

3.5 Conclusion

In surveying the main available technologies applicable for NIR PAC applications, it is clear that a very wide range of analyser and sampling interface technologies can be brought to bear on any specific proposed process analytical application. This spread of techniques allows for a remarkable flexibility. The careful use of this flexibility to select the optimum analyser and sampling configuration for the target application is one key to success. The other keys to success in a NIR PAC application lie outside the scope of this chapter, but are no less important and are covered elsewhere. They include key issues such as calibration modelling development techniques, project management, application support and ongoing performance audit, validation and project maintenance. However, with all these issues correctly addressed, NIR technologies allow a unique opportunity to provide the highest-quality real-time process analytical information ideally suited to process control and optimisation techniques. It is a hallmark of modern NIR process analysers, with their improved reliability, sampling methods, and suitable process-oriented software and communication capabilities that they are more and more a normal and indispensable part of process unit control strategies.

Bibliography

Koch, K.H., *Process Analytical Chemistry: Control, Optimisation, Quality and Economy*; Springer-Verlag; Berlin, 1999.

Chalmers, J. (ed.), *Spectroscopy in Process Analysis*; Sheffield Academic Press; Sheffield, 2000.

McClellan, F. and Kowalski, B.R. (eds), *Process Analytical Chemistry*; Blackie Academic and Professional; London, 1995.

Williams, P. and Norris, K. (eds), *Near-Infrared Technology in the Agricultural and Food Industries*, 2nd Edn; American Association of Cereal Chemists, Inc.; St Paul, MN, 2001.

Wetzel, D.L., Contemporary Near-Infrared Instrumentation. In Williams, P. and Norris, K. (eds), *Near-Infrared Technology in the Agricultural and Food Industries*, 2nd Edn; American Association of Cereal Chemists; St Paul, MN, 2001; pp. 129–144.

Brimmer, P.J., DeThomas, F.A. and Hall, J.W., Method Development and Implementation of Near-Infrared Spectroscopy in Industrial Manufacturing Processes. In Williams, P. and Norris, K. (eds), *Near-Infrared Technology in the Agricultural and Food Industries*, 2nd Edn; American Association of Cereal Chemists; St Paul, MN, 2001; pp. 199–214.

Miller, C.E., Chemical Principles of Near-Infrared Technology. In Williams, P. and Norris, K. (eds), *Near-Infrared Technology in the Agricultural and Food Industries*, 2nd Edn; American Association of Cereal Chemists; St Paul, MN, 2001; pp. 19–38.

McClure, W.F., Near-Infrared Instrumentation. In Williams, P. and Norris, K. (eds), *Near-Infrared Technology in the Agricultural and Food Industries*, 2nd Edn; American Association of Cereal Chemists; St Paul, MN, 2001; pp. 109–128.

Burns, D.A. and Ciurczak, E.W. (eds), *Handbook of Near-Infrared Analysis*, 2nd Edn; Marcel Dekker, Inc.; New York, 2001.

Ciurczak, E.W., Principles of Near-Infrared Spectroscopy. In Burns, D.A. and Ciurczak, E.W. (eds), *Handbook of Near-Infrared Analysis*, 2nd Edn; Marcel Dekker, Inc.; New York, 2001; pp. 7–18.

McCarthy, W.J. and Kemeny, G.J., Fourier Transform Spectrophotometers in the Near-Infrared. In Burns, D.A. and Ciurczak, E.W. (eds), *Handbook of Near-Infrared Analysis*, 2nd Edn; Marcel Dekker, Inc.; New York, 2001; pp. 71–90.

Workman, J.J. and Burns, D.A., Commercial NIR Instrumentation. In Burns, D.A. and Ciurczak, E.W. (eds), *Handbook of Near-Infrared Analysis*, 2nd Edn; Marcel Dekker, Inc.; New York, 2001; pp. 53–70.

Siesler, H.W., Ozaki, Y., Kawata, S. and Heise, H.M. (eds), *Near-Infrared Spectroscopy: Principles, Instruments, Applications*; Wiley-VCH; Weinheim (Germany), 2002.

Wilson, J. and Hawkes, J.F.B., *Optoelectronics: An Introduction*; Prentice-Hall International; London, 1983.

Griffiths, P.R. and de Haseth, J., *Fourier Transform Infrared Spectrometry*; Wiley-Interscience; New York, 1986.

Hendra, P., Jones, C. and Warnes, G., *Fourier Transform Raman Spectroscopy*; Ellis Horwood; Chichester, 1991.

Mirabella, F.M. (ed.), *Modern Techniques in Applied Molecular Spectroscopy*; John Wiley & Sons; New York, 1998.

Brown, C.W., Fibre Optics in Molecular Spectroscopy. In Mirabella, F.M. (ed.), *Modern Techniques in Applied Molecular Spectroscopy*; John Wiley & Sons; New York, 1998; Chapter 10. pp. 377–402.

Chalmers, J.M. and Griffiths, P.R. (eds), *Handbook of Vibrational Spectroscopy*; John Wiley & Sons; New York, 2002.

Wetzel, D.L., Eilert, A.J. and Sweat, J.A., Tunable Filter and Discrete Filter Near-Infrared Spectrometers. In Chalmers, J.M. and Griffiths, P.R. (eds), *Handbook of Vibrational Spectroscopy*, vol 1; John Wiley & Sons; New York, 2002, pp. 436–452.

Wetzel, D.L. and Eilert, A.J., Optics and Sample Handling for Near-Infrared Diffuse Reflection. In Chalmers, J.M. and Griffiths, P.R. (eds), *Handbook of Vibrational Spectroscopy*, vol 2; John Wiley & Sons; New York, 2002, pp. 1163–1174.

Kruzelecky, R.V. and Ghosh, A.K., Miniature Spectrometers. In Chalmers, J.M. and Griffiths, P.R. (eds), *Handbook of Vibrational Spectroscopy*, vol 1; John Wiley & Sons; New York, 2002, pp. 423–435.

Miller, C.E., Near-Infrared Spectroscopy of Synthetic and Industrial Samples. In Chalmers, J.M. and Griffiths, P.R. (eds), *Handbook of Vibrational Spectroscopy*, vol 1; John Wiley & Sons; New York, 2002, pp. 196–211.

DeThomas, F.A. and Brimmer, P.J., Monochromators for Near-Infrared Spectroscopy. In Chalmers, J.M. and Griffiths, P.R. (eds), *Handbook of Vibrational Spectroscopy*, vol 1; John Wiley & Sons; New York, 2002, pp. 383–392.

Stark, E.W., Near-Infrared Array Spectrometers. In Chalmers, J.M. and Griffiths, P.R. (eds), *Handbook of Vibrational Spectroscopy*, vol 1; John Wiley & Sons; New York, 2002, pp. 393–422.

Goldman, D.S., Near-Infrared Spectroscopy in Process Analysis. In Meyers, R.A. (ed.), *Encyclopedia of Analytical Chemistry*, vol 9: *Process Instrumental Methods*; John Wiley & Sons; New York, 2000, pp. 8256–8263.

Coates, J.P. and Shelley, P.H., Infrared Spectroscopy in Process Analysis. In Peterson, J.W. (ed.), *Encyclopedia of Analytical Chemistry*, vol 9: *Process Instrumental Methods*; John Wiley & Sons; New York, 2000, pp. 8217–8239.

Gies, D. and Poon, T.-C., Measurement of Acoustic Radiation Pattern in an Acousto-Optic Modulator; Proceedings of IEEE SoutheastCon 2002, pp. 441–445.

Kauppinen, J., Heinonen, J. and Kauppinen, I., Interferometers Based on Rotational Motion; *Applied Spectroscopy Reviews* 2004, 39, 99–129.

Hoffman, U. and Zanier-Szydlowski, N., Portability of Near-Infrared Spectroscopic Calibrations for Petrochemical Parameters; *J. Near Infrared Spectrosc.* 1999, 7, 33–45.

Dallin, P., NIR Analysis in Polymer Reactions; *Process Contr. Qual.* 1997, 9, 167–172.

Chapter 4

Infrared Spectroscopy for Process Analytical Applications

John P. Coates

Abstract

Infrared (IR) spectroscopy offers many unique advantages for measurements within an industrial environment, whether they are for environmental or for production-based applications. Historically, the technique has been used for a broad range of applications ranging from the composition of gas and/or liquid mixtures to the analysis of trace components for gas purity or environmental analysis. The instrumentation used ranges in complexity from the simplest filter-based photometers to opto-mechanically complicated devices, such as Fourier transform infrared (FTIR) instruments. Simple non-dispersive infrared (NDIR) instruments are in common use for measurements where well-defined methods of analysis are involved, such as the analysis of combustion gases for carbon oxides and hydrocarbons. In complex measurement situations it is necessary to obtain a greater amount of spectral information, and either full-spectrum or multiple wavelength analyzers are used. Of the analytical techniques available for process analytical measurements, IR is the most versatile, where all physical forms of a sample may be considered – gas, liquids, solids, and even mixed phase materials. Over the past two decades a wide range of sample interfaces (sampling accessories) have been developed for IR spectroscopy and many of these may be adapted to either near-line/at-line production control or on-line process monitoring applications. Another aspect that is sometimes ignored is sample condition as it exists within the process, and issues such as temperature, pressure, chemical interferants (such as solvents), and particulate matter need to be addressed. In the laboratory these are handled within sample preparation; for process applications this has to be handled by automated sampling systems.

Instrumentation used within an industrial environment normally requires special packaging to protect the instrument from the environment, and to protect the operating environment from the instrument, relative to fire and electrical safety. This places special constraints on the way that an instrument is designed and the way that it performs. This places important demands on the mechanical and electrical design, providing the reliability and stability necessary to accommodate 24/7 operational performance. Traditional methods of IR measurement require significant adaptation to meet these needs, and in most cases, the instrumentation must be developed from the ground up to ensure that the performance demands are met. In the past 5 years a number of new technologies have become available, many as a result of the telecommunications developments of the late 1990s. While not necessarily directly applicable now, many of these technologies lend a way to enhance performance of IR measurement devices, and are expected to provide new generations of smaller and lower-cost instrumentation or even spectral sensors.

4.1   Introduction

Infrared (IR) spectroscopy is undoubtedly one of the most versatile techniques available for the measurement of molecular species in the analytical laboratory today, and for applications beyond the laboratory. A major benefit of the technique is that it may be used to study materials in almost any form, and usually without any modification; all three physical states are addressed – solids, liquids, and gases.

Before going any further it is important to define IR spectroscopy in terms of its measurement range. It has been suggested that based on the information content, which is assigned to molecular vibrations, the operational wavelength range for IR spectroscopy is from 800 to 200 000 nm (0.8–200 μm), or from 12 000 to 50 cm^{-1}, in frequency units. This can be separated functionally into what is defined as fundamental and overtone spectroscopy. This leads to the classical view of IR spectroscopy, which is based on the terms 'mid-infrared' (mid-IR) and 'near-infrared' (NIR). NIR, as traditionally applied to process monitoring applications, is discussed in a chapter elsewhere in this publication. However, NIR applications over the years have been viewed in very well-defined areas, notably agricultural and food, chemical and petrochemical, and pharmaceutical industries. Applications that were once considered to be classical mid-IR applications, such as gas monitoring applications, are now becoming popular within the NIR spectral region, primarily because of laser-based technologies and other technologies developed from the telecoms industries. Such applications of NIR will be mentioned later in this chapter.

The mid-IR region covers the fundamental vibrations of most of the common chemical bonds featuring the light- to medium-weight elements. In particular, most gases (with the exception of diatomic molecular gases such as oxygen and nitrogen) and organic compounds are well represented in this spectral region. Today, the mid-IR region is normally defined in terms of the normal frequency operating range of laboratory instruments, which is 5000–400 cm^{-1} (2–25 μm). Note that today, for most chemical-based applications, it is customary to use the frequency term 'wavenumber' with the units of cm^{-1}. Some of the early instruments, which utilized prism-based dispersion optics, fabricated from sodium chloride, provided a scale linear in wavelength (microns or more correctly, micrometers, μm). These instruments provided a spectral range of 2.5–16 μm (4000–625 cm^{-1}). With the wavelength (λ) described in microns, the relationship to the frequency (ν) in wavenumbers is provided by $\nu = 10\,000/\lambda$. The upper limit is more or less arbitrary, and is chosen as a practical limit based on the performance characteristics of laboratory instruments where all fundamental absorptions occur below this limit. The lower limit, in many cases, is defined by a specific optical component, such as a beamsplitter with a potassium bromide

(KBr) substrate which has a natural transmission cut-off just below $400\,\mathrm{cm}^{-1}$. An understanding of this spectral range is important because it dictates the type of optics, optical materials and optical functional components, such as sources and detectors, that have to be used.

IR spectroscopy is one of the oldest spectroscopic measurements used to identify and quantify materials in on-line or near-line industrial and environmental applications. Traditionally, for analyses in the mid-IR, the technologies used for the measurement have been limited to fixed wavelength NDIR filter-based methods and scanning methods based on either grating or dispersive spectrophotometers or interferometer-based FTIR instruments. The last two methods have tended to be used more for instruments that are resident in the laboratory, whereas filter instruments have been used mainly for process, field-based and specialist applications, such as combustion gas monitoring.

Only a few classical scanning dispersive instruments have been used for mid-IR applications in the non-laboratory applications, and no further commentary in regard to their application outside of the laboratory will be made here. However, there are new technologies emerging where non-scanning dispersive instruments can play an important role, and these will be covered. Originally, similar constraints applied to FTIR instruments, although in the past 15 years there has been a strong desire to move instruments away from the laboratory and toward point of use applications, either at-line, or on-line or field-based. These are primarily associated with the great flexibility associated with measurements made on an FTIR instrument, and the performance characteristics of spectral resolution, speed and optical throughput, when compared to traditional grating instruments.

Practical FTIR solutions have been developed by paying attention to the fundamental design of the instrument. Moving an FTIR instrument out of the benign environment of a laboratory to the more alien environment of either a process line or a portable device is far from trivial. Major efforts have been placed on the instrument design in terms of both ruggedness and fundamental reliability of components. Furthermore, issues such as environmental contamination, humidity, vibration, and temperature are factors that influence the performance of FTIR instruments. As a consequence, FTIR instrumentation that is to be successfully applied in process and field-based application must be designed to accommodate these factors from the start. Some of the main issues are packaging and making the system immune to effects of vibrations and temperature variations, the need to protect delicate optical components, such as beamsplitters, and the overhead associated with making a high-precision interferometric measurement.

Vibrational spectroscopy, in the form of mid-IR and more recently NIR spectroscopy, has been featured extensively in industrial analyses for both quality control (QC) and process monitoring applications.[1–3] Next to chromatography, it is the most widely purchased instrumentation for these measurements and analyses. Spectroscopic methods in general are favored because they are relatively straightforward to apply and implement, and are often more economical in terms of service, support, and maintenance. Furthermore, a single IR instrument in a near-line application may serve many functions, whereas chromatographs (gas and liquid) tend to be dedicated to only a few methods at best.

The first industrial applications of IR spectroscopy were for quality and production control in the petrochemical industries, primarily for the analysis of fuels, lubricants, and polymers. Early instruments were designed only for mid-IR absorption measurements, and this was primarily dictated by the available methods for sample interfacing, which were limited to simple transmission cells. In the early days of analytical vibrational spectroscopy, NIR was never really considered to be an analytical technique with any practical value, and in fact it was originally addressed as an extension of the visible spectral region, rather than the IR. However, in the 1970s, with applications targeted at agricultural and food products the true value of this spectral region was appreciated, in particular for quantitative measurements. It is worth noting that in this case, non-invasive approaches to sample interfacing were used in the form of diffuse reflectance. This latter sample-handling technique is seldom used for practical mid-IR-based process applications.

Today, the value, functionality, and benefits offered by mid-IR and NIR spectroscopy techniques are largely uncontested. Taking into account all of the techniques, virtually any form of sample can be handled, including gases, aerosols, liquids, emulsions, solids, semi-solids, foams, slurries, etc. This is also a major factor when comparing chromatographic methods, which are traditional for process analysis, with spectroscopic methods. In a traditional QC environment, a single spectrometer can be configured for a broad range of sample types merely by changing the method used for sample handling/sample interfacing. The nature of the sample plays an important role in the decision process, and in many ways it is the most demanding element for consideration when assembling a measurement system for an industrial application. This is especially the case if we extend the application to become a full on-line process measurement, where the sample extraction and conditioning can be more expensive than the spectrometer itself.

Infrared spectroscopy, in the forms of both NIR and mid-IR spectroscopy, has been hailed as a major growth area for process analytical instruments.[4,5] This view assumes that traditional laboratory instruments are outside of those covered by the areas concerned. Overall, this has been difficult to define in terms of the boundaries between the laboratory and the process; where they start and end. Some of the confusion arises by the term 'process analysis' itself. In the most liberal form, it can be considered to be any analysis that is made within an industrial environment that qualifies a product. This ranges from the QC of incoming raw materials, to the control of the manufacturing process through its various intermediate stages, and on to the analysis of the product. How this is viewed is very much industry-dependent.

The ideal is for process analysis to provide feedback to the process control system, thereby ensuring optimum production efficiency. For some industries product quality is the most important controlling parameter, and this requires analytical controls through-out critical stages of production. This is a view shared by the current process analytical technologies (PATs) initiative that is being endorsed by the pharmaceutical industry. Other examples include high-tech manufacturing, such as in the semiconductor industry, and specialty or value-added chemical production. In the petrochemical industries, which are dominated by continuous processes, sources of information that help to control the process itself and/or maintain production efficiency are the important attributes of a process analytical system. Considering all these factors and requirements, it is important to review the physical implementation, giving rise to terms such as 'on-line,' 'off-line,' 'near-line,' 'at-line,' and even remote and portable methods of analysis. In general, these descriptions are self-explanatory and the only question tends to be: Which is the best way to implement an analysis? The deciding factors are the turn-around time required for the analysis and the cost and ease of implementation. Whatever approach is used, it is important to address the instrumentation as a complete package or system. Depending on the requirements of the analysis and the technique selected, a system consists of several key components; the sampling system (how the sample is extracted from the process and how it is conditioned), the sample interface, the basic spectrometer (including how it is packaged), and the way that the system communicates back to the process. Relative to this last point, if the measurement system is not permanently on-line, then the issue can be the form of human interface.

4.2   Basic IR spectroscopy

As noted, for the mid-IR the spectral region is in the range of $5000–400\,\mathrm{cm}^{-1}$, where the region below $400\,\mathrm{cm}^{-1}$ is classified as the *far IR* and is assigned to low-frequency vibrations. The classical definition of NIR is the region between the IR and the visible part of the spectrum, which is nominally from 750 to 3300 nm ($0.75–3.3\,\mu\mathrm{m}$, approximately $13\,300–3000\,\mathrm{cm}^{-1}$). Originally, the NIR spectrum was serviced by extensions to ultraviolet-visible (UV-vis) instrumentation. In the late 1970s, NIR was recognized to be useful for quantitative measurements for a wide variety of solid and liquid materials, and as a result a new class of instruments was designed and dedicated to this spectral region.

There are pros and cons when considering the use of mid-IR versus that of NIR for process monitoring applications. The biggest advantage of mid-IR is its broad applicability to a multitude of applications based on the information content of the mid-IR spectrum.[6,7] Analytically, the fundamental information content of the NIR is limited and varies throughout the spectral region, with different regions used for different applications. In all cases, the measurements utilize absorptions associated with overtones and combinations from the fundamental vibrations that occur in the mid-IR region.[8] For the most part, these regions are dominated by absorptions from the 'hydride' functionalities —CH, OH, NH, etc. Three groups of information are observed, and these are nominally assigned to first, second, and third overtones of the hydride fundamentals, combination bands, plus some coupling from vibrational frequencies of other functional groups. Note that many functional groups are measured directly in the mid-IR region, but are only measured by inference in the NIR. Also, there are sensitivity advantages in the mid-IR, especially for gas-phase measurements, where intensities are as much as 10–1000 times greater.

One important factor that favors NIR as a method of measurement over mid-IR for process monitoring applications is the issue of sampling. The overtone and combination bands measured in the NIR are between one and three orders of magnitude weaker than the fundamental absorptions. This has an important impact on the way that one views the sample and the method used for the sample interface. Sample thickness or pathlength is a critical part of a quantitative measurement. In the mid-IR, pathlength is generally measured in micrometers, or tenths of millimeters, whereas millimeter and centimeter pathlengths are used in the NIR to measure comparable levels of light absorption. The longer pathlength simplifies sampling requirements for all condensed-phase materials, and in particular for powdered or granular solids. It also favors more representative sampling. On the downside, the presence of particulate matter, which leads to light scattering and loss of energy, has a larger impact with shorter wavelengths and with larger pathlengths. Optical construction materials, as used in cell windows, and also for light conduits (including optical fibers) are important to consider. Common silicate materials such as quartz and glass are transparent throughout the NIR, in contrast to most of the mid-IR region (note that anhydrous quartz is transparent to around 2500 cm^{-1} per 4 μm). Often hygroscopic or exotic materials are used for sampling and instrument optics within the mid-IR region.

For most common IR applications (near-, mid- or far-IR) the calculations are made from data presented in a light absorption format with the assumption that the traditional Beer-Lambert law holds (see Equation 4.1). In practice, this involves comparing light transmission through the sample with the background transmission of the instrument. The raw output, normally expressed as percent transmittance, is I/I_0, where I is the intensity (or power) of light transmitted through the sample, and I_0 is the measured intensity of the source radiation (the background) illuminating the sample. The actual absorption of IR radiation is governed by the fundamental molecular property, absorbance, which in turn is linked to molar absorptivity for a light transmission measurement by Beer's law:

$$A = \alpha \cdot c \tag{4.1}$$

Or more generally, by the Beer-Lambert-Bouguer law:

$$A = \alpha \cdot b \cdot c \tag{4.2}$$

where α is *molar absorptivity*, b is sample thickness or optical pathlength, and c is the analyte concentration.

Each spectral absorption can be assigned to a vibrational mode within the molecule, and each has its own contribution to the absorptivity term, α. Absorbance, in turn is determined from the primary measurement as:

$$A = \log_{10}\left(\frac{1}{T}\right) \text{ or } \log_{10}\left(\frac{I_0}{I}\right) \tag{4.3}$$

A majority of traditional NIR measurements are made on solid materials and these involve reflectance measurements, notably via diffuse reflectance. Likewise, in the mid-IR not all spectral measurements involve the transmission of radiation. Such measurements include internal reflectance (also known as attenuated total reflectance, ATR), external reflectance (front surface, 'mirror'-style, or specular reflectance), bulk diffuse reflectance (less common in the mid-IR compared to NIR), and photoacoustic determinations. In these cases more complex expressions may be required to describe the measured spectrum in terms of IR absorption. In the case of photoacoustic detection, which has been applied to trace-level gas measurements, the spectrum produced is a direct measurement of IR absorption. While most IR spectra are either directly or indirectly correlated to absorption, it is also possible to acquire IR spectra from emission measurements. This is particularly useful for remote spectral measurements, and in an industrial environment this is especially important for fugitive emissions and stack gas emissions.

Although IR spectroscopy and IR instrumentation may be discussed elsewhere in this publication, it is important to cover the basic principles of the measurement. This is important for instruments dedicated to specific measurements and targeted applications because it dictates the optimum technologies and materials that are best suited for the measurement system. For a dedicated process IR system, the measurement techniques used need to be defined in terms of their spectral range, their general applicability to the operating environment, and their ease of implementation. In other words, it is essential to match the characteristics of an individual technology to the performance required for the target application. Too often, new technologies are arbitrarily assigned to target applications without due consideration of all measurement and/or environmental factors.

4.3 Instrumentation design and technology

In its most basic form, IR spectroscopy is a technique for the measurement of radiation absorption. By convention the measurement is made on a relative basis, where the absorption is derived indirectly by comparing the radiation transmitted through the sample with the unattenuated radiation obtained from the same energy source. For most measurements, the resultant signal is expressed either as a percentage of the radiation transmitted through the sample or reflected from the sample. For many process measurements featuring simple filter-based NDIR analyzers, the measurement may be made at a nominal single wavelength. However, in common with most laboratory instruments, many modern IR process analyzers utilize multi-wavelength information, often acquired from a scanning spectrometer. Irrespective of the actual style of instrument used, the main components of an IR analyzer can be summarized in the form of a block diagram (Figure 4.1). Fundamentally, this is a generic format that features an energy source, an energy analyzer, a sampling point or interface, a detection device, and electronics for control, signal processing, and data presentation.

Although not indicated in Figure 4.1, the sample conditioning and the nature of the optical interface used to interact with the sample are as important, if not at times more important, to the success of the measurement as the actual instrument technology itself. Note that the generic format used in the illustration also applies to emission measurements, where the sample is in essence the source. As illustrated, the configuration implies that the radiation passes through the sample, and this only applies to transmission measurements. For reflection applications, the surface of the sample is illuminated, and the reflected radiation is collected from the same side. Another variant on the theme (as illustrated) is the remote sampling approach where the sample is located away from the main spectrometer and the radiation to and from the sample is carried via optical fibers or some form of light conduit. Figure 4.2 illustrates some of the possible component-sample configurations that are used for IR spectral measurements.

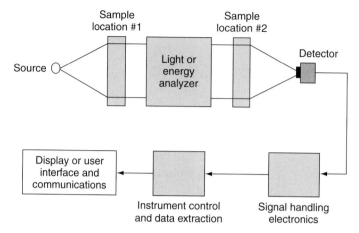

Figure 4.1   Generic block diagram for a spectral analysis system.

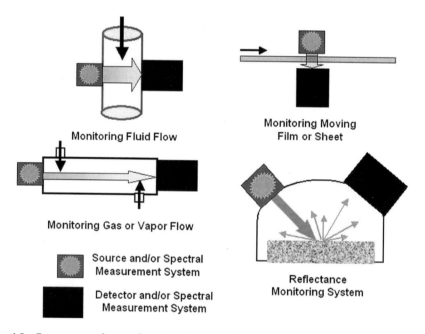

Figure 4.2   Common sampling configurations for spectral process analyzer systems.

There are several measurement techniques that can be considered for use in IR-based instrumentation in the NIR and the mid-IR spectral regions, and these are summarized in Table 4.1. Some of these techniques are classical, as in the case of optical filter-based instruments and scanning monochromators. Others, such as acousto-optically tunable filter (AOTF), have been considered for 15+ years, but are mainly applied to NIR

Table 4.1 Measurement technologies available for NIR and mid-IR instrumentation

Measurement technology	NIR	Mid-IR
Fixed and variable optical filters	X	X
Correlation filters (e.g. gas correlation filters)		X
Scanning slit monochromator	X	X
Fixed polychromators	X	X
IR detector arrays	X	X
Photodetector arrays/CCD arrays	X	
Broadband tunable filters (e.g. AOTF, LCTF)	X	
Narrow-band tunable filters	X	X
Diode Lasers or tunable lasers	X	X
Interferometers (FTIR)	X	X

measurements.[9,10] References to the use of lasers have been made for some time, and recent work in the telecoms-related industries has resulted in the availability of solid state devices that could lend themselves to spectroscopic applications in both NIR and mid-IR spectral regions. Tunable lasers are slowly becoming available, and some have been used for specialized gas measurement applications. At the time of writing, many of the devices were expensive; however, it was expected that in the future there will be an opportunity for these devices to become available at a relatively low cost.[11–16]

There are two ways to view instrumentation: in terms of the method used to separate the IR radiation into its component parts as used for the analysis, or in terms of the enabling technologies used to selectively capture the relevant radiation required to perform the analysis. The selection of the approach used for the measurement, in line with the techniques outlined in Table 4.1, tends to dictate which view is the most relevant. To explain this we can consider the following examples:

An FTIR instrument: The three critical components (excluding the sample) are the source, the detector, and the interferometer. In terms of enabling technology it is the interferometer that is critical to the measurement.

A detector array spectrograph: The three critical components (excluding the sample) are the source, the light separation technique that can be a variable filter or a standard diffraction grating-based spectrograph, and the detector array. In terms of the enabling technology, it is the detector array that is critical in the way the instrument operates. Figure 4.3 provides typical configurations for array-based spectrometers. Typically, they are simple in construction, and they feature no moving parts.

A diode laser spectrometer: In this case, if the laser is a tunable laser, there are only two critical components; the tunable laser and the detector. Typically, the enabling technology is the laser, which in this mode acts as the light source and the wavelength selection device.

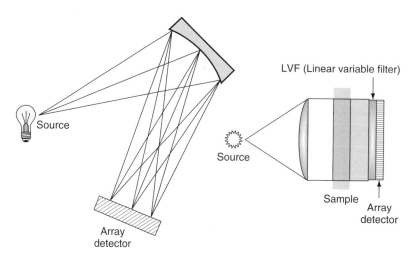

Figure 4.3 Typical configurations for array-based spectral analyzers.

Note that the examples above are over-simplifications to some extent, and other critical elements, such as optics, the associated electronics and the sampling interfaces are also important for a final implementation. The latter is particularly the case for laser instruments, where attention to optical interfacing is essential and often features specialized optical fiber-based optical coupling.

Next, when instrumentation is implemented in an industrial environment there are many other parameters associated with the final instrument packaging and the electrical and fire hazard safety that need to be addressed. Often these issues are not trivial and can have a profound impact on both the design and the final performance of an instrument. In the following sections, instrumentation and component technology will be reviewed, and where appropriate the impact of operation in an industrial environment will be addressed.

4.4　Process IR instrumentation

It takes a lot more than just connecting an IR instrument to a pipe to make the instrument into a process analyzer. The system not only has to be constructed correctly and be capable of flawless operation 24 hours a day, but it also has to be packaged correctly, and be able to access the stream correctly and without any interference from process-related disturbances such as bubbles, water, particulates, etc. Finally, the data acquired from the instrument has to be collected and converted into a meaningful form and then transmitted to the process computer, where the information is used to assess the quality of the product and to report any deviations in the production. Beyond that point, the process computer, typically not the process analyzer, handles any decisions and control issues. There are a few exceptions where robotics are integrated with the analyzer, and good vs. bad product quality are assessed at the point of measurement. The use of

FTIR analyzers for semiconductor wafer quality testing is an example, as mentioned below. Today we speak of process analytical instrumentation (PAI), and we use the term 'PAI' which relates to the totally integrated system, implemented on-line, and made up of some or all of the following components.

The term 'process analyzer' sometimes conjures up the image of plumbing and a highly ruggedized instrument in a heavy-duty industrial enclosure. Not all analyzers used for process applications have to follow that format. For example, in the semiconductor industry, the traditional format analyzers are used as on-line gas and water sensors and are considered as process monitors. But product monitoring analyzers, which are configured more as automated laboratory analyzers, are viewed differently. The semiconductor wafer, the product, is monitored at various stages of the process, effectively *in-line* or *off-line* by FTIR analyzers. It should be pointed out that FTIR is the most commonly accepted method for the QC of the product, examples being the ASTM Standard methods F1188/F1619 for interstitial oxygen measurement, and F1391 for substitutional carbon content. In a process that can involve up to 32 critical steps, and where a high yield at each step is essential, it is unreasonable not to consider FTIR as a true process analysis tool, in spite of its laboratory-style analyzer format. The application is made practical by the use of high-precision wafer-handling robotics, which are integrated into the analyzer.

IR photometers have in the past been considered as the standard instrument format for IR measurements. These are used as on-line systems and are mainly used in the environmental market (combustion gases) and for the measurements of standard process gases such as hydrocarbons. Today, there are a relatively large number of on-line process stream analyzers for liquids, solids, and gases (mid- and NIR), which feature either filter (discrete) or full-spectrum analyzers. Many industries are rethinking the role of process analyzers, and applications featuring 'near-line' or 'at-line' are now considered to be appropriate especially when the timescale does not require real-time measurements. Many of these analyses can be legitimately viewed as process measurements as the direct on-line analogues. Such instruments are, however, often easier and much less expensive to implement than an on-line instrument. The final implementation and format are usually based on the nature of the process, the needs of the end-user, the overall economics, and the required speed of analysis. In many cases this comes down to whether the process itself is operated in a batch mode or is continuous. A good example of a laboratory platform being adapted for near-line and even mobile applications is the TravelIR (SensIR/Smiths Detection), as illustrated in Figure 4.4 in a mobile mode of operation.

4.4.1 Commercially available IR instruments

As noted, IR instruments fall into several categories, from simple photometers (single- or multiple-filter devices) to relatively complex full-spectrum devices, such as FTIR instruments. Today's process engineers often prefer optical methods of measurement rather than the traditional chromatographs because of perceived lower maintenance and ease of implementation. However, the final selection is often based on the overall economics and the practicality of the application.

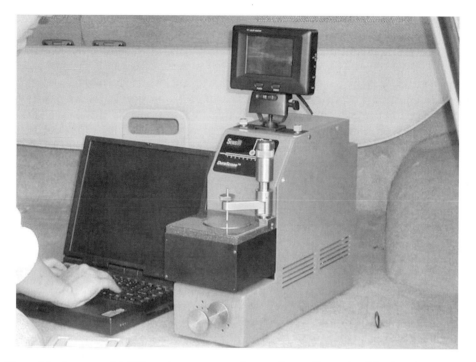

Figure 4.4　A commercial FTIR spectrometer system for near-line, at-line and mobile IR analysis.

4.4.1.1　A historical perspective

For the measurement of a single chemical entity, analyzers in the form of NDIR have been in use for decades for both industrial and environmental monitoring applications. These feature one or more wavelength-specific devices, usually optical filters, customized for the specific analyte, such as CO, CO_2, hydrocarbons, moisture, etc. This type of instrument is normally inexpensive (on average around US$5000) and relatively easy to service, dependent on its design. This class of instrument is by far the most popular if one reviews the instrumentation market in terms of usage and the number of analyzers installed.

Filter-based instruments are often limited to applications where there is simple chemistry, and where the analytes can be differentiated clearly from other species or components that are present. Today, we may consider such analyzers more as sensors or even meters, and the analytical instrument community typically does not view them as true instruments. Since the late 1980s a new focus on instrumentation has emerged based on the use of advanced measurement technologies, and as such is considered to be more of the consequence of an evolution from laboratory instruments. Some of the first work on full-spectrum analyzers started with an initial interest in NIR instruments. The nature of the spectral information obtained in the NIR spectral region is such that an analyzer capable of measuring multiple wavelengths or preferably a full spectrum is normally required. Based on this understanding an entire market for NIR instruments developed in the late 1980s through to the 1990s for scanning NIR instruments. In some cases filter instruments were used, but because of the need to measure multiple wavelengths the instruments were more complex than traditional filter IR photometers.

In parallel to the development of full-spectrum NIR analyzers, work started on the use of mid-IR spectrometers. By the mid-1980s, FTIR became the de facto method for measuring the mid-IR spectrum in the laboratory. For many production applications, FTIR was viewed as being too complex and too fragile for operation on the factory floor or for implementation on a process line. Some pioneering FTIR companies, notably Bomem, Nicolet and Analect, proved that the concept would work with adaptations to the design of laboratory instruments. These companies, starting around the mid-1980s, and later others made the decision to focus on industrial applications, and designed instruments for operation in more demanding environments.

Today most of the major FTIR companies offer some form of instrumentation that is used in process monitoring applications. Some are application-specific; others are industry-specific, the pharmaceutical industry being a good example. Companies such as Hamilton Sundstrand (originally Analect), Midac and ABB (Bomem) tend to be 100% focused on process applications, whereas a company such as Brüker supports both laboratory and process applications. All of these companies offer on-line and off-line implementations of their products. Other FTIR companies, notably Perkin Elmer, Digilab (now Varian) and Thermo (Nicolet/Mattson), tend to lean toward the off-line applications, and largely in the consumer and pharmaceutical or the 'high-tech' application areas. The companies mentioned are the ones that have formed the core of the traditional instrument business. There are many smaller or less well-known companies that provide FTIR-based analyzers for specialized applications. It must also be noted that today with FTIR technology there are no clear distinctions made between NIR and mid-IR because FTIR instruments can be readily configured for any or both of these spectral regions.

The implementation costs of FTIR instrumentation, especially for on-line applications can be very high, sometimes in excess of US$100 000. Also, there are issues relative to size, performance and still the physical reliability issues that make FTIR less than acceptable for many areas of application. Furthermore, there is a general desire to make instrumentation much smaller, and even handheld and portable (one example is shown in Figure 4.4). As a result, today we are seeing a number of new competing technologies. In some cases these are direct competition to FTIR, but in others there are often compromises between size and/or performance, and overall versatility. FTIR is a very versatile approach to spectral measurement. However, such versatility is often not required for specific applications, and it is in such cases that alternative technologies can often excel.

In the ensuing sections the basic instrument technologies will be discussed as used in commercial instrumentation. A section will be included, which reviews some of the emerging technologies that are used as either replacements or substitutes for the standard technologies, often for very specific applications.

4.4.1.1.1 Photometers

One of the oldest forms of process analyzers are the filter photometers, sometimes also known as filtometers.[17] They have simple construction, and in the simplest form they consist of a source, an optical filter (or filters), a sample cell and a detector (Figure 4.5 provides example presentations). For many applications they are single-channel devices,

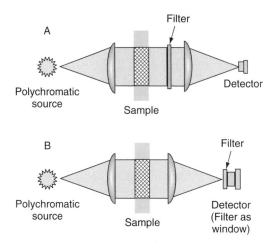

Figure 4.5   Block diagram for simple filter photometer analyzers.

and consequently operate as a single-beam instrument. Miniaturization has led to the production of smaller more integrated devices that essentially behave as single-species sensors. In each case, the selected filter, the source and the detector define the analytical spectral region. For the simplest versions, the filters can be integrated with the detectors, where the filter forms the window for the detector (as indicated in Figure 4.5b). The filters are either bandpass or correlation (gas- or material-based), dependent on application – the latter being used for highly specific low-concentration applications, and most often gas-phase applications. Most process photometers are designed for single component measurement, with many featuring just a single filter selected to transmit at the analytical wavelength/frequency, as indicated in Figure 4.5. For some applications, where a reference wavelength is required, a second channel may be implemented with a filter set to the reference wavelength. In some modern implementations, two-color detectors are employed. These are special detector packages, which contain two detector elements, internally separated, and where each element is separately illuminated from two separate filter channels. The most practical of these implementations features two separate windows above the detector elements constructed from the two different filters (analytical and reference wavelengths). Two color mid-IR detectors are commercially available.

   With a traditional photometer a baseline is established with a zero sample (contains none of the analyte), and the span (range) is established with a standard sample containing the anticipated maximum concentration of the analyte. Although the electronics are simple, and designed for stability and low noise, it is usually necessary to establish zero and span settings on a regular basis, often daily, and sometimes more frequently. This can result in significant calibration overhead costs for even the simplest photometer. Note that often the lower instrument cost is negated by higher operational costs.

   Conventional wisdom leans toward a simple measurement system featuring one or two optical elements and with a simple construction. In practice there are many forms of filter assemblies that can be used. These include simple interference filters, filter clusters or arrays (multiple wavelengths), correlation filters (customized to the analyte), continuously variable filters, such as the circular variable filter (CVF) and the linear variable filter (LVF), and tunable filters (both broadband and narrowband).

Variable filters, in the form of CVFs and LVFs, are not new and their use in scanning instruments featuring single-element detectors has been reported for at least 30 years. Note that commercial instruments based on these filters were available in the 1970s, the best examples being the Miran® analyzers, originally offered by Wilks. Variable filters can be expensive to produce and alternatives such as clusters of filters or filter arrays have been used for NIR instruments and for some mid-IR instruments, again in combination with single-element detectors. For all of these applications, a mechanical device is required to drive the filter assembly. The reproducibility and reliability of these devices, real and perceived, limits the performance and utility of these approaches. Today, detector arrays are becoming available, and filter clusters and continuously variable filters can be integrated with these arrays to form a multi-wavelength spectral measurement system. Such a combination leads to compact instrumentation for simple transmission measurements on liquids or gases, and removes all problems associated with moving parts.

Tunable filters in the form of AOTF devices, liquid crystal tunable filter (LCTF) and also tunable cavity Fabry-Perot étalon (FPE) devices have been considered in non-moving part instrument designs for many years. Today, the AOTF and the LCTF devices are used in the NIR spectral region.[9,10] Originally, designs were also proposed for mid-IR AOTF devices, but these have not become available, mainly because of fabrication issues (cost and material purity). Tunable FPE devices, which are really just variable cavity interference filters, have been developed for the telecommunications industry. While these have been primarily used in the NIR, in most cases they can be fabricated to work also in the mid-IR, the latter being only an issue of material/substrate selection.

Tunable filters provide the benefit of rapid spectral tuning, providing full or partial range spectral acquisition in about 100 ms, sometimes less. Most of these tunable devices can be step-scanned where the device is tuned to specific wavelengths. This is an extremely fast process, and emulates a multi-wavelength filter system, with near-instantaneous output at each wavelength. Over the years since their introduction, there have been various issues relative to the use of tunable filters, ranging from cost to poor measurement reproducibility. However, stringent demands from the telecommunications industry have led to the development of highly reliable tunable devices.

The largest application segment for filter photometers is in the area of combustion gases analysis, primarily for CO, CO_2, hydrocarbons, SO_2, etc. Other major areas of application include the petrochemical industry, with natural gas and other hydrocarbon process gas streams being important applications. As measurements become more complex, there is the need for more advanced instrumentation. Variable or tunable filter solutions (as described above) or full-spectrum FTIR or NIR instruments are normally considered for these applications, primarily in terms of overall versatility. Now that array-based systems are becoming available, there is a potential for an intermediate, less expensive, and more compact solution. Note that compact instrumentation tends to be environmentally more stable, and is well suited for industrial applications.

4.4.1.1.2 *Monochromators and polychromators*

Traditional optical spectrometers for both mid-IR and NIR were based on a scanning monochromator. This design features a single source and detector, and a mechanically scanned dispersion element in combination with optical slits for spectral line separation. The dispersion element, either a refracting prism or a diffraction grating, and the slits were driven mechanically by relatively complex mechanisms. These instruments are now referred to as dispersive instruments. Although this arrangement worked adequately in a laboratory environment for years, it is now considered to be inappropriate for most applications, particularly the case for a heavy industrial setting. Reliability, calibration, and overall performance have been factors that limit the use of a monochromator. Today, a few commercial NIR instruments still feature monochromators that are digitally synchronized, but without exception the interferometer-based FTIR instruments have replaced dispersive mid-IR instruments.

Monochromators, theoretically, provide energy of one wavelength (or practically a few wavelengths) at a time; the device scans the spectrum sequentially. In contrast, the polychromator allows multiple wavelengths to be measured simultaneously. A polychromator is used in instruments that feature detector arrays or other arrangements that can handle multiple wavelengths at a time via unique methods of wavelength encoding. In a simplest form of polychromator, a concave diffraction grating spatially separates the source energy into the analytical wavelengths and provides the necessary imaging on the array. The advantages of array-based instruments are obviously based on the fact that there are no moving parts, they are mechanically simple, and good wavelength stability is usually achieved leading to highly reproducible measurements in a factory or plant environment. Currently, commercial instruments, often of a very compact form, are available based on this concept. While most current commercial instruments are limited to the NIR region, new mid-IR arrays are becoming available, and it is reasonable to expect commercial instruments featuring these arrays will soon become available. Note that a commercial system was introduced in the UK for the pharmaceutical industry (original instrument sponsored by Astra-Zeneca) based on an integrated ATR probe coupled to a mid-IR-based polychromator system.

An instrumentation technique that utilizes the output from a monochromator, and that provides some of the benefits of multiplexed data acquisition, is known as a Hadamard transform spectrometer. This class of instrument can feature either a monochromator or a polychromator (equipped with a detector array). Hadamard transform instruments are available as custom-made devices, but none have been fully commercialized.

4.4.1.1.3 *Interferometers*

During the 1980s most mid-IR instruments moved away from dispersive methods of measurement (monochromator-based) to interferometric measurements based on the widespread introduction of FTIR instrumentation.[18,19] These instruments provided the performance and flexibility required for modern-day mid-IR applications. At the heart of an FTIR instrument is a Michelson-style of interferometer. The critical elements in this style of instrument are the beamsplitter and two mirrors: one fixed and the other moving (as illustrated in Figure 4.6). This type of measurement requires the production of an

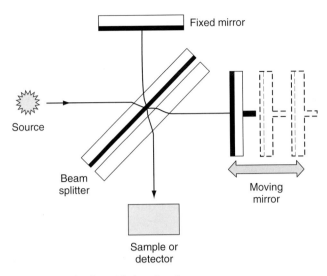

Figure 4.6 Basic components for the Michelson interferometer.

optical interferogram, which is generated at high optical precision from the mirror–beamsplitter combination. The mechanical and optical tolerances on the interferometer design are extremely demanding, a factor that leads to many challenges for instruments placed in an industrial environment.

The drive mechanism of the moving mirror assembly, which eventually yields the spectral information, has to be critically controlled and must be made immune to external interferences such as temperature and vibration. Both can have significant impact on the performance of an interferometric measurement system. In the mid-IR region, the beamsplitter is an environmentally fragile optical element, which is normally constructed from a hygroscopic substrate such as potassium bromide. This substrate has to be maintained to high optical flatness, and is coated with multiple layers of high-index materials, such as germanium (Ge). This critical element must be environmentally protected, in particular from dust and humidity. At minimum the instrument needs to be hermetically sealed, dry gas-purged, or protected by desiccant. In an industrial environment, purging can be difficult, and so sealing and desiccation are normally required. For demanding industrial applications, non-hygroscopic materials, such as barium fluoride and zinc selenide, have been used for beamsplitter construction. Temperature is another major issue, and unless well designed and/or protected the interferometer will be inherently sensitive to changes in ambient temperature. In acute situations the interferometer can lose optical alignment, with consequent losses in stability and energy throughput. Instruments designed for process applications often feature thermal compensation in the beamsplitter design, and will also feature some form of controlled mirror alignment, to compensate for thermally induced changes in the interferometer geometry.

It is possible to construct a rugged interferometer suitable for use in a factory or production environment with careful attention to design details. Manufacturers specializing in process applications have focused on these issues and a range of ruggedized systems are available, suitable for on-line process monitoring applications. Operational issues can arise, mainly because of the need to maintain good optical alignments and the fact that an FTIR instrument still features critical moving parts. An alternative design of interferometer, known as a polarizing interferometer, has no moving parts, and at least two commercial designs have been reported, mainly for operation in the NIR region. One negative aspect of most interferometers, even for those designed for alien environments, is the relatively high cost and the overall size and complexity. However, at the time of publication, the FTIR instrument is still a viable option for the mid-IR spectral region.

4.4.2   *Important IR component technologies*

This section is focused primarily on source and detector technologies. For some applications the source or the detector actually defines the entire measurement technology, for example tunable lasers (source) and array spectrographs (detector). There are other important technologies to consider, especially in the area of data acquisition, control, computer, and communication technologies. These are rapidly changing areas and, if viewed generically, service all forms of instrumentation. Where practical, companies tend to use standard platforms. But for certain applications where performance is critical, there is still a case for proprietary solutions.

Beyond performance optimization, issues relative to packaging and the need for compliance with certain safety and electronics regulatory codes are cited as reasons for a customized solution. In the latter case, a systems approach is required, especially when attempting to meet the code or performance requirements for compliance with European Certification (CE) mark or electrical and fire safety codes such as National Fire Prevention Association (NFPA) and ATEX (replaces the old Cenelec requirements). Off-the-shelf electronics may provide the necessary performance characteristics for generic applications, and their use eliminates large expenses related to product development, plus the associated time delays. Photonics-related components are solely addressed in this section because they are used to customize instruments for application-specific systems.

Source and detector selections are interrelated, where the output of the source is matched to the sensitivity range of the detector. However, the exact nature of the source is also dependent on the type of sample(s) under consideration, the intended optical geometry, the type of measurement technique, and the final desired performance. The bottom line is that adequate source energy must reach the detector to provide a signal-to-noise performance consistent with the required precision and reproducibility of the measurement. This assumes that the detector is well matched, optically and performance-wise, and is also capable of providing the desired performance.

4.4.2.1   *Broadband sources*

Most optical spectral measurements, where the measurement of multiple wavelengths is required, will feature some type of polychromatic or broadband light source. There are a few exceptions here, such as tunable laser sources and source arrays. In such instances, the source is effectively *monochromatic* at a given point in time. These sources are covered separately under monochromatic sources.

Most sources intended for IR measurements are thermal sources, and feature some form of electrically heated element, providing a characteristic blackbody or near black-body emission. There are essentially two types of broadband sources used in IR instrumentation: the open element source and the enclosed element source. The latter is effectively a light bulb, with tungsten lamp with a quartz envelope being one of the most important. Note that this source does not transmit energy much below 5.5–5.0 μm ($2000 \, cm^{-1}$). A variant on the *light bulb* is an enclosed electrically heated element within a standard electronics package, such as a TO-5 style can. The unit is sealed with a low-pressure atmosphere of a protective inert gas, such as nitrogen, by an IR-transmitting window, such as potassium bromide (KBr), zinc selenide (ZnSe), sapphire or germanium (Ge). The thin filaments used in bulbs and bulb-like sources have a low thermal mass, and this provides the opportunity to pulse the source. A variant of this source, provided in similar packaging, is MEMS-fabricated and has a diamond-like surface coating, which provides exceptionally good thermal and oxidative stability. All of the above sources are particularly useful for process-related applications, including handheld monitoring systems because they have low power, high efficiency, and can be electrically modulated.

The other forms of blackbody sources are adaptations of those used widely in conventional laboratory-style IR instruments, which feature exposed electrically heated elements. Various designs have been used, with metal filaments made from Kanthral® and Nichrome being simple solutions in lower-cost laboratory instruments. They tend to be mechanically fragile, especially when subject to vibrations while hot. Electrically heated ceramic sources are currently popular and these are more robust and reliable than simple wire sources. These include sources made from silicon carbide, with the Globar® and the Norton sources being common examples. The Norton source is used extensively for modern low-to-medium cost instruments and for most instruments used for process applications. It is a simple, rugged source, which was originally used as a furnace igniter. Typical operating temperatures are in the 1100–1300 K range, with the Globar® being an exception at around 1500 K. Note that the Globar® usually requires some form of assisted cooling and is seldom used for process applications.

Most detector systems require that the IR beam be modulated, where the source energy is adequately differentiated in the measured signal from the ambient background. One of the traditional approaches is to use some form of mechanical 'chopper,' usually in the form of a rotating sector wheel, which modulates the beam by blocking the radiation in one or more sectors during a rotation. While this has been used in the past on older-style dispersive instruments, it is considered to be unsuitable today for most process-related applications. The issue here is that moving parts are considered as failure sources, and are generally considered to be undesirable. An alternative, practical approach is to pulse the source by modulating the electric power to the source, thereby eliminating the need for any form of mechanical chopping device, an important factor in dedicated and process-style instruments. This approach works with low-mass sources, such as wire filament, light bulb style or MEMS wafer sources as described earlier. Certain measurement techniques, such as interferometry as used in FTIR and EP-IR (encoded photometric IR) instruments, modulate the radiation as a consequence of their mode of operation, and do not require independent modulation of the source.

4.4.2.2 Monochromatic sources

The most common forms of 'monochromatic' source are the light emitting diode (LED) and the laser. Today, these types of source are becoming popular for a wide range of spectral measurements for both instruments applied to specific applications and dedicated chemical sensors. LEDs are traditionally limited to NIR spectral measurements, but lasers are now available operating in the mid-IR, from 3 to 30 μm. Tunable lasers have always been the dream of spectroscopists. The concept of having a small, high-powered, narrow-beam monochromatic IR light source that can be tuned rapidly to other wavelengths is considered the ultimate for many process-related applications. Today, solid-state lasers and certain types of LEDS are tunable over narrow wavelength ranges throughout the NIR and mid-IR spectral regions. While the tuning range is typically very small, often no more than a few nanometers, it is sufficient to be able to scan through the narrow line spectra of certain molecular gas species. The devices are useful for the measurement of low concentrations of specific gases and vapors, with detection limits down to parts-per-trillion (ppt) being reported. The concept of using an array containing several laser elements is feasible for handling gas mixtures. Tunable laser sources have been used for monitoring species such as hydrogen fluoride, light hydrocarbons, and combustion gases.

Other non-continuum sources which are effectively multi-wavelength 'monochromatic' (because of line width) or multiple line sources, such as gas-discharge lamps, have been used and are often considered to be useful for calibration purposes. Another novel source, which is polychromatic in character but is monochromatic or species-specific in behavior, is the molecular equivalent of the electrodeless discharge lamp. The source utilizes a low-pressure gas, the same as the analyte species, and the molecules of the gas are excited within a radio-frequency field. The excitation includes transitions in the vibrational energy states of the molecule, and the source emits radiation in the IR region at the characteristic, resonance wavelengths of the molecular species. In this manner, a spectrometric measurement system can be developed where the analytical wavelengths of the analyte are identical and resonantly coupled to that of the source – comparable in nature to an atomic absorption measurement. This is similar in concept to a correlation filter photometer, except that the source provides the filter function. Commercial instrumentation has been developed based on this technology, for medical applications, but it has potential for use in other applications, including the measurement of combustion gases in an industrial environment.

4.4.2.3 Single-element detectors

Most standard IR spectrometers currently operating in the mid-IR spectral region feature single-element detectors. Exceptions are analyzers that are equipped with multiple detectors dedicated to 'single' wavelength measurements, either within a spectrograph arrangement where the detectors are placed at the physical analytical wavelength locations, or where multiple-element detectors are used with dedicated filters. The selection of detector type is dependent on the spectral region being covered.[11,12,20] There are two main types of detector in use for mid-IR measurements – thermal detectors and photon detectors. In the mid-IR region, thermal detectors are used for most routine applications. These are preferred for most process applications because they operate at room temperature. The most common thermal detectors are the pyroelectrics, such as deuterated triglycine sulfate (DTGS) and lithium tantalate. Lithium tantalate detectors are used widely for low-cost sensing devices (including fire alarms), where the detector element can cost as little as US$10. They are also used in filter-based instruments and in low-cost scanning instruments. Most process FTIR instruments feature the more expensive and more sensitive DTGS detector, which provides between one and two orders of magnitude increase in sensitivity over lithium tantalate, dependent on implementation. However, for general purpose process analytical applications it is necessary to thermally stabilize the DTGS detector, typically at temperatures around 25–28°C. This is an optimum temperature providing maximum stability, especially important for this particular detector, which becomes inoperable at temperatures above 41°C.

Thermal detectors by nature have a relatively slow response, and are not efficient for fast data-acquisition rates. Also, the same applies in certain measurements that feature low energy throughput, as in some reflectance measurements, high-resolution gas analyses, and mid-IR fiber-optic-based measurements or applications where a light-pipe conduit is used for interfacing an instrument to a sampling point. Traditionally, for such applications featuring low light levels or high-speed acquisition, the MCT (mercury–cadmium–telluride) detector is used. If measurements are to be made within 4000–$2000\,cm^{-1}$ spectral region, a lead selenide detector can be used. For many applications the detector can be used at room temperature. A sensitivity increase can be obtained by reducing the detector temperature with a single-stage thermo-electric (TE) cooler.

The MCT detector, which can operate at longer wavelengths, operates at sub-ambient temperatures, and is usually cryogenically cooled to liquid nitrogen temperatures. When operating with liquid nitrogen cooling, it is necessary to manually fill a Dewar cold-trap. These normally have an 8-hour capacity but extended fill versions are available, providing up to 24-hour usage. Obviously, the requirement to manually fill the Dewar, and the limited capacity of the Dewar pose limitations on the usage of this style of detector in a process environment. Although a commercial system exists for auto-filling the Dewar, the detector system tends to be limited to laboratory-based applications. While this type of detector system has been used in some specialized industrial applications, it is mainly limited to high-resolution gas monitoring, chemical reaction studies, and specialized semiconductor analyses.

Alternatives to cryogenic cooling include thermoelectric cooling (multi-stage Peltier devices) and thermo-mechanical cooling devices, such as Sterling coolers, recirculating micro-refrigerators and Joule–Thompson (J–T) coolers. None of these solutions are ideal: Peltier coolers have to be operated as three- or four-stage devices, Sterling engine coolers have limited mechanical lifetimes, and J–T coolers often exhibit vibrational coupling problems. Of these, the thermo-electric coolers are often the most straightforward to implement, and can be purchased as standard detector packages. However, they do not achieve liquid nitrogen temperatures (77 K), which are necessary in order to gain optimum sensitivity. Typically, the devices cool down to about −75°C (198 K). This limits the spectral range to a bandwidth of around $1000 \, \text{cm}^{-1}$ in the mid-IR and reduces the detector's performance closer to the DTGS detector. However, the high-speed data-acquisition benefits of the detector are retained. Both the three- and four-stage Peltier-cooled detectors and Sterling cycle-engine cooled detectors have been used for process and environmental monitoring applications with dedicated FTIR instruments.

4.4.2.4   Array detectors

At the time of writing, the only detector array elements in common use were the silicon photodiode arrays, and these were initially used for UV-vis measurements, and later with extended performance into the short-wave NIR. Indium-Gallium-Arsenide (InGaAs) detector arrays operating in the mid-range NIR, which were initially developed for military imaging applications, have been used extensively in telecoms applications. Originally, these devices lacked consistency (pixel-to-pixel), were very expensive, and were limited to 128- and 256-element arrays. Recently, arrays based on lead salts (lead sulfide and lead selenide) have been introduced, and these have provided extended performance into the mid-IR region – down to around $5 \, \mu\text{m}$ ($2000 \, \text{cm}^{-1}$) – which is adequate for measuring the 'hydride' fundamental (OH, NH, and CH) vibrations, for cyano compounds, and for certain combustion gases.

Mercury-cadmium-telluride arrays have been used by the military for night-vision imaging for many years, and versions of the detector have been available for imaging-based applications (hyper-spectral imaging). These are limited to such highly specialized applications and the price is cost-prohibitive for any process-related measurements. Recently, other array detectors devices operating in the mid-IR spectral region have become available. These are MEMS-based structures that function as pyroelectric devices. Two European suppliers (one UK-based, the other Swiss) have introduced arrays featuring 64 and 128 pixels respectively. Both companies offer the devices with the associated electronics and with the option for an integrated linear variable filter. The latter device, available with spectral ranges of 1.5–$3 \, \mu\text{m}$, 3–$5 \, \mu\text{m}$ and 5.5–$11 \, \mu\text{m}$, provides the equivalent of a solid-state spectrometer. At least one process spectrometer has been developed on one of these platforms, with an integrated internal reflectance probe. The development of this system was sponsored by a major pharmaceutical manufacturing company.

4.4.3   New technologies for IR components and instruments

There is a constant move toward the improvement of instrumentation technology. This ranges from new detector and source technologies, to new ways to make IR measurements and new ways to fabricate instruments. In the area of components, there are new MEMS-based sources, which operate at low power and provide extended source lifetimes, new tunable laser devices, new tunable filter devices, both broad and narrow range, and new detector arrays. Many of these devices are currently relatively high cost. However, most are made from scalable technologies, such as semiconductor fabrication technologies, and as such if the demand increases the cost can come down. Many can follow Moore's law of scaling. In the future, for a dedicated IR sensing system, it is feasible to consider costs in the US$50–100 price range.

In the area of instrument design there are examples of small interferometers being designed for portable applications. In some cases there is a compromise in resolution, but this is more than compensated by the convenience of the smaller size. One other new technology worthy of mentioning is EP-IR. Conceptually, this technique can be considered as a hybrid approach, providing the performance and flexibility benefits of FTIR, while providing the simplicity, ease of implementation, and low cost associated with an NDIR-based system. The basic concept layout of an EP-IR spectral measurement system is provided in Figure 4.7.

In its current form the technology is applied in a simple photometric analyzer where the incoming IR beam from the sample is imaged into a diffraction grating-based spectrograph. The dispersed radiation from the grating is imaged across an aperture above the surface of a rotating encoder disk. The encoder disk has a series of reflective tracks, which are spatially located within the dispersed grating image to correspond to the wavelengths and wavelength regions used for the analysis (as illustrated in Figure 4.7). Each track has a pattern that produces a reflected beam with a unique sinusoidal modulation for each individual wavelength. The reflected beams are brought to a common image point on a single detector, which generates a signal that forms a discrete interferogram. The intensity contribution for each wavelength component is obtained by applying a Fourier transform to this interferogram. The results, in terms of analyte

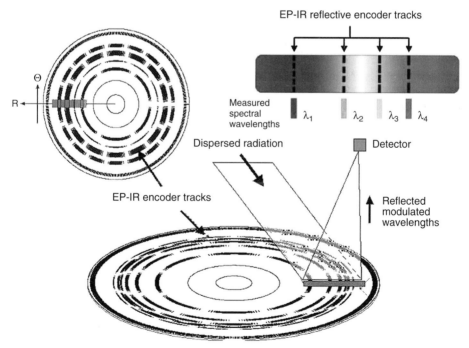

Figure 4.7 The EP-IR spectrometer system.

concentrations, can be determined from the measured intensities by the application of a classical least squares or other chemometric methods.

The heart of the system is the encoder disk, which has the appearance of a CD with a series of concentric barcode-like bands. The system is currently implemented with a spectral range from 3.0 to 5.0 μm by the use of a thermo-electrically cooled lead selenide detector. The analyzer has a baseline noise level of better than 10^{-5} absorbance, and, as an example, is capable of providing sub-ppm detection limits for hydrocarbon-based species from a 1-meter sample pathlength. In future incarnations the system will be configured with a dual spectrograph and will provide example spectral ranges of 3–5 μm and 7–14 μm. With the combined ranges the analyzer can handle most process analytical applications within a package that will be around the same price point as a simple NDIR analyzer.

4.4.4   Requirements for process infrared analyzers

The operating environment for instrumentation designed for industrial process-related measurements is understood, and guidelines (written and otherwise) exist for their implementation. While there may be compelling reasons to adapt traditional instrumentation technology used in the laboratory for process analytical applications, there are also many reasons not to do so. A common assumption is that it is a packaging issue and all that is needed for a process analyzer is to repackage the laboratory instrument in an industrial-grade enclosure. While this concept may succeed for early prototyping, it seldom works for a final process analyzer. At issue is the fact that there are more factors in making a process analyzer than simply the repackaging of an existing instrument. Important issues are the measurement philosophy, the overall system design philosophy, and changes in the operational concepts. It is necessary to classify the mode of operation and the operating environment of the analyzer, and also it is beneficial to examine the term 'process analysis.'

One important distinction is a logical separation between the terms *instrument* and *analyzer*. An instrument is totally open-ended and flexible in terms of sampling and operation, and typically generates data that requires detailed interpretation. This is particularly the case with IR instrumentation. An analyzer is a total system, which normally includes automated sampling, operation with predefined instrument settings, a defined user interface, and a prescribed method for the measurement and calculation of results. Process instrumentation tends to be the world of analyzers. Process analysis can be defined as an analytical procedure that is performed to provide information on raw material quality (screening), the progress of a process – whether it is batch or continuous – and composition/quality of the final product. This can range from dedicated laboratory-based off-line testing to continuous on-line measurements. Several clearly defined platforms of process analyzer exist and these can be delineated as follows.

4.4.4.1   Laboratory-based analyzers

For many applications, laboratory-based analyses may be developed on standard laboratory instruments. While these are not always the most ideal, they do provide a starting point; and once well defined, a custom-designed or dedicated analyzer can be produced. For IR methods of analysis the main issues are the ease of sample handling, preferably without modification of the sample, a simple operator interface, and the automated calculation and reporting of results. One of the benefits of an IR analysis is that a single instrument is multifunctional, and can be configured for a wide range of samples. As a consequence, analytical methods tend to be *recipe-based* and are usually developed by a support group. Most instrument companies provide macro-generation software that allows the procedures to be automated with little operator input, and provide a documented audit trail in the event that a measurement is questioned. Once standardized a dedicated analyzer can be produced just for that measurement. This is an important requirement for regulated analytical procedures. While there are minimum instrument performance requirements for most analyses, the most common sources of variability and non-reproducibility originate from sample-handling errors. Where practical, automated methods of sample handling are preferred to eliminate or reduce this source of error.

4.4.4.2   Analyzers for near-line or at-line plant measurements

Plant-based analyzers often share a common heritage with laboratory-based instruments, but their design requirements are more rigorous and demanding. Laboratory-based instruments are normally packaged in a molded plastic or sheet metal housing, with standard power and data cable connections. For operation in a plant environment, it is important to protect the instrument from the operating environment, and to ensure that the instrument conforms to factory and worker safety mandates. These can have a significant impact on the way that an instrument/analyzer is packaged and the type of electrical connections used for power and data transmission.

Most process analyzers are required to function 24 hours a day, and must work reliably for all work shifts. It is unacceptable for an analyzer to fail unexpectedly especially late at night. This requirement places a demand for a high degree of reliability, which in the end is a design issue – normally achieved by minimizing critical moving parts, and by using high-quality components with long mean-time-between-failure (MTBF). This favors newer methods of measurement that have minimized or eliminated the need for critical moving parts, and as a result the traditional FTIR instrument can be viewed as being less desirable.

The requirement for simplicity of sampling and analyzer operation is even more essential within a plant environment. Sampling must be highly reproducible and require only a minimum of operator skill. One must not second-guess an operator when designing either the sample–analyzer interface or the user interface of the analyzer. Keystrokes for any operation must be kept to a minimum and, where practical, devices such as bar code readers should be implemented. All visual displays must be unambiguous, and any cautions, warnings or error messages must be clearly displayed.

Most analyzers incorporate a personal computer (PC) for data acquisition, control, data analysis, presentation/display of results, and communications. Many are integrated via Ethernet-based networks. Two approaches are used: either the computer is integrated into the analyzer package, including the visual display and keyboard/control panel (an abbreviated keyboard or a preprogrammed equivalent), or a separate industrial-grade computer is used. In the first approach high-performance single-board computers are available based on a passive back plane mounted inside the analyzer. Alternatively, rack-mounted PCs and/or environmentally hardened PCs are used in systems requiring separate computers. The focus on computer hardware includes keyboards and input devices, which may feature integrated touch-sensitive displays screen.

If the analyzer is required to operate in a true production environment then it may be necessary to mount it within an environmental enclosure which is sealed to provide protection from water and dust. Not all manufacturing plants offer a controlled operating environment and so the analyzer must be designed to operate over a wide range of humidity (from 0 to 100%, non-condensing) and over a wide temperature range, at minimum from around 0–40°C, sometimes wider. If there is a risk of exposure to flammable vapors, either incidentally or continuously, then the analyzer must conform to fire safety codes. One standard approach is to place the analyzer within a purged and pressurized stainless steel enclosure. As technologies change, and as miniaturization takes place, these requirements can be reduced by paying attention to the way that the analyzer is designed, and by understanding the requirements for intrinsically safe electronics.

Designing a process analyzer does not stop at the enclosure and the internal components. The electronics, the spectrometer and all associated optics must be made to be insensitive to the operating environment. This is particularly important for IR instruments that feature sensitive optics and also high-temperature energy sources. This may be achieved by design, where all components must operate reliably independent of the operating environment of the instrument/analyzer, which can include wide variations in humidity and temperature, and the presence of vibration, over a broad range of frequencies.

4.4.4.3 On-line process analyzers

The basic requirements of an on-line analyzer are very similar to those defined above for an off-line process analyzer. The main difference is that the analyzer usually operates 24 hours a day and unattended. This places higher dependence on reliability and plant safety issues. Any maintenance must be planned in advance and is ideally carried out during a scheduled plant shutdown. Components, such as sources, that are known to have a limited lifetime must be replaced regularly during scheduled maintenance. Premature replacement reduces the likelihood of an unexpected failure. Replacement of serviceable items must be simple, and must not require any sophisticated alignment procedures. An example is the laser used in an FTIR instrument, which must not require any tedious alignment, typically associated with laser replacement in laboratory-based instruments. Also, an average electrical technician or maintenance engineer must be able to replace this component.

Most on-line analyzers are installed as permanent fixtures. Bearing this in mind, environmental issues associated with temperature and vibration become more critical, and the requirement for conformance to safety standards may be enforced by the need for system certification. The latter requires an assessment of the working environment, and the potential for fire and/or explosion hazard, relative to the anticipated presence of flammable vapors or gases. Safety hazard and local electrical design code compliance, which includes CE Mark for Europe, Canadian Standards Association (CSA) for Canada, and Underwriter's Laboratory Association (UL) for the USA, may be required for both off-line and on-line analyzers for certain operating environments. Certification is usually conducted by recognized organizations such as factory mutual (FM, USA) or TÜV (Germany). Figure 4.8 is an example of a fully configured FTIR-based analyzer, equipped with on-board sample handling and on-board data processing, that complies with the standard safety codes for the USA and Europe (originally Cenelec, now ATEX). The analyzer features integrated thermo-electric cooling and also substantial anti-vibration mounting. The inset picture to the right shows the sealed and purged sample-handling chamber.

System installation in a permanent location may require a sample conditioning system featuring some degree of automation, such as automatic cleaning and outlier sample collection and the need to interface with an existing control system process computer. The latter may require that the system operates with a standardized communications protocol, such as Modbus, for the chemical industry. Certain specialized industries use different protocols, such as the semiconductor industry which uses SECS and SEC-II protocols. A 'standardized' approach designated the *Universal* Fieldbus as another method/protocol for process analyzers which is being supported by certain hardware manufacturers.

One approach where the analyzer is housed within a more benign environment is based on the use of an optical fiber-based sample interfacing. Using this type of interface one is able to place the analyzer within the plant control room, possibly within a standard 19-inch electronics rack. With this type of configuration the sample cell and sample conditioning system are placed within the hazardous location, and the instrument is located in a general purpose, or a safe, area where the optical fibers provide the analyzer–sample cell/sampling system interface.[21] Note that there is only limited use of optical fibers in the mid-IR region as covered by most FTIR analyzers. Currently, common choices are zirconyl fluoride glasses, which provides a spectral window down to around $2000\,\mathrm{cm}^{-1}$ ($5.0\,\mu\mathrm{m}$) and the chalcogenide glasses, which may be used from around $3300\,\mathrm{cm}^{-1}$

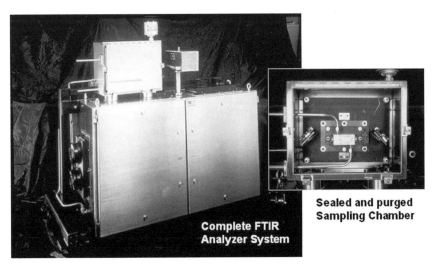

Complete FTIR Analyzer System

Sealed and purged Sampling Chamber

Figure 4.8　Fully configured and certified FTIR process analyzer system.

(3.03 μm) to 800 cm^{-1} (12.5 μm). Chalcogenides are highly attenuating and the length of fiber that can be used is limited to a few meters. Note for some applications, a practical alternative to optical fibers is light-pipe technology (optical conduit), where the sampling interface is interconnected to the instrument via a series of light pipes with flexible connections. At least two manufacturers offer this approach as an option. For such installations, there are fewer requirements for the analyzer to conform to safety standards, and the rack-mounted format is quite suitable, assuming that a rack assembly is available to house the analyzer.

The functional specifications of most on-line process analyzers are similar to those used for off-line applications, and consequently the design requirements are similar. A very practical design concept is to use common instrument measurement technology throughout, providing all three analyzer platforms as discussed above. This enables laboratory-based methods development with the ability to handle both off-line and on-line plant applications as required. A factor that is important from an implementation point of view and for ongoing serviceability is the requirement to be able to support instrument calibration off-line, by providing a practical approach to calibration and method transferability between analyzers. This can make the final analyzer prohibitively expensive for certain applications – FTIR is a case in point. One solution is to consider the use of simpler technology, including a filter-based or EP-IR instrument for the final installation, where all of the initial work is performed on a more sophisticated laboratory instrument. Once completed, the calibrations are moved over to the simpler instrument by the use of a suitable calibration transfer function.

In the design of a process spectrometer there are many attributes and design features to consider. Adapting FTIR technology to an industrial environment can provide some of the greatest challenges. Consequently, FTIR can be considered as a suitable model for the discussion of the issues involved. In this case, the main issues are safety, reliability and serviceability, vibration, temperature, optics, sources and detectors, sample interfacing and sampling, and calibration.

Analyzer safety: There are several potential safety hazards, and the source is one – providing localized temperatures in the range of 1100–1500 K. The laser and its power supply, which is a potential spark hazard, are other examples. Methods for thermal isolation and instrument purging can address the source issues. The controlled discharging of any charge within the capacitance circuitry of the laser power supply addresses the spark hazard issue.

Reliability and serviceability: FTIR instruments have not been known for long-term reliability, based on lab experience. Attention must be paid to optical stability, and the reliability of key components, such as optics (non-hygroscopic preferred), electronics, lasers, and sources. If a component has to be replaced, then a non-specialist technician must be able to perform the task.

Vibration: Vibration can be a major disruption and destructive influence for any form of instrumentation. FTIR is a special case because the fundamental measurement is vibration-sensitive. Care must be taken to ensure that the mirror drive and the associated optics and components are immune to vibration. Note that mirror mounts, and even pyroelectric detectors can be a source of vibrational interference. From a more basic standpoint, constant vibration can loosen components, and so the extensive use of a screw-bonding adhesive is strongly recommended.

Temperature: Temperature changes can result in dimensional changes, which inevitably cause problems if not addressed, for opto-mechanical assemblies within an instrument. Temperature compensation is usually required, and careful attention to the expansion characteristics of the materials of construction used for critical components is essential. This includes screws and bonding materials. If correctly designed, the optical system should function at minimum over typical operating range of 0–40°C. Rapid thermal transients can be more problematic because they may result in thermal shock of critical optical components. Many electronic components can fail or become unreliable at elevated temperatures, including certain detectors, and so attention must be paid to the quality and specification of the components used.

Optics: Issues relative to hygroscopic optics and the need to pay attention to mirror mounts relative to vibration and/or thermal effects have already been addressed. Zinc selenide is an important alternative material, especially when anti-reflection coated. If potassium bromide absolutely has to be used for its lower transmission range (down to $400\,cm^{-1}$) then a protective coating must be used. Most process analyzers use protective windows between the spectrometer and the sample interface. If used, back reflections from the window surfaces into the interferometer must be avoided because these will cause photometric errors.

Sources and detectors: Specific discussions of sources and detectors have been covered elsewhere in this chapter. The issues here are more service- and performance-related. Most sources have a finite lifetime, and are service-replaceable items. They also generate heat, which must be successfully dissipated to prevent localized heating problems. Detectors are of similar concern. For most applications, where the interferometer is operated at low speeds, without any undesirable vibrational/mechanical problems, the traditional lithium tantalate or DTGS detectors are used. These pyroelectric devices operate nominally at room temperature and do not require supplemental cooling to function, and are linear over three or four decades.

Certain applications requiring higher spectrum acquisition rates or higher sensitivity may require a higher-performance detector, such as an MCT detector or the lead selenide detector (5000–$2000\,cm^{-1}$). The MCT detector is capable of high-speed acquisition and has a very high sensitivity; however, the main issues here are practical in terms of implementation. The detector ideally must be cooled down to liquid nitrogen temperatures ($77\,K$), and the detector exhibits non-linearity with absorbances greater than 1 (unless electronically linearized).

The DTGS detector does not require cooling to operate; however, it will stop functioning as a detector at temperatures above 40°C because of depolarization. Note that internal instrument temperatures can be up to 10° higher than the external ambient temperature, and so there is a need for concern about the DTGS detector. A solution is to thermally stabilize the DTGS detector in a process analyzer, typically with a one-stage Peltier (thermoelectric) cooling device.

Transfer optics: The interface of the IR beam with the sample, especially in a classified hazardous environment, can be a major challenge. Gas samples are not too difficult, although it is important to pay attention to the corrosivity of the gases, relative to the windows and any internal optics, as with a folded pathlength cell. Liquids offer different challenges. For on-line applications, users prefer to minimize the use of valves, bypass streams and auxiliary pumps, especially over long distances between the stream and the analyzer. At times there is a benefit to sample the stream either directly or as close as possible to the stream, and in such cases there is the need for transfer optics between the sample cell and the spectrometer/analyzer.

There are a couple of quite elaborate schemes for 'piping' the light from the spectrometer to the sample and back. While these do work, they are typically cumbersome, and they can be awkward to implement, and maintaining alignment and optical throughput can be a problem. The use of fiber optics has already been mentioned, and it is worthwhile to consider their use in context because there are practical limitations as previously mentioned. Recent work has involved hollow capillary light fiber guides, which provide a better transmission range than chalcogenide mid-IR glasses, and with much lower attenuation losses. In this implementation, the hollow fiber guides have much of the flexibility of fiber optics, but with most of the transmission characteristics of a light pipe interface.

Diagnostics: There are many sources of error that can develop within a process analyzer. Component failure is usually relatively easy to diagnose (as long as it is not the computer) and so on-line diagnostics can be straightforward. More insidious problems are caused by the analysis itself, or by electronic component aging (does not fail but operates out of specification), or when a sampling device gives problems, such as corrosion or the coating of windows of a cell. Many of these problems are application-dependent, but often the failure mode is anticipated, and preventative measures implemented to reduce the impact. Note that the FTIR instrument is a full-spectrum device, providing a significant amount of spectral over-sampling, and as such is very beneficial for application diagnostics.

Calibration: Most process analyzers are designed to monitor concentration and/or composition. This requires a calibration of the analyzer with a set of prepared standards or from well-characterized reference materials. The simple approach must always be adopted first. For relatively 'simple' systems the standard approach is to use a simple linear relationship between the instrument response and the analyte/standard concentration.[22] In more complex chemical systems, it is necessary to either adopt a matrix approach to the calibration (still relying on the linearity of the Beer–Lambert law) using simple regression techniques, or model the concentration and/or composition with one or more multivariate methods, an approach known as chemometrics.[23–25]

Standardization: The instrument response function can vary from analyzer to analyzer. If calibration transfer is to be achieved across all instrument platforms it is important that the instrument function is characterized, and preferably standardized.[26] Also, at times it is necessary to perform a local calibration while the analyzer is still on-line. In order to handle this, it is beneficial to consider an on-board calibration/standardization, integrated into the sample conditioning system. Most commercial NIR analyzers require some form of standardization and calibration transfer. Similarly, modern FTIR systems include some form of instrument standardization, usually based on an internal calibrant. This attribute is becoming an important feature for regulatory controlled analyses, where a proper audit trail has to be established, including instrument calibration.

Overall, most of the requirements for a process spectrometer/analyzer are straightforward to implement, but they do require attention at the design level. Another important area, which is FTIR-specific, is the user interface and the need to provide for industry standard data communications. Standard software packages do exist for process instrumentation. For prototype development, and even for the front-end interface in a stand-alone mode of operation, software products, such as Microsoft's Visual Basic and National Instruments' LabView, are also important instrumentation development tools. Note that National Instruments also provides important computer-based electronics and hardware that meet most of the computer interfacing, and system control and communications needs for modern instrumentation.

The design issues discussed so far are based on the requirements of process analyzers operating in a traditional industrial environment that are governed by the need to comply with established electrical, fire and safety standards. For analyzers operating in the high-technology fabrication industries (semiconductors, optics, vacuum deposition, composites, etc.), consumer-associated industries, such as the automotive after market, and environmental, safety, and hygiene applications (United States Environmental Protection Agency, United States Occupational Safety and Health Administration, etc.), the needs and requirements are often very different. Instrument reliability and performance are always important, and operator safety requirements are much the same in all industries although the actual standards may differ, depending on the regulatory agencies involved. For operation in some specialized areas of application, the need to follow well-defined operating procedures and operating protocols is essential, and full documentation and audit trails are a requirement. For some applications, ISO 9000 certification of the analyzer company may be a prerequisite.

4.4.5 Sample handling for IR process analyzers

Many references have been made regarding the importance of the sample, how it is handled and the manner in which it is presented to the IR analyzer. Sample handling for IR spectroscopy has become a major topic of discussion for more than 20 years. and is well documented in standard texts.[27-32] IR spectroscopy, when compared to other analytical techniques, tends to be unique relative to the large number of sample-handling technologies that are available. This is in part due to the fact that samples can be examined in any physical state. Selection of the most appropriate sampling technique is extremely important relative to the overall success of a process-related application. Although the modes of operation may be very different, this applies equally to both off-line and on-line styles of analyzers. The only difference for on-line applications is that the sample-handling method must be sufficiently robust to ensure optimum performance during 24-hour operation. Attention to sample corrosion, and pressure and temperature effects is essential. This often requires the special consideration of the materials of construction used for cell fabrication. The use of specialized and even exotic materials may be required for both the optical elements and the cell construction. Traditional IR optics (Table 4.2) are generally unsuitable, and optical materials such as sapphire, cubic zirconia, and even diamond are required for demanding applications; in these cases the material selection is based on the optical transmission range that is required for the measurement (Table 4.3). Likewise, materials such as nickel, Monel®, and Hastelloy® are used for cell construction when corrosive materials are analyzed.

One of the most important factors in the selection of the sample-handling technique is to attempt to analyze the sample, as it exists, without any form of chemical or physical modification. For gases and certain liquids, simple transmission cells, often with a flow-through configuration meet these requirements.

However, many liquid samples have a high IR absorption cross section, and strong IR absorption occurs even for short pathlengths of the material. This situation causes problems for mid-IR measurements in a process application. In the laboratory, it is sometimes possible to reduce the sample pathlength for such samples down to a capillary film. This is impractical for monitoring applications that would require a flow-through configuration. This problem does not apply in the NIR, where sample pathlengths are typically one or two orders of magnitude larger (typically between 1 and 100 mm) than in the mid-IR spectral region.

A solution to this pathlength issue is to use the technique known as ATR or, simply, just internal reflectance. During the 1980s and 1990s there was a focus on the ATR for both off-line and on-line sampling.[30–32] There are several different configurations available that provide for manual sample loading, continuous flow methods, and dip and insertion probes. The technique is based upon the principle of internal reflectance within a transparent medium of high refractive index. The medium is cut or shaped to provide the equivalent of facets, or reflective surfaces, and the final component is called an internal reflectance element (IRE). As the IR radiation reflects from the internal surfaces, there is an interaction at the surface with any material (the sample) in intimate contact with that surface.

Table 4.2 Common IR-transmitting window materials

Material	Useful range, cm^{-1} (transmission)	Refractive index at 1000 cm^{-1}	Water solubility (g/100 ml, H$_2$O)
Sodium chloride, NaCl	40 000–590	1.49	35.7
Potassium bromide, KBr	40 000–340	1.52	65.2
Cesium iodide, CsI	40 000–200	1.74	88.4
Calcium fluoride, CaF$_2$	50 000–1140	1.39	Insoluble
Barium fluoride, BaF$_2$	50 000–840	1.42	Insoluble
Silver bromide, AgBr	20 000–300	2.2	Insoluble
Zinc sulfide, ZnS	17 000–833	2.2	Insoluble
Zinc selenide, ZnSe	20 000–460	2.4	Insoluble

Table 4.3 Common IR-transmitting materials used for IR optics and ATR internal reflectance elements (IREs)

Material	Useful range, cm^{-1} (transmission)[a]	Refractive index at 1000 cm^{-1}	Water solubility (g/100 ml, H$_2$O)
Zinc sulfide, ZnS	17 000–833	2.2	Insoluble
Zinc selenide, ZnSe	20 000–460	2.4	Insoluble
Cadmium telluride, CdTe	20 000–320	2.67	Insoluble
AMTIR[b]	11 000–625	2.5	Insoluble
KRS-5[c]	20 000–250	2.37	0.05
Germanium, Ge	5 500–600	4.0	Insoluble
Silicon, Si	8 300–660	3.4	Insoluble
Cubic zirconia, ZrO$_2$	25 000–~1600	2.15	Insoluble
Diamond, C	45 000–2 500, 1 650–<200	2.4	Insoluble
Sapphire	55 000–~1800	1.74	Insoluble

Notes: Materials such as cubic zirconia, diamond, and sapphire may be used for transmission windows for special applications.
[a] The usable range is normally less than the transmission range because of the higher optical attenuation that results from the extended optical pathlength of an IRE.
[b] AMTIR: IR glass made from germanium, arsenic, and selenium.
[c] Eutectic mixture of thallium iodide/bromide.

The radiation effectively penetrates in the region approximately 1–3 µm into the surface of the contact material (the sample). Within this penetration depth the equivalent of absorption occurs at the characteristic frequencies of the sample, and as a consequence a spectrum of the sample is generated. The actual final 'effective pathlength' is the accumulated sum of the absorptions as a function on the number of internal reflections. There is a range of materials available, which are suitable for the construction of the IRE, and the final selection is based on the index of refraction of the IRE material, the effective range of optical transparency, and the surface strength and inertness of the IRE, relative to the sample. A selection of suitable materials is included in Table 4.3, and this includes diamond, which along with sapphire and cubic zirconia is particularly applicable to process monitoring applications.

ATR offers many benefits for the measurement of samples in the mid-IR because it can handle a wide range of materials and sample characteristics. It can be applied to liquids, powders, slurries, foams, and solids. It is particularly well suited to liquids because a high degree of intimate contact with the surface can be maintained with a liquid. An issue that one needs to be aware of is the potential for surface contamination and fouling, especially for on-line, flow-through applications. Fouling of the surface not only causes interferences and/or potential energy losses, but can also change the photometry of the experiment by modifying the refractive index of the surface. The propensity for fouling is application-dependent, and is also a function of the surface characteristics of the IRE and the chemical nature of the sample. A material such as zinc selenide has polar sites at the surface (it is effectively a salt, and zinc and selenide ions can exist on the surface) and ionic interactions with certain sample types can occur. In such cases, it is important to try to implement algorithms for detecting material build-up on the surface. Cleaning schemes can be incorporated, and by definition this requires some form of automated sampling system. For some applications, a simple flushing or back-flushing with a cleaning solvent is all that is required. Certain commercial analyzer systems incorporate an ultrasonic cleaning option into the sampling area for cases where there is an incidence of strong material adhesion to the surface.

It is a general perception that sample handling is easier in the NIR region compared to the mid-IR. To some extent this is true because longer pathlengths may be used for liquid sample handling, and direct reflectance methods, such as diffuse reflectance, may be used for convenient sampling of solids. Diffuse reflectance is used for mid-IR applications in the laboratory, but in general it is difficult to implement for on-line applications. Certain applications have been documented where mid-IR diffuse reflectance has been applied to mineral characterization, and this has been used successfully as a field-based application. In practice, spectral artifacts generated by anomalous surface reflections tend to limit the use of reflection techniques for solids in process applications.

4.4.6 Issues for consideration in the implementation of process IR

4.4.6.1 Low maintenance and turnkey systems

The average process engineer is extremely conservative and is hard to convince when it comes to the implementation of process analyzers, and process IR is no exception. Unless there is an agreement to use a process analyzer in an experimental role, it is normally a requirement to supply the analyzer system as a turnkey product. Also, the product is expected to operate 'out-of-the-box,' with minimum technical support required at the time of installation – very much like installing a modern refrigerator at home. Once installed, the analyzer system must operate with minimum operator attention. Maintenance, as noted earlier, must fit into a scheduled maintenance program. Failure modes must be understood, and the recommended time frame for the replacement of parts must be determined, with an adequate supply of spare parts. The process engineer is responsible for downtime and lost production. An analyzer company must not be perceived as the cause of such problems. With an FTIR system, components such as sources and lasers are often the most likely to fail, and knowledge of the MTBF for these is important.

An on-board embedded computer is welcomed in a process analyzer; however, fragile components such as integrated hard disk drives are not recommended. The best approach for on-board storage is solid-state storage, such as flash random access memory (RAM) or battery-backed RAM. Moving parts in analyzers are always subject to critical assessment, and technologies that minimize the need for moving parts are preferred. If FTIR is the only option for a process analyzer, then it is important to re-enforce design features and attributes that address system stability and reliability. There is a trend toward miniaturization with no moving parts. It is early days yet for some of these technologies, but with time they are expected to provide practical alternatives to FTIR.

Maintenance issues are important to address, and high-cost replacement parts must be kept to a minimum, unless the frequency of replacement is only once every 3–4 years. Today, process FTIR analyzers are not necessarily a problem, as long as components are simple to replace. The normal replacement items are the source and laser, and the replacement of these must be kept simple. If an interferometer alignment is required, then this must be automated via an integrated mechanized auto-align procedure.

4.4.6.2   Cost per analyzer point

One aspect that has limited the use and introduction of process analyzers, and in particular FTIR analyzers, has been the high cost of the analyzer, and the resultant high cost per analysis point. Many process analyzer installations are financially evaluated in terms of cost per analysis or analysis point, as well as a payback. Typically, this needs to be kept as low as possible.

The use of fiber optics and fiber-optic multiplexing can increase the number of analysis points, and hence can reduce the overall costs related to a single analyzer. This approach has been used successfully with NIR instrumentation, where typically up to 8 points can be handled. As noted earlier, the use of fiber-optics with mid-IR Fourier transform instruments has been limited. Although in the past commercial multiplexers have been available for mid-IR fiber systems, their use has been minimal.

The opportunity to reduce the cost per analysis point can be very important for certain applications. One option with FTIR is to use sample stream multiplexing with a manifold system. For gas-based systems this is very straightforward, and does not impose high overhead costs. Liquid systems are more complex, and are dependent on the nature of the stream involved, and material reactivity, miscibility, and viscosity are important factors to consider. As noted previously, with the introduction of new, miniaturized technologies, the hope is that newer, less-expensive devices can be produced, and these can provide the needed multiplicity to reduce the cost per analysis point to an acceptable level.

4.4.6.3   Overall cost considerations

The cost per analysis point and the projected payback are two metrics used to evaluate the feasibility of installing a process analyzer. If a spectrometer replaces an existing service-intensive analyzer then that is a positive situation. Also, because of the flexibility of IR analyses, it is feasible that a mid-IR analyzer might replace several existing analyzers; it is a case where direct chemical information from IR can be more important than certain inferential methods that only provide indirect information. Also, if the only other option is lengthy off-line laboratory analyses, then it is easier to justify installing a process IR analyzer. However, if it is necessary to maintain complex chemometric methods, as with some of the NIR applications, then that can be a large negative factor. Cost of ownership is an important metric and is often estimated over 5- or 10-year time frames. Analyzers operating in the mid-IR, or even in the first overtone (FTNIR), are typically much easier to calibrate than traditional NIR analyzers. Often a chemometrics-based calibration is unnecessary in the mid-IR. Consequently, cost-benefits associated with mid-IR analyzers may be a positive selling point compared to an NIR solution.

4.5   Applications of process IR analyzers

Up to this point IR spectroscopy has been discussed relative to the requirements of a process analyzer, and what is needed to implement IR technologies. It is also important to evaluate the IR instrumentation relative to specific applications. IR provides important information that relates directly to the molecular structure and composition of both organic and inorganic materials. Virtually all classes of chemical compounds have functional groups that provide characteristic absorptions throughout the classical mid-IR spectral region. Similarly, many parallel applications exist for inorganic compounds, which typically also provide very characteristic and often very intense IR signatures. A detailed discussion of potential and actual applications, both industrial and otherwise, is too long to cover in any detail here. Table 4.4 indicates strong candidate applications of mid-IR that can be extended to process, process-related, or QC measurements.

Table 4.4   Example applications for process IR spectroscopy

Application area	Materials and products	Techniques and applications
Refinery production	Distillates, crude oils, reforming and cat cracking, and blending	Liquid and gas – transmission Composition monitoring
Fuels	Natural gas, LPG, propane/butane, distillates, gasoline, diesel, fuel oil, syn-fuels, additives	Liquid and gas – transmission Composition monitoring, Octane and cetane number measurements
Solvents	High-purity products, mixtures	Composition monitoring Vapor analysis
Oils and lubricants	Mineral and synthetic oils, greases, coolants, hydraulic fluids	Blending analysis Additive chemistry Oil degradation monitoring
Specialty gas products	High-purity gases, gas mixtures	Composition monitoring Impurity monitoring
General chemicals	Organic/inorganic, liquids and solids	Composition monitoring Impurity monitoring
Chemical reaction monitoring	Polymerizations, esterifications and other condensation reactions, diazo reactions, oxidation, and reduction	On-line and dip-probe applications
Plastics and polymers	Monomers, homopolymers and copolymers, and formulated products	Composition monitoring and physical properties, such as crystallinity, conformation, monomer units, and end groups
Polymer products	Adhesives, adhesive tapes, sealants, latex emulsions, rubber materials, plastic fabrication, etc.	Composition monitoring Rate of cure monitoring Product QC
Consumer products	Liquid detergents and soaps Personal care products Cosmetics Polishes – waxes, etc.	Blending control Composition monitoring Raw materials screening QC

Table 4.4   (Continued)

Application area	Materials and products	Techniques and applications
Specialized products	Water treatment products Dyes and pigments Textiles Pulp and paper Agrochemicals	Production control Composition monitoring Raw materials screening QC
Environmental	Solid and liquid waste Drinking and waste water Priority pollutants	Trace analysis Vapor analysis
Combustion gases	Regulatory applications, fire hazard assessment	Composition monitoring – gas Trace gas and vapor analysis
Ambient air monitoring	Workplace, ventilation systems	Composition monitoring – gas Trace gas and vapor analysis
Food products	Dairy products, General food additives Natural oils and fats Beverages Flavors and fragrances Fermentation reactions	Blending control Composition monitoring Raw materials screening QC
Pharmaceutical products and development	Prescription drugs Dosage forms Active ingredients Product synthesis	Reaction monitoring Dryer monitoring – vapors Blending control Composition monitoring Raw materials screening QC
Aerosol products	Packaging, propellants, final product assay	Raw materials screening QC Gas mixture analysis
Packaging	Paper, cardboard, plastics, films, adhesives	Raw materials screening Physical properties, such as crystallinity – polymers QC
Refrigeration	Automotive and industrial, monitoring of refrigerants	Gas composition analysis
Semiconductors	Process gases, plasma gases, substrate analysis for contaminants	Gas composition analysis Raw materials screening Trace analysis QC

4.6   Process IR analyzers: A review

Process IR spectrometers can be used throughout a large number of industry segments, ranging from their use in traditional chemicals and petrochemicals manufacturing to fine chemicals and specialized manufacturing. The latter includes semiconductor, pharmaceutical, and consumer-oriented products, such as packaged goods and food products. A potential end-user has to balance the pros and cons of mid-IR vs. NIR. In recent years NIR has become accepted in many industry segments as a near-line/at-line tool, and as an on-line continuous monitoring technique. However, there are areas not well covered by NIR, such as gas-phase monitoring applications, especially for the detection of low concentrations of materials. For these, mid-IR spectroscopy, used in the form of either traditional analyzers or FTIR instrumentation, has become the technique of choice.

Gas and vapor applications go beyond product manufacture, and include industrial health and safety, hazard monitoring, and, as appropriate, the needs of various government agencies, including those that focus on the environment (United States Environmental Protection Agency), energy (Department of Energy), and homeland security (Department of Defense). In some cases the monitoring devices are not required by the agency directly, but are used as tools of compliance for industry as a whole.

The petrochemical and chemical industries are the traditional industries for IR analyzers. The needs are extremely diverse and the versatility associated with process IR spectral measurements make the analyzers a good fit. Products manufactured by the petrochemical and chemical industries range from commodity materials (basic chemicals and raw materials) to fully formulated or compounded end-use products. Value-added products (formulated products) often require an advanced spectral analyzer, such as a process FTIR, whereas commodity chemicals are often served by a traditional IR photometer.

It has always been assumed that manufacturing companies want to make plants more efficient and wish to maximize profitability during production. While this is true, it is not an automatic assumption that a company will install expensive equipment to make manufacturing more efficient. In general, IR analyzers are viewed as cost-savers providing information rapidly, often capable of providing it in 'real-time' and with a minimal amount of operator intervention. In many cases, mid-IR analyzers can be easier to justify than other forms of process analyzers. Adding new technology might be just a small part of an overall modernization scheme – it depends on the industry, and on the company. Note that there is a requirement for a return on investment (ROI) and this can vary significantly from industry to industry: Payback ranges from within 6 months to 4 or 5 years.

Beyond the ROI issues, one of the premises for the use of advanced process analyzers is improved product quality. For many industries, and in particular industries close to consumers, there is the need to define procedures and installation qualification/operational qualification/performance qualification (IQ/OQ/PQ) issues, good laboratory practice and good manufacturing practice (GLP/GMP), audit trails, etc. Such requirements open the door for the use of improved analytical methodology, which can be readily provided by the implementation of good process analyzer technology.

At times there has been a focus on a specific application, and instruments have been developed based on either a new technology or existing ones. While such devices may appear to be limited to the application at hand, with some modification, either to the optics or the sampling, they can be applied more generally. Many instrument developments are either sponsored by the federal government (for military, aerospace, energy, and environmental applications) or are spawned from projects undertaken by engineering-consulting firms for major manufacturing companies. One example is a refrigerant analyzer that was developed for the automotive industry for the determination of Freon®/halocarbon type in vehicle air conditioning systems – a requirement initiated from an EPA mandate. In this particular case, specific filters were used to define the analytes, and the packaging was small and inexpensive. This type of system can lend itself to many process applications with only minor adaptations. Examples such as this are applications-driven, and the products are often developed quite independently of the traditional process analyzer business. Today, we are seeing a similar phenomenon with the telecommunications industry, where IR and NIR spectral devices were designed to monitor laser throughput in fiber-optic networks. With some adaptation, many of these devices can be expanded to meet industrial and environmental applications. The benefits to such technologies are small size, low cost potential, environmental ruggedness, and high reliability.

4.7 Trends and directions

Over time, a large number of traditional laboratory instruments have been morphed to meet industrial needs for QC applications. Example applications include raw material, product QC and also some environmental testing. In such scenarios laboratory instruments appear to work adequately. Having said that, there are issues: the need for immediate feedback and the need for smaller, cheaper, and more portable measurements. There is a growing interest in the ability to make measurements in almost any area of a process, with the idea that better production control can lead to a better control of the process and of the quality of the final product. The cost of implementation of today's (2004) process analyzers is still too high, and it is impractical to implement more than a couple of instruments on a production line. Also, there is growing concern about the operating environment, worker safety, and environmental controls.

The ideal scenario would be to have the power of a traditional IR analyzer but with the cost and simplicity of a simple filter device, or even better to reduce the size down to that of a sensor or a simple handheld device. This is not far-fetched, and with technologies emerging from the telecommunications industry, the life science industry, and even nanotechnology, there can be a transition into analyzer opportunities for the future. There is definitely room for a paradigm shift, with the understanding that if an analyzer becomes simpler and less expensive to implement then the role of analyzers/sensor can expand dramatically. With part of this comes the phrase *good enough is OK* – there is no need for the ultimate in versatility or sophistication. Bottom line is that even in process instrumentation, simple is beautiful.

Expanding on this theme, spectrometer devices with lower resolution, shorter spectral ranges, and multiple wavelength selection can be customized to a wide range of analyses. Devices already exist, sometimes implemented in different arenas, in particular for aerospace, military and defense applications, and certain medical applications. Technologies outlined in this article can be modified and changed with this understanding. Specialized enabling technologies exist, such as thin-film optical coatings, chip-scale devices, micro-engineering and micro-machining, and alternative light transmission and detection technologies. Most come from outside of the traditional instrument industry (as already suggested) and, in the future, new generations of instruments and sensor devices will be introduced from companies other than the established instrument vendors. An excellent example is shown in Figure 4.9, for a miniature process analyzer system, based on a super-luminescent broadband LED source or a tunable laser, which is a product from

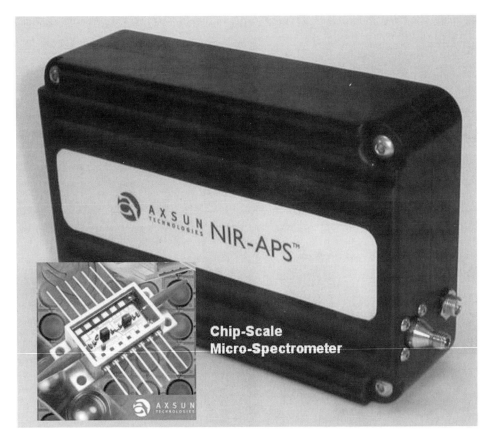

Figure 4.9 A micro-spectral analyzer system based on a telecommunications chip-scale spectrometer package.

a player in the telecommunications industry. While this technology is currently limited to the NIR range, in principle the platform can be expanded in the future to include mid-IR. Note the inset picture which depicts the chip-based spectrometer at the heart of the analyzer.

References

1. Burns, D.A. & Ciurczak, E.W. (eds), *Handbook of Near-Infrared Analysis*; Practical Spectroscopy Series; Marcel Dekker; New York, Vol. 13, 1992.
2. Dent, G. & Chalmers, J.M., *Industrial Analysis with Vibrational Spectroscopy*; Royal Society of Chemistry; Cambridge, 1997.
3. Coates, J.P., Vibrational Spectroscopy: Instrumentation for Infrared and Raman Spectroscopy; *Appl. Spectrosc. Rev.* 1998, 33(4), 267–425.
4. Workman, J.J., Jr *et al.*, Process Analytical Chemistry; *Anal. Chem.* 1999, 71, 121R–180R.

5. Workman, J.J., Jr, Review of Process and Non-Invasive Near-Infrared and Infrared Spectroscopy; *Appl. Spectrosc. Rev.* 1999, 34, 1–89.

6. Colthrup, N.B.; Daly, L.H. & Wiberley, S.E., *Introduction to Infrared and Raman Spectroscopy*; Academic Press; San Diego, 1990.

7. Lin-Vien, D.; Colthrup, N.B.; Fateley, W.G. & Grasselli, J.G., *Infrared and Raman Characteristic Group Frequencies of Organic Molecules*; Academic Press; San Diego, 1991.

8. Workman, J.J., Jr, Interpretive Spectroscopy for Near Infrared; *Appl. Spectrosc. Rev.* 1996, 31, 251–320.

9. Wang, X.; Soos, J.; Li, Q. & Crystal, J., An Acousto-Optical Tunable Filter (AOTF) NIR Spectrometer for On-line Process Control; *Process Contr. Qual.* 1993, 5, 9–16.

10. Wang, X., Acousto-Optic Tunable Filters Spectrally Modulate Light; *Laser Focus World* 1992, 28, 173–180.

11. Hudson, R.D., Jr, Infrared System Engineering. In Ballard, S.S. (ed.); *Pure and Applied Optics*; Wiley-Interscience; New York, 1969.

12. Wolfe, W.L. & Zissis, G.J., *The Infrared Handbook*; Revised Edition; The Infrared Information and Analysis (IRIA) Center; Environmental Research Institute of Michigan; Office of Naval Research; Dept. of the Navy; Washington, DC, 1985.

13. Schlossberg, H.R. & Kelley, P.L., Infrared Spectroscopy Using Tunable Lasers. In Vanasse, G.A. (ed.); *Spectrometric Techniques*, Vol. II; Academic Press; New York, 1981, pp. 161–238.

14. Wright, J.C. & Wirth, M.J., Lasers and Spectroscopy; *Anal. Chem.* 1980, 52, 988A–996A.

15. Sauke, T.B.; Becker, J.F.; Loewenstein, M.; Gutierrez, T.D. & Bratton, C.G., An Overview of Isotope Analysis Using Tunable Diode Laser Spectrometry; *Spectroscopy* 1994, 9(5), 34–39.

16. Curl, R.F. & Tittel, F.K., Tunable Infrared Laser Spectroscopy; *Annu. Rep. Prog. Chem.*, Sect. C 2002, 98, 1–56.

17. Wilks, P.A., High Performance Infrared Filtometers for Continuous Liquid and Gas Analysis; *Process Contr. Qual.* 1992, 3, 283–293.

18. Griffiths, P.R. & de Haseth, J.A., *Fourier Transform Infrared Spectrometry*; Wiley-Interscience; New York, Vol. 83, 1986.

19. Ferraro, J.R. & Krisnan, K., *Practical Fourier Transform Infrared Spectroscopy: Industrial and Chemical Analysis*; Academic Press; San Diego, 1989.

20. Lerner, E.J., Infrared Detectors Offer High Sensitivity; *Laser Focus World* 1996, 32(6), 155–164.

21. Ganz, A. & Coates, J.P., Optical Fibers for On-line Spectroscopy; *Spectroscopy* 1996, 11(1), 32–38.

22. Coates, J.P., A Practical Approach to Quantitative Methods of Spectroscopic Analysis. In George, W.O. & Willis, H.A. (eds); *Computer Methods in UV, Visible and IR Spectroscopy*; Royal Society of Chemistry; Cambridge, UK, 1990, pp. 95–114.

23. Workman, J.J., Jr; Mobley, P.R.; Kowalski, B.R. & Bro, R., Review of Chemometrics Applied to Spectroscopy: 1985–1995, Part 1; *Appl. Spectrosc. Rev.* 1996, 31, 73–124.

24. Standard Practice for Infrared, Multivariate Quantitative Analysis; The American Society for Testing and Materials (ASTM) Practice E1655–00 (2003); ASTM Annual Book of Standards; West Conshohocken, Vol. 03.06, 2003.

25. Beebe, K.R.; Pell, R.J. & Seasholtz, M.-B., *Chemometrics: Practical Guide*; John Wiley; New York, 1998.

26. Workman, J.J., Jr & Coates, J.P., Multivariate Calibration Transfer: The Importance of Standardizing Instrumentation; *Spectroscopy* 1993, 8(9), 36–42.

27. Porro, T.J. & Pattacini, S.C., Sample Handling for Mid-Infrared Spectroscopy, Part I: Solid and Liquid Sampling; *Spectroscopy* 1993, 8(7), 40–47.

28. Porro, T.J. & Pattacini, S.C., Sample Handling for Mid-Infrared Spectroscopy, Part II: Specialized Techniques; *Spectroscopy* 1993, 8(8), 39–44.

29. Coleman, P.B. (ed.); *Practical Sampling Techniques for Infrared Analysis*; CRC Press; Boca Raton, 1993.
30. Coates, J.P., A Review of Sampling Methods for Infrared Spectroscopy in Applied Spectroscopy. In Workman, J.J., Jr & Springsteen, A. (eds); *A Compact Reference for Practitioners*; Academic Press; San Diego, 1998, pp. 49–91.
31. Harrick, N.J., *Internal Reflection Spectroscopy*; Harrick Scientific Corporation; Ossining, New York, 1987. (Original publication by John Wiley & Sons, New York, 1967.)
32. Mirabella, F.M., Jr (ed.); *Internal Reflection Spectroscopy: Theory and Applications*; Marcel Dekker; New York, Vol. 15, 1993.

Chapter 5
Process Raman Spectroscopy

Nancy L. Jestel

5.1   How Raman spectroscopy works

Raman spectroscopy is based on an inelastic scattering process involving an energy transfer between incident light and illuminated target molecules. A small fraction of the incident light is shifted from its starting wavelength to one or more different wavelengths by interaction with the vibrational frequencies of each illuminated molecule. The molecule's vibrational energy levels determine the size of the wavelength shift and the number of different shifts that will occur. This amounts to a direct probe into the state and identity of molecular bonds, and this is why Raman spectroscopy is so useful for chemical identification, structure elucidation, and other qualitative work.

Stokes Raman shift occurs if energy is transferred from the light to the molecule, resulting in red-shifted scattered light, with a longer wavelength and lower energy. In the molecule, a ground state electron is excited from a lower vibrational energy level to a higher one. Anti-Stokes Raman shift occurs when energy is transferred to the light from the molecule. The observed light is blue-shifted, with a shorter wavelength and higher energy. A ground state electron in the molecule is excited from a higher vibration energy level through a virtual state and ends at a lower level. The molecular bonds vibrate at a lower frequency. In either Stokes or anti-Stokes scattering, only one quantum of energy is exchanged. The absolute energy difference between incident and scattered radiation is the same for Stokes and anti-Stokes Raman scattering.

The Raman effect is weak. Only approximately one in 10^8 incident photons is shifted. Of these few Raman-shifted photons, even fewer (relative fraction at room temperature) are anti-Stokes-shifted because fewer molecules start in the required excited state. The Boltzmann distribution describes the relationship between temperature and the fraction of molecules in an excited state. As temperature increases, the relative proportion of ground and excited states changes and the Stokes-to-anti-Stokes intensity proportion changes accordingly. The rest of those 10^8 photons are known as Rayleigh scatter. Since the electrons start and finish at the same vibrational energy level, the photons also start and finish at the same wavelength. Everything is unchanged by the encounter with a molecule. A representative vibrational energy level diagram is shown in Figure 5.1.[1] The challenge of Raman spectroscopy is to detect the Raman 'needle' in the Rayleigh 'haystack.'

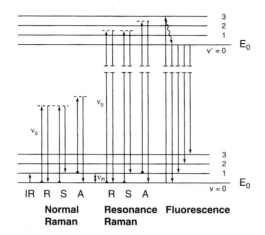

Figure 5.1 Energy level diagram illustrating changes that occur in IR, normal Raman, resonance Raman, and fluorescence. Notation on the figure stands for Rayleigh scattering (R), Stokes Raman scattering (S), and anti-Stokes Raman scattering (A). Reprinted from Ferraro *et al.* (2003)[1] with permission from Elsevier.

A monochromatic light source such as a laser is used to generate enough Raman scattered photons to be detected. A complete Raman spectrum is symmetric about the wavelength of the incident light, with the Stokes portion typically presented on the left side and anti-Stokes on the right side, when plotted on an increasing wavenumber axis. However, the Stokes and anti-Stokes portions will not be mirror images of each other because of their relative intensity differences. A complete Raman spectrum of carbon tetrachloride is shown in Figure 5.2.[2] Since classical Raman spectroscopy typically has been done at or near room temperature and since the Raman effect is inherently weak, most practitioners collect only the Stokes portion of the spectrum. Just as in infrared spectroscopy, the normal *x*-axis units for Raman spectra are wavenumbers, or inverse centimeters, which enables Raman spectra to be compared regardless of the wavelength of the incident light.

Energy transfer actually occurs as a two-photon event:

(1) one photon interacts with the electron cloud of the molecule, inducing a dipole moment and modifying the vibrational energy levels of electrons; and
(2) this in turn provokes emission of a second photon in response to those changes.

However, because this occurs through one quantum mechanical process, it should not be considered to be separate absorption and emission steps.[3] Raman instruments count the number of photons emitted as a function of their wavelength to generate a spectrum. This differs substantially from IR spectroscopy where incident light is actually absorbed by the vibrational energy transition of the molecule. IR instruments measure the loss of intensity relative to the source across a wavelength range. Raman and IR spectroscopy both probe the vibrational energy levels of a molecule, but through two different processes.

Accordingly, the selection rules for Raman and IR spectroscopy are different. In Raman spectroscopy, there must be a change in polarizability of the molecule upon excitation, whereas a change in dipole moment is required for IR. A dipole moment is the magnitude

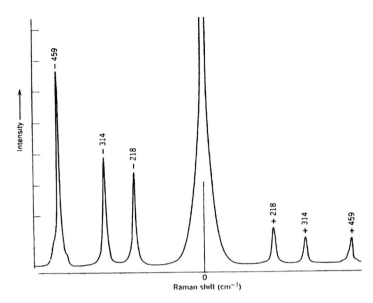

Figure 5.2 Complete Raman spectrum of carbon tetrachloride, illustrating the Stokes Raman portion, Rayleigh scattering, and the anti-Stokes Raman portion. Reprinted from Nakamoto (1997)[2] and used by permission of John Wiley & Sons, Inc.

of the electronic force vector between the negative and the positive charges or partial charges on a molecule. A permanent dipole moment exists in all polar molecules. Changes in dipole moment mean that the magnitude of the charge or the distance between the charges has changed. When the oscillating electric field of a light source interacts with a molecule, the positively charged nuclei and the negatively charged electrons move in opposite directions, creating a charge separation and an induced dipole moment.[2] Polarizability reflects the ease with which this occurs.[4] Changes in polarizability, or the symmetry of the electron cloud density, can occur for polar or non-polar molecules. Since polyatomic molecules can move in any combination of x, y, or z axes, there are 3^2, or 9, possible directions of motion to evaluate. A change in any of them means Raman scattering can occur.[2] The mutual exclusion principle states that there are no shared transitions between IR and Raman for molecules with a center of symmetry. Both Raman and IR bands can be observed for the many molecules without a center of symmetry, though the relative intensities will be quite different. Raman overtones and combination bands are forbidden by the selection rules (symmetry) and are generally not observed.

Using these concepts, it is possible to establish some rough rules to predict the relative strength of Raman intensity from certain vibrations.[2]

- Totally symmetric ring breathing >> symmetric > asymmetric
- Stretching > bending
- Covalent > ionic
- Triple > double > single
- Heavier atoms (particularly heavy metals and sulfur) > lighter atoms

A practical rule is that Raman active vibrations involve symmetric motions, whereas IR active vibrations involve asymmetric motion. For example, a symmetric stretching or bending mode would be Raman active, but an asymmetric stretching or bending mode is IR active. Thus, it is clear why highly symmetric molecules, particularly homonuclear diatomic species, such as $-C-C-$, $-C=C-$, $-N=N-$, or $-S-S-$, generate such strong Raman scatter and are correspondingly weak in the IR. Some classic strong Raman scatterers include cyclohexane, carbon tetrachloride, isopropanol, benzene and its derivatives, silicon, and diamond. By comparison, some strong IR absorbers are water, carbon dioxide, alcohols, and acetone.

Under constant experimental conditions, the number of Raman scattered photons is proportional to analyte concentration. Quantitative methods can be developed with simple peak height measurements.[5] Just as with IR calibrations, multiple components in complex mixtures can be quantified if a distinct wavelength for each component can be identified. When isolated bands are not readily apparent, advanced multivariate statistical tools (chemometrics) like partial least squares (PLS) can help. These work by identifying all of the wavelengths correlated to, or systematically changing with, the levels of a component.[6] Raman spectra can also be correlated to other properties, such as stress in semiconductors, polymer crystallinity, and particle size because these parameters are reflected in the local molecular environment.

5.2   When Raman spectroscopy works well and when it does not

5.2.1   Advantages

Raman spectra encode lots of information about molecular structure and chemical environment. Fortunately, that information can be accessed in many ways because Raman spectroscopy is an extremely flexible technique, both in what it can measure and in how the measurements can be made.

5.2.1.1   Wide variety of acceptable sample forms

Samples for analysis may be solids, liquids, or gases, or any forms in between and in combination, such as slurries, gels, and gas inclusions in solids. Samples can be clear or opaque, highly viscous or liquids with lots of suspended solids. While this variety is easy for Raman spectroscopy, it would be challenging for mid-IR and near-infrared (NIR) spectroscopy and numerous non-spectroscopic approaches.

5.2.1.2   Flexible sample interfaces

Raman spectroscopy offers a rich variety of sampling options. Fiber optics separate the measurement point from the instrument, up to about 100 meters. Multiple points can be measured, simultaneously or sequentially, with one instrument. Measurements can be made non-invasively or in direct contact with the targeted material. Non-contact and contact probes are discussed and pictured in Section 5.3.4. Non-invasive or non-contact probes mean that there are several inches of air between the front lens of the probe and the sample. These are particularly valuable for very dangerous or radioactive compounds, for corrosives, and for sterile or packaged items. Clear glass or plastic bottles, bags, blister packs, or ampoules are suitable package forms. Powders are most easily measured with a non-contact probe.

Contact or immersion probes can be placed directly in the process. The fiber optics are protected behind a suitable window, such as sapphire, quartz, or glass, but the process material touches the window. Slip streams are not required, but could be used. Available probes have been engineered to meet almost any specified chemistry or condition, including high or low temperature, high or low pressure, caustic or corrosive, viscous, slurries, or submerged. The same sample interface does not have to be used at each point. The flexible sampling configurations and long fiber lengths possible give Raman spectroscopy considerable advantages over mid-IR, and even NIR, spectroscopy.

5.2.1.3   No sample preparation, conditioning, or destruction

Generally, little or no sample preparation is required for Raman experiments, which is particularly valuable in a process installation. Most process implementations are designed to require no sample conditioning. The sample is not destroyed during the experiment, except for a few cases like black or highly fluorescent materials or materials that absorb enough laser light to be thermally degraded.

5.2.1.4   Flexible sample size

There is no inherent sample size restriction, large or small, but is fixed by the optical components used in the instrument. The diffraction limit of light, roughly a few cubic micrometers depending on the numerical aperture of the optics used and the laser's wavelength, sets the lower bound.[7] In a process application, the type of fiber optics used also affects sample volume examined. Macroscopic to microscopic samples can be measured with the appropriate selections of laser wavelength, laser power, and optics.

5.2.1.5   Attractive spectral properties

Raman spectra with sharp, well-resolved bands are easily and quickly collected over the entire useful spectral range, roughly 50–4000 cm^{-1}, though the exact range depends on instrument configuration. Bands can be assigned to functional groups and are easily interpreted. For quantitative work, often simple univariate calibration models are sufficient. Calibration models can range from 0.1 to 100% for most compounds, and can reach detection limits of 100 parts per million (ppm) for strong or resonance-enhanced Raman scatterers.

5.2.1.6   Selection rules and acceptable solvents

The differences in selection rules between Raman and IR spectroscopy define ideal situations for each. Raman spectroscopy performs well on compounds with double or triple bonds, different isomers, sulfur-containing and symmetric species. In contrast to IR, the Raman spectrum of water is extremely weak so direct measurements on aqueous systems are easy to do. In general, quality IR spectra cannot be collected from samples in highly polar solvents because the solvent spectrum overwhelms that of the sample; typically the Raman spectrum of the solvent is weak enough not to present similar problems.

5.2.1.7   High sampling rate

Acquisition times commonly vary from seconds to minutes, often with negligible time between acquisitions, even with multiple simultaneous measurement locations (multiplexing). The dedication of different areas on the charge coupled device (CCD) detector to each point makes this possible. The different detectors used for NIR and mid-IR instruments cannot be multiplexed in the same fashion and must measure multiple samples sequentially.

5.2.1.8   Stable and robust equipment

Most process Raman instruments have few moving parts, if any, and thus are quite stable and robust. This also reduces the number of parts likely to need replacement. The laser and detector shutter are the most common service items. Special utilities and consumables are not required. Proper enclosure selection allows instruments to survive harsh environments, such as installation outdoors or around vibrations. An instrument built to accurate specifications and well installed is unlikely to require much care or daily attention. However, it is important to remember that there is no such thing as a maintenance-free instrument. Limited additional maintenance will ensure robust performance for years.

5.2.2   Disadvantages and risks

Like all techniques, Raman spectroscopy has its limitations and disadvantages.

5.2.2.1   High background signals

Fluorescence is one of the biggest challenges to collecting quality Raman spectra, but does not impact mid- or near-infrared techniques. There are numerous approaches to reducing the problem, including using longer laser wavelengths and summing numerous short acquisitions. However, it remains one of the first issues a feasibility study must address.[8]

It is all but impossible to collect Raman spectra of black materials. The sample is usually degraded, burned, or otherwise damaged from absorption of the intense laser energy. Any Raman signal is masked by the strong blackbody radiation. Many of these samples are equally challenging for mid- or near-infrared. Raman spectra of highly colored materials can be similarly difficult to collect, though the technique has been used to great advantage in examinations of paintings and pottery.[9,10]

5.2.2.2 Cost

In a chapter on process Raman spectroscopy, Pelletier wrote that the 'analyzer may cost from US$50 000 to US$150 000, and the integration of the analyzer into the production facility can cost as much as the analyzer itself.'[11] The cost varies so much because there is no standard configuration. Differences include the number of measurement points required and distance of each from the analyzer, the kinds of probes used, the demands on an enclosure from weather and electrical classification, the system used to communicate with a facility's host computer, and required redundancies. In general, Raman instruments are more expensive than mid- and near-infrared units. In addition, lasers need frequent replacement and are expensive. Popular lasers, such as diode lasers at 785 nm and Nd:YAG at 532 nm, have estimated lifetimes of just 1–1.5 years. Though lasers can sometimes be repaired, the usual replacement cost is US$10 000–20 000. While a NIR process unit might require the light source to be changed every 6–12 months, even the most expensive light bulb is practically free compared to a laser. However, the costs of reduced manufacturing yield, poor product quality, or unsafe sampling or operating conditions can easily justify the laser replacement expenses for many installations. No matter how a firm calculates its required return on investment, such initial and ongoing expenses demand that an installation delivers substantial benefits. Until prices fall, it is probable that many applications easily handled by Raman spectroscopy will be done with cheaper but more technically challenging techniques like NIR spectroscopy.

5.2.2.3 Stability

Subtle changes to the position and shape of sharp Raman bands are indicative of small changes in the local chemical environment. This makes the technique very sensitive and suitable for challenging chemical problems; however, it also puts substantial demands on instrument stability. Small changes in laser wavelength or instrument environment could appear as wavenumber shifts and be mistaken for chemical changes. Besides using the highest-quality, most stable instrument available for such applications, there are other approaches that can be used to reduce the potential for problems.

One of the most common approaches is to mathematically align reference or pseudo-reference peaks, such as from a solvent or unchanging functional groups, prior to use in a model. In other cases, it can help to mathematically reduce peak resolution slightly, analogous to using peak areas instead of peak heights. If an instrument has not yet been purchased, changes to the grating to increase resolution by sacrificing some spectral coverage may help minimize sensitivity to the fluctuations. While these approaches may seem like hurdles, they are analogous to learning to handle NIR spectra with multiplicative scatter correction or second derivatives. Similar problems plague NIR units, but are less noticeable on the broad, ill-defined peaks. While Raman instrument stability is excellent for many applications, users must remain alert to the possibility of a problem in long-term use and develop strategies to mitigate this risk.

5.2.2.4 Too much and still too little sensitivity

Raman spectroscopy's sensitivity to the local molecular environment means that it can be correlated to other material properties besides concentration, such as particle size or polymer crystallinity. This is a powerful advantage, but it can complicate the development and interpretation of calibration models if not recognized in advance. For example, if a model is built to predict composition, it can appear to fail if the sample particle size distribution does not match what was used in the calibration set. Some models that appear to fail in the field may actually reflect a change in some other aspect of the sample, rather than a problem with the instrument. This issue is not unique to Raman spectroscopy, but users may not be as aware of potential problems as for other techniques. Good understanding of the chemistry and spectroscopy of the system being modeled along with well-designed calibration experiments can help prevent these problems. The power of the technique can be a hindrance if not recognized and managed.

Raman spectra are very sensitive indicators of changes in hydrogen bonding, even though the water spectrum is weak. Water spectra can be measured, but it is quite possible to accidentally turn the spectrum and instrument into an extremely expensive thermometer. The ability to obtain composition and temperature from a single spectrum is a substantial advantage for many applications. However, it is a disadvantage in others and must be taken into consideration during the development of a calibration model.

Conversely, Raman spectroscopy has been described as an insensitive technique. Except with resonance-enhanced materials or very strong scatterers, Raman spectroscopy is generally considered to have a detection limit around 0.1%.[11] The technique is often described as being poorly suited to detecting small changes in concentration and this has limited the technique's use in some applications. It is hard to generalize since the limit depends on the situation and the equipment configuration. It is important to test sensitivity and detection limits for target applications during feasibility studies and to recognize the demands those requirements put on equipment.

5.3 What are the special design issues for process Raman instruments?

Common characteristics of all viable process Raman instruments include the ability to measure at multiple locations, simultaneously or sequentially and in a variety of electrically classified or challenging environments with no special utilities. The instrument must be reasonably rugged and compact, and not require much attention, servicing, or calibration.[12] Modern process Raman instruments essentially are turnkey systems. It is no longer necessary to understand the optical properties and other instrument details to successfully use the technique. The general design options will be described briefly, but more information is available elsewhere.[13,14] This section will focus primarily on special instrumentation considerations for process installations.

Any Raman instrument consists of a laser as a monochromatic light source, an interface between sample and instrument, a filter to remove Rayleigh scatter, a spectrograph to separate the Raman scattered light by wavelength, a detector, and a communications system to report analysis results. Purchasers must make decisions about laser wavelength and stability, appropriate sampling mechanism, spectrometer design, and a communications system. The filter and detector choices generally follow from these decisions. Generally, electricity is the only utility required; gases, liquid nitrogen, and cooling water are not required. Normally, plugging-in fiber optics and an electrical cord are the only connections required to get the equipment running. Few adjustments or alignments are required.

5.3.1 Safety

Safety must be the first consideration of any process analytical installation. Electrical and weather enclosures and safe instrument–process interfaces are expectations of any process spectroscopy installation. The presence of a powerful laser, however, is unique to process Raman instruments and must be addressed because it has the potential to injure someone. Eye and skin injuries are the most common result of improper laser exposure. Fortunately, being safe is also easy. Because so many people have seen pictures of large industrial cutting lasers in operation, this is often what operations personnel erroneously first envision when a laser installation is discussed. However, modern instruments use small footprint, comparatively low-power lasers, safely isolated in a variety of enclosures and armed with various interlocks to prevent accidental exposure to the beam.

It is important and worthwhile to spend time educating project participants about the kinds of lasers used, their power, safety classifications, proper handling and operation, and risks.[15–17] Ample free educational resources are available on the Internet. Operations and maintenance personnel must receive regular laser safety training commensurate with their expected level of interaction with the instrument. This might range from general awareness of a potential hazard in the area to standard operating procedures to lock-out/tag-out procedures when working on the instrument, laser, or fiber optics. Government laser-safety regulations have specific requirements depending on the laser wavelength and power.[16,18–20] Common requirements include restricting access to the laser and its emission points, both to prevent accidental laser exposure and to ensure uninterrupted operation, and wearing wavelength-specific safety glasses whenever the laser beam is exposed. Appropriate safety warning signs and labels must be used.

Vendors who sell process Raman equipment have implemented many safety features as part of their standard instrument configurations. However, it remains the responsibility of the purchaser to ensure a safe environment and compliance with their company, industry, and governmental safety regulations. Anything involved in an industrial process has associated hazards that must be controlled to eliminate or reduce risks. Those originating from the laser in a Raman instrument are readily managed when given due consideration and must not be ignored.

5.3.2 Laser wavelength selection

Raman signal intensity is proportional to the fourth power of the inverse of the incident wavelength. As Table 5.1 illustrates, the use of a 785-nm laser instead of one at 480 nm means that only 14% of the possible photons could be generated. It is desirable to use the shortest laser wavelength possible to maximize the number of Raman photons available for detection.

Fluorescence is a common interference and its severity tends to increase with shorter wavelengths. By using longer-wavelength lasers, many fluorescence problems can be avoided, but this comes at the expense of greatly reduced Raman intensity. CCD detectors have limited response functions at longer wavelengths. Lasers and detectors should be selected together in order to minimize limitations on spectral coverage. Project feasibility studies are an excellent time to evaluate different laser wavelengths. The selection of a laser wavelength to balance the trade-offs between maximizing Raman signal and minimizing fluorescence is a critical decision.

A dramatic enhancement in Raman signal can be obtained when the laser wavelength falls in the sample's electronic absorption band. This is referred to as resonance Raman spectroscopy and it greatly increases sensitivity and reduces detection limits. All vibrational bands in the spectrum are not enhanced equally since the resonance is only with the chromophoric functional group. It makes the system highly selective, though that could be a disadvantage if more than one component needs to be predicted.[2] Resonant systems are prone to local sample heating so care must be taken to ensure that samples are not exposed to the laser too long or that there is an efficient sample-cooling mechanism.

5.3.3 Laser power and stability

Higher laser power is preferred since acquisition time (signal strength) is directly related to laser power for a given system. Lower-power lasers are used when samples are not suitably robust or for budget considerations. The maximum sampling rate in a process application is limited by acquisition time, which depends on the molecule's Raman

Table 5.1 Decrease in Raman intensity as a function of laser wavelength

Wavelength λ (nm)	Intensity $(1/\lambda)^4$ (% relative to nm)
480	100
532	62.27
633	33.06
785	13.98
840	10.66
1064	4.14

scattering efficiency. High laser power can cause local sample heating, which may cause damage to static samples unless heat dissipation is addressed. The ready availability of quality lasers has made process Raman instruments possible and further improvements will translate directly to increased sensitivity and power of Raman spectroscopy.

Laser stability is a key differentiator between commercial instruments. Some vendors choose to use the most stable lasers (constant power, wavelength, and mode) available whereas others use less robust lasers but continuously correct for fluctuations with math and programming. The first option is more expensive but conceptually easier to understand; the second is less expensive but quite challenging for the instrument designer and programmer, and sometimes the user. Both are valid approaches, though many Raman spectroscopists prefer the first approach. However, even with very stable lasers, model robustness often benefits from some implementation of the second approach to eliminate the effects of long-term fluctuations and changes. The choice between these approaches will depend on the application's spectral sensitivity, analysis requirements, and budget.

5.3.4   Sample interface/probes

Fiber-optic cables up to approximately 100 meters connect each measurement point to the instrument. The cables typically contain four single fibers: a primary pair with one to carry laser light to the sample and another to return the Raman signal to the spectrometer, and a spare pair. If a fiber is broken, light will not reach the sample and the spare fiber must be used instead.

Silica Raman scatter always is generated when the laser travels through fiber optics. It will overwhelm any analyte signal if it is not removed before the laser reaches the sample. For similar reasons, the laser must be blocked from the return fiber. The optical path of one commercial probe design is shown in Figure 5.3. Though this task can be

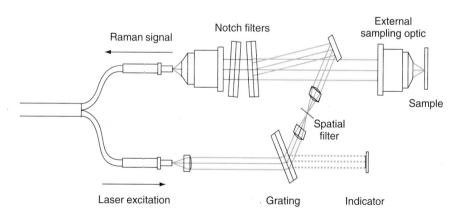

Figure 5.3   Optical path of a commercial Raman probe. Adapted, with permission, Copyright © 2004 Kaiser Optical Systems, Inc.

accomplished in different ways, all probes use optics which must be protected from dirt and weather with a suitable enclosure or housing.

Most probes use a 180° backscatter configuration, which refers to the angle between the sample and the Raman scatter collection optics. Because the illumination and collection optics occupy essentially the same space, Raman spectroscopy can be performed using a single probe inserted directly in a process stream. As previously discussed, the probe may be in contact with the sample or at a distance away from it. In a non-contact system, the laser usually comes to a focus several inches away from the probe's optical face. A clear, process-compatible sight glass may be placed in between the probe and the focal point if necessary. In a contact system, the probe is inserted through a fitting, and its optical window touches the targeted material. The choice between the approaches is usually made by which implementation would be the easiest, safest, and most robust.

The ability to separate the fiber-optic tip and sample by as much space as required makes Raman sampling extremely flexible. For example, 15-foot immersion probes to insert in the top of chemical reactors have been made. The laser beam travels down the 15-foot probe shaft, finally coming to a focus just outside of the optical window at the probe's tip. A photo of such a probe is shown in Figure 5.4(a). Non-contact probes could also be made to come to a focus at similar distances, but then the beam must be enclosed to address safety concerns.

Probes can be engineered for almost any environment, but these requirements must be clearly defined before probes are built. No probe can survive all environments so there may be some design trade-offs. Many immersion probes have a second window as a safety backup to ensure the process remains contained in the event of a window seal failure. Probes have been designed with motorized focus adjustments, automatic solvent or steam washes or air jets to keep the window clear, and sealed in water- and pressure-safe containers. Three commercial probes, two immersion and one non-contact, are pictured in Figure 5.4. Different optics can be attached to the face of the non-contact probe, much like conventional cameras using different lenses.

Raman spectroscopy's flexible and minimally intrusive sampling configurations reduce sample pretreatment demands. Since the sample does not have to pass through a narrow gap, it is easy to analyze viscous streams or ones with high solids content. The main sampling challenge is ensuring that the window remains clear enough for light to pass through.

5.3.5 Spectrometer

As with all process spectroscopy instruments, it is essential to use the right environmental enclosure. In general, process Raman instruments must be protected from weather and kept reasonably thermally stable. A variety of manufacturers offer process-ready systems that have been specifically engineered to handle common environmental challenges, such as temperature fluctuations, vibrations, weather, and electrical classification. Similarly, a convenient, automated mechanism must be included for calibrating the spectrometer wavelength, laser wavelength, and optical throughput or intensity axes.[21]

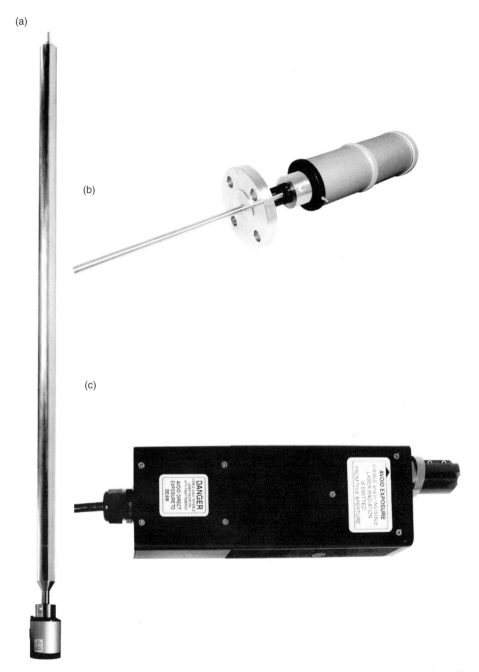

Figure 5.4 Collection of commercial Raman probes designed for different installations: (a) fifteen-foot immersion probe; (b) typical immersion probe, approximately one-foot long; and (c) probe body for a non-contact probe with fitting to accept different focal length optics. Adapted, with permission, Copyright © 2004 Kaiser Optical Systems, Inc.

Key spectrometer design evaluation criteria include optical throughput or efficiency, spectral coverage, spectral resolution, instrument stability and robustness, and instrument size. Three major spectrometer designs are commonly available: axial transmissive, concave holographic, and Fourier transform. A fourth design based on Echelle gratings is used for field applications, but is not sold as a dedicated process instrument. A fifth design, Fabry-Perot interferometer, has been discussed for process applications but is not readily available.[22] The axial transmissive, concave holographic, Echelle, and other designs used only for laboratory units, such as Czerny-Turner, are all considered dispersive designs.

Most dispersive Raman systems use a silicon-based CCD as a detector. The system's resolution is a function of the grating, spectral range, detector width in pixels, and pixel size. The detector width is usually the limiting factor. Volume holographic gratings or grisms (combination of grating and prism) are used in axial transmissive systems. In both cases, the light travels along a single axis, passes through the grating, and is dispersed on the detector. Additionally, since the grating and all optics are mechanically fixed, these designs have excellent stability and a small footprint. Because there are few optical surfaces, the optical throughput is high. Simultaneous measurement of multiple locations is accomplished by dedicating a location on the detector for each.

The holographic grating enables a complete spectrum to be split in half, usually around a low information portion of the spectrum like $1750\,cm^{-1}$, with each half directed to a different portion of the detector. This ensures resolution comparable with other designs, around $5\,cm^{-1}$, and simultaneous acquisition of a complete spectrum. The Rayleigh scatter is removed with a holographic notch filter. Such filters have become standard in the industry.[23,24] The grism enables numerous fibers to be imaged onto the detector simultaneously, thus offering the possibility of monitoring tens of channels simultaneously.

In a related dispersive approach, concave holograms can be used as both the grating and the imaging optics to build a one-element spectrograph. Since there are so few components, these instruments are compact, inexpensive, and stable. Over long wavelength ranges, some image-quality problems may be observed.[13]

Fourier transform Raman (FT-Raman) systems are mainly used to avoid substantial fluorescent backgrounds by exciting the sample with long laser wavelengths. The systems generally use a 1064-nm laser, though a system is available at 785 nm. As with dispersive systems, the longer excitation wavelength greatly reduces signal intensity. The poor sensitivity of these systems has limited their process utility, but may be the best choice when fluorescence is severe. The instruments typically use Indium-Gallium-Arsenide (InGaAs) or germanium detectors since the photoelectric response of silicon CCDs at these wavelengths is negligible.

5.3.6 Communications

In order for the data generated by the analyzer to be useful, it must be transferred to the operation's centralized control or host computer and made available to process control algorithms. Vendor packages manage instrument control and can do spectral interpretation and prediction or pass the data to another package that will. Most vendors support a variety of the most common communications architectures. In addition to reporting predictions, the software also typically reports a variety of possible instrument errors.

5.3.7   Maintenance

Outside of the occasional system calibration and model verification tests, the routine maintenance burden of a Raman system is quite low. Optical windows may need to be cleaned, though automatic window cleaning systems can be implemented if it is a known issue. The most likely maintenance activity is laser replacement. Some systems have a backup laser that turns on automatically if the primary laser fails. This lessens the impact of a failure if the unit is being used for closed-loop process control. With a quality instrument and well-developed models, process Raman installations need less maintenance than many competing techniques.

5.4   Where Raman spectroscopy is being used

The diversity of potential applications of Raman spectroscopy and the environments where it could be deployed are staggering. It is being used in many arenas, including macro-, micro-, and nano-materials science, chemicals and polymers, pharmaceuticals, biomedical and medical, semiconductors, space exploration, ceramics, marine science, geology, gemology, environmental monitoring, forensics, and art and archeology applications.[25–30] More and more scientists without formal training in Raman spectroscopy are using it to solve their problems. The following extensive, but not exhaustive, collection of examples is organized by fundamental purpose rather than by industry. They focus on process analysis and were selected to demonstrate the viability and flexibility of the technique, to stimulate brainstorming on possible uses of Raman spectroscopy, and to guide the reader to a better understanding of technical considerations.

The examples are taken from technical and patent literature. Many firms patent their work but never publish in peer-reviewed literature. Despite the absence of peer-review and pertinent technical details, patents can offer important insights into how technology is being applied.

5.4.1   Reaction monitoring

Many spectroscopic techniques are increasingly used for chemical reaction monitoring because of their speed and detailed chemical data.[31] Process Raman spectroscopy is well suited for monitoring the progression of many types of batch, semi-batch, or continuous reactions. The reaction scale can be large, such as a polymer produced in a 25 000-gallon vessel, or small, such as the curing of an epoxy coating on a tiny substrate. Common motivations are to ensure the correct raw materials are used, to monitor reaction progress to a targeted product yield, to detect side-product generation, and to ensure product quality. Though a few systems are studied to elucidate reaction mechanism, the information usually is intended to trigger a quick, automatic corrective action by a process control system.

5.4.1.1   Bioreactors

From the earliest batches of mead to modern beer and wine, controlling microorganisms to produce the desired product has long been an art. As the biochemical and bioprocess fields progress, the science behind the art is being identified and playing an increasingly important role.[32] Bioprocesses keep growing in industrial importance, producing everything from food to alcohols to medicines.[33] It is often difficult to collect representative samples without contaminating the vessel. Off-line analysis is too slow for these rapidly changing reactions. Bioengineers are looking for non-invasive, non-destructive, and fast measurement systems that can measure every component simultaneously without contaminating the vessel or altering the reaction profile.[34] With process analyzers, sampling issues are reduced or eliminated, while still providing data more frequently than off-line testing. This facilitates investigations of mechanisms and component relationships, optimized production conditions, and high product quality.

Carotenoids are commercially important as a food supplement and are being studied as a treatment for a variety of diseases including macular degeneration, cataracts, heart disease, and cancer.[35] Production by traditional chemical synthesis is giving way to bioreactors to help meet the US$935 million demand projected by 2005.[36] Engineers need to quickly identify nearly optimal production conditions to accurately screen the many organisms being considered as biological generators. As the manufacturing processes mature, such motivations will turn to controlling and maximizing output.

The use of Raman spectroscopy to quantify the production of the carotenoid astaxanthin from the yeast *Phaffia rhodozyma* and from the alga *Haematococcus pluvialis* in a bioreactor has been described.[36] The bioreactor was operated normally for more than 100 hours per reaction. The calibration curve covered total intracellular carotenoid concentrations from 1 to 45 mg/l. The precision was better than 5% and the calibration model transferred across batches. Three experimental designs were used, but no data were provided on the number of replicate batches studied. The data suggest that calibration curves also could be built for other reactor components such as glucose. Though other carotenoids were generated, totaling roughly 20–25% of the stream, no attempt was made to spectrally differentiate between the various species. The continuous measurements via in-line probe were easy, sufficiently fast, and accurate.

This example illustrates several important lessons. First, this application takes advantage of Raman signal enhancements to measure concentrations normally considered too low for Raman spectroscopy. Extremely strong Raman signal is a hallmark of carotenoid measurements. Traditionally, this enhancement has been considered a resonance effect but others believe it is due to coupling between π-electrons and phonons.[37] It can be beneficial to use such enhancements whenever possible.

Second, the results of off-line experiments do not necessarily predict the success or failure of on-line experiments. Off-line experiments had previously produced poor results due to excessive sample heating and photobleaching. These problems arose from limited sample volumes and poor stirring. The in-line experiments used a large, well-mixed tank and were correspondingly easier to perform. It is important to identify any differences between laboratory and plant conditions and perform a series of experiments to test the impact of those factors on the field robustness of any models.

Third, it can be difficult to match the results from samples removed for off-line analysis (reference values) with the right spectra. If the system is changing rapidly, removed samples may be quenched incompletely and continue to react. This will introduce bias into reference values. Careful procedures must be developed to minimize bias between on-line and off-line results.

Fourth, while fluorescence was not a problem in this system, some bioreactor components may be highly fluorescent. Until more experience is gained, this makes it difficult to make generalizations about the likelihood of success with new systems. Finally, water is a weak Raman scatterer and can be hard to track. While Ulber *et al.* considered the challenge in obtaining a spectrum of the aqueous cell culture broth to be a disadvantage, others consider it an advantage since these solutions overwhelm mid- or near-infrared detectors.[33,34,36]

Lee *et al.* determined the concentration of glucose, acetate, formate, lactate, and phenylalanine simultaneously in two *E. coli* bioreactions using Raman spectroscopy.[33] The bioreactor was modified to have a viewing port, and spectra were collected through the port window. This enabled researchers to sterilize the reactor and ensure that it would never be contaminated by contact with the instrument.

These experiments differed from many others because the models developed were based on spectra of the pure raw materials, products, and the expected interferents. Reference samples were collected and measured in order to independently evaluate the model, but were not used in model development. This approach promises greater efficiency and productivity for situations where many different reactions are being studied, instead of the more common industrial case of the same reaction run repeatedly. However, the researchers were not satisfied with their models and identified several probable limiting factors that serve as good reminders for all process analytical projects.

Experimental differences between calibration work and routine use can introduce model error. The sapphire viewing port was used as a convenient, built-in reference sample. The calibration models were built before the reactor was sterilized. Unfortunately, the window material changed slightly during sterilization. The resulting subtle differences in the reference spectra collected before and after sterilization hurt the model's predictive ability. Reference samples are useful only if they are stable.

The authors list other possible error sources, including minor differences between the spectra of pure components and mixtures, temperature effects, and missing minor components. Again, it is usually best to mimic the true process conditions as much as possible in an off-line set-up. This applies not only to physical parameters like flow rate, turbulence, particulates, temperature, and pressure, but also to minor constituents and expected contaminants. The researchers anticipate correcting these issues in future models, and expect to achieve approximately 0.1-mM detection limits. The models successfully accommodated issues with specular scattering from the biomass and air bubbles and from laser-power fluctuations.

5.4.1.2   Emulsion polymerization

Emulsion polymerization is used for 10–15% of global polymer production, including such industrially important polymers as poly(acrylonitrile-butadiene-styrene) (ABS), poly(styrene), poly(methyl methacrylate), and polyvinyl acetate.[38] These are made from aqueous solutions with high concentrations of suspended solids. The important components have unsaturated carbon–carbon double bonds. These systems are ideal for Raman spectroscopy and a challenge for other approaches, though NIR spectroscopy has been used.

Bauer *et al.* describe the use of a non-contact probe coupled by fiber optics to an FT-Raman system to measure the percentage of dry extractibles and styrene monomer in a styrene/butadiene latex emulsion polymerization reaction.[39] Simple PLS models were developed and easily accounted for the non-linear effects of temperature variation and absorption of NIR radiation by water (self-absorption effect). Though these factors presented no problem in this example, it is important to consider their effect whenever a new model is being developed, and design calibration experiments to verify their impact. The models worked over wide calibration ranges: 2–50% dry extractibles and 0–70 000 ppm styrene monomer. The earliest samples in the reaction cycle were inhomogeneous and difficult to analyze by the reference method, gas chromatography (GC), thus limiting the quality of the styrene model. As observed for the bioreactions, obtaining valid reference values is one of the most common problems with developing models for many process spectroscopy techniques, including Raman.

Wenz and colleagues at Bayer Polymers Inc. describe the use of Raman spectroscopy to monitor the progress of a graft emulsion polymerization process, specifically the manufacture of ABS graft copolymer, in order to select the appropriate reaction termination point.[40] Early termination reduces product yield and results in an extra product purification step; termination too late reduces product quality. As Figure 5.5 illustrates, the reaction composition over time is not smooth and predictable, making it unlikely that off-line analysis would ever be fast enough to ensure correct control decisions.

McCaffery *et al.* discuss the use of a low-resolution Raman spectrometer to directly monitor a batch mini-emulsion polymerization.[41] While this kind of equipment is unlikely to be installed in an industrial facility, the article raises several important points. In order to compensate for laser-power fluctuations, a functional group present in both the reactants and the product, the phenyl ring in styrene, was used as an internal standard. Since internal standards cannot be added to industrial reactions, this approach can be quite helpful. However, scientists must be certain that the internal standard will remain unchanged by the reaction and that changes in its signal only reflect laser-power fluctuations.

The authors used an immersion optic but reported difficulty keeping the polymer from coating the optical window. Eventually a protective coating for the window was identified. Though probe designs continue to improve and strive to reduce window fouling, this is a major issue to address during instrument selection. It occasionally may be advantageous to begin testing the performance of typical window materials in the process even before actual Raman experiments begin. As noted in previous cases, McCaffery *et al.* found it challenging to obtain accurate reference values to correlate with spectral changes. Since the reaction proceeds dynamically, it is extremely difficult to quench the reaction fast enough to avoid introducing a bias between the spectroscopic and the off-line reference values.

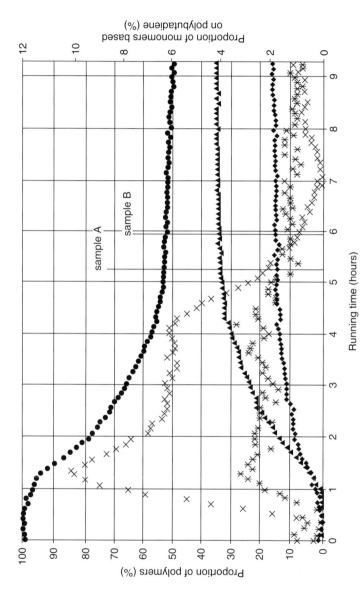

Figure 5.5 Concentration profiles of polystyrene, polyacrylonitrile, polybutadiene, styrene, and acrylonitrile as a function of time since the reaction started. Reprinted from Wenz et al. (2001).[40]

5.4.1.3 Halogenation: Chlorination and iodination

The use of Raman spectroscopy to monitor the reaction of chlorine gas and elemental phosphorous to produce phosphorous trichloride has been extensively described in patents and journals.[42–47] This work was first reported in 1991, making it one of the oldest examples of process Raman spectroscopy and an excellent study on the evolution of an installation.[43] The project initially used an FT-Raman spectrometer to minimize fluorescence, but that instrument's high laser power led to coating formation on the directly inserted, optical probe's window. The probe was moved to a flowing slip stream, which reduced but did not eliminate the problem. Since reducing laser power sacrificed too much measurement performance and time, the equipment was changed to an axial transmissive dispersive spectrometer. This instrument's higher optical throughput made it possible to use a shorter wavelength laser to improve scattering efficiency, even at a lower laser power. This also enabled the probe to be returned to its original location directly in the process. Though the sample fluorescence that originally mandated the use of the FT instrument was observed at the end of the reaction, it did not interfere with process control.[45]

Small details can impact how successfully an installation is viewed. The unforeseen window coatings and associated maintenance were not sufficient for the plant to stop using the analyzer, but they probably reduced the personnel's satisfaction with the technology. Since it is simply not possible to predict and investigate every possible issue during feasibility studies, owners of any kind of on-line analyzer must be prepared to consider the need to make changes to the system over time.

Malthe-Sorenssen *et al.* of Nycomed Imaging filed a PCT (world) patent application, which describes monitoring the production of an organic iodinated X-ray contrast agent with IR, Raman, or NIR spectroscopy. Raman spectroscopy is described as 'particularly sensitive to detection of molecular species with large numbers of polarizable electrons, such as iodinated X-ray contrast agents.' Raman spectroscopy was specified to monitor the concentration of final product in effluent and in cleaning rinse water. The calibration models were estimated to have a detection limit of approximately 1-mg product/mL or 2-mg product/mL. A strong Raman band at approximately $170\,cm^{-1}$ was the foundation of the multivariate calibration model.[48]

5.4.1.4 Calcination

Another well-documented process Raman application is monitoring the efficiency of converting titanium dioxide from anatase to rutile form by calcining.[44,49,50] Monsanto reported that it has four Raman process analyzers installed globally and each measures three reactors via non-contact probes coupled to fiber optics as long as 100 meters. The units have been operating since 1994 and were described as reliable, valuable process control tools. One motivation for the project was the need to obtain process composition data quickly enough to enable more efficient process control algorithms. Anatase and rutile titanium dioxide have dramatically different spectra, enabling simple univariate band ratio models to be used. The uncertainty at the 95% confidence level in the mean estimate was 0.1%, roughly four times more precise than the X-ray diffraction (XRD) reference method.

This example provides several valuable lessons, some of which are the same as observed in previous examples. First, the experimental conditions and compositions need to be the same for calibration and routine operation. In this example, the initial on-line measurements were biased relative to the reference values. This was traced to the temperature differences between the on-line samples (100–200°C) and the room temperature samples used to build the calibration models. Raman band intensity is a known function of temperature, but affects high- and low-frequency bands unequally. A separate temperature measurement enabled the results to be corrected.

Second, all possible sources of variation should be included in the calibration set. Sample parameters such as crystal size and shape can influence the position, shape, and intensity of Raman bands. Extensive experimentation was necessary to prove that these parameters did not influence this system.

Third, the reference method results should be compared with measurements on an equivalent amount of material, particularly if the sample is not homogeneous. In many instruments, Raman spectra are collected only from the sample located in a small focal volume. If that material is not representative of the bulk, then the Raman results will appear to be biased or erroneous. To avoid this problem, multiple sequential spectra are added together to represent an effectively larger composite sample. Alternatively, a larger area could be sampled if the instrument design permits it. If it were desired to study within-sample inhomogeneity, short acquisition times could be used.

Fourth, the process window clarity is related to measurement quality. If the window becomes substantially fouled or obscured, the signal will be attenuated or the coating material itself will be measured repeatedly, depending on where the focal point of the probe is relative to the window. In this example, the spectra are collected with a non-contact probe looking through a window at material being conveyed by the screw of an auger. A purge and air-blast window cleaning apparatus and protocol were developed to ensure that material was not building up and sticking to the window. A thin coating of very fine powder quickly recoats the window and slightly attenuates the signal, but not enough to reduce measurement quality significantly.

Finally, the impact of possible calibration instability on the prediction quality must be understood. In this case, spectrograph and laser wavelength calibrations were very stable and easy to update, but of minor concern because of the broad bands being used. However, since the intensity calibration function is non-linear, any changes in it could unequally affect bands used in the calibration ratio and introduce error in the prediction. Newer equipment offers easy intensity calibration routines but this can be difficult to use automatically with immersion probes since it requires that they be removed from the process.

5.4.1.5 Hydrolysis

Super- or near-critical water is being studied to develop alternatives to environmentally hazardous organic solvents. Venardou *et al.* utilized Raman spectroscopy to monitor the hydrolysis of acetonitrile in near-critical water without a catalyst, and determined the rate constant, activation energy, impact of experimental parameters, and mechanism.[51] The study utilized a miniature high-temperature, high-pressure probe for experiments between 275–350°C and 150–350 bar. The probe could have gone to higher temperatures, but fluorescence began to mask critical bands. The authors did not speculate on the responsible species. Understanding if it arose from the decomposition of an impurity or from a product of the main reaction would help determine the best approach to solve the problem. Related studies are also being performed.[52]

5.4.1.6 Acylation, alkylation, esterification, and catalytic cracking

Li *et al.* discuss the use of on-line Raman spectroscopy to dynamically model the synthesis of aspirin, one of the most documented and well-understood reactions in organic chemistry. That makes it an excellent choice for building confidence in the sampling interface, Raman instrumentation, and analysis procedures. The researchers used wavelets during analysis to remove fluorescent backgrounds in the spectra and modeled the concentrations with multiple linear regression.[53]

Gosling of UOP LLC patented the use of Raman spectroscopy to control a solid catalyst alkylation process.[54] Based on the measured composition of the stream, specific process parameters are adjusted to bring the composition back to a targeted value. Multiple probes are placed in the process, for example, near a feed stock inlet, in the reaction zone, or after the reactor. In the process discussed, the catalyst can be deactivated faster than an on-line GC can produce a reading. This reason, coupled with obtaining greater process efficiency, is strong motivation for a fast on-line system like Raman.

McGill *et al.* compared the accuracy, between-run precision, and rate constant predictions of in-line NIR, Raman, and UV-vis spectroscopy to measure the concentration of 2-butyl crotonate during the acid-catalyzed esterification of crotonic acid and butan-2-ol.[55] To compensate for slight fluctuations in the laser intensity, the spectra were normalized to the intensity of the toluene solvent peak. Using an unchanging solvent peak for normalization is a common and useful approach. The authors note that the spectra showed evidence of slight *x*-axis instability, which would translate into greater uncertainty in intensity values and thus predicted values. Mathematical alignment of spectra, such as aligning solvent peaks during pre-processing, might improve results. Univariate and multivariate model results were similar. Raman values agreed well with GC results, and were more precise than from UV-vis, although not as precise as NIR.

United States patent 5,684,580 by Cooper *et al.* of Ashland Inc. discusses monitoring the concentration of benzene and substituted aromatic hydrocarbons in multiple refinery process streams and using the results for process control.[56] Xylene isomers can be differentiated by Raman spectroscopy, making it technically preferable to NIR. This patent is a good example of process Raman spectroscopy and subsequent process control.

5.4.1.7 Microwave-assisted organic synthesis

Pivonka and Empfield of AstraZeneca Pharmaceuticals describe the continuous acquisition of Raman spectra of an amine or Knoevenagel coupling reaction in a sealed microwave reaction vessel at elevated temperatures and pressures.[57] A microwave synthesizer was modified to accommodate a non-contact Raman probe. This approach helps elucidate the chemical mechanism and makes reaction kinetics instantaneously and continuously available. Other reported advantages included safety, sufficient sensitivity, ease, and the ability to work with solution-containing solids.

The authors note that the use of little or no solvent in the reactions helped mitigate Raman spectroscopy's problems with sensitivity. Subtraction of the initial spectrum from each subsequent spectrum in a complete time series removed the signal from unchanged species, such as solvent and other additives. Though no problems were observed from elevated temperature, blackbody radiation, or the duty cycle of the microwave, the authors caution that more experiments on those factors are needed.

5.4.1.8   Oxidative dehydrogenation (ODH)

In situ Raman spectroscopy is being used to measure the structure of a catalyst during field-deployed reaction conditions, in combination with on-line GC for activity measurements. This is intended to help develop a molecular level structure-activity/selectivity model and to understand catalyst-poisoning mechanisms. Cortez and Banares are using this approach to study propane ODH using a supported vanadium oxide catalyst.[58] Stencel notes that the scattering cross-sections of molecules adsorbed on noble metals are greatly enhanced, making it possible to measure sub-monolayer concentrations.[59] Another advantage was the ability to simultaneously monitor the chemical changes in the catalyst supports to understand their role.

5.4.1.9   Polymer curing

While high-throughput screening (HTS) is traditionally associated with the pharmaceutical industry, Schrof *et al.* of BASF Corporation discuss a similar concept to quickly measure the extent of cure in many different lacquer formulations during the product development process.[60] Their patent describes the pipetting and curing of thousands of formulations on a substrate in a grid pattern. Changes in the concentration of reactive functional groups can be correlated to the degree of cure, gloss, and yellowing. The motivation for this work was the ability to quickly and inexpensively screen formulations with little energy consumption and few environmentally damaging by-products. Van Overbeke discussed monitoring epoxy resin cure, neat and in blends, with Raman spectroscopy.[61] The approach of Schrof *et al.* could be applied to different systems, such as epoxy formulations, if desired.

5.4.2   In-process aid or quality-monitoring tool

5.4.2.1   Composition

5.4.2.1.1   Blending raw materials
The pharmaceutical industry must be able to prove that blends of active and excipient (binding) ingredients are uniformly distributed in order to ensure delivery of the intended dosage. Traditional manual sampling and laboratory composition analysis is

error-prone, laborious, and slow. Sample collection tools can create segregation simply by touching the process. The ability to use a non-contact spectroscopic system to acquire continuous data without any disruption to normal mixing action is appealing. NIR spectroscopy has been used for this and other pharmaceutical activities. However, the use of Raman spectroscopy is increasing because the sharp bands in Raman spectra make interpretation similar to Fourier transform infrared (FTIR) spectroscopy and reduce the need for chemometrics.[28]

The use of FT-Raman spectroscopy with fiber-optic sampling to assess pharmaceutical blend homogeneity has been demonstrated.[62] The approach is calibration-free since it is based on the mean square difference (MSD) between two sequential spectra, an approach also used in NIR.[63] This significantly reduces the burden to install and maintain the method. It was easy to detect the homogenization time on the plot of MSD vs. time, as shown in Figure 5.6. The study confirmed a previous hypothesis that particle size and mixing speed have significant effects on homogenization time. The collected Raman spectra could also be used to provide chemical proof that the correct raw materials were used in the formulation. This work demonstrates that mixing protocols and mixing blade designs can be developed much more quickly than by manual sampling, while also quickly and automatically confirming successful mixing.

Blending also plays a role in the chemical industry. The use of Raman spectroscopy to control gasoline blending, particularly oxygenated species such as methanol, ethanol, propanols, and butanols, is discussed in US patent 5,596,196.[64] The system can be operated in a feed-forward and/or feedback control scheme. In feed-forward, the composition of each of six remote feed tanks is measured sequentially. A control computer then

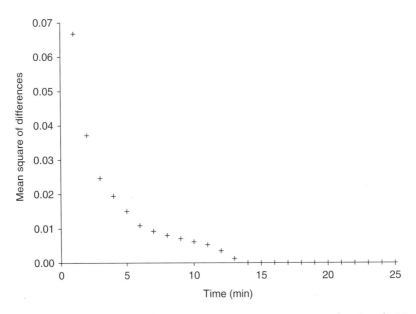

Figure 5.6 Mean square of differences between consecutive Raman spectra as a function of mixing time of a binary system. Reprinted from Vergote *et al.* (2004)[62] with permission from Elsevier.

calculates dynamically the extent to which the valve from each tank should be opened to obtain the targeted formulation. Alternatively, in feedback mode, the final composition of the blended product can be measured and the control computer can back-calculate the change required in the valve-setting of a particular stream. Multivariate calibration models were used. This patent was filed in 1995, making it another long-standing example of process Raman spectroscopy.

5.4.2.1.2 Polymer melt composition

Polymer extrusion is a raw-material blending process where the materials change form during the process. There are several reports on composition monitoring in an extruder.[65] Coates *et al.* compared the predicted value and sensitivity of mid-IR, NIR, Raman, and ultrasound to step-changes in the polymer melt composition of simple polypropylene–polyethylene blends.[66] Exceptional agreement was found between the techniques, though all of the vibrational spectroscopy techniques had difficulty resolving a composition change of 1 wt%. Each technique was deemed capable and worth pursuing, though NIR was the most accurate. The authors discuss the difficulty of doing quantitative Raman measurements on-line without the presence of an internal standard to normalize the signal. One alternative is to normalize to one of the components and predict composition in terms of that component. The work is further complicated by the presence of additives, stabilizers, dyes and pigments in addition to turbidity and elevated temperatures and pressures, all of which can generate substantial backgrounds. These complications remain substantial hurdles to routine industrial use.

5.4.2.1.3 Purification: Distillation and separation

Distillations are perfect candidates for Raman process monitoring since many potential fluorescent or interfering species are reduced or removed. The Institut Francais du Petrole holds two relevant patents discussing the use of Raman spectroscopy to control the separation of aromatics by adsorption on simulated moving beds and the separation of isomers, particularly aromatic hydrocarbons with 8–10 carbon atoms such as *para*-xylene and *ortho*-xylene.[67,68]

Dow Corning has reported monitoring the concentration of a chlorosilane distillation stream with Raman spectroscopy. On-line GC was used, but the 20–60-minute response time proved too slow to detect important column dynamics. The on-line GCs also required extensive maintenance due to the hydrochloric acid in the stream. The Raman system is used to predict the concentration of nine species. Many could not be detected or distinguished by NIR.[69]

5.4.2.2 Corrosion

In situ Raman spectroscopy is being used to investigate corrosion products from zinc in a humid atmosphere and sodium chloride[70] and from Type 304L stainless steel in aerated water at elevated temperatures and pressures.[71] The changes in detected species over time helped identify possible corrosion mechanisms and the effect of different variables on corrosion rates and mechanisms.

5.4.2.3   Waste stream upset

The feasibility of Raman spectroscopy for monitoring of nitrate and nitrite ions in wastewater treatment process reactors was evaluated. The study used an ultraviolet laser, operating at either 204 or 229 nm, to obtain resonance enhancement of the nitrate and nitrite ions and to avoid fluorescence from other components.[72] The Raman signal was collected at ~135° backscatter rather than with the more common 180° configuration. The 1640 cm^{-1} water band was used as an internal standard. The detection limit for both species was estimated to be less than 14 μM (200 ppb). Other work is being done to use Raman spectroscopy for environmental monitoring, including at contaminated sites and for potable water analysis. Sampling and lower detection limits have been major challenges.[72–75]

5.4.2.4   Physical form or state

5.4.2.4.1   Polymorphs or crystal form

The United States Food and Drug Administration (US FDA) now advocates the use of process analytical technology in the pharmaceutical industry.[76] This opinion change is expected to result in a rapid increase in process Raman applications, not just for reaction monitoring, but also as part of the drug development process. Drug discovery relies on HTS, which is a special category of process monitoring. Just as in traditional process monitoring applications, small changes in HTS operational efficiency can translate into millions of dollars saved. One aspect of HTS focuses on identifying the crystal forms, or polymorphs, of a drug under different conditions.[77]

Scientists at the Merck Research Laboratories have used in situ Raman spectroscopy to follow the kinetics of polymorph conversion from one form to another in a series of experiments designed to mimic manufacturing upset conditions.[78] Hilfiker *et al.* at Solvias used carbamazepine (CBZ) as a model compound to describe the use of Raman micros-copy to characterize crystal forms.[79] The ability to collect spectra directly in a microtiter plate with or without solvent and during evaporation is a major advantage over many other techniques. Several vendors sell systems preconfigured for automated analysis of microtiter plates. Another advantage cited was the ability to collect spectra from multiple locations within a single sample well. Spectra were acquired automatically except for the evaporation experiments, which required an operator to focus on a specific crystal. Such manual intervention could be eliminated with appropriate image-processing routines interfaced with the spectrometer software. The spectra were processed into clusters by spectral similarity. The authors note that all published and several new crystal forms were identified during the study. Solvias' HTS uses a specific set of crystallization protocols that have tended to produce new polymorphs. Hilfiker notes that Raman microscopy is 'an ideal analytical tool for high-throughput discrimination between crystal structures.'[80]

Anquetil *et al.* used optical trapping to isolate a single microcrystal in a 35-μL cell and collect Raman spectra.[81] The concentration and the polymorphic form of the CBZ crystal(s) were easily determined without handling or consuming the sample. The spectra were normalized to the intensity of the methanol solvent band. Spectra were collected continuously as the temperature in the cell varied. Either of two forms could be formed selectively depending on the temperature profile used. The authors note that this is an ideal HTS technique because it uses very little sample, reaches equilibrium quickly, is done in situ, and is chemically distinct.

5.4.2.4.2 Orientation, stress, or strain

Paradkar *et al.* used on-line Raman spectroscopy to measure the molecular orientation of polypropylene fibers and crystallinity of high-density polyethylene fibers. Birefringence[82] and differential scanning calorimetry (DSC)[83] were the reference methods. In both cases, the work was done on a pilot-scale melt spinning apparatus and spectra were collected at different distances from the spinneret. The 841/809-cm^{-1} band ratio was used to monitor the polypropylene. The polyethylene studies used the integrated intensity of the 1418-cm^{-1} band normalized to the area of the 1295–1305-cm^{-1} bands. These bands have been shown to be insensitive to the crystallization state of the material and thus are a suitable built-in internal standard. The identification of appropriate analysis bands is critical and often is supported by many laboratory-based studies, as was the case here. Additionally, the equipment was slightly modified to eliminate crystallinity calculation errors from laser polarization. The biggest experimental problem faced in both studies was keeping the fiber from moving in and out of the focal plane, resulting in unacceptably long acquisition times. The authors predict this problem could be overcome.

Ergungor describes the application of on-line Raman spectroscopy and neural networks to the simultaneous prediction of temperature and crystallinity of nylon-6 nanocomposites as a function of cooling rate. The authors prefer their neural network approach because they make use of information in the entire spectrum rather than from a few bands as most studies have done.[84] Van Wijk *et al.* of Akzo Nobel obtained a patent on the use of a Raman spectrum of a polymeric fiber to determine dye uptake and other structural or mechanical properties based on previously developed models.[85]

Voyiatzis and Andrikopoulos discuss adding an orientation-sensitive, resonant Raman additive to a polymer mixture prior to drawing in order to calculate molecular orientation continuously. The approach was tested with poly(vinyl chloride) (PVC), isotactic polypropylene (iPP), and poly(vinylidene fluoride) (PVF2). The ratio of a band from the additive to the orientationally insensitive CH$_2$ bending mode at ~1435 cm^{-1} in the polymer was computed. While the addition of such an additive is unreasonable for many industrial processes, the authors note a favorable alternative for industrial PVC samples.

The all-trans conjugated double-bond segments, formed in PVC as a result of processing-induced thermal degradation, are sensitive to orientation and could be used instead of the additive. These bonds are resonance-enhanced with 532-nm excitation and can be detected at very low concentrations.[86]

Wright of Advanced Micro Devices, Inc. discusses the use of Raman spectroscopy to measure the integrity of a film on semiconductor wafers during manufacture in US patent 6,509,201.[87] The Raman measurements are made during the manufacturing process and can be considered an on-line system. Unlike many process Raman installations, this one is based on micro-Raman, where a microscope is used to focus the laser beam to a spot only a few micrometers in diameter. The Raman data is combined with other measurements, such as scatterometry, to calculate a stress level and compare it to predetermined control limits. All of the data are matched to which wafer and region they came from since the measurements are not necessarily acquired at the same time. The data are input to a manufacturing process control transfer function or model. Based on the calculations, corrective adjustments are made to critical variables, such as etch rate or exposure time, later in the production process in a classic feed-forward loop. This kind of process control system improves yield on the first pass, reducing scrap generation by providing a tailored repair for each part (via feed-forward control). Raman spectroscopy has a well-established place in the semiconductor industry for this and other applications.[88]

5.4.2.5   Film or layer thickness

Adem of Advanced Micro Devices, Inc. was granted a patent on the use of Raman spectroscopy to monitor the thickness, crystal grain size, and crystal orientation of polysilicon or other films as they are deposited on semiconductor wafers via low-pressure chemical vapor deposition (CVD).[89] The spectra are acquired with a non-contact probe through a suitably transparent window in the loading door. A feedback scheme is discussed. When the thickness has achieved the targeted value, the deposition is stopped. If the crystal grain size or orientation is deemed unsuitable, the deposition temperature is adjusted accordingly.

Mermoux *et al.* discuss the application of Raman spectroscopy to monitoring the growth of diamond films prepared under assorted CVD conditions. The authors synchronized a pulsed laser and gateable detector in order to increase the Raman signal intensity enough to be detectable in the presence of the hydrogen emission spectrum from the plasma and blackbody radiation from the hot substrate. The signal enhancement was close to the duty cycle of the laser. Spectra were collected without interruption during 24 hours of deposition. The spectra enable simultaneous measurement of the deposition temperature, the development of non-diamond phases, and stress in the film. A previously measured spectrum of pure diamond at elevated temperatures and pressures was used as a reference to detect changes in key bands.[90]

The use of a fiber-optic Raman probe to monitor the thickness and composition of a thin film on a substrate is discussed in US patent 6,038,525.[91] The patent claims that even 0–1000-Å films can be detected. A spectrum of the uncoated substrate is collected. The attenuation of a particular Raman band is correlated to the film thickness and monitored to decide if process parameters need to be adjusted. The spectrum changes over time from that of the substrate to the coating. Other calibration models are used to determine if the film coating has the desired properties. The example discussed is the coating of lanthanum aluminum oxide ($LaAlO_3$) with yttrium barium copper oxide and includes specific bands to use in the models. In a follow-up patent, the system is modified to collect Raman spectra from multiple points to form a chemical image in order to identify physical defects in a film during deposition.[92] Though Raman imaging is a valuable research tool, difficulties in illuminating a target area with enough laser power or in stepping through individual points fast enough have prevented it from being used in many potential process control applications.

5.4.3   Product properties

Several patents discuss the use of Raman spectroscopy to determine the properties of finished products.[93,94] For reformulated gasoline, some of these properties include sulfur, olefin, benzene, volatile organic carbon (VOC), nitrogen oxides (NOx), aromatic contents, total air pollutants (TAPs), Reid Vapor Pressure (RVP), distillation properties, motor and research octane numbers, and drivability. For the octane numbers, the accuracy of the Raman method was limited by errors in the reference method.

5.4.4   Mobile or field uses

There also are many mobile or field applications of Raman spectroscopy. While these do not control a process in the traditional sense, the repeated analyses still trigger an action, whether it is the confiscation of illegal drugs or the diagnosis of a medical condition. The environments and requirements are at least as challenging as in any industrial plant but have a different focus. Space exploration programs continue to investigate Raman spectroscopy as a possible analytical tool for geological identifications to include in future planetary exploration missions.[95] Portable Raman spectroscopy was subjected to a blind field test of 58 unknown samples to evaluate its suitability for use by emergency response and forensics teams and achieved a satisfactory 97% correct. Development work was continued based on the results.[96]

Vibrational spectroscopy continues to make progress as a medical diagnostic tool.[97–100] Examples include probes that fit in standard endoscopes to classify diseased and normal esophageal tissue,[101] and catheters to identify atherosclerotic disease which is not easily recognized visually.[102,103] Models are being developed to measure and track the concentration of water in the brain as a measure of brain swelling.[104] Other researchers are trying to monitor blood gases and glucose concentrations through the skin.[105,106] A patent was issued for a spectroscopic device to be placed on an intravenous medication line to verify that a hospital patient is receiving the correct medication and concentration; one version of the instrument is based on Raman spectroscopy.[107] While it will be years before these are viable and common clinical tools, the results of these studies help improve instrumentation and modeling techniques.

5.5   What is the current state of Raman spectroscopy?

The previous section included many examples of the application of Raman spectroscopy to process monitoring. The collection clearly reveals a thriving process analytical field. Then again, did it? Many articles end with a phrase like '... demonstrates the feasibility of using Raman spectroscopy for online monitoring of an important industrial process...' or 'Raman spectroscopy appears to be the ideal online method.' However, there are relatively few articles actually discussing industrial implementation. If one limits the examples to peer-reviewed articles presenting detailed spectroscopic information, showing graphs of composition vs. time, and linked to process feedback loops, the list is very short indeed. Why?

What is the state of process Raman spectroscopy right now? Is it still slowly building? Is it a niche market and unlikely to expand? Is it thriving and flourishing? Since almost all industrial applications will use a commercial instrument, the instrument vendors are the only ones who can really answer these questions but are likely to be bound by confidentiality agreements with their customers. In the absence of publications or other public disclosures, it is useful to examine the factors that may be limiting or appearing to limit the use of Raman spectroscopy for process control.

5.5.1   Publication reluctance

Process spectroscopy is, almost by definition, done to measure and control an industrial process. Almost all of the work is driven by business needs, such as improving profits or product quality. In competitive business environments, firms preserve every advantage possible by protecting valuable measurement systems as trade secrets. Thus, firms are often reluctant to reveal process spectroscopy applications, whether successful or not. Notable exceptions to this include the desire for positive publicity around improved safety or to direct the regulatory environment. Often, companies will patent the work and will not publish in a scientific journal until after the patent is filed, if ever. Many applications, such as the classic titanium oxide–monitoring paper, are revealed only years after implementation. As a consequence, the current state of the art in the literature is quite likely far out of date.

Educated inferences suggest that the field is thriving despite the limited publications on full process control installations. The number of Raman vendors is increasing, along with mergers and acquisitions activity. Additionally, the list of companies that have or are willing to publicly acknowledge their use of process Raman spectroscopy or advocate its utility is impressive. Some of these are listed in Table 5.2.

Table 5.2 Companies publicly acknowledging or willing to acknowledge their use or advocacy of process Raman spectroscopy

Advanced Micro Devices (AMD)[89]	Dow Corning Corporation[109]	Intel[110]
Akzo Nobel[109]	DuPont[45]	Monsanto[109]
Ashland Oil[56]	Ford Motor Company[45]	Pharmacia[109]
Bayer Polymers[40]	French Institute of Petroleum[68]	Pfizer[109]
BorgWarner Morse TEC[a]	General Electric	Procter & Gamble Pharmaceuticals[45]
BP Amoco[109]	Glaxo-SmithKline[45]	Sematech[111]
Colgate-Palmolive[45]	Hoeganaes Corp.[b]	Solvias[79,109]
Compagne Europeanne des Accumulateurs[45]	ICI Chemicals[109]	Timken[c]
Dow Chemical[109,112]	ICI Polymers[109]	

Notes:
[a] Personal communication (2004). Raman gas analysis during heat treatments.
[b] Personal communication (2004). Raman gas analysis on a powdered metal annealing line.
[c] Personal communication (2004). Raman gas analysis for advanced heat treatment atmosphere assessment.

5.5.2 Technique maturity and long-term performance

Raman spectroscopy was recognized over 75 years ago, but has only been a viable process tool for about a decade. Industrial chemists and production engineers rarely want the newest technology and the lowest serial number instrument. Many experienced scientists remember old rules about Raman spectroscopy having poor sensitivity, being difficult to do, and used for qualitative applications only. Modern equipment has changed much of this. There have been enormous advances in every aspect of performing Raman spectroscopy: new lasers, new sampling interfaces, new spectrometer designs, new detectors, and cheap fast computers with powerful control, analysis, and communications software.[99,108] Though fluorescent backgrounds can still be a problem, judicious equipment selection – including different laser wavelengths, dispersive versus FT platforms, and carefully balancing acquisition time versus number of exposures – can help manage much of the problem. To experienced practitioners, the instrumentation challenges of the past have been overcome and it is simply time to start practicing process Raman spectroscopy.

For others, Raman spectroscopy is a new or exotic technique. Compared to other process analytical techniques, there are still relatively few published examples of process Raman spectroscopy, and no published studies, as yet, of its long-term performance. This kind of information is imperative for new users to be able to accurately assess the role Raman spectroscopy could play in their organizations. It also helps supplement and balance information received from vendors. While the technique is long past its infancy, the community of potential users may perceive the situation differently.

5.5.3 Lack of widespread knowledge and experience

Differences in perception of the technique's maturity may originate from a simple lack of widespread knowledge about the approach. There are only a few short courses offered. Many students learn of it only as a side note in a physical chemistry textbook and never have hands-on training to use it. Laboratory-based Raman instrumentation is not as ubiquitously available for familiarization and casual experimentation as FTIR. While most vendors will arrange to do feasibility studies and preliminary trials with potential customers, these kinds of short experiments make it difficult for new users to build a solid familiarity with the technique. If a group has years of experience with NIR, it can be difficult to remember to re-examine Raman spectroscopy for each new project to see if it would be a better technical choice and ultimately easier to do.

References

1. Ferraro, J.R.; Nakamoto, K. & Brown, C.W.; *Introductory Raman Spectroscopy*; 2nd Edition; Academic Press; Boston, 2003.
2. Nakamoto, K.; *Infrared and Raman Spectra of Inorganic and Coordination Compounds*. Part A: *Theory and Applications in Inorganic Chemistry*; 5th Edition; Wiley-Interscience (John Wiley & Sons, Inc.); New York, 1997.

3. Nafie, L.A.; Theory of Raman Scattering. In: Lewis, I.R. & Edwards, H.G.M. (eds); *Handbook of Raman Spectroscopy: From the Research Laboratory to the Process Line*; 1st Edition; Marcel Dekker, Inc.; New York, 2001; pp. 1–10.

4. Gardiner, D.J.; Introduction to Raman Scattering. In: Gardiner, D.J. & Graves, P.R. (eds); *Practical Raman Spectroscopy*; 1st Edition; Springer-Verlag; New York, 1989; pp. 1–12.

5. Pelletier, M.J.; Quantitative Analysis Using Raman Spectroscopy; *Appl. Spectrosc.* 2003, 57, 20A–42A.

6. Shaver, J.M.; Chemometrics for Raman Spectroscopy. In: Lewis, I.R. & Edwards, H.G.M. (eds); *Handbook of Raman Spectroscopy: From the Research Laboratory to the Process Line*; 1st Edition; Marcel Dekker, Inc.; New York, 2001; pp. 275–306.

7. Treado, P.J. & Morris, M.D.; Infrared and Raman Spectroscopic Imaging. In: Morris, M.D. (ed.); *Microscopic and Spectroscopic Imaging of the Chemical State*; 1st Edition; Marcel Dekker, Inc.; New York, 1993; pp. 71–108.

8. Pelletier, M.J.; Raman Instrumentation. In: Pelletier, M.J. (ed.); *Analytical Applications of Raman Spectroscopy*; 1st Edition; Blackwell Science Ltd; Malden, MA, 1999; pp. 53–105.

9. Edwards, H.G.M.; Raman Spectroscopy in the Characterization of Archaeological Materials. In: Lewis, I.R. & Edwards, H.G.M. (eds); *Handbook of Raman Spectroscopy: From the Research Laboratory to the Process Line*; 1st Edition; Marcel Dekker, Inc.; New York, 2001; pp. 1011–1044.

10. Perez, F.R.; Applications of IR and Raman Spectroscopy to the Study of Medieval Pigments. In: Lewis, I.R. & Edwards, H.G.M. (eds); *Handbook of Raman Spectroscopy: From the Research Laboratory to the Process Line*; 1st Edition; Marcel Dekker, Inc.; New York, 2001; pp. 835–863.

11. Pelletier, M.J.; New Developments in Analytical Instruments: Raman Scattering Emission Spectrophotometers. In: McMillan, G.K. & Considine, D.M. (eds); *Process/Industrial Instruments and Controls Handbook*; 5th Edition; McGraw-Hill; New York, 1999; pp. 10.126–10.132.

12. Chauvel, J.P.; Henslee, W.W. & Melton, L.A.; Teaching Process Analytical Chemistry; *Anal. Chem.* 2002, 74, 381A–384A.

13. Slater, J.B.; Tedesco, J.M.; Fairchild, R.C. & Lewis, I.R.; Raman Spectrometry and its Adaptation to the Industrial Revolution. In: Lewis, I.R. & Edwards, H.G.M. (eds); *Handbook of Raman Spectroscopy: From the Research Laboratory to the Process Line*; 1st Edition; Marcel Dekker, Inc.; New York, 2001; pp. 41–144.

14. Chase, B.; A New Generation of Raman Instrumentation; *Appl. Spectrosc.* 1994, 48, 14A–19A.

15. Laser Institute of America; www.laserinstitute.org.

16. OSHA Safety & Health Topics: Radiation – Laser Hazards; http://www.osha.gov/SLTC/laserhazards/index.html.

17. Harvey, S.D.; Peters, T.J. & Wright, B.W.; Safety Considerations for Sample Analysis Using a Near-Infrared (785 nm) Raman Laser Source; *Appl. Spectrosc.* 2003, 57, 580–587.

18. Rockwell, R.J., Jr & Parkinson, J.; State and Local Government Laser Safety Requirements; *J. Laser Appl.* 1999, 11, 225–231.

19. ANSI. Z136 Laser Safety Standards Series: Z136.1-2000 American National Standard for Safe Use of Lasers. 2000.

20. OSHA Guidelines for Laser Safety and Hazard Assessment (Std 01-05-001-Pub 8-1.7). 1991; http://www.osha.gov/pls/oshaweb/owadisp.show_document?p_table=DIRECTIVES&p_id=1705.

21. Gemperline, P.J.; Rugged Spectroscopic Calibration for Process Control; *Chemom. Intell. Lab. Syst.* 1997, 39, 29–40.

22. Kotidis, P.; Crocrombe, R. & Atia, W.; Optical, Tunable Filter-Based Micro-Instrumentation for Industrial Process Control. In: *Federation of Analytical Chemistry and Spectroscopy Societies* (FACSS), Orlando, FL USA, 2003.

23. Battey, D.E.; Slater, J.B.; Wludyka, R.; Owen, H.; Pallister, D.M. & Morris, M.D.; Axial Transmissive f/1.8 Imaging Raman Spectrograph with Volume-Phase Holographic Filter and Grating; *Appl. Spectrosc.* 1993, 47, 1913–1919.

24. Everall, N.; Owen, H. & Slater, J.; Performance Analysis of an Integrated Process Raman Analyzer Using a Multiplexed Transmission Holographic Grating, CCD Detection, and Confocal Fiberoptic Sampling; *Appl. Spectrosc.* 1995, 49, 610–615.

25. Pelletier, M.J. (ed.); *Analytical Applications of Raman Spectroscopy*; 1st Edition; Blackwell Science Ltd; Malden, MA, 1999.

26. Lewis, I.R. & Edwards, H.G.M. (eds); *Handbook of Raman Spectroscopy: From the Research Laboratory to the Process Line*; 1st Edition; Marcel Dekker, Inc.; New York, 2001.

27. Lehnert, R.; Non-Invasive Analysis of Ceramics by Raman Spectroscopy; *CFI-Ceram. Forum Int.* 2002, 79, E16–E18.

28. Vankeirsbilck, T.; Vercauteren, A.; Baeyens, W. *et al.*; Applications of Raman Spectroscopy in Pharmaceutical Analysis; *TrAC, Trends Anal. Chem.* 2002, 21, 869–877.

29. Brewer, P.G.; Malby, G.; Pasteris, J.D. *et al.*; Development of a Laser Raman Spectrometer for Deep-Ocean Science; *Deep Sea Res.* (I Oceanogr. Res. Pap.) 2004, 51, 739–753.

30. Balas, K. & Pelecoudas, D.; Imaging Method and Apparatus for the Non-Destructive Analysis of Paintings and Monuments; US Patent Application Publication 2003/0117620 A1; Filed in 2000.

31. Gurden, S.P.; Westerhuis, J.A. & Smilde, A.K.; Monitoring of Batch Processes Using Spectroscopy; *AIChE J.* 2002, 48, 2283–2297.

32. von Stockar, U.; Valentinotti, S.; Marison, I.; Cannizzaro, C. & Herwig, C.; Know-How and Know-Why in Biochemical Engineering; *Biotechnol. Adv.* 2003, 21, 417–430.

33. Lee, H.L.T.; Boccazzi, P.; Gorret, N.; Ram, R.J. & Sinskey, A.J.; In Situ Bioprocess Monitoring of *Escherichia coli* Bioreactions Using Raman Spectroscopy; *Vib. Spectrosc.* 2004, 35, 131–137.

34. Ulber, R.; Frerichs, J.-G. & Beutel, S.; Optical Sensor Systems for Bioprocess Monitoring; *Anal. Bioanal. Chem.* 2003, 376, 342–348.

35. Edge, R. & Truscott, T.G.; Carotenoids-Free Radical Interactions; *Spectrum* 2000, 13, 12–20.

36. Cannizzaro, C.; Rhiel, M.; Marison, I. & von Stockar, U.; On-Line Monitoring of *Phaffia Rhodozyma* Fed-Batch Process with In Situ Dispersive Raman Spectroscopy; *Biotech. Bioeng.* 2003, 83, 668–680.

37. Parker, S.F.; Tavender, S.M.; Dixon, N.M.; Herman, H.K.; Williams, K.P.J. & Maddams, W.F.; Raman Spectrum of β-Carotene Using Laser Lines from Green (514.5 nm) to Near-Infrared (1064 nm): Implications for the Characterization of Conjugated Polyenes; *Appl. Spectrosc.* 1999, 53, 86–91.

38. van den Brink, M.; Pepers, M. & Van Herk, A.M.; Raman Spectroscopy of Polymer Latexes; *J. Raman Spectrosc.* 2002, 33, 264–272.

39. Bauer, C.; Amram, B.; Agnely, M. *et al.*; On-Line Monitoring of a Latex Emulsion Polymerization by Fiber-Optic FT-Raman Spectroscopy. Part I: Calibration; *Appl. Spectrosc.* 2000, 54, 528–535.

40. Wenz, E.; Buchholz, V.; Eichenauer, H. *et al.*; Process for the Production of Graft Polymers; US Patent Application Publication 2003/0130433 A1; Assigned to Bayer Polymers LLC; Filed in 2002. Priority Number DE 10153534.1 (2001).

41. McCaffery, T.R. & Durant, Y.G.; Application of Low-Resolution Raman Spectroscopy to Online Monitoring of Miniemulsion Polymerization; *J. Appl. Polym. Sci.* 2002, 86, 1507–1515.

42. Feld, M.S.; Fisher, D.O.; Freeman, J.J. *et al.*; Process for Preparing Phosphorus Trichloride; US 5,310,529; Assigned to Monsanto Company; Filed in 1992.

43. Feld, M.S.; Fisher, D.O.; Freeman, J.F. *et al.*; Apparatus for Preparing Phosphorus Trichloride; US 5,260,026; Assigned to Monsanto Company; Filed in 1991.

44. Leugers, M.A. & Lipp, E.D.; Raman Spectroscopy in Chemical Process Analysis. In: Chalmers, J.M. (ed.); *Spectroscopy in Process Analysis*; 1st Edition; Sheffield Academic Press; Sheffield, England, 2000; pp. 139–164.

45. Lewis, I.R.; Process Raman Spectroscopy. In: Lewis, I.R. & Edwards, H.G.M. (eds); *Handbook of Raman Spectroscopy: From the Research Laboratory to the Process Line*; 1st Edition; Marcel Dekker, Inc.; New York, 2001; pp. 919–974.

46. Freeman, J.J.; Fisher, D.O. & Gervasio, G.J.; FT-Raman On-Line Analysis of PCl_3 Reactor Material; *Appl. Spectrosc.* 1993, 47, 1115–1122.

47. Gervasio, G.J. & Pelletier, M.J.; On-Line Raman Analysis of PCl_3 Reactor Material; *At-Process* 1997, III, 7–11.

48. Malthe-Sørenssen, D.; Schelver Hyni, A.C.; Aabye, A.; Bjørsvik, H.R.; Brekke, G. & Sjøgren, C.E.; Process for the Production of Iodinated Organic X-ray Contrast Agents; World Intellectual Property Organization Patent Cooperation Treaty International Application WO 98/23296; Assigned to Nycomed Imaging; Filed in 1997. Priority Number GB 9624612.9 (1996).

49. Clegg, I.M.; Everall, N.J.; King, B.; Melvin, H. & Norton, C.; On-Line Analysis Using Raman Spectroscopy for Process Control During the Manufacture of Titanium Dioxide; *Appl. Spectrosc.* 2001, 55, 1138–1150.

50. Besson, J.P.; King, P.W.B.; Wilkins, T.A.; McIvor, M.C. & Everall, N.J.; Calcination of Titanium Dioxide; EP 0 767 222 B1; Assigned to Tioxide Group Services Ltd; Filed in 1996. Priority Number GB 9520313 (1995).

51. Venardou, E.; Garcia-Verdugo, E.; Barlow, S.J.; Gorbaty, Y.E. & Poliakoff, M.; On-Line Monitoring of the Hydrolysis of Acetonitrile in Near-Critical Water Using Raman Spectroscopy; *Vib. Spectrosc.* 2004, 35, 103–109.

52. Belsky, A.J. & Brill, T.B.; Spectroscopy of Hydrothermal Reactions. 12. Acid-Assisted Hydrothermolysis of Aliphatic Nitriles at 150–260 °C and 275 bar; *J. Phys. Chem. A* 1999, 103, 3006–3012.

53. Li, W.H.; Lu, G.W.; Huang, L.; Gao, H. & Du, W.M.; Aspirin Synthesis On-Line Measurement by Raman Spectroscopy; *Acta Phys.-Chim. Sin.* 2003, 19, 105–108.

54. Gosling, C.D.; *Control of Solid Catalyst Alkylation Process Using Raman Spectroscopy*; US 6,528,316 B1; Assigned to UOP LLC; Filed in 2000.

55. McGill, C.A.; Nordon, A. & Littlejohn, D.; Comparison of In-Line NIR, Raman and UV-Visible Spectrometries, and At-Line NMR Spectrometry for the Monitoring of an Esterification Reaction; *Analyst* 2002, 127, 287–292.

56. Cooper, J.B.; Flecher, J.P.E. & Welch, W.T.; Hydrocarbon Analysis and Control by Raman Spectroscopy; US 5,684,580; Assigned to Ashland Inc.; Filed in 1995.

57. Pivonka, D.E. & Empfield, J.R.; Real-Time In Situ Raman Analysis of Microwave-Assisted Organic Reactions; *Appl. Spectrosc.* 2004, 58, 41–46.

58. Cortez, G.G. & Banares, M.A.; A Raman Spectroscopy Study of Alumina-Supported Vanadium Oxide Catalyst During Propane Oxidative Dehydrogenation with Online Activity Measurement; *J. Catal.* 2002, 209, 197–201.

59. Stencel, J.M.; *Raman Spectroscopy for Catalysis*; 1st Edition; Van Nostrand Reinhold; New York, 1990.

60. Schrof, W.; Horn, D.; Schwalm, R.; Meisenburg, U. & Pfau, A.; Method for Optimizing Lacquers; US 6,447,836 B1; Assigned to BASF Aktiengesellschaft; Filed in 1999. Priority Number DE 19834184 (1998).

61. Van Overbeke, E.; Devaux, J.; Legras, R.; Carter, J.T.; McGrail, P.T. & Carlier, V.; Raman Spectroscopy and DSC Determination of Conversion in DDS-Cured Epoxy Resin: Application to Epoxy-Copolyethersulfone Blends; *Appl. Spectrosc.* 2001, 55, 540–551.

62. Vergote, G.J.; De Beer, T.R.M.; Vervaet, C. *et al.*; In-Line Monitoring of a Pharmaceutical Blending Process Using FT-Raman Spectroscopy; *Eur. J. Pharma. Sci.* 2004, 21, 479–485.

63. Blanco, M.; González Bañó, R. & Bertran, E.; Monitoring Powder Blending in Pharmaceutical Processes by Use of Near Infrared Spectroscopy; *Talanta* 2002, 56, 203–212.

64. Cooper, J.B.; Wise, K.L.; Welch, W.T. & Sumner, M.B.; Oxygenate Analysis and Control by Raman Spectroscopy; US 5,596,196; Assigned to Ashland Inc.; Filed in 1995.

65. Fischer, D. & Pigorsch, E.; Process Monitoring of Polymer Melts by In-Line Near IR and Raman Spectroscopy in Extruders; *Abstr. Pap. Am. Chem. Soc.* 2001, 221, U323.

66. Coates, P.D.; Barnes, S.E.; Sibley, M.G.; Brown, E.C.; Edwards, H.G.M. & Scowen, I.J.; In-Process Vibrational Spectroscopy and Ultrasound Measurements in Polymer Melt Extrusion; *Polymer* 2003, 44, 5937–5949.

67. Couenne, N.; Duchene, P.; Hotier, G. & Humeau, D.; Process for Controlling a Separation System with Simulated Moving Beds; EP 0 875 268 B1; Assigned to Institut Francais du Petrole; Filed in 1998. Priority Number FR 9705485 (1997).

68. Cansell, F.; Hotier, G.; Marteau, P. & Zanier, N.; Method for Regulating a Process for the Separation of Isomers of Aromatic Hydrocarbons Having from 8 to 10 Carbon Atoms; US 5,569,808; Assigned to Institut Francais du Petrole; Filed in 1994.

69. Lipp, E.D. & Grosse, R.L.; On-Line Monitoring of Chlorosilane Streams by Raman Spectroscopy; *Appl. Spectrosc.* 1998, 52, 42–46.

70. Ohtsuka, T. & Matsuda, M.; In Situ Raman Spectroscopy for Corrosion Products of Zinc in Humidified Atmosphere in the Presence of Sodium Chloride Precipitate; *Corrosion* 2003, 59, 407–413.

71. Maslar, J.E.; Hurst, W.S.; Bowers, W.J. & Hendricks, J.H.; In Situ Raman Spectroscopic Investigation of Stainless Steel Hydrothermal Corrosion; *Corrosion* 2002, 58, 739–747.

72. Ianoul, A.; Coleman, T. & Asher, S.A.; UV Resonance Raman Spectroscopic Detection of Nitrate and Nitrite in Wastewater Treatment Processes; *Anal. Chem.* 2002, 74, 1458–1461.

73. Collette, T.W.; Williams, T.L. & D'Angelo, J.C.; Optimization of Raman Spectroscopy for Speciation of Organics in Water; *Appl. Spectrosc.* 2001, 55, 750–766.

74. Collette, T.W. & Williams, T.L.; The Role of Raman Spectroscopy in the Analytical Chemistry of Potable Water; *J. Environ. Monit.* 2002, 4, 27–34.

75. Bowen, J.M.; Sullivan, P.J.; Blanche, M.S.; Essington, M. & Noe, L.J.; Optical-Fiber Raman Spectroscopy Used for Remote In-Situ Environmental Analysis; US 4,802,761; Assigned to Western Research Institute; Filed in 1987.

76. Guidance for Industry. PAT – A Framework for Innovative Pharmaceutical Manufacturing and Quality Assurance. Draft. 2003; http://www.fda.gov/cder/guidance/5815dft.htm.

77. Yu, L.X.; Lionberger, R.A.; Raw, A.S.; D'Costa, R.; Wu, H. & Hussain, A.S.; Applications of Process Analytical Technology to Crystallization Processes; *Adv. Drug Deliv. Rev.* 2004, 56, 349–369.

78. Starbuck, C.; Spartalis, A.; Wai, L. *et al.*; Process Optimization of a Complex Pharmaceutical Polymorphic System Via In Situ Raman Spectroscopy; *Cryst. Growth Des.* 2002, 2, 515–522.

79. Hilfiker, R.; Berghausen, J.; Blatter, F. *et al.*; Polymorphism – Integrated Approach from High-Throughput Screening to Crystallization Optimization; *J. Therm. Anal. Calorim.* 2003, 73, 429–440.

80. Hilfiker, R.; Berghausen, J.; Blatter, F.; De Paul, S.M.; Szelagiewicz, M. & Von Raumer, M.; High-Throughput Screening for Polymorphism; *Chimica Oggi – Chemistry Today* 2003, 21, 75.

81. Anquetil, P.A.; Brenan, C.J.H.; Marcolli, C. & Hunter, I.W.; Laser Raman Spectroscopic Analysis of Polymorphic Forms in Microliter Fluid Volumes; *J. Pharm. Sci.* 2003, 92, 149–160.

82. Paradkar, R.P.; Sakhalkar, S.S.; He, X. & Ellison, M.S.; On-Line Estimation of Molecular Orientation in Polypropylene Fibers Using Polarized Raman Spectroscopy; *Appl. Spectrosc.* 2001, 55, 534–539.

83. Paradkar, R.P.; Sakhalkar, S.S.; He, X. & Ellison, M.S.; Estimating Crystallinity in High Density Polyethylene Fibers Using Online Raman Spectroscopy; *J. Appl. Polym. Sci.* 2003, 88, 545–549.

84. Ergungor, Z.; Batur, C. & Cakmak, M.; On Line Measurement of Crystallinity of Nylon 6 Nanocomposites by Laser Raman Spectroscopy and Neural Networks; *J. Appl. Polym. Sci.* 2004, 92, 474–483.

85. Van Wijk, R.J.; De Weijer, A.P.; Klarenberg, D.A.; De Jonge, R. & Jongerden, G.J.; Technique for Measuring Properties of Polymeric Fibres; World Intellectual Property Organization Patent Cooperation Treaty International Application WO 99/12019; Assigned to Akzo Nobel N.V.; Filed in 1998. Priority Number NL 1006895 (1997).

86. Voyiatzis, G.A. & Andrikopoulos, K.S.; Fast Monitoring of the Molecular Orientation in Drawn Polymers Using Micro-Raman Spectroscopy; *Appl. Spectrosc.* 2002, 56, 528–535.

87. Wright, M.I.; Method and Apparatus for Monitoring Wafer Stress; US 6,509,201 B1; Assigned to Advanced Micro Devices, Inc.; Filed in 2001.

88. Wolf, I.D.; Semiconductors. In: Pelletier, M.J. (ed.); *Analytical Applications of Raman Spectroscopy*; 1st Edition; Blackwell Science Ltd; Malden, MA, 1999; pp. 435–472.

89. Adem, E.; Method of In Situ Monitoring of Thickness and Composition of Deposited Films Using Raman Spectroscopy; US 6,667,070; Assigned to Advanced Micro Devices, Inc.; Filed in 2002.

90. Mermoux, M.; Marcus, B.; Abello, L.; Rosman, N. & Lucazeau, G.; In Situ Raman Monitoring of the Growth of CVD Diamond Films; *J. Raman Spectrosc.* 2003, 34, 505–514.

91. Maguire, J.F.; Busbee, J.D.; Liptak, D.C.; Lubbers, D.P.; LeClair, S.R. & Biggers, R.R.; Process Control for Pulsed Laser Deposition Using Raman Spectroscopy; US 6,038,525; Assigned to Southwest Research Institute; Filed in 1997.

92. Maguire, J.F.; Busbee, J.D. & LeClair, S.R.; Surface Flaw Detection Using Spatial Raman-Based Imaging; US 6,453,264 B1; Assigned to Southwest Research Institute; Filed in 1999.

93. Welch, W.T.; Bledsoe, R.R.; Wilt, B.K. & Sumner, M.B.; Gasoline RFG Analysis by a Spectrometer; US 6,140,647; Assigned to Marathon Ashland Petroleum; Filed in 1997.

94. Cooper, J.B.; Bledsoe, J.R.R.; Wise, K.L.; Sumner, M.B.; Welch, W.T. & Wilt, B.K.; Process and Apparatus for Octane Numbers and Reid Vapor Pressure by Raman Spectroscopy; US 5,892,228; Assigned to Ashland Inc. and Old Dominion University Research Foundation; Filed in 1996.

95. Sharma, S.K.; Lucey, P.G.; Ghosh, M.; Hubble, H.W. & Horton, K.A.; Stand-Off Raman Spectroscopic Detection of Minerals on Planetary Surfaces; *Spectroc. Acta Pt. A-Molec. Biomolec. Spectr.* 2003, 59, 2391–2407.

96. Harvey, S.; Vucelick, M.; Lee, R. & Wright, B.; Blind Field Test Evaluation of Raman Spectroscopy as a Forensic Tool; *Forensic Sci. Int.* 2002, 125, 12–21.

97. Mantsch, H.H.; Choo-Smith, L.-P. & Shaw, R.A.; Vibrational Spectroscopy and Medicine: An Alliance in the Making; *Vib. Spectrosc.* 2002, 30, 31–41.

98. Workman, J., Jr; Koch, M. & Veltkamp, D.J.; Process Analytical Chemistry; *Anal. Chem.* 2003, 75, 2859–2876.

99. Workman, J., Jr; Creasy, K.E.; Doherty, S. *et al.*; Process Analytical Chemistry; *Anal. Chem.* 2001, 73, 2705–2718.

100. Crow, P.; Wright, M.; Persad, R.; Kendall, C. & Stone, N.; Evaluation of Raman Spectroscopy to Provide a Real Time, Optical Method for Discrimination between Normal and Abnormal Tissue in the Bladder; *Eur. Urol. Suppl.* 2002, 1, 80.

101. Boere, I.A.; Bakker Schut, T.C.; van den Boogert, J.; de Bruin, R.W.F. & Puppels, G.J.; Use of Fibre Optic Probes for Detection of Barrett's Epithelium in the Rat Oesophagus by Raman Spectroscopy; *Vib. Spectrosc.* 2003, 32, 47–55.

102. Kittrell, C.; Cothren, J.R.M. & Feld, M.S.; Catheter for Laser Angiosurgery; US 5,693,043; Assigned to Massachusetts Institute of Technology; Filed in 1990.

103. Kittrell, C. & Feld, M.S.; Catheter System for Imaging; World Intellectual Property Organization Patent Cooperation Treaty International Application WO 89/02718; Assigned to Massachusetts Institute of Technology; Filed in 1988. Priority Number US 100714 (1987).

104. Wolthuis, R.; van Aken, M.; Fountas, K.; Robinson, J.; Joe, S.; Bruining, H.A. & Puppels, G.J.; Determination of Water Concentration in Brain Tissue by Raman Spectroscopy; *Anal. Chem.* 2001, 73, 3915–3920.

105. Feld, M.S.; Biomedical Raman Spectroscopy: Basic and Clinical Applications. In: Eastern Analytical Symposium; Somerset, NJ USA; 19 November 2003.

106. Berger, A.J.; Brennan, I.I.I.J.F.; Dasari, R.R. *et al.*; Apparatus and Methods of Raman Spectroscopy for Analysis of Blood Gases and Analytes; US 5,615,673; Assigned to Massachusetts Institute of Technology; Filed in 1995.

107. Allgeyer, D.O.; Device and Method for Qualitative and Quantitative Determination of Intravenous Fluid Components; US Patent Application Publication 2003/0204330 A1; Filed in 2002.

108. Farquharson, S.; Smith, W.W.; Carangelo, R.M. & Brouillette, C.R.; Industrial Raman: Providing Easy, Immediate, Cost Effective Chemical Analysis Anywhere; *Proc. SPIE-Int. Soc. Opt. Eng.* 1999, 3859, 14–23.

109. Selected Successful Raman Applications. http://www.kosi.com/raman/analyzers/ramanrxn2.html.

110. Merrit, R.; Intel Applies Chip-Making Prowess to Cancer Research – MPU Giant Hopes Raman Analyzer Speeds Detection; *Electronic Engineering Times* 2003, 10272003: 6.

111. Pelletier, M.J.; Davis, K.L. & Carpio, R.A.; Shining a Light on Wet Process Control; *Semiconductor International* 1996, 19, 103–108.

112. Lipp, E.D. & Leugers, M.A.; Applications of Raman Spectroscopy in the Chemical Industry. In: Pelletier, M.J. (ed.); *Analytical Applications of Raman Spectroscopy*; 1st Edition; Blackwell Science Ltd; Malden, MA, 1999; pp. 106–126.

Chapter 6
UV-Vis for On-Line Analysis

Lewis C. Baylor and Patrick E. O'Rourke

6.1   Introduction

The use of ultra-violet (UV) spectroscopy for on-line analysis is a relatively recent development. Previously, on-line analysis in the UV-visible (UV-vis) region of the electromagnetic spectrum was limited to visible light applications such as color measurement, or chemical concentration measurements made with filter photometers. Three advances of the past two decades have propelled UV spectroscopy into the realm of on-line measurement and opened up a variety of new applications for both on-line UV and visible spectroscopy. These advances are high-quality UV-grade optical fiber, sensitive and affordable array detectors, and chemometrics.

Non-solarizing (or at least solarization resistant) optical fibers make analyses at wavelengths shorter than 280 nm possible by fiber-optic spectroscopy. Prior to this improvement, optical fibers quickly developed color centers when exposed to intense UV radiation from either a deuterium lamp or a xenon flash lamp. The light transmitting ability of the fiber degraded quickly, often in a matter of a few minutes. Current optical fibers maintain their light transmission at or near that of a fresh fiber for months or years. The length of the fiber run, nevertheless, must be kept fairly short if the analytical work is to be done in the deep UV range of 190–220 nm. This is due to the decreased transmission efficiency of the fibers at these wavelengths relative to those in the range of 220–280 nm. Fiber-optic probes built with non-solarizing fiber make possible in situ sampling of a process, at the same time allowing the detection equipment to be positioned safely away from the process hazards.

The emergence of sensitive and affordable array detectors in 1980s and 1990s has also improved measurement capability in the UV-vis. Commercially available UV-vis instrumentation with photodiode-array (PDA) detectors appeared in the mid-1980s. This made it possible to produce a UV-vis spectrophotometer with no moving parts, with the exception of a shutter for the lamp. PDAs work well in high-light applications, such as absorption spectroscopy. Charge coupled device or CCD detectors began to appear in commercial process instrumentation in the 1990s. These detectors offer improved sensitivity over the PDA and are two dimensional, rather than just line arrays. Front-illuminated CCDs may be used in the UV if they are coated by a chromophore that absorbs UV radiation and re-emits the energy as a visible photon. Back-thinned CCDs directly sense

UV photons and are about ten times more sensitive in the UV than a front-illuminated CCD or PDA.

Finally, the development of chemometrics over the past 20 years has also aided in the use of UV-vis technology for more complicated chemical matrices than was possible at earlier times. Chemometrics allows large quantities of spectral data to be analyzed and reduced to a useful bit of information such as the concentration of a chemical species. Contributions from overlapping absorption features may be separately analyzed to determine the concentrations of more than one chemical species. In addition, through analysis of residuals, it is possible to detect when something unexpected occurs in a process.

6.2   Theory

6.2.1   Chemical concentration

Most liquid- and gas-phase UV-vis spectroscopic measurements rely on the well-known Beer's Law[1] (Equation 6.1) to relate the amount of light absorbed by a sample to the quantity of some chemical species that is present in that sample.

$$A = \varepsilon b C \tag{6.1}$$

where A is absorbance, ε is the molar absorption coefficient in units of mol/L-cm, b is the pathlength of the measurement in units of cm, and C is the concentration in units of mol/L.

While spectra are sometimes recorded in units of transmittance (T) or percent transmittance ($\%T$), these units are not suitable for determining chemical concentration. The reason for this comes from the relationship between absorbance and transmittance[1] as shown in Equation 6.2.

$$A = \log(1/T) \tag{6.2}$$

It can easily be seen from Equations 6.1 and 6.2 that only absorbance is linear with concentration.

Without going into great detail on any of the different algorithms, chemometric models can be divided into a few simple categories. Single-wavelength models simply track the magnitude of the absorbance at a chosen wavelength and correlate that to concentration by Beer's Law. This method works well if there is only one absorbing species present (at the analytical wavelength) and if there is a well-behaved baseline that does not introduce any offsets to the data. If baseline offsets are present but the baseline remains flat even when offset, then a simple model that correlates the difference in the absorbance values at the analytical wavelength from the baseline wavelength to the concentration of the analyte, again by Beer's Law, will work well. There must be no interfering absorptions at either the analytical or the baseline wavelengths. For more complex spectra other methods are required. These cases involve overlapping absorption bands from two or more components and/or sloping baselines in the data. More sophisticated methods of background subtraction include first and second derivatives, as well as a multi-segment background fit and subtraction. Chemometric techniques such as principle component regression (PCR), partial least squares (PLS), and multiple linear regression (MLR) are used to correlate spectral response to analyte concentration.

While baseline offsets are usually not a problem in a laboratory cuvette, they can be a significant issue in on-line measurements. There are several sources of such offsets. Bubbles scatter light in a sample and induce baseline offsets. Scattering has a greater effect the farther into the UV region (shorter wavelengths) one goes. If too much light is scattered, not enough light will be transmitted through the sample and a reliable measurement cannot be made. If possible, minimization of bubbles in the optical cavity of a probe or cell is desirable. Particles also cause the scattering of light in a process sample and should be minimized if possible. Filtering probes and cells are available, which can eliminate most particles and bubbles from the optical path and make reliable process measurements possible for difficult sample matrices. In some cases, the sample will deposit a film onto the optical surfaces of a probe or cell. This film might lead to a bias in the analytical results, and eventually could reduce light transmission to make the analysis impossible. Film build-up requires some action be taken to clean the cell or the probe, either cleaning in place or by removing the probe or cell for cleaning by hand. The movement of a fiber-optic probe and/or the fibers leading to/from a probe or cell can also induce baseline shifts in the recorded spectra. These shifts are usually minor and can be eliminated with proper pre-processing of the data (e.g. derivatives).

6.2.2 Color

Transmittance and reflectance data are used in color measurements. Transmittance spectra are used for liquid color measurements, while reflectance spectra are used on solid samples (powders, surfaces) and on opaque liquids (paint). Spectra of solid samples are usually recorded in the units of reflectance (R) or percent reflectance (%R), which is analogous to percent transmittance in that reflectance equals the ratio of the reflected radiation to the incident radiation. A large number of color scales are in use today.[2] Some are specific to a particular industry, while others enjoy broader application. The most common scales are the CIE L*a*b* (Commission Internationale de L'éclairage, 1976) and the L*C*h°. For L*a*b* coordinates, L* refers to the brightness of a sample, a* is the red-green coordinate, and b* is the yellow-blue coordinate. For L*C*h° coordinates, L* is again the brightness, while C* is the chromaticity and h° is the hue. Spectral data from 380 to 720 nm are normally used for color calculations, as this is the wavelength range to which the human eye is sensitive. While the number of different color scales may seem daunting, it is important to remember that all color coordinates are calculated from the same basic spectral data. The only difference is in the wide variety of mathematical transforms that are then applied to that data to extract a particular useful bit of color information. Whiteness index (ASTM E313-73) and yellowness index (ASTM D1925-70) are two more examples of different color scales.

6.2.3 Film thickness

The thickness of thin film layers separated by uniform, parallel interfaces can be determined from optical interference patterns that result. These measurements can be made from about 400 nm out through the visible spectrum and on into the near-infrared (NIR) region. Since film thickness measurements rely not on the absolute magnitude of the reflected light, but on the variation of that signal with wavelength, the choice of units is less important. Typically %R is used, but in some cases raw intensity is also satisfactory. We will treat thickness determinations in more detail in the applications section of this chapter.

6.3 Instrumentation

Instrumentation for UV-vis process analysis falls into four categories: scanning instruments, diode-array instruments, photometers, and fiber-optic diode-array and CCD instruments. The former two are more typically encountered in at-line or near-line applications, whereas the latter two are better suited to actual on-line analyses.

Conventional scanning UV-vis spectrophotometers come in two configurations: single beam and double beam.[1] In general, scanning UV-vis spectrophotometers offer superior stray light performance to an array-type instrument. Their chief disadvantages for process analysis are the presence of moving parts and the difficulty in getting the sample to the instrument for analysis.

Single-beam instruments are found in at-line or near-line analyses where a sample vial or cuvette may be inserted into the instrument. While these instruments can collect data over a broad wavelength range, they are not capable of sophisticated data analysis and an analysis based upon one or two wavelengths is performed. A separate reference vial or cuvette is usually required. In most cases, a sample without any interference (bubbles, particles, multiple absorbing components) is required. Double-beam instruments are more typically encountered in the laboratory environment. They offer the advantage of a separate reference channel and, in general, better optical performance than a single-beam instrument.

Diode-array spectrophotometers have no moving parts, or at most a shutter that moves; thus they are more suited to locations near the process. However, these instruments still require the introduction of the sample into the instrument's sample cavity, as well as the necessity for a separate reference measurement. These instruments are usually only found in a single-beam configuration.

Photometers are relatively simple devices that make use of a light source and one or more filters to present a narrow band of wavelengths to a sample and then to a photodetector.

One, two, or even three wavelength analyses are possible with these devices. In addition, photometers are designed for location on-line, with samples generally plumbed into a sample cavity in the analyzer.

Fiber-optic-coupled spectrophotometers (single beam and double beam) are the best choice for on-line analyses. The advent of non-solarizing optical fiber has made possible on-line analyses in which the spectrophotometer may be located remotely from the process and light is carried to/from the process by the optical fiber. A rugged probe or flow cell provides the sample interface.

Photodiode arrays are almost always used in single-beam instruments. PDAs provide high signal-to-noise ratio (S/N) in high-light applications such as absorbance/transmission measurements. They are less expensive than CCDs and come with 512, 1024, or even 2000+ pixel arrays. CCDs are more costly than the PDA; however, they offer superior sensitivity and versatility to the PDA. With proper electronic processing, CCDs work well in both high- and low-light (emission, fluorescence) applications. They offer lower dark noise than PDAs and greater sensitivity over the entire UV-vis range. For use in the UV, the CCD must be either coated with a material that absorbs UV light and emits visible radiation, or used in the back-thinned configuration.

Because a CCD is a two-dimensional array of pixels, it is possible to create a double-beam instrument with a single detector. Coupled in an optical design with an imaging grating, the output of two fibers may be imaged simultaneously and separately onto the CCD and both spectra readout by the electronics.

Both PDA and CCD fiber-optic spectrophotometers may be combined with a fiber switch or multiplexer to access multiple probes or cells in a sequential fashion. Multiplexers are available to select from among 2, 6, 9, or even more fibers/probes.

6.4   Sample interface

Design and selection of the sample interface is vital to provide the best-quality data for an analysis. The sample interface may be located in the sample cavity of a spectrophotometer, as in the cases of laboratory cuvettes, vials, and flow cells. The sample interface may also be fiber-coupled and located closer to the process. Fiber-optic sample interfaces include flow cells, insertion probes, and reflectance probes.

6.4.1   Cuvette/vial

In a few cases, analyses may be done on the process floor using a conventional scanning spectrophotometer and the usual laboratory cuvette or a specially made vial that fits into the sample cavity of the instrument. Cuvettes may be made from quartz (various grades depending upon UV transmission requirements), or from plastic (disposable) when UV wavelengths are not required for the measurements. In most cases these at-line analyzers have lower performance characteristics than dedicated laboratory instruments, but may be quite adequate for the task at hand. However, their use still requires manual sampling of the process, which itself brings on problems such as operator exposure to the process and sample integrity. In some automated applications such as drug dissolution, samples are pumped through a flow cell (cuvette) located in the instrument's sample cavity.

6.4.2 Flow cells

Flow cells are used in fiber-optic applications. They may be plumbed directly into a process, or onto a side-sampling loop. They typically have two opposing lenses with a sample gap in between. The flow path is usually the 'Z' configuration (Figure 6.1). The volumes of the flow cells may vary widely depending upon the sampling requirements.

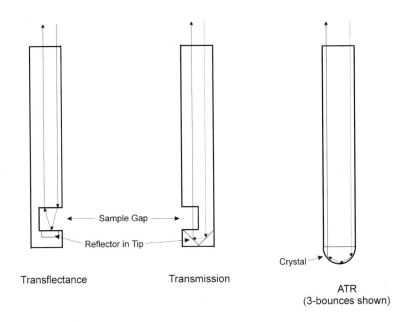

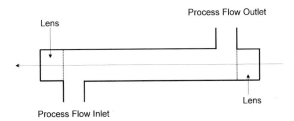

Figure 6.1 Schematic of basic probe designs: transmission cell, immersion probe (retroreflecting and transmission), attenuated total internal reflection.

Standard flow cells with a single-pass optical design have pathlengths from 1 mm to 1 m for liquid-phase applications. Alignment of optics and optical throughput become problematic at greater lengths. These flow cells usually have flow path diameters from 1 mm up to several inches. Flow cells with small flow path diameters have very low cell volumes and are ideal for use in applications where there is very little sample available. Examples of these applications include high-performance liquid chromatography (HPLC), and various combinatorial chemistry techniques.

Short pathlength flow cells may be used in lower volume, or lower flow situations. The extremely small gap between the optics of fibers limits them to these types of applications. It is possible to make flow cells with sample gaps as small as 25 microns, so that they may be used with highly absorbing species.

A filtered flow cell offers the advantage of removing particles or bubbles from the optical path. This makes reliable process measurements possible in difficult environments where particles and/or bubbles abound. A fermenter is one example of such an environment.

6.4.3 Insertion probe

When it is not possible to plumb a cell into the process, an insertion probe offers a second method of accessing the sample. These probes may be put in through a port in a pipe or vessel. It is also possible to design the probe port to allow for removal for cleaning while the process continues to run.

Transflection probes (see Figure 6.1) work by sending a beam of light into a sample gap from a collimating lens and then bouncing the light off a mirror at the other side. As such, these probes have a double-pass optical design. These probes may be manufactured with sample gaps from 0.5 mm to 5 cm. At gaps smaller than 0.5 mm (1-mm pathlength) it becomes difficult to reliably move sample through the probe's measuring cavity.

In a transmission insertion probe (see Figure 6.1) light makes only a single pass through the sample gap. In these probes either the fiber is forced into a 'U' bend at the end of the probe, or corner-cube reflectors must be used to turn the light path 180°. Again, the smallest practical gap is 0.5 mm (0.5-mm optical path).

Like the filtered flow cell, the filtered immersion probe (either transflection or transmission) may be used in sample environments in which bubbles or particles make the use of unfiltered samples impossible.

Attenuated total internal reflection (ATR) probes offer several advantages over other probe types. ATR is a phenomenon that relies on a difference in the index of refraction of a crystal and that of the solution with which it is in contact to prevent light from escaping the crystal. Only the evanescent wave of the light interacts with the solution layer at the crystal face. The result is an optical pathlength of only a few microns. Typical designs make use of faceted crystals or hemispheres (see Figure 6.1). The most common ATR material in the UV-vis is sapphire. In rare cases, fused silica may be used. ATR allows spectra to be taken of neat samples with optical density (OD) of 500–1000 or more. ATR is largely immune to particle or bubble effects in the solution. A reference spectrum of an ATR probe is also easily obtained, as a clean probe in air may serve as the reference: once a vessel is flushed, it is not necessary to fill it with a reference solution in order to re-reference the probe. The major disadvantage to ATR is that quantitation of the detected absorption is complicated by the fact that the strength of the absorption band will vary with the refractive index of the sample. The closer the index of refraction of the sample is to that of the crystal (cut-off index) the larger the absorption peak (longer effective pathlength), and more quickly the absorption cross section will change for a given change in refractive index. The refractive index may change because of concentration changes or because of temperature changes in the sample. ATR probes make ideal insertion probes because there is no mirror or sample gap that would require cleaning. So long as the crystal is rinsed clean, the probe is ready for use. ATR probes are also generally immune to the effects of bubbles and particles in the sample.

6.4.4 Reflectance probe

A diffuse reflection probe is used to measure the light reflected from a solid surface or powder. This probe is immune to specular reflections from the sample, and detects only diffusely reflected light.

As differentiated from diffuse reflection, we use the term 'backscatter' to mean a probe that detects both specular and diffuse light. The most common designs of these probes incorporate a number of illumination fibers (usually six) in a ring around a central detection fiber. This probe is useful in a slurry or solution high in particulates (e.g. crystallization).

6.5 A complete process analyzer

We have mentioned many of the components of a complete UV-vis process analyzer system, but have not put them together as a single installation to this point. An example of such a system based upon fiber optics is shown in Figure 6.2. The UV-vis analyzer contains a spectrophotometer and a light source. A fiber multiplexer may be added if multiple sampling points are required. Modern analyzers will also contain a control computer, display and some kind of data I/O interface to the process. These interfaces can include analog current outputs, digital alarm signals, or numerical streams sent to the plant's distributed control system (DCS). One or more probes are placed into the process. If the measurements made by the analyzer indicate that the process is drifting towards the control limits, the DCS may now take some action to change the operation of the process and bring it back to the optimal range. For example, the amount of reactant being fed may be adjusted, the temperature may be changed, or mixing speeds may be altered.

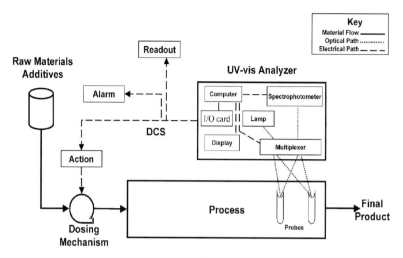

Figure 6.2   Example of an on-line process analysis.

6.6   Applications

UV-vis technology finds use in a wide variety of applications that cut through a cross section of industry. Applications exist for gas, liquid, and solid phase materials and range from chemical concentration, to color measurement, to film thickness determinations. We will discuss only a few of these in detail here.

6.6.1   Gas analysis – toluene

A number of important pollutant gases and vapors have signatures in the UV region of the electromagnetic spectrum. Among these are NO_x, SO_x, CO, benzene, toluene, and the xylenes. This application (Dr William Napper, Ciba Specialty Chemicals, McIntosh, Alabama) focused on the monitoring of toluene in a chemical manufacturing plant for the purpose of preventing the formation of an explosive mixture of toluene and air. This is a critical safety analysis for the plant and must be accurate and reliable at all times. In this particular case, sulfur compounds in the gas stream acted to poison electrochemical sensors, so the decision was made to use a UV analyzer. The analyzer (Equispec™ Chemical Process Analyzer, Equitech Int'l Corporation, New Ellenton, SC) consisted of a fiber-optic UV-vis CCD spectrophotometer, a xenon flash lamp, fiber multiplexer, three 100-cm transmission cells and a control computer. The presence of the multiplexer allowed for the use of multiple measurement cells with a single analyzer. The goal of the analysis was to keep the toluene level below the lower explosive limit (LEL) of 12 000 ppm. The alarm level for the analyzer was set to 60% of the LEL. Typical toluene concentrations were in the 1000–2500-ppm range. As toluene concentrations rose, nitrogen was bled into the system to keep the mixture inert. If the alarm level was reached (~7000 ppm), the process went into shutdown. A further advantage of the UV analyzer was that while the transmission cells and optical fibers were in an explosion hazard area, all the electrical components of the analyzer were kept in a general-purpose area. Another key aspect of this analysis was sample preparation. The sample stream was at ambient temperature, so the transmission cells were heated to 45°C to prevent condensation on the optical surfaces inside. Prior to entry into the cell, the sample was sent through a 100-mL 'knock out pot' to catch any droplets in the gas stream, and then through a 5-micron filter to remove any particulates. The careful preparation of the sample stream for the gas cells was one of the keys to the success of this application. The sample was drawn into the cell by a vacuum venturi; the negative pressure in the process was created by a blower system on the carbon bed that was used to collect the toluene.

The calibration spectra from one of the gas cells are shown in Figure 6.3. Duplicate samples at the same concentrations have been omitted for clarity. These spectra have been pre-processed by subtracting the baseline offset (average response over the 360–370-nm range) from each spectrum. A second-order MLR (quadratic) fit to the data at 266 nm gave satisfactory results, with a correlation coefficient of 0.99875 and a standard error of 11.4%. A plot of the known versus predicted toluene concentrations from the calibration set is shown in Figure 6.4. While the standard error of the fit may seem high, it was quite adequate for the purposes of this application. Extremely accurate toluene concentrations were not required since the alarm level for process shutdown was set at 60% of the LEL, well outside of the error bars on the high-concentration calibration points.

This instrument ran very reliably while sampling every thirty seconds for 4 years from installation until the process was discontinued, with the need for only one lamp

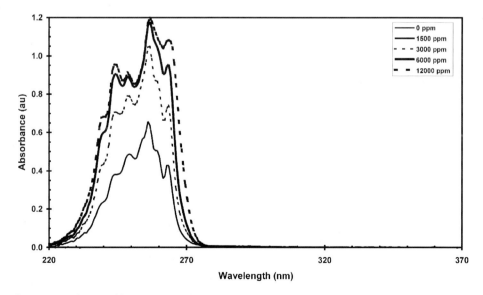

Figure 6.3 Toluene calibration spectra.

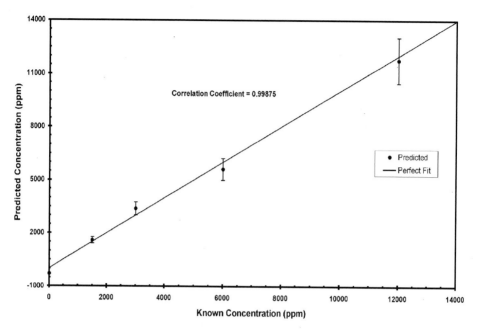

Figure 6.4   Known vs. predicted concentrations of toluene.

replacement and one replacement of the CPU cooling fan. Keeping the downtime to a minimum was essential, as this was a safety analysis; if the analyzer was down, the process did not run.

6.6.2   *Liquid analysis – nickel*

The power of on-line technology is the ability to monitor a critical point in the process for an event of some kind that intermittent manual sampling and off-line analysis would miss. One such event is column breakthrough on an ion-exchange column, where, for example, recovery of a catalyst may have both economic and environmental impacts. In terms of plant efficiency, on the one hand, it is desirable to use the maximum capacity of the column and then stop or divert the flow only when column breakthrough is detected. On the other hand, it is also desirable to minimize the amount of breakthrough if it has an impact on permitted plant emissions. The authors have successfully implemented a UV-vis analyzer (Equispec^{TM} Chemical Process Analyzer, Equitech Int'l Corporation, New Ellenton, SC) for the detection of nickel breakthrough on an ion-exchange column. Once breakthrough is detected in this particular installation, process flow is diverted into a second ion-exchange column while the first one is being regenerated. The analyzer consists of a fiber-optic UV-vis CCD spectrophotometer, a xenon flash lamp, a long-path high-flow transmission cell and a control computer. The analyzer runs in a two-channel (dual beam) configuration in which a reference fiber-loop serves as real-time compensation for lamp intensity fluctuations. The methanol carrier has no signal in the wavelength range of the analysis; however, the nickel peak does have a wavelength shift due to changes

Column Breakthrough

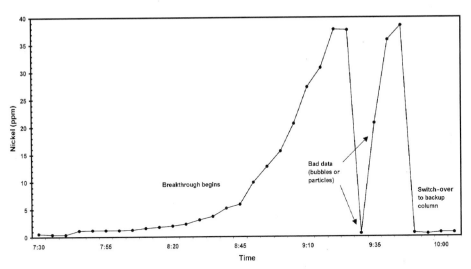

Figure 6.5 Data showing nickel breakthrough.

in the pH of the process flow. In order to account for this shift, a chemometric model is used to look for the nickel peak within a 'window' along the wavelength axis. A first-order MLR model was built based upon the area of the nickel peak. Baseline correction is necessary due to the potential for bubbles in the process flow. The larger inner diameter of this flow cell (5 cm) allows for the presence of some bubbles without negatively affecting light transmission through the cell. Nickel is being monitored in the range of 0–20 ppm using a 50-cm pathlength transmission cell. Figure 6.5 is a plot showing an example of breakthrough on the column, followed by switch-over to the backup column.

6.6.3   Solid analysis – extruded plastic color

Proper control of the color of extruded plastic is a key quality issue for many consumer products, including vinyl siding, fencing and decking materials and flooring. Tradition-ally, samples of the product are pulled from the process line and analyzed for color properties in the plant laboratory. This can cause a delay on the order of 10–30 minutes before notification of out-of-spec material might reach the manufacturing floor. During that lag period between sampling and results, quite a lot of bad products will have been made, which now must be at best, re-ground and at worst, scrapped. The addition of an on-line color analyzer can greatly improve efficiency by reducing scrap and can also improve product quality by allowing for control of the color to a tighter specification. The hardware configuration (Equispec^{TM} Color Analyzer, Equitech Int'l Corporation, New Ellenton, SC) for this application is identical to that for the above nickel application, with the exception of the fiber-optic probe and the software algorithms. In this case, the probe is a specular-excluded reflectance probe with a 45/0° (illumination/detection) geometry that is mounted a short distance (<0.5 cm) above the moving sample web. Such probes can be extremely rugged and may be placed only a short distance outside of the extruder head. Again, the analyzer runs in a two-channel (dual beam) configuration in which a reference fiber-loop serves as real-time compensation for lamp intensity fluctuations. A white reference tile serves as the instrument's color calibration; this tile is checked about once a week for accuracy. In a vinyl siding installation, it is possible to reliably measure ΔE of 0.3 with this type of UV-vis system (ΔE is a measure of the distance between two points (colors) in color space). Gloss changes in the product do not affect the measured color coordinates. As long as product embossing depth is ~50% or less than the optimum probe-sample distance, it too does not affect the measured color coordinates.

6.6.4   Film thickness – polymer

With the appropriate fiber-optic probe and data processing techniques, a UV-vis analyzer may be used to determine the optical thickness of a transparent thin film (Equispec™ Film Thickness Analyzer, Equitech Int'l Corporation, New Ellenton, SC). It is possible to simultaneously measure thickness of different layers in a multi-layer structure as long as each layer falls within the analysis range of the instrument. Typically, this means layers in the 0.5–150-micron range. A further constraint on this technique is that the layer structure of the film must be smooth on the scale of the spot size of the fiber-optic probe. Another constraint is that neighboring layers must have different indices of refraction in order for them to appear as distinct layers to the analyzer. The preferred light source for this analysis is the tungsten-halogen bulb, for it offers a smooth emission envelope and high stability. The highly structured nature of a xenon source's emission spectrum makes it unsuitable for film thickness measurements. White light from the fiber-optic probe strikes the sample, and reflected light is collected from each layer interface in the film. For example, a two-layer film returns signals representing each individual layer as well as the total film thickness (assuming all are within the range of the instrument) (Figure 6.6). Depending upon the particular sample, all or a portion of the spectrum in the 400–1000 nm wavelength range may be used. In contrast to a color measurement or a chemical concentration measurement, maximum light return from the sample does not guarantee the best results. It is better to optimize the probe placement based upon the ringing pattern in the data, rather than upon the reflected light intensity. The signal returned from the sample represents the optical thickness of the sample's layer or layers. This optical thickness may be converted to a true physical thickness by dividing the result by the index of refraction of the layer in question. In this application, a vis-NIR spectrophotometer is used to measure the clear top layer on a co-extruded polymer film. The bottom layer is pigmented to an opaque white color and its thickness may not be determined by this method. Prior to the installation of the fiber-optic spectroscopy system, film samples were measured manually in the laboratory by a subtractive scheme. First, the total thickness of a sample was measured on a manual profilometer. The top layer of the polymer was removed with methylene chloride in a hood. The sample was

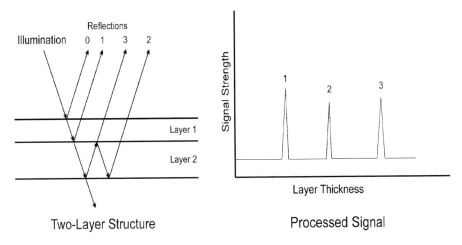

Figure 6.6 Diagram of reflections from a two-layer film.

then repositioned on the profilometer as closely as possible to the originally measured spot and the thickness of the second white layer was determined. The thickness of the top layer was then determined by difference. There were a number of problems with this analytical scheme: difficulty in measuring the same physical spot before and after solvent treatment, use of a hazardous solvent and generation of an additional waste stream for the lab. The fiber-optic spectroscopy system eliminates all three of these problems, at the same time providing more accurate and more rapid analysis of the samples. The film thickness analyzer consists of a fiber-optic CCD spectrophotometer, tungsten-halogen light source, and a diffuse reflectance probe to interrogate the sample. The tip of the probe is held at a constant distance from the film sample by means of a sample-holder jig. A comparison of the optically determined thickness to the manual laboratory method is shown in Figure 6.7.

6.6.5   *Dissolution testing*

Dissolution testing is a mandated step in the production of all tablets and extended release products in the pharmaceutical industry. There are well-established United States Pharmocopeia (USP) guidelines for performing all dissolution testing. Although none of them are written specifically for fiber-optic methods, they do make allowances for new technologies that are proven to be of equal or greater sensitivity or accuracy than the existing methods.[3,4] A dissolution testing apparatus consists of a set of six or eight thermostatted, stirred vessels of an established geometry and volume from the USP guidelines. The dissolution apparatus provides a means to dissolve each sample, but does not provide a means to determine the concentration of the active ingredient in the bath. In the most well established scheme, sipper tubes withdraw samples from each dissolution vessel and send them through a multi-port valve to a flow cell sitting in the sample chamber of a UV-vis spectrophotometer. Most often, this is a diode-array spectrophotometer that has

Film Thickness Data

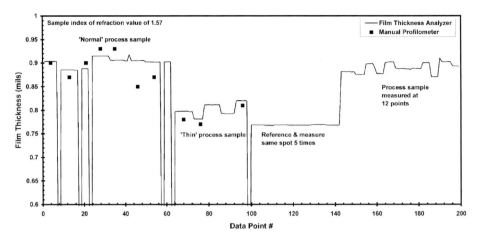

Figure 6.7   Comparison of optical thickness measurements to a physical technique.

been integrated with the control of the dissolution testing apparatus (Agilent 8453 UV-Visible Spectrophotometer, Agilent Technologies, Palo Alto, CA). In recent years, moves have been made to make in situ measurements in the dissolution baths by means of fiber-optic probes. There are three possible probe implementations: in situ, down shaft, and removable in situ. A recent review article evaluates various probe and instrument designs.[5]

An in situ probe is a slender probe of either transflectance or transmission design that is permanently inserted into each dissolution bath in an apparatus. This has the advantage of allowing measurements at the same physical location in the vessel where the sipper tube had been positioned, according to USP guidelines. A disadvantage is the disturbance of the laminar flow in the stirred vessel – thus there has been an effort to make this probe with as small a diameter as possible (1/8" or less).

A second design positions the probe down the hollow shaft of the stirring blade in each vessel. The design eliminates the problem of flow disturbance at the sampling point; however, it does not sample the active at the depth prescribed by the USP guidelines, and the effectiveness of the flow through the shaft is difficult to determine. Again, these probes may be either of tranflectance or transmission design.

In a third configuration, a probe is dipped into the vessel only when it is time to make a measurement in the vessel. Once the data have been collected, the probe is withdrawn. This minimizes the flow disturbance in the vessel, while still allowing sampling at the same location as the sipper tube. A disadvantage of this method is that a film may form on the optical surfaces of the probe as they dry while the probe is suspended in air over the dissolution bath.

There are three instrument designs in use with fiber-optic probes. One system makes simultaneous measurements on up to eight dissolution bath probes using a CCD-based spectrophotometer with xenon flash lamp source. The signals from all eight probes are simultaneously imaged onto a single CCD detector and concentrations determined in real time on all vessels (Opt-Diss Fiber Optic UV System, LEAP Technologies Inc., Carrboro, NC). This method has the advantage that all measurements use the same light source and detector, eliminating any variation that might arise from the use of multiple detectors and light sources. The biggest disadvantage of this design is the crosstalk between neighboring fibers on the detector surface. Crosstalk is the spilling over of light from one fiber onto the pixels designated to receive light from a neighboring fiber. This effect can be minimized by careful design of the grating and choice of pixels on the CCD itself.

The second instrument design makes use of an array of eight separate CCD spectrophotometers, each with its own light source (Rainbow Dynamic Monitor, Delphian Technology LP, Woburn, MA). The advantage of this design is the complete elimination of crosstalk between probes. Disadvantages of this approach include cost (eight detectors and eight light sources) and the needed correction for variation in performance among the eight different detectors.

The third design employs a mechanical fiber multiplexer (fiber switch) to allow one spectrophotometer to interrogate sequentially each of up to eight probes (IO Fiber Optic Dissolution System, C Technologies Inc., Cedar Knolls, NJ). Like the second design, there is no crosstalk between the different fiber probes. Like the first design, there is a single light source and detector in use for all of the probes. The key to success of this design is that the mechanical multiplexer is both very accurate and very precise in its repeated movements to each probe.

6.6.6 Liquid analysis – vessel cleaning

Another excellent application of fiber-optic UV-vis spectroscopy is in the monitoring of a tank or vessel during a cleaning in place operation (Nova, SpectrAlliance Inc., St Louis, MO; Cleanscan, Carl Zeiss, Germany). This is important in industries where product quality and purity is tantamount: pharmaceuticals, foods, and beverages. Once a batch reaction is completed and the vessel drained, it must be cleaned before the next batch of product can be made. This next batch may be the same product, or a different one. In either case, it is vital that the cleanliness of the vessel be verified. If an off-line analysis is to be used, the cleaning process is delayed while laboratory results are obtained. In other cases, a tank may be flushed for a length of time that has been determined to be satisfactory through trial and error over time at the plant. In fact, much less cleaning may be necessary and the ingrained method in use is overkill – wasting production time and money. An on-line analyzer can pinpoint when the vessel is adequately clean and may be able to decrease turnaround times for the next reaction batch. Fortunately, in the pharmaceutical and food/beverage industries, many products possess strong UV or visible signatures and the reaction solvent is likely to be transparent to the UV-vis analyzer. It is not even necessary to calibrate for a particular analyte, so long as the decay of a signal peak may be followed over time as the tank is washed.

Fiber-optic UV-vis spectrophotometers are well suited to a clean-in-place application. For analytes with signals in the deep UV, best results will be obtained with a deuterium source and short fiber runs. Further out in the UV and visible regions, a xenon flash lamp is the best choice and fiber runs may be much longer. If fiber runs must be kept short, as is likely in pharmaceutical applications, then either the sample must be plumbed into a cell located at or inside of the analyzer enclosure or the analyzer must be located within a few meters of the flow cell or probe mounted in the process piping. Because this application relies on following the degradation of a signal to determine cleanliness, it is important to attain the lowest detection limit possible. This may be achieved by improving the signal-to-noise ratio through longer-path flow cells and probes (5 cm and up), signal averaging, and using a low-noise detector.

One of the difficulties in quantitative spectroscopic analysis in any part of the electromagnetic spectrum is that a separate calibration model must be maintained for nearly every different analyte. However, in the case of cleaning-in-place applications this is not necessarily true. Since the goal of this measurement is to verify when no signal is present – that is when the sample spectrum matches the spectrum of the neat solvent(s) – the model may be a simple wavelength type (peak minus background) and the signal peak selected automatically by a peak search algorithm. Alternatively, the integrated area under a peak might be tracked. Another means would be to simply compare the stored solvent spectrum to each sample spectrum, calculate the residual and continue cleaning and making measurements until it reaches a pre-set minimum level.

References

1. Willard, H.O., Merritt, L.L., Jr, Dean, J.A. & Settle, F.A., Jr, *Instrumental Methods of Analysis*, 6th Edition; Wadsworth Publishing Company; Belmont, 1981.
2. Hunt, R.W.G., *Measuring Color*, 3rd Edition; Fountain Press Ltd; Oxfordshire, 2001.
3. General Notices, USP 26, 2003; pp. 6–7.
4. General Chapter <711> Dissolution, USP 26, 2003; p. 2156.
5. Lu, X., Lozano, R. & Shah, P., *In-Situ* Dissolution Testing Using Different UV Fiber Optic Probes and Instruments; *Dissolution Technologies*; November 2003, 6–15.

Chapter 7
Near-Infrared Chemical Imaging as a Process Analytical Tool

E. Neil Lewis, Joseph W. Schoppelrei, Eunah Lee, and Linda H. Kidder

Conventional near-infrared (NIR) spectroscopy is a very rugged and flexible technique that can be generally adopted for use in a wide range of chemical analyses in many different research and industrial (process) applications. The additional information provided by the spatial perspective of NIR spectroscopic imaging offers access to greater understanding and therefore control of the manufacturing process of complex composite materials and products. While the focus of this work is on pharmaceuticals and their manufacture, the reader can easily incorporate these ideas for other industries and products, including general chemical, polymer, agricultural, food, cosmetic, biological, or any other manufacturing environments where organic materials are processed to form a finished product.

7.1 The process analytical technology (PAT) initiative

The PAT initiative of the US Food and Drug Administration (US FDA) strongly advocates the use and development of new measurement technologies for pharmaceutical manufacturing.[1] The ultimate goal of the initiative is to maintain a stable process throughout manufacturing using real-time monitoring of critical process parameters, and proposes the most significant changes in pharmaceutical manufacturing in four decades. This goal cannot be reached immediately, as first the manufacturing process must be well characterized, and parameters that impact its stability must be identified. For example, what effect, if any, do small changes in the blending, drying, pressing, coating or other manufacturing steps have on the final dosage form? And once the critical parameters are identified, analytical techniques that can monitor those parameters must be identified, optimized for process applications, validated and deployed. To encourage their use, the FDA is streamlining the mechanism for adopting new technologies in pharmaceutical manufacturing. A successful deployment of the initiative will result in significant investment in analytical technology by the pharmaceutical industry, both in proven and in novel techniques, and a much more efficient and reliable manufacturing process for pharmaceuticals.

7.2   The role of near-infrared chemical imaging (NIR-CI) in the pharmaceutical industry

7.2.1   *Characterization of solid dosage forms*

Solid dosage forms (tablets or capsules) are not composed of a single material, rather they are carefully designed mixtures that follow a recipe called 'formulation.' A typical formulation may include one or more active pharmaceutical ingredients (APIs), fillers, binders, disintegrants, lubricants, and other materials. Each of these materials is chosen to provide desirable characteristics for manufacturing, storage, handling, and eventual release of the therapeutic agent. Throughout formulation development and the scale-up process, the primary goal is to produce a formulation and a manufacturing procedure that is robust and consistent, such that every tablet or capsule produced will yield identical therapeutic characteristics.

A basic problem in pharmaceutical manufacturing is that even a relatively simple formulation may produce widely varying therapeutic performance depending on the distribution of ingredients in the final matrix. While one usually thinks of a tablet as a homogeneous block of material, this is rarely the case. Microscopic inspection of a typical solid dosage form reveals that some, or all, of the individual ingredients are not distributed uniformly, but reside in discrete particles or agglomerations. The compositions, sizes, spatial relationships, and consistency of these localized domains comprise the entire formulation and reflect the individual unit operations used to produce the product.

Modern formulations are further increasing the demands on product uniformity. Products containing more potent APIs may be formulated at quantities ranging from a few to even less than one milligram per dose, yet the finished tablet must still be large enough for convenient handling. Therefore, the therapeutic agent may represent significantly less than 1% of the bulk form, and maintaining content uniformity is absolutely crucial. Pharmaceutical makers are also developing advanced tablets for drug dosage management that can provide longer, flatter and complex bloodstream profiles. A wide range of approaches can be used to produce these effects including barrier layers, cored tablets, selective-release microspheres, and even osmotic pumps. These types of tablets are really highly engineered drug delivery systems in which the physical structure is as critical as the chemical composition.

Purity and potency are two of the metrics that pharmaceutical quality assurance (QA) departments emphasize to determine if a batch may be released. To do this they typically employ HPLC and/or mass spectrometry to determine gross composition, and to verify the absence of any contaminants. However, only a small subset of tablets in a batch is tested because the tests destroy the sample. While these methods offer some insight into sample consistency, they provide no information on the distribution of the components in an individual finished form. Dissolution testing is used to indicate the manner and duration of component release but cannot elucidate the cause of failure for tablets outside of the allowed dissolution profile.

In addition to requiring the destruction of samples, the standard QA techniques for pharmaceutical release testing are laboratory-based and time-consuming, making it difficult or impossible to quickly trace and correct the sources of variations or failures. A number of spectroscopic techniques such as mid-infrared (MIR), NIR and Raman enable rapid, non-destructive sample analysis, and these methods can be employed at a number of points in the pharmaceutical development and manufacturing process. In particular, NIR spectroscopy is quickly becoming a workhorse technique for the industry due to its high information content and flexibility of implementation.[2–4] Currently, it is widely used for the rapid characterization of raw materials,[5,6] and has also been used in applications such as blend homogeneity,[7–9] moisture measurement, and the analysis of intact tablets.[10,11]

7.2.2 'A picture is worth a thousand words'[12]

As the pharmaceutical manufacturing process becomes better understood, it is apparent that the metrics of potency and purity are no longer sufficient. Also, in the process of creating this increased understanding, traditional analytical technologies currently employed cannot provide the direct and timely information necessary to fully characterize a solid dosage form, or the process that was used to create it. Manufacturing processes that rely on testing only after the production process is complete are understandably inefficient. This need is addressed in part by NIR-CI that offers the ability to obtain high fidelity, spatially resolved pictures of the chemistry of the sample. The ability to visualize and assess the compositional heterogeneity and structure of the end product is invaluable for both the development and the manufacture of solid dosage forms.[13] NIR chemical images can be used to determine content uniformity, particle sizes and distributions of all the sample components, polymorph distributions, moisture content and location, contaminations, coating and layer thickness, and a host of other structural details.[14–18]

The underlying tenant of the PAT initiative is to develop a better product through understanding of the manufacturing process. NIR-CI can be used to identify the elusive critical control parameters that will impact the performance of the finished product. The technique is fast and non-destructive and can be used independently, or in concert with other techniques such as dissolution analysis, to rapidly diagnose potential production problems. Incorporating NIR-CI analysis into the pre-formulation and formulation development phases of manufacturing can improve testing turnaround, limit scale-up difficulties, and reduce the time to market. This presents a more robust process, imparting significant economic benefits to the pharmaceutical manufacturer.

Near-infrared chemical imaging instrumentation is rugged and flexible, suitable for both the laboratory and the manufacturing environment. Therefore analysis methods developed in the laboratory can often be tailored for implementation near-line or at-line. NIR-CI is also a massively parallel approach to NIR spectroscopy, making the technique well suited for high-throughput applications.

7.3   The development of imaging spectroscopy

7.3.1   *Spatially resolved spectroscopy – mapping*

The development of spatially resolved molecular spectroscopy has its beginnings in 1949 with a publication in *Nature* that described the use of a microscope coupled with an IR spectrometer, collecting the first spatially resolved IR spectra.[19] While this was not an imaging instrument, since it only collected single-point IR spectra, it proved the feasibility of collecting spectral information from a spatially defined location. Taking this idea one step further, the first chemical maps were published in 1988 by Harthcock and Atkin.[20] In this work, spectra were collected from an inhomogeneous sample using a Fourier transform infrared (FTIR) microscope fitted with a moving stage. This experiment consisted of collecting a series of spectra from adjacent locations by moving the sample at regular intervals between each measurement. The resulting 'pictures,' while somewhat crude by today's standards, demonstrated the concept of spatially localizing chemical species within the sample using infrared (IR) spectroscopy. The value of this approach was swiftly realized in the analytical community.

Since that time, improvements in optical design and advancements in computers, software, and automation technology have enhanced the speed, performance, and utility of spectroscopic mapping instruments. Subsequently, FTIR and Raman microscope mapping systems are now found in many research laboratories. However, these instruments still utilize a moving sample stage and the 'step-and-acquire' acquisition mode, and because of this limitation, data collection progresses relatively slowly. Additionally, the fact that moving parts are not optimal for at-line or on-line instrumentation limits the usefulness of this type of system for PAT applications.

7.3.2   *The infrared focal-plane array (FPA)*

Unlike mapping, the detector used for infrared chemical imaging is a two-dimensional detector array. The IR FPA detector, the long-wavelength analog of a typical digital camera, was originally developed for military applications such as missile detection, targeting and night-vision devices. It is only within the last decade or so that they have been 'discovered' and used by the general IR spectroscopy community.[21] Originally, only small format arrays (64 × 64 pixels) were available and these arrays were constructed from material such as indium antimonide (InSb) or mercury-cadmium-telluride (MCT). They were prone to mechanical failure, required cryogenic cooling, and were quite expensive (>US$10 per pixel). More recently, as the technology has begun to shift from military applications to higher-volume commercial uses, more economical, large format (320 × 256 pixels) FPAs made from InSb, MCT, or Indium-Gallium-Arsenide (InGaAs) have entered the market at much lower cost (<US$1 per pixel). Many of these cameras operate uncooled or incorporate solid-state cooling, and exhibit much improved operability and durability. More recently the IR camera market has also seen the emergence of the uncooled microbolometer array using low-cost CMOS technologies.[22] These arrays promise to set whole new price performance levels for IR FPAs.

7.3.3 Wavelength selection

While the two-dimensional format of an FPA provides the ability to spatially analyze a sample, a spectroscopic imaging instrument must still incorporate a method of wavelength discrimination. Several approaches have been described in the literature[14,23–27] and, in some cases, subsequently implemented by instrument manufacturers. Each technology has its own benefits and limitations but it is beyond the scope of this work to explore each of these in detail. MIR and some NIR imaging instruments utilize the Fourier transform (FT) scheme and as a result, function in much the same way as traditional FTIR mapping systems; however, if a large-format FPA is employed, the need for moving the sample with an automated stage is eliminated. NIR and Raman imaging instruments have been constructed using solid-state tunable filter technologies, which offer several distinct advantages for industrial applications, including mechanical simplicity and flexibility for industrial applications.

7.3.4 The benefits of NIR spectroscopy

As stated in the introduction to this chapter, NIR spectroscopy has been shown to be an excellent tool applied to the solution of a wide range of pharmaceutical applications.[2–4] The high information content coupled with ease and reliability of sampling is a primary attraction of the technique over other spectroscopies. While it is arguably true that fundamental modes observed in the MIR may offer higher specificity and sensitivity than the overtone and combination bands measured in the NIR region, one must balance the value of this 'additional' information against both the increased sampling constraints of the MIR experiment and the actual needs of the application at hand.

The high sensitivity of the MIR technique arises from the highly absorbing nature of most organic materials in this region. Samples must be prepared in such a way as to limit the amount of material that interacts with the probing radiation (e.g. microtomed sections or KBr dilutions). The use of reflection techniques such as specular (SR) and diffuse reflectance (DRIFTS) in the MIR is highly dependent on the optical properties and morphology of the surface of the sample.[28] The data from these measurements often require care in interpretation, significant post-collection corrections, and can be difficult to acquire reproducibly.[29,30] Even attenuated total reflectance (ATR) techniques require an extremely flat or deformable sample surface to achieve intimate contact with the internal reflection element. Subsequently, the quality and even the information content of MIR data will typically have at least some dependence on the technique and skill of the operator and the analyst. These sampling constraints become even more significant for a chemical imaging experiment. Preparing a sample suitable for large field interrogation without compromising its original spatial integrity is a formidable task, greatly limiting the applicability and speed of the MIR experiment.

On the other hand, the overall absorbance of NIR radiation exhibited by most organic materials is reduced by a factor of 10–100 compared to that of MIR light. Therefore the constraints placed upon sample presentation are significantly relaxed, even for optically 'thick' materials. The majority of solid samples can be measured non-invasively with little or no preparation using diffuse reflectance techniques. While the optical properties and physical uniformity of the sample surface will still influence the spectral data, their overall impact is less significant and more easily separated from the spectral information.[31] Therefore, NIR methods significantly reduce the sample handling burden and the dependence on operator technique to gather high-quality, reproducible data. Relaxed sampling requirements increase the efficiency and widen the applicability of the technique, especially for non-laboratory-based measurements. For imaging, this also means that the native spatial integrity of the sample need not be disturbed.

The trade-off is that the broader overtone and combination bands observed in the NIR spectrum can be more difficult to interpret; that is, to assign to discrete vibrations or functionalities of the interrogated material. This is the source of the (largely undeserved) reputation of inferior specificity that has been assigned to the NIR technique. While interpretation problems may arise if one has no *a priori* information as to the chemical nature of the material being analyzed, this is seldom the case for pharmaceutical applications. In these samples, the component materials are known and generally well characterized. The pure components are readily available and can be used to acquire the independent spectral references, if necessary. The real information content coupled with enhanced reproducibility has proved NIR spectroscopy to be sensitive to subtle compositional changes, even in complex samples, making it preferable for a wide range of both qualitative and quantitative pharmaceutical applications.[2–4] This aspect of NIR spectroscopy makes it especially well suited for use in chemical imaging analysis where the added value is largely realized through the comparison of a large number of spatially resolved measurements rather than the examination of individual spectra.

7.3.5 NIR imaging instrumentation

In addition to the analytical and sampling benefits of NIR spectroscopy, implementing this spectral range in a chemical instrument also poses several distinct advantages. First, as described earlier, the most dramatic improvements in commercial FPA technology have been seen for detectors that operate in this region, spurred by applications such as night-vision devices and telecom diagnostic tools. The spectroscopy communities benefit from these advances in the wider choice of reliable, larger format, and more reasonably priced 'cameras' for the chemical imaging application. The breadth of illumination source options for work in the NIR region is also excellent. DC-powered quartz-halogen lamps are typically used as sources, and are widely available in a range of powers and configurations. These sources are relatively inexpensive and flexible, and do not pose any significant safety concerns.

An ideal PAT imaging analyzer should be flexible, robust and amenable to the manufacturing environment. Figure 7.1 depicts a generalized schematic of an NIR imaging instrument incorporating a large-format FPA and a liquid crystal tunable filter (LCTF) for wavelength discrimination. This relatively simple experimental configuration offers many advantages over other chemical imaging techniques. One tremendous benefit is the more

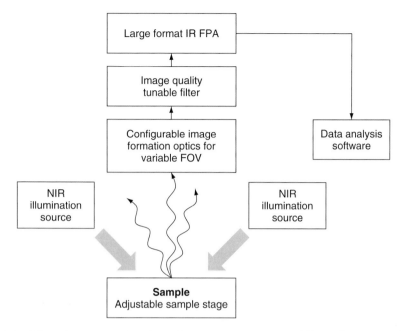

Figure 7.1   Schematic representation of NIR-CI instrument operating in diffuse reflectance mode. Radiation from the illumination source interacts with the sample. Light reflected off of the sample is focused onto a NIR sensitive 2D detector, after passing through a solid-state tunable wavelength selection filter.

relaxed demands for optics relative to MIR and Raman techniques. MIR requires the use of IR transmissive optics manufactured from exotic materials, thereby making them difficult to manufacture and typically very expensive. IR imaging systems using reflective optics are a possibility, but they are bulky and non-trivial to configure for large field of view (FOV) applications. Optics for Raman imaging applications demand high light-gathering efficiencies because of the relative weakness of the Raman signal. This makes it difficult to work with samples that are not perfectly flat. For global imaging implementations, curved or rough samples will not be in focus across the working FOV.

Working in the NIR permits the use of relatively simple refractive achromatic optics with long working distances. A standard laboratory system can be quickly configured by the operator for microscopic applications (<10 microns per pixel) or macroscopic applications (approximately 500 microns per pixel). The result is a chemical image data set comprising tens of thousands of spectra of a single pharmaceutical granule or an entire blister pack collected rapidly from the same instrument. Instruments that image even larger areas can be purpose-built for process, high throughput, or other specialized applications. Because the source used for the experimental configuration is a readily available quartz-halogen lamp, the area to be illuminated can be simply and cost-effectively scaled up.

The tunable filter technology also has distinct PAT advantages. Rapid tuning of discrete wavelengths through software control and no moving parts enables the collection of data sets comprising approximately 80 000 spectra in a matter of a couple of minutes. For most research and problem-solving experiments it may be useful to use the tunable filter to collect data over its entire NIR spectral tuning range. However, finished process applications may only require data collection over a narrow spectral range or even use just a few analytically relevant wavelengths. Under these circumstances the use of a random access tunable filter offers compelling advantages by minimizing the size of the data sets, and dramatically speeding up the data collection process. Further, once a data collection method has been streamlined and validated, it can be readily 'bundled' such that collection and analysis are integrated and accomplished in 'real-time.' These chemical imaging protocols can proceed with little or no operator intervention.

Another significant benefit of the tunable filter approach is its amenability for use in a 'staring' configuration whereby the image, not the source, is spectrally tuned or modulated. This has the benefit of resulting in an instrument that is compact and has a highly configurable FOV. When coupled with a two-dimensional IR FPA, the solid-state optical design and compact size of the LCTF allows it to be deployed in a wide range of situations. Standard commercial instruments are normally mounted in a downward-looking orientation, on an optical rail with an illumination source and a sampling stage. These instruments are suitable for installation in either the laboratory or the plant floor because they have no moving parts, and require no external utilities with the exception of power. This basic design also permits the camera, LCTF, and imaging optics to be unified into a single compact wavelength-selectable chemical imaging module which can also be mounted in side-looking or inverted orientations for more specialized applications.

In summary, the technology readily enables a broad selection of optical configurations to be used, providing ease of switching between highly disparate fields of view and the flexibility to chemically image irregular and uneven samples. Furthermore, this is achieved with a system that can collect >80 000 moderate-resolution NIR spectra in several minutes or >80 000 lower resolution spectra in a few seconds, all with no moving parts. This suggests that the generalized instrument design depicted by Figure 7.1 is possibly the most rugged and robust chemical imaging approach available.

7.4   Chemical imaging principles

Imaging, in the broad sense, refers to the creation of a representative reproduction of an object or a scene. Images are created to convey and preserve certain types of information about the scene. For example, an artist will use pigments to capture the colors of a landscape in a painting. The artist may apply shading and perspective to depict three-dimensions in a two-dimensional rendering. In the artistic example, much of the information conveyed is subjective. In the realm of science, objective information is preferred, but the painting example is useful in demonstrating that the information contained within an image is much greater than the sum of its individual spatial elements.

Conventional digital imaging is the process of reproducing the spatial information of a scene onto a two-dimensional optical detector. Typical visible grayscale or color images are collected over a broad range of optical wavelengths. On the other hand, a single *spectral* image is usually collected over a narrow wavelength range. In the IR or NIR spectral region it can be used to reveal the chemical composition of the sample through absorption by one or more chemical species within the sample at a particular 'diagnostic' wavelength. The result is a spatially resolved chemical map of the sample. Chemical (also called *hyperspectral*) imaging is the acquisition of images over a larger, usually contiguous, series of narrow spectral bands comparable to traditional (single-point) spectroscopic techniques. This is the fundamental concept of chemical imaging – a rapid analytical method that simultaneously delivers spatial, analytical (chemical), structural, and functional information.

7.4.1   The hypercube

Chemical image data sets are 'visualized' as a three-dimensional cube spanning one wavelength and two spatial dimensions called a hypercube (Figure 7.2). Each element within the cube contains the intensity-response information measured at that spatial and spectral index. The hypercube can be treated as a series of spatially resolved spectra (called

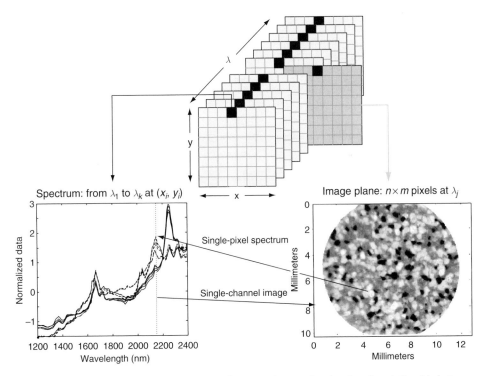

Figure 7.2   Schematic representation of spectral imaging hypercube showing the relationship between spatial and spectral dimensions. The image displayed is a single-wavelength image, and its contrast is based on the intensity of IR absorption at that wavelength. Pixels containing sample components with a strong absorption at that wavelength will appear bright. The spectra shown demonstrate how the spectral features of different sample components can be used to derive image contrast.

pixels) or, alternatively, as a series of spectrally resolved images (called image planes or channels). Selecting a single pixel will yield the spectrum recorded at that particular spatial location in the sample. Similarly, selecting a single image plane will show the intensity response (absorbance) of the object at that particular wavelength.

7.4.2   Data analysis

Hypercubes collected from a chemical imaging experiment can be quite large. For example, a moderate-format NIR FPA (320×256) collects data using $>80\,000$ individual detector elements (pixels). A full-range spectrum may contain more than 100 wavelength channels resulting in a cube with more than $8\,000\,000$ individual elements. The corresponding information content of the entire cube is enormous. It is impractical (and not particularly useful) to attempt to manually interpret each element independently. Therefore the analysis software used to mine and portray the salient information from the hypercube becomes an integral part of the chemical imaging experiment. The package must be specifically designed to handle these large, multidimensional data sets, and ideally incorporates a complete set of image and spectral processing options, as well as chemometric analysis procedures, and image visualization tools.

Armed with a complete set of software tools, there is a seemingly endless range of possible processing paths that can be employed, and there is no set method that is universally applicable for all data sets. However, there are four basic steps that are customarily used for the analysis of chemical imaging data. They are: spectral correction, spectral pre-processing, classification, and image analysis. The first two steps are somewhat universal for all comparative spectroscopic analyses. Classification techniques are also common in spectral analyses, but their application in imaging may differ from more familiar methods. Image analysis serves as the departure from 'spectral' thinking and is used to evaluate and quantify the heterogeneity (contrast) developed in the image from the first three steps. There are no firm requirements to use all of the steps – they merely serve as a guideline for the general flow of data work-up. Later in this chapter we will present two 'case studies,' showing some typical data work-ups. For each case study, the desired and derived information is different, and thus illustrates strategies for data processing.

Due to the large information content of an image cube, there is usually no way to investigate and present all of the details simultaneously. As is true in approaching any large set of analytical data, efficient analysis of chemical images requires at least some general goals with regard to the information being sought. These will direct the analyst to choose certain processing tools that are appropriate for the desired analysis. For poorly characterized samples, it is often necessary to use an iterative approach, using the initial results as a guide to develop more appropriate processing and analysis methods. Also, because typically there is more than one processing path that can be used to elucidate particular details, getting the same result with a different method can be used as a tool to confirm results and conclusions.

While this is not the place for a complete description of the theory, uses, and ramifications of all possible processing treatments that can be applied to chemical imaging data, the majority of NIR chemical images recorded to date have been obtained from solid or powder samples, measured predominantly by diffuse reflectance. As such, the following discussions may serve as a general outline of the more common procedures and their application to these types of imaging data.

7.4.3 Spectral correction

As for all absorbance-based spectral measurements, the intensity data represented in a raw (single beam) chemical image are a combination of the spectral response of both the instrument and the sample. In order to remove the instrument response component, it is necessary to ratio the data to a background reference. For reflection measurements, the background is a separate data cube typically acquired from a uniform, high-reflectance standard such as Spectralon^{TM} (Labsphere, Inc., North Sutton, NH) or white ceramic. Since the filter-based NIR-CI instruments operate in a true 'staring' or DC mode, it is also necessary to subtract the 'dark' camera response from the image cube. The correction to reflectance (R) is therefore:

$$R = (\text{Sample} - \text{Dark})/(\text{Background} - \text{Dark})$$

Further processing is also usually performed to transform the reflectance image cube to its $\log_{10}(1/R)$ form, which is effectively the sample 'absorbance.' This results in chemical images in which brightness linearly maps to the analyte concentration, and is generally more useful for comparative as well as quantitative purposes. Note that for NIR measurements of undiluted solids the use of more rigorous functions such as those described by Kubelka and Munk[32] are usually not required.[31]

7.4.4 Spectral pre-processing

The primary objective of pre-processing treatments is to remove the non-chemical biases from the spectral information. Most diffuse reflectance measurements will carry some physical and/or optical influence from the interrogated sample. Scattering effects induced by particle size or surface roughness may lead to offsets or other more complex baseline distortions. Differences in sample density or the angle of presentation may induce overall intensity variations as well as additional baseline distortions. Most samples do not present perfectly uniform surfaces on the microscopic scale, and due to the highly localized nature of the individual measurements, these effects may dominate the initial contrast observed in an un-processed chemical image. In some cases this contrast may provide useful information to the analyst; however, it is generally removed in order to focus on the chemical information.

For physically uniform samples, simple techniques such as offset or polynomial baseline correction may suffice to minimize the physical/optical artifacts from the spectral data, but there are also a number of more advanced techniques available, which are quite effective. For example, it is common to convert the spectral data to a Savitzky-Golay[33] first- or second-order derivative form. This conversion minimizes the baseline and random noise contributions while highlighting subtle band shape and wavelength position information, often aiding in the resolution of overlapping bands. It also has the benefit of preserving the relative band intensity information necessary for further quantitative treatments of the data. For powder samples multiplicative scatter correction (MSC)[34] can be very effective in correcting baseline as well as any intensity artifacts due to scattering. It may be used by itself, or as a preliminary step to taking spectral derivatives.

Spectral normalization techniques are another set of powerful tools for pre-processing chemical imaging data. The most common of these is mean-centering and/or normalizing the spectral intensities to unit variance. Such treatment will virtually eliminate baseline offsets and physical intensity variations. By design, it will rescale the spectral intensities to a common range. This is useful in comparing spectra from different pixels and may enhance the chemical contrast in samples that contain materials exhibiting widely varying overall absorbance properties. However, rescaling will also distort the quantitative information contained in the spectral intensities, and this pre-processing approach should be used with caution if quantitative assessments of the individual spectra is required during further analysis.

7.4.5 Classification

Classification is the process of using spectral differences of the chemical species that comprise a sample to highlight regions of spectral, and therefore chemical, similarity within that sample. The result is a chemical image that visualizes these distributions. In some cases, simply selecting an image plane at a wavelength of a characteristic absorption band for a particular species of interest will provide sufficient contrast to readily indicate the spatial distribution and relative abundance of that material. This is called a *univariate* approach and is the simplest and most direct way of probing the sample. This data interrogation method is fast and intuitive, but it requires that the spectra from each of the various species display a region of relatively unique absorbance. Univariate analysis often falls short of discerning finer distribution details in more complex samples, as examination of the disassociated image planes is wasteful of the total information content of the complete hypercube. Since the cube is composed of full spectra, the data are well suited for *multivariate* analysis techniques. These methods utilize all or distinctive regions of the spectral range in order to better distinguish and deconvolve the overlapping spectral responses of the individual species, even in very complex samples.

Multivariate analysis of single-point NIR spectra has become a mainstay for a wide range of pharmaceutical applications.[3,4] These same techniques are also applicable to NIR-CI data, albeit with a somewhat different emphasis. Single-point methods are generally based on a relatively small number of individually collected measurements. Much of the work in constructing a robust method actually depends on producing and measuring a sufficient number of individual reference samples. NIR-CI data sets, on the other hand, represent tens of thousands of individual measurements acquired simultaneously from a single sample. This difference has several important ramifications. In single-point applications, the value of a multivariate model lies in its ability to extract highly accurate information from a single-sample spectrum. In imaging applications where the goal is to generate chemical contrast, the focus is shifted from the individual predictions to the relationships between the vast numbers of predictions representing the image. In this case selectivity is the primary concern. Even in more 'quantitative' applications, the sheer statistical superiority of the large imaging data sets enhances the relative accuracy and robustness of most multivariate methods. Additionally, the high spatial resolution of the individual pixel measurements often will yield relatively pure spectra that may be used as 'internal' references for the analysis.

Multivariate analysis tools are available in many commercial software packages and, in general, one does not have to be an 'expert' to utilize them effectively. However, a general familiarity with the basic assumptions and limitations of the algorithms at hand is useful in both the selection of appropriate methods and the interpretation of the results. Many good texts are available which discuss these methods from the fundamental mathematical origins,[35,36] through general spectroscopic[37–40] and imaging applications.[41]

The multivariate tools typically used for the NIR-CI analysis of pharmaceutical products fall into two main categories: pattern recognition techniques and factor-based chemometric analysis methods. Pattern recognition algorithms such as spectral correlation or Euclidian distance calculations basically determine the similarity of a sample spectrum to a reference spectrum. These tools are especially useful for images where the individual pixels yield relatively 'unmixed' spectra. These techniques can be used to quickly define spatial distributions of known materials based on external reference spectra. Alternatively, they can be used with internal references to locate and classify regions with similar spectral response.

Factor-based chemometric methods are often more appropriate for the analysis of images where there is significant spectral mixing in the individual pixels, as these algorithms are specifically designed to account for the varying spectral contributions of multiple materials in individual (pixel) measurements. Principal components analysis (PCA) is an 'undirected' algorithm meaning that it does not use a pre-defined set of references or a spectral library to interrogate the sample spectra set. Instead, this method defines the significant sources of variance within the sample set as a series of ranked components or factors and assigns each spectrum (pixel) a score (brightness) based on the relative contribution of each factor. PCA analysis is often a very effective method for highlighting subtle spectral variations between different objects or regions in a sample scene. It can also serve as an efficient preliminary analysis step by pointing the user to significant spectral and spatial regions that may be further analyzed by other methods.

Directed chemometric algorithms such as principal component regression (PCR) and partial least squares (PLS) analyses utilize a pre-defined set of reference spectra (a library) to discriminate discrete spectral contributions attributable to each of the materials. The resulting models are then applied to the sample image data to produce score predictions that relate to the abundance of each of the materials in the individual measurements. Traditional quantitative NIR applications employing these methods are typically calibrated using a large set of reference samples with a range of varying component concentrations. More qualitative 'classification' models are easily developed from a simpler library composed of pure component spectra. While either approach is applicable to NIR-CI data, the 'classification' mode is generally preferred. This is not only because it is easier, but also because the qualitative classification of the individual pixels generally produces the contrast necessary to effectively characterize the individual component distribution. Additionally, imaging data enjoy a certain 'statistical robustness' due to the large number of individual pixels representing the sample. Therefore, it is often possible to extract bulk quantitative information from the mean value of the distribution of qualitative score predictions.

7.4.6   Image processing

Image processing refers to the techniques used to convert the contrast developed in the image into information which is of value to the analyst. One of the goals of image analysis is to produce a 'picture' depicting the distribution of one or more components. After the chemical contrast between the individual pixels has been revealed using the techniques described previously, it can be displayed more effectively through the choice of an appropriate color map and intensity scaling of that map. For example, a gray-scale map can be used to simply convey a range of IR absorbances, or a pseudo-color scheme may aid the eye in associating different regions of similar intensities. The visualization quality of the image may also be improved by masking or truncation to remove irrelevant regions of the image. Conversely, concatenation of multiple images can assist in the direct comparison of samples collected in different measurements. Another useful technique for the visualization of the distribution of multiple components simultaneously is to produce a red-green-blue (RGB) composite image. In this method each color of the map represents the contrast developed for a separate component.

One aspect of sample analysis that is somewhat unique to imaging data is the ability to statistically characterize the distributions within the sample through the interrogation of the histogram of intensities associated with the image. The large number of individual measurements represented by a single NIR chemical image produces statically significant histogram distributions, often providing objective and quantitative information unobtainable through standard spectral analysis methods. The image histogram can be analyzed at many different levels depending on the nature of the sample at hand. For high-contrast images, the histogram will contain two or more clearly defined modes. Simply comparing the number of pixels contained within the modes (often called pixel-counting) can provide a direct method for the determination of bulk 'abundance,' an estimate of the sample concentration. When image contrast is based on quantitative or semi-quantitative classification techniques such as PCR or PLS modeling, the mean of the resulting score distribution can provide accurate abundance estimations. Non-normal distribution of the histogram can also be characterized mathematically and may provide more detailed insight into the nature of the distribution of particular materials within the sample, often useful for defining subtle differences between similar samples.[16] A more thorough discussion of image histogram analysis is provided in one of the case studies that follows in this chapter.

Another image-processing method that is especially applicable to many pharmaceutical sample images is particle statistics analysis. This technique works well when the individual component distribution appears in the form of discrete domains in the processed image. Such domains may represent actual particles of material, but more often are indications of agglomerations of the components either intentionally or inadvertently produced by the manufacturing process. The particle analysis tools permit the analyst to quickly characterize the statistical distribution of the domain sizes based on particle area or diameter, and can also be used to characterize domain shapes, orientations, and relative separation distances. This information can be vital in the elucidation of product performance differences between samples that have identical chemical assays and can be used to relate the impact of the different unit operations on the composition of the final dosage form.

7.5   PAT applications

In this section we would like to focus on the attributes of NIR-CI that make it such a powerful tool for PAT applications. Chemical imaging approaches provide tremendous capabilities due to the high-throughput nature of the technique. For example, 'self-calibrating' content uniformity measurements can be performed rapidly, counterfeit products can be identified amongst real products, intact blister packs can be screened for QA, and trace amounts of contaminants, including different polymorphic forms, can be identified and visualized within finished products. This is in addition to performing 'typical' imaging applications such as investigating the spatial and chemical structure of dosage forms for root-cause analysis of manufacturing problems.

7.5.1   'Self-calibrating' high-throughput content uniformity measurements

In this example, the FOV of the experiment encompasses a large area of approximately 9×7 cm. This large FOV enables a matrix of samples (20 tablets and 3 pure components) to be simultaneously evaluated. The goal for this example is inter-sample comparison, using the pure component samples as a built-in calibration set. The capability to simultaneously view both samples and pure components is unique to array detector-based chemical imaging instrumentation. Data were collected using a Sapphire^{TM} (Spectral Dimensions, Inc., Olney, MD, USA) chemical imaging system in the spectral range of 1100–2450 nm at 10-nm increments with 16 frames co-adds. The FOV covers approximately 9×7 cm, with each pixel sampling an area of approximately 275×275 µm (81 920 pixels).

To demonstrate the ability to evaluate inter-sample variations, tablet groups from two different manufacturers in an over-the-counter (OTC) pain relief medication were compared. Pure acetaminophen, aspirin, and caffeine samples are obtained in either tablet form or powder compact and included within the same FOV as the tablets to provide simultaneous reference materials for the tablet samples. The tablets and pure components were arranged as shown in Plate 1a. This FOV contains 20 tablets and the three API pure components. Measurements on all samples were made simultaneously. Tablet groups labeled A and B contain three APIs each: acetaminophen, aspirin, and caffeine. Tablet A samples from one manufacturer have a reported label concentration of 37, 37, and 10% for the three API components, respectively. Tablet B samples from the second manufacturer contain the same three APIs, at label concentrations of 39, 39, and 10%, respectively. In addition to these samples, tablet C samples are included in the array of tablets. These samples contain only acetaminophen as the API with a reported label concentration of 79%, and are made by the manufacturer who produces tablet A. The remaining mass of all three tablet types represents the excipient (binder, disintegrant, and lubricant) materials.

The raw images were dark corrected as described previously, then background corrected using a data cube collected from Spectralon^{TM} (Labsphere, Inc., North Sutton,

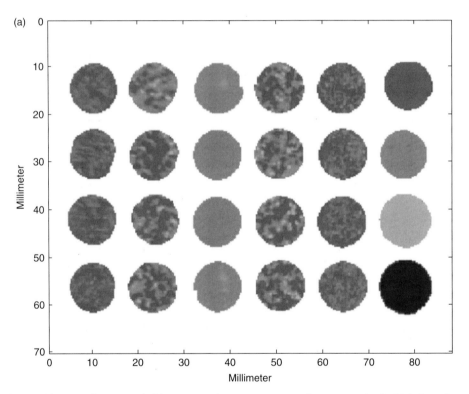

Plate 1 (a) Key to placement of tablet matrix and pure component reference samples for high-throughput inter-tablet comparison. Tablet groups A and B are all tablets of an OTC pain relief medication, with A and B from two different manufacturers. The tablets contain three APIs, acetaminophen, caffeine, and aspirin. Group C is composed of tablets that have only one of the three APIs contained in tablet groups A and B. The API pure components are positioned in the right column of the tablet matrix. (b) Summary image in which the distribution of the three APIs are shown simultaneously. Acetaminophen is shown in red, caffeine in green, and aspirin in blue, and the three color channels are overlaid. (This figure is produced in color in the color plate section, which follows page 204.)

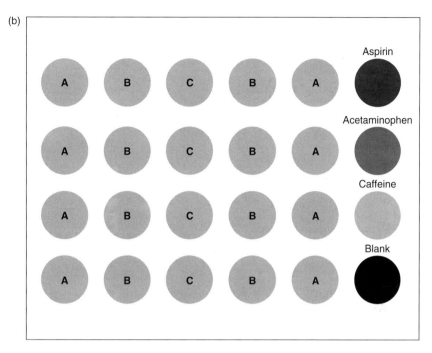

Plate 1   (Continued).

NH) as the reference, and converted to $\log_{10}(1/R)$. A Savitzky-Golay second derivative was performed to eliminate contrast due to baseline offset and slope, and the resulting spectra mean-centered and normalized to unit variance. A spectral reference library for the experiment was created from the sample data set by selecting pixels from regions of the sample matrix containing the pure component materials. A PLS type two (PLS2) model was constructed from the reference library and used to generate PLS score images for each of the library components.

Plate 1b shows the RBG image created from the PLS score images, with acetaminophen assigned to the red channel, aspirin to blue, and caffeine to green. The channels are scaled independently, and only represent qualitative information about the component distributions. This image highlights the advantages of having reference samples in the same FOV. The results can immediately be visually confirmed – the acetaminophen pure sample is solid red – indicating that the PLS score is accurately identifying that component. The same holds true for aspirin (blue), and caffeine (green). Additional confirmation is that the samples labeled tablet C show all red with no blue or green, confirming that the model is correctly classifying acetaminophen. Creating single data sets that contain both the sample and the reference materials simplifies many aspects of the experiment. The data set is 'internally' consistent because the reference and the sample data are measured simultaneously by the same instrument – temporal variations in instrument response are automatically accounted for. Also since the relevant data always remain together in a single image data file, the sample and the reference data are always processed together in identical ways and cannot be separated. The reference samples may be left in place when new samples are introduced, and therefore a new experiment simultaneously results in new reference spectra. Any changes that effect instrument response or data collection parameters result in those changes being reflected in the reference sample spectra. This obviates the need to apply transfer of calibration procedures to analyze new samples. Moving beyond qualitative information, quantitative information can be gleaned by examining the statistical properties of these tens of thousands of spectra collected in a single data set, to allow many inter-tablet and intra-tablet quantitative comparisons to be made. Even by using low (high-throughput) magnification it is possible to glean intra-tablet uniformity information since approximately 1100 NIR spectra are recorded for each of the 24 tablets.

7.5.2   High-throughput applications: Counterfeit screening/quality assurance

Figure 7.3 demonstrates another example of utilizing NIR-CI as a 'high throughput' technology for use in the pharmaceutical manufacturing process. As in the previous example, the highly flexible FOV and robust imaging capability of the technique is used to great value to spatially analyze a larger array of samples simultaneously. In this example, a sealed blister pack approximately 7×11 cm in size containing ten visually identical white tablets is analyzed. Prior to the experiment, one of the tablets was carefully removed and replaced with one of identical form factor containing a different API. Figure 7.3a is a visible image of the adulterated package. The NIR spectral image of the complete sample scene was collected without disturbing the samples or the packaging. The PCA score image of the arrangement in Figure 7.3b clearly identifies the rogue tablet containing the different active ingredient. While the result of this measurement is an image, the data can also be considered as simply a series of NIR spectra taken in parallel. We can

(a)

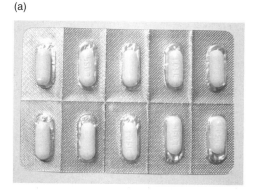

Figure 7.3 Adulterated pharmaceutical blister pack containing a single 'rogue' tablet. (a) Visible image showing that visually the tablets are virtually indistinguishable; and (b) NIR PCA score image highlighting the single tablet that has a different chemical signature.

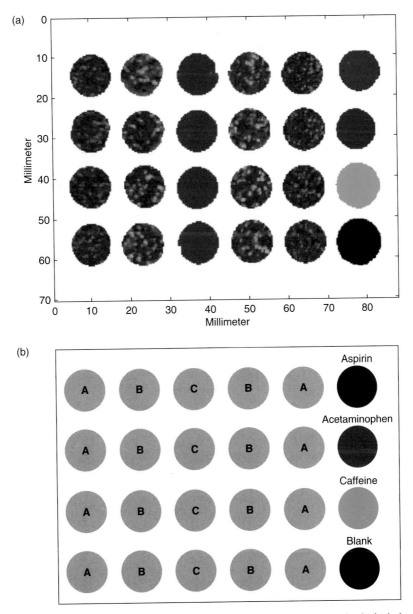

Plate 1 (a) Key to placement of tablet matrix and pure component reference samples for high-throughput inter-tablet comparison. Tablet groups A and B are all tablets of an OTC pain relief medication, with A and B from two different manufacturers. The tablets contain three APIs, acetaminophen, caffeine, and aspirin. Group C is composed of tablets that have only one of the three APIs contained in tablet groups A and B. The API pure components are positioned in the right column of the tablet matrix. (b) Summary image in which the distribution of the three APIs are shown simultaneously. Acetaminophen is shown in red, caffeine in green and aspirin in blue, and the three color channels are overlaid.

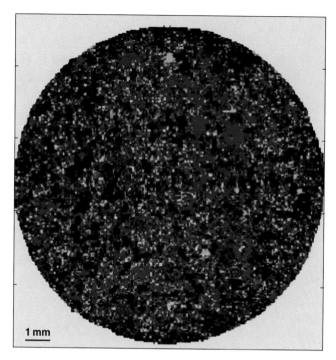

Plate 2 Trace analysis of polymorphic form 1 of ranitidine. The RGB summary image highlights the distribution of sample components: polymorphic form 2 of ranitidine in red, polymorphic form 1 of ranitidine in green and the excipients in blue.

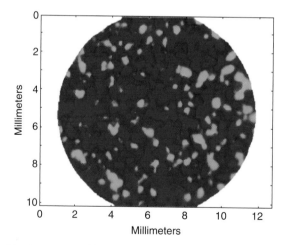

Plate 3 Red-Green-Blue image created from PLS score images of three APIs of an OTC analgesic tablet (Excedrin™). Red highlights acetaminophen-rich areas, green highlights caffeine-rich areas, and blue highlights aspirin-rich areas.

(b)

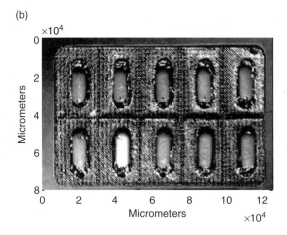

Figure 7.3   (Continued).

therefore interpret the data as a quality control measurement performed on all ten tablets simultaneously, with the number of 'objects' that can be simultaneously addressed in this manner limited only by magnification and the number of pixels on the IR array. Simple, bulk quantitative information can also be derived: 90% of the objects are correct for that blister pack, 10% of the objects are incorrect.

There are numerous applications where such a high-throughput system could simply replace single-point or multiplexed NIR spectrometers, by providing significant advantages in sample throughput and handling. These advantages are operative over a range of distance scales and sample shapes that are difficult or impossible for current single-point instrumentation to address. One such example is the validation of the filling of clinical trial blister packs in which a series of variable dosage and/or placebo dosage forms are handled and packaged in pre-defined geometric arrangements. The intent of such an exercise is to provide a highly controlled dosage regimen in a manner which is 'blind' to the patient and also, in the case of double-blind trials, to the dispensing physician. In order for this to be feasible, the physical size and shape of all types of dosage forms must be indistinguishable. Errors can void a clinical trial, imposing significant costs in both immediate resources and lost time to market for a new drug. The ability to rapidly verify the accuracy of this complex packaging process after the product is sealed therefore offers significant benefit. In many cases this array-based approach can be optimized so that only a few carefully selected wavelengths are needed for sample identification and confirmation, yielding an inspection time for a complete sealed package on the order of seconds. The ability to perform this measurement close to real time is critical if 100% inspection is the goal. In addition, the large, flexible FOV of an array-based approach enables a variety of package shapes and orientations to be analyzed with little or no reconfiguration. Figure 7.4 shows the ease with which difficult packaging problems are handled with an imaging approach. Unlike a single-point or multiplexed system, the redundancy in the number of detector elements makes the system relatively insensitive to complex spatial patterns or sample placement reproducibility.

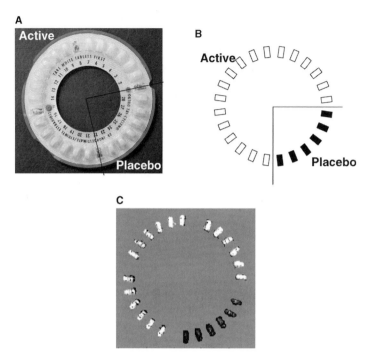

Figure 7.4 Contraceptive contained in the original package, demonstrating high-throughput classification possibilities. (a) visible image; (b) description of sample positioning; and (c) NIR PCA score image showing contrast between the placebo and the active-containing tablets.

7.5.3 Defeating sample dilution: Finding the needle in the haystack

In addition to high-throughput capabilities, chemical imaging also has the ability to detect low concentrations or weakly absorbing species. It is often not possible to detect these species in the mean spectrum obtained through bulk, macro-scale spectroscopy. This is because the distinguishing spectral features are simply lost in the average of the stronger signals obtained from the rest of the sample. Microscopic approaches can provide highly localized spectra, effectively increasing the relative concentration (and thus, the detectivity) of a scarce component in the sampled area. However, this requires the selection of the 'correct' location for detection. Therefore, either one must have prior knowledge of the location of the impurity, or it must be found by time-consuming mapping techniques. NIR-CI collects spatially organized microspectra from large areas of the sample in a single experiment. The full area of a typical 1-cm diameter tablet sample can be interrogated with a spatial resolution of about $40 \times 40\,\mu m$ in only a few minutes, providing both enhanced sensitivity and location information. Even if a rouge material is detected but is unidentifiable by its NIR spectrum alone, the location information derived from the imaging experiment can greatly assist in the rapid analysis of the suspect region by other analytical methods.

To illustrate this point, two examples are presented: the first identifying a low level of impurity in a tablet, and the second detecting low levels of a particular polymorphic form. Figure 7.5 summarizes the NIR-CI analysis of a tablet that contains a single API (furosemide) in a mixture of excipients. Furosemide has been shown to degrade with exposure to light,[42] resulting in a finished product that also contains a low level of contaminant – the degradation product. A PLS2 model was constructed using library spectra of pure components, acquired separately. The resulting PLS score image for furosemide is shown in Figure 7.5a. The NIR spectra of the API and the degradation

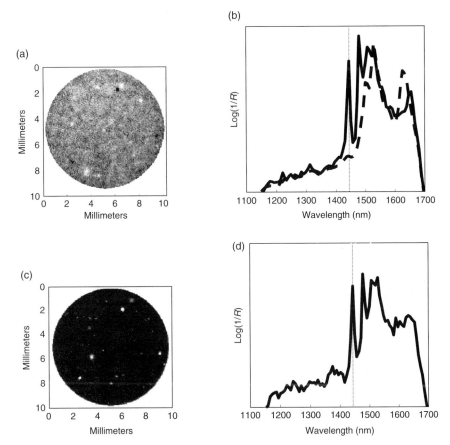

Figure 7.5 Trace contaminant analysis of furosemide and its breakdown product in a tablet. (a) PLS score image highlighting furosemide-rich areas in tablet; (b) Dashed line spectrum is the pure API spectrum. Solid line spectrum is a spectrum of the degradation product. The dotted vertical line on the graph shows that the largest spectral difference occurs at 1444 nm; (c) Normalized and scaled single-channel image at 1444 nm highlighting the impurity rich areas. The impurity concentration derived from sample statistics is ∼0.61%; and (d) Solid line spectrum is a residual microspectrum observed at the center of one of the bright spots after subtracting the variance due to the predominant chemical component (the excipients) from the entire data set. This spectrum compares well to the spectrum of the degradation product in panel 7.6b.

product are overlaid in Figure 7.5b. The latter exhibits a prominent spectral feature at 1444 nm, which readily distinguishes it from both the API and the excipient materials. Since the impurity is expected to occur in very low concentrations, the spectral contribution from the drug material was enhanced in the image through a scaled subtraction of the spectrum of the bulk (excipient) material from each pixel in the original hypercube. The corrected single-channel image at 1444 nm of the residual data clearly shows several small areas of increased intensity distributed across the sample area (Figure 7.5c). Single-pixel residual spectra from these regions closely resemble the known spectrum of the breakdown product (Figure 7.5d) and verify its presence in the sample tablet. The overall abundance of the impurity in the sample was calculated through pixel-counting statistics and estimated to be approximately 0.61% of total tablet area.

In other experiments, we have also used this approach with some success to determine the existence and localization of low levels of different polymorphic forms within finished products. Plate 2 shows an image of a sample containing both the polymorphic forms (1 and 2) of the API ranitidine. The overall composition of the tablet is 75% ranitidine (both forms) and 25% excipients. Plate 2 highlights the form 1 domains in green, whose abundance is estimated to be approximately 1.2% of the total tablet weight. This information is derived from the analysis of a single tablet.

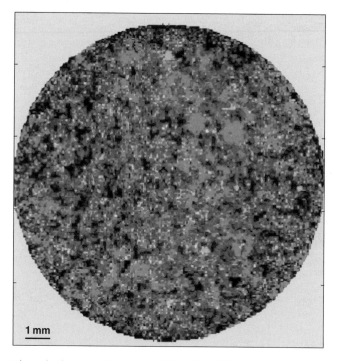

1 mm

Plate 2 Trace analysis of polymorphic form 1 of ranitidine. The RGB summary image highlights the distribution of sample components: polymorphic form 2 of ranitidine in red, polymorphic form 1 of ranitidine in green and the excipients in blue. (This figure is produced in color in the color plate section, which follows page 204.)

7.5.4   Advanced dosage delivery systems

The NIR-CI analysis of an extended-release dosage form is shown in Figure 7.6. This product is a capsule filled with small microspheres containing the therapeutic agent to affect the extended-release profile. A sample of the microspheres was removed from a capsule and imaged with the Sapphire™ NIR-CI system. The corrected single-channel image at 2130 nm (Figure 7.6a) indicates that there are two 'types' of spheres present in the sample, each having a distinctly different spectral response (Figure 7.6b). The solid line spectrum is from a 'bright' bead, and the dashed line spectrum is from a 'dark' bead. Quantitative information about the dosage form was obtained from the image by simply

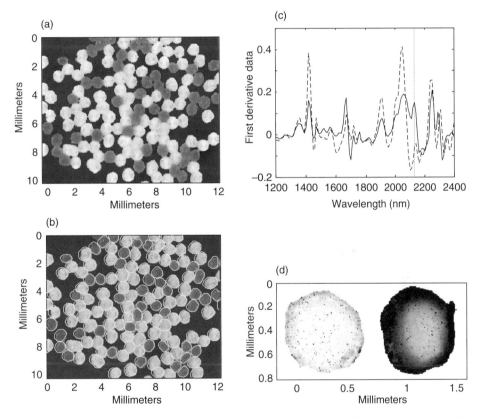

Figure 7.6   Microbeads from extended-release dosage form. Whole beads at low magnification showing two distinct 'types.' The contrast is derived from differences in their NIR spectra. (a) Single-channel image at 2130 nm highlights differences between beads; (b) Single-pixel microspectra of the two bead 'types.' A dashed line spectrum is from a 'dark' bead and a solid line spectrum from a 'bright' bead. The dotted vertical line on the graph shows that the largest spectral difference occurs at 2130 nm; (c) Image showing intensity-based particle contours from which bead size statistics may be determined; and (d) High-magnification chemical images showing spatial differences in the composition of the two 'types' of beads identified in Figure 7.6a. The left- and the right-hand beads correspond respectively to the 'dark' and the 'bright' beads from panel 7.6a.

setting the appropriate intensity level so as to define each of the observed spheres as discrete particles (Figure 7.6c). Automated image analysis indicated that there were twice as many 'bright' (91) as 'dark' (45) spheres in this particular FOV, and that the 'bright' spheres are, on average, approximately 5% larger in diameter.

The release mechanism of the form was elucidated through further microanalysis of the individual sphere types. One of each 'type' of sphere was cut in half and the cut faces were chemically imaged at higher magnification. The resulting images (Figure 7.6d) reveal that the 'dark' sphere is relatively homogeneous while the 'bright' sphere is made up of a coating that surrounds a core composed of the same material found in the 'dark' spheres. This coating presumably comprises the extended-release agent.

It is readily apparent that the spatial information provided by NIR-CI is invaluable in both the development and the manufacture of complex delivery systems such as this product. Note the relatively uneven nature of the coating observed in the image of the 'bright' sphere in Figure 7.6d. It ranges in thickness from approximately 150–250 μm. Imaging analysis during the development and scale-up of the process which produces the coated spheres can provide a clear picture of this component and allow the 'critical control parameters' to be quickly identified, improving quality and reducing the risk of costly performance failures. During manufacturing, chemical imaging can be applied to monitor the mean and the standard deviation of coating thickness uniformity, the mean and the standard deviation of the microsphere diameter, as well as the relative composition of mixtures of different types of microspheres. These data can be used to enable the desired composition to be optimized and maintained.

7.6 Processing case study one: Estimating 'abundance' of sample components

One of the unique capabilities of chemical imaging is the ability to estimate the 'abundance' (an approximation of concentration) of sample components that comprise a heterogeneous pharmaceutical sample *without* developing a standard concentration calibration model. NIR-CI can sample an entire tablet surface with >80 000 spectral data points. For a spatially heterogeneous sample, a precise calibration model may not be necessary *because* of this density of sampling. The classification of 80 000 spectra across the sample surface provides a statistically robust measurement of the distribution of sample components. For tablets that are extremely heterogeneous relative to the experimental magnification (typically approximately 40 μm/pixel), sample components appear to aggregate in domains of pure components. A single-pixel spectrum can be examined, and classified as belonging to one species or another. For pharmaceutical samples that are well blended, a better approach is to use a multivariate algorithm to determine the mixture of components for each pixel in the data set. When summarized across all pixels, the contribution of each sample component can be determined. In this example, the abundance of three APIs is calculated for an OTC pain relief medication. A multivariate analysis gave the best results when comparing the estimated composition to the label concentrations.

7.6.1 Experimental

The imaging data were collected over the spectral range of 1200–2400 nm at 10-nm increments using a Sapphire™ NIR-CI system (Spectral Dimensions, Inc., Olney, MD, USA). The size of FOV was 13 × 10 mm, each pixel sampling an area of approximately 40 × 40 μm. The concentrations of the active ingredients as reported from the manufacturer's label are 37% acetaminophen, 37% aspirin, and 10% caffeine. The remainder of the tablet mass represents the excipient (binder, disintegrant, and lubricant) materials. Pure acetaminophen, aspirin, and caffeine samples are obtained in either tablet form or powder compact and are used to obtain reference spectra of pure components.

7.6.2 Spectral correction and pre-processing

The data cube was converted to $\log_{10}(1/R)$ as described in the section on data analysis, using background and dark-cube correction.

As described above, pre-processing steps minimize effects due to optical or physical differences in the sample. Differences in packing density or particle sizes create baseline offset and slope differences, just as for single-point spectroscopy. In chemical imaging techniques, other physical issues such as non-uniform lighting also can influence the data. For chemical imaging, physical contributions to spectra impact the contrast in resulting images. The naïve user that does not perform appropriate pre-processing can be fooled into thinking that image contrast is chemical in nature, when actually it is physical. In this example, to eliminate baseline offset and slope effects, a series of pre-processing algorithms were applied. First, a Savitzky-Golay second derivative was taken; second, the data were mean-centered. Finally, the spectra were normalized to unit variance. It is worth noting that the strong NIR positive-going bands in $\log_{10}(1/R)$ spectra become strong negative-going bands in second derivative data.

7.6.3 Analysis

Traditional macro-scale NIR spectroscopy requires a 'calibration' set, made of the same chemical components as the target sample, but with varying concentrations that are chosen to span the range of concentrations possible in the sample. A concentration matrix is made from the known concentrations of each component. The PLS algorithm tries to create a model that best describes the mathematical relationship between the reference sample data and the concentration matrix, and applies that transformation model to the 'unknown' data from the target sample to estimate the concentration of sample components. This is called 'concentration mode PLS.'

These same analysis techniques can be applied to chemical imaging data. Additionally, because of the huge number of spectra contained within a chemical imaging data set, and the power of statistical sampling, the PLS algorithm can also be applied in what is called 'classification mode.' In this case, the reference library used to establish the PLS model is made from pure component spectra of each sample component in the target sample, and does not include mixture examples. Because the library is composed of pure component spectra, the membership matrix is made of 1s and 0s: a '1' indicates membership to a certain class (or chemical component) and a '0' indicates no similarity to that class. The PLS algorithm works the same as it does with a real concentration matrix with genuine mixture spectra. When the model is applied to the target sample data, each spectrum is scored relative to its membership to a particular class (i.e. degree of purity relative to a chemical component). Higher scores indicate more similarity to the pure component spectra. While these scores are not indicative of the absolute concentration of a chemical component, the relative abundance between the components is maintained and can be calculated. If all sample components are accounted for, the scores for each component can be normalized to unity, and a statistical assessment of the relative abundance of the components made.

A PLS2 analysis was employed, based on a library of the three API pure components. Applying the model in 'classification mode' to the sample data set results in PLS score images that show the spatial distribution of the three API components. The PLS score images for acetaminophen, aspirin, and caffeine highlighting high concentration areas for each chemical component are shown in Figures 7.7a–c. Contrast in these score images is based on the value of the PLS score of each pixel relative to the respective component. A pixel with high aspirin concentrations will have a high aspirin score, and will appear bright in the aspirin score image. To verify that the model is working in a qualitative sense, a comparison is made between pure component spectra of the APIs and single-pixel spectra from acetaminophen-, aspirin-, and caffeine-rich areas shown in their respective score images (Figures 7.7d–f).

Qualitatively, acetaminophen and aspirin appear to be more abundant than caffeine, which concurs with the label concentration values. Caffeine shows very distinctive areas of localized high concentration domains while aspirin is relatively evenly distributed on the spatial scale of the image. Acetaminophen appears to be somewhat in the middle, showing up as large domains that blend into one another.

To quickly see how the three components overlap spatially, it is possible to create a three-color RGB image in which each component is assigned to a separate color channel. This is a useful technique to correlate the spatial distributions of components. Often pharmaceuticals are formulated so that particular components or combinations of components form spatial aggregates. Imaging can verify that this association of specific components is taking place as designed. Acetaminophen is assigned to the red channel, aspirin to blue, and caffeine to green. To create the composite, each of the individual color channels is scaled independently, so any quantitative information is lost. When overlapped, they form the RGB image shown in Plate 3, in which none of the three components show overlapping or associative spatial distributions.

It is extremely useful to move beyond a subjective and qualitative analysis of the spatial distribution of sample components, and to begin to explore the quantitative information contained within chemical imaging data sets. One of the most powerful statistical representations of an image does not even maintain spatial information. A chemical image can be represented as a histogram, with intensity along the x-axis and the number of pixels with that intensity along the y-axis. This is a statistical and quantitative representation of a chemical image. Since each image contains tens of thousands of pixels from across the sample, the statistical sampling is quite reliable.

For the acetaminophen and aspirin PLS score images, histograms show distinctive bi-modal normal distribution curves (Figure 7.8a,b, respectively). The bright pixels in the image are represented in the high-intensity populations in the histograms. A threshold can be applied below which pixels are estimated not to belong to the class of interest. These thresholds are set at the inflection point between the bi-modal distributions. The threshold in the acetaminophen score image was set at a score value of 0.43. By counting the pixels above this threshold, 22 813 pixels were estimated to be acetaminophen-containing. The threshold for aspirin in its corresponding score image was 0.36, and 23 058 pixels were classified as aspirin-containing pixels. The number of pixels that are classified as acetaminophen or aspirin divided by the total number of pixels in the tablet yields an estimation of abundance.

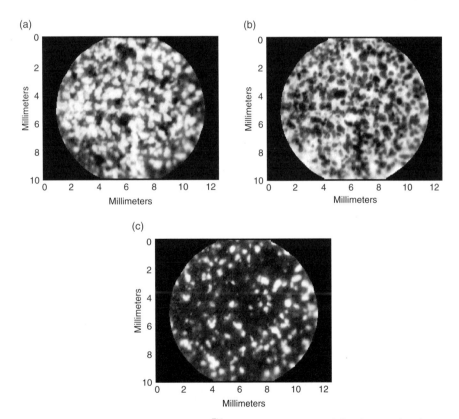

Figure 7.7 An OTC analgesic tablet (Excedrin™) with three APIs. Spatial distribution of each API was obtained using PLS analysis. (a–c) PLS score images of acetaminophen, aspirin, and caffeine, respectively; and (d–f) Single-pixel microspectra (solid line) in comparison with pure component acetaminophen, aspirin, and caffeine spectra (dashed line), respectively.

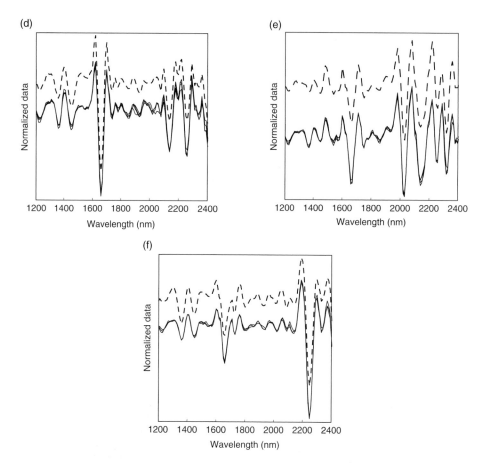

Figure 7.7 (Continued).

The histogram representation of the caffeine score image (Figure 7.8c) does not clearly show a bi-modal distribution of pixels, even though the image shows the clearly distinctive caffeine domains. In other words, morphologically it is clear which pixels are contained within caffeine domains and which are not, but these populations are not clearly separated in the histogram. Obviously, a different approach is needed to estimate caffeine abundance. Returning to the image, caffeine domains can be selected by connecting pixels that have the same PLS score. Rather than choosing a threshold based on a statistical distribution, morphology in the image is used as the metric. Using standard image-processing procedures, polygons are drawn around the caffeine domains (Figure 7.9). A less subjective threshold is now determined by looking at the image, and iteratively choosing a PLS score value so that the bright areas in the image are included in caffeine domains, and the darker pixels are excluded. In this example, pixels with a score above 0.45 are classified as caffeine-containing, totaling 6755 pixels. From further investigation of the image statistics for the particles, one can determine the following

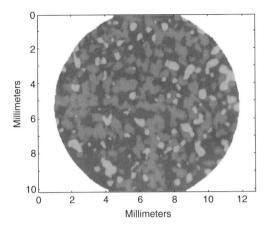

Plate 3 Red-Green-Blue image created from PLS score images of three APIs of an OTC analgesic tablet (Excedrin™). Red highlights acetaminophen-rich areas, green highlights caffeine-rich areas, and blue highlights aspirin-rich areas. (This figure is produced in color in the color plate section, which follows page 204.)

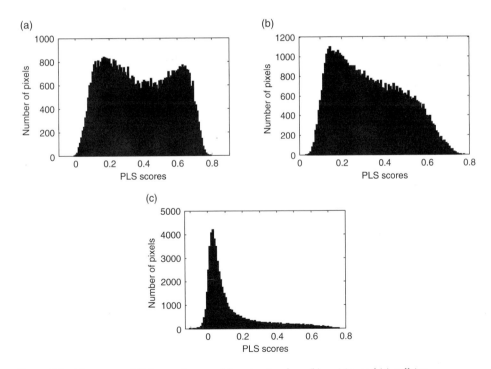

Figure 7.8 Histograms of PLS score images. (a) acetaminophen; (b) aspirin; and (c) caffeine.

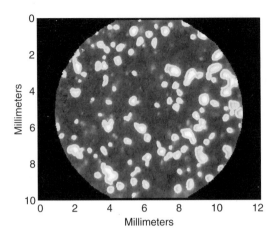

Figure 7.9 Chemical image analysis tools use domain morphology to highlight caffeine domains in the tablet.

information: there are 112 individual caffeine particles with a mean area of 0.097 mm², and a mean nearest neighbor distance of 0.728 mm.

In addition, taking the ratio of pixels calculated to belong to each component to the total number of pixels yields the following abundance estimates: 41% for acetaminophen, 42% for aspirin, and 12% for caffeine. There are other components in the tablet, composing the remaining 16% of the tablet. These excipients were not included in the initial model, so the results for the three API components need to be normalized to their expected contribution to the whole tablet. When normalized to a total contribution of 84%, the abundance estimations are 36% for acetaminophen, 37% for aspirin, and 11% for caffeine. These values compare favorably with the manufacturers' label claim of 37% for acetaminophen, 37% for aspirin, and 10% for caffeine.

The ability to determine the domain size of the individual components comprising a finished pharmaceutical product as shown above is not readily attainable by other analytic techniques. It is reasonable to assume that many of the physical properties of a tablet, such as dissolution and hardness, will be directly or indirectly impacted by how the sample components are packed – the size of a typical component domain, and how these domains interact with each other. Much of the 'art' of formulation is a function of tailoring the interactions of the sample components to bring about the desired physical characteristics of the finished product. NIR-CI can easily track the effects of changed manufacturing conditions on domain sizes and inter- and intra-component affinities. In our experience in examining finished pharmaceutical tablets, the typical domain size of the individual components is often much larger than the manufacturers generally assume them to be. This does not imply that the tablets do not have the desired physical characteristics, but rather that the final domain sizes that give the desired properties are larger than would be estimated by knowing the size of particles of the pure components before blending.

A final 'quantitative' approach to estimating absolute abundance is through normalized mean PLS scores. This approach is totally objective, unlike the previous examples in which thresholds need to be determined. While a classification mode PLS model is not designed to estimate absolute concentration, the relative abundance of the components is maintained: the higher the PLS scores, the higher the concentration. As with the abundance estimations performed using histogram thresholds, if the PLS scores are normalized then they give objective estimates of absolute abundance. In this example, as has already been shown, the three API components account for 84% of the total tablet. The abundance estimations based on normalized PLS scores are 33% for acetaminophen, 38% for aspirin, and 12% for caffeine, compared to the true value of 37, 37, and 10%, respectively. If the complete set of ingredients was used for the PLS calculation, the normalization step would be unnecessary.

7.6.4   Conclusions

This example shows a variety of processing approaches that can be used to estimate the abundance of the three API components using a single tablet. The statistics available when collecting over 80 000 spectra per imaging sample not only enable a 'standard' analytical assay to be performed, but also provide real information on domain size, morphology, and distribution of the individual components comprising the finished product. By statistically characterizing domain sizes through morphological data processing, even more insight can be gained into what sample characteristics determine tablet physical properties, and ultimately product performance.

7.7   Processing case study two: Determining blend homogeneity through statistical analysis

As discussed in the previous case study, the physical properties of finished pharmaceutical forms may be affected by the interactions and the domain sizes of the individual sample components. Because the blending process can determine the formation of these component domains and component interactions, how the blending proceeds may have a significant impact on the performance of the finished form. For example, improperly blended powders may result in a final product where the release of therapeutic materials occurs at less than optimal rates. Determination of the blend quality in both powder-blend samples and finished forms has posed a particular problem to the analyst. Traditional spectroscopic and chromatographic methods can only yield the bulk content relationship of the components, and dissolution testing on the finished form can only indicate that a problem exists but cannot provide direct information as to its source. NIR-CI is well suited to rapid and direct determination of the blend uniformity for all of the chemical species comprising the formulation, both for powder blends and for finished tablets. This example will focus on examining blend homogeneity of finished forms.

7.7.1   *Experimental*

In this application, a series of six tablets representing a wide range of blending homogeneity are analyzed. Visually, and in terms of total chemical composition, all of the
tablets are identical, composed of exactly the same proportion of materials: 80 mg
furosemide (API) and 240 mg excipient mix (99.7% Avicel PH 102 and 0.33% magnesium
stearate). Five of the samples were tableted in the laboratory with homogeneity controlled
by varying the blending time. These tablets range from nearly unmixed to fully blended.
The sixth tablet is a high-quality 'commercially' blended example. All of the tablets were
formed by direct compression method; the laboratory examples with a Carver press, and
the commercial tablet on an automated tableting press.

Each tablet was imaged with a MatrixNIR^{TM} chemical imaging system (Spectral
Dimensions, Inc., Olney, MD) using a pixel resolution of $40 \times 40\,\mu m$ over the wavelength
range of 1000–1700 nm at 10-nm spectral resolution. The time required to acquire each
image is approximately 2 minutes. The raw images were background-corrected using a
data cube collected from Spectralon^{TM} (Labsphere, Inc., North Sutton, NH) as the
reference, converted to $\log_{10}(1/R)$, and baseline-corrected. Images were also collected
from samples of pure furosemide and Avicel. These data were processed in the same
manner as the tablet images and used to create a pure component reference library for the
system. A PLS2 model was constructed from the reference library and used to generate
PLS score images from each of the tablet data cubes. The resulting score images are shown
in Figure 7.10 in which 'white' denotes the location of the API component.

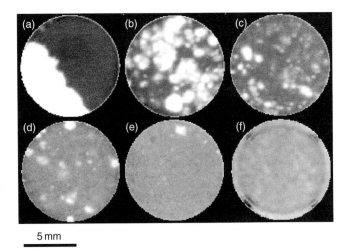

5 mm

Figure 7.10   Chemical image analysis of six tablets composed of 80 mg of furosemide (API) mixed with
240 mg of the excipient mix (99.7% Avicel PH 102 and 0.33% magnesium stearate). Tablets A–E were
created in the laboratory using increasing blending time, tablet F is an example of a commercial product.
These PLS score images of the tablets highlight the location of the API component (bright pixels), and show
the gross blend-quality differences.

7.7.2   Observing visual contrast in the image

The quality of the blending in each of these samples is readily apparent in the PLS score images. In the poorly blended tablets (A and B) there is sharp contrast between the domains of nearly pure API against a black background of the nearly pure excipient materials. As the blend quality improves in tablets C and D, the contrast softens as the images start to become more 'gray,' indicating more mixing of the white and the dark, that is API and excipients. The API domains are still present, although they are smaller and less distinct. Tablet E is nearly uniform gray indicating good blending, although one small API-rich domain is still observed near the top of the image. The commercial tablet F is extremely well blended and shows almost no contrast at all when looking at the API distribution.

7.7.3   Statistical analysis of the image

While visual inspection of the chemical images gives a qualitative indication of the blend quality in the samples, more quantitative metrics are available through statistical analysis of the images using histogram representations. The histogram is the graphical representation of the distribution of intensities in the image. Figure 7.11 shows corresponding histograms for the six individual tablet score images. In these graphs the abscissa represents the range of intensities (in this case, the individual PLS score values) and the

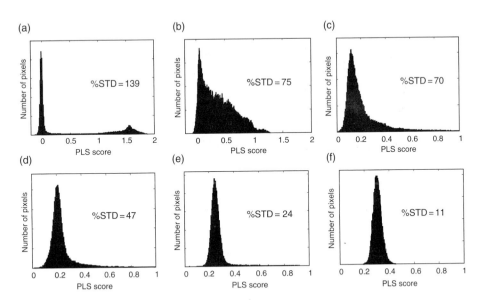

Figure 7.11   (a–f): Histograms of the PLS score images showing the statistical distribution of the API class. Heterogeneity in the blend is indicated by deviations from a normal distribution and can be expressed as the percent standard deviation (%STD) metric calculated by dividing the standard deviation by the mean value of the histogram distributions.

ordinate denotes the number of individual pixels within the image possessing that value. A highly homogeneous image will produce a narrow Gaussian distribution such as that seen in the histogram of the commercial tablet (F). On the other hand, a starkly hetero-geneous image will yield multi-modal distributions. For example, the histogram for the un-blended tablet (A) contains two distinct modes, the larger population at the left depicts the pure excipient area with near-zero PLS score values, and the smaller population on the right represents the pure API region with large score values. The histograms from the images of the intermediate tablets indicate heterogeneity as varying degrees of distortion from the normal distribution.

For a statistical analysis to adequately characterize the distribution of a component, the data should first be processed to obtain the optimum selectivity for that component. In this application the PLS model produces a score image that effectively separates the spectral response of the API from the excipients. Even though there is very little observable contrast in the images of the well-blended samples, the results from the poorly blended samples convey confidence that the method is effective at tracking the API distribution.

It should be noted that the relationship between domain sizes of the components in the matrix and the magnification employed will influence the image statistics. For the well-blended samples, the domain sizes of the API and excipient materials are much smaller than the individual image resolution elements. Therefore, each pixel should correspond to a mixture spectrum denoting the relative amounts of the components at that location in the sample. The multivariate PLS modeling approach produces a measure of the relative API concentration at each individual pixel in the image. The resulting histogram repre-sentations of the PLS score images represent the distribution of concentrations at the individual pixel locations. On the other hand, if the real domain sizes were larger than the pixel size, each pixel would record a relatively pure component spectrum, regardless of the actual blend quality. In this case, the distributions obtained from the image may appear bi- or multi-modal even for a highly homogeneous mixture (in the bulk sense). In these cases, it may be advantageous to use other image analysis tools, such as particle statistics as shown in the previous example, to access the bulk uniformity of the blend. Alternatively, it is also possible to adjust the magnification of the NIR-CI instrument to increase the area interrogated by the individual pixels.

Four metrics are typically used to characterize the shape of the distribution in the histogram. The mean is the average value of the intensities within the distribution. The standard deviation describes the dispersion about the mean, that is the width of the distribution. Skew and kurtosis denote deviation from a normal distribution. Skew measures the asymmetrical tailing of the distribution: positive skew indicates tailing toward higher values, negative skew indicates tailing toward lower values. Kurtosis measures the symmetrical tailing or 'peakedness' of the distribution. Positive kurtosis describes a sharper peak with longer tails (as in a Lorentzian shape), while negative kurtosis indicates a flatter peak with smaller tails (more like the classic 'mouse-hole' shape). Both skew and kurtosis are equal to zero for a normal distribution.

The interpretation of these metrics for a chemical image is somewhat dependent on the overall morphology of the sample. For a highly to moderately homogeneous sample (e.g. tablets D–F), the mean value indicates the bulk abundance of the component and the standard deviation is a measure of the overall heterogeneity in the sample. As the sample becomes more heterogeneous, skew and kurtosis illustrate the distortion of the normal distribution and may indicate the localization of one component or another. For example, positive skew suggests domains of increased concentration and conversely, 'holes' in the distribution will be represented by negative skew values.

7.7.4 Blend uniformity measurement

For the blend uniformity application, the primary concern is overall heterogeneity within the sample. This is directly indicated by the width of the distribution and is quantitatively measured as the percent standard deviation relative to the mean (%STD). The %STD value obtained from the PLS scores distribution from each of the six imaged tablets is noted on the histograms in Figure 7.11. The value of %STD consistently decreases as the blending quality of the sample improves.

The range of blend quality represented by these six samples is much greater than would normally be encountered in an actual manufacturing process. However, the sensitivity of the %STD metric is demonstrated through the comparison of the results obtained from the commercial tablet (F) and the best-blended laboratory tablet (E). The general lack of contrast in the visual images suggests that both of these tablets are, on the whole, extremely well blended. The significantly larger %STD values obtained from tablet E likely stem from the isolated domain of increased concentration observed near the top edge of the sample in the PLS score image. This hypothesis was tested by truncating the PLS score image from these samples to a 100×100 pixel region from the center region of the tablets. The enlarged region is approximately 20% of the whole tablet area and is still large enough to be representative of the overall blend quality of the tablet. The truncated images and the corresponding histograms are shown in Figure 7.12a and b respectively. In this analysis, the %STD value for the tablet (F) changes only very slightly while that of tablet (E) decreases significantly to a value only slightly larger than that yielded by tablet (F). This indicates that, except for a single localized region, the general blend uniformity of both these tablets is quite similar.

This result also indicates the value of examining as much of the entire sample as possible when characterizing blend homogeneity. If only a small region is analyzed, it is possible that isolated 'hot-spots' would escape detection. NIR-CI has the advantage of the ready availability of large, high-density array detectors as well as flexible fields of view. This means that the configuration can be easily tailored such that the entire sample surface is analyzed rapidly and simultaneously. If one is concerned that the surface may not be representative of the entire thickness of the tablet, the sample can be turned over or easily scraped (e.g. with a razor blade or microtome device) to reveal interior surfaces. Subsequent images of the revealed interior surfaces can then be collected and analyzed in the same manner to yield a more complete, volumetric analysis of the blend uniformity. In either case, the ability to rapidly generate statistically relevant blending data from single or multiple tablets is a critical aspect of its value to the PAT initiative.

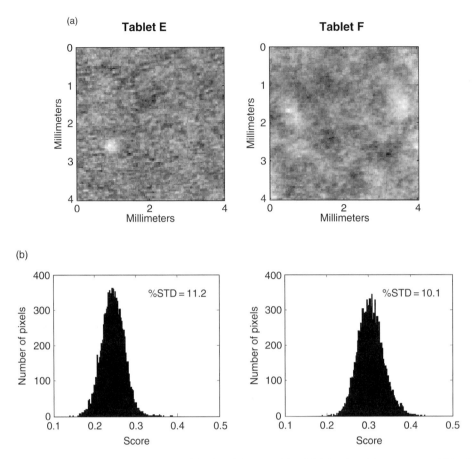

Figure 7.12   Analysis of selected areas of the best laboratory blend (tablet E) and the commercial blend (tablet F). (a) truncated PLS score images for the API component (depicted as white) from the central region of the tablets showing that on a more local scale, the blending in both samples is more uniform; and (b) Histograms and the resulting %STD of the distribution from the truncated images. Note that both examples display nearly normal distribution.

7.7.5   Conclusions

The spatial aspect of NIR-CI provides the information necessary to develop a fast and accurate approach for the quantitative characterization of the homogeneity of a solid dosage form. This information cannot be directly accessed by any other traditional compendial or single-point spectroscopic technique. Access to such data is of great value in the development of the manufacturing process, monitoring finished products, and root cause analysis of manufacturing problems. For any given pharmaceutical product, the ultimate goal of the manufacturer may be to achieve a product that is as homogeneous as possible. In this case, the relative width (%STD) of the API distribution shown in the histogram of the chemical image provides an objective, accurate, and very sensitive metric for blend quality. Other components of a complex formulation could also be measured in the same manner. The data collection and processing can also be easily built into a 'turnkey' method so that the entire procedure can be automated.

7.8 Final thoughts

Near-infrared-CI provides analytical capabilities that are not achievable with standard single-point spectroscopic methods, or even other chemical imaging techniques. Deriving spatially resolved spectral information, such as domain size and component interactions, throughout the various pharmaceutical manufacturing steps is anticipated to provide much of the core understanding of critical process parameters that ultimately drive product performance. The ability to provide a simplified, turnkey data collecting and processing solution also enables this application to move from an R&D to a manufacturing environment. In addition, the flexible FOV and built-in redundancy of detector elements in the FPA make it possible to utilize 'built-in' calibration sets, potentially obviating the need for cumbersome transfer of calibration methods. In the particular hardware configuration emphasized in this chapter, a system with no moving parts, flexible FOV and modular approach to data collection and processing can be integrated to provide high-throughput capabilities for QA processes. What and how much we may be able to learn about the complexities of pharmaceutical manufacturing from NIR-CI is still relatively unknown, but early results suggest that it might be a great deal.

Acknowledgements

We would like to thank Dr Robbe C. Lyon of the FDA for providing the furosemide tablet samples.

References

1. See the internet site: http://www.fda.gov/cder/OPS/PAT.htm.
2. McClure, W.F., Near-Infrared Spectroscopy: The Giant is Running Strong; *Anal. Chem.* 1994, 66, 43A–53A.
3. Bakeev, K.A., Near-Infrared Spectroscopy as a Process Analytical Tool. Part I: Laboratory Applications; *Spectroscopy* 2003, 18(11), 32–35.
4. Ciurczak, E.W. & Drennen, J.K. III (eds), *Pharmaceutical and Medical Applications of Near-Infrared Spectroscopy*, Practical Spectroscopy Series, Volume 31; Marcel Dekker; New York, 2002 – and references therein.
5. Gimet, R. & Luong, A.T., Quantitative Determination of Polymorphic Forms in a Formulation Matrix Using the Near Infrared Reflectance Analysis Technique; *J. Pharm. Biomed. Anal.* 1987, 5, 205–211.

6. Norris, T.; Aldridge, P.K. & Sekulic, S.S., Determination of End-Points for Polymorph Conversions of Crystalline Organic Compounds Using On-Line Near-Infrared Spectroscopy; *Analyst* 1997, 122, 549–552.

7. Sekulic, S.S.; Ward, H.W.; Brannegan, D.R.; Stanley, E.D.; Evans, C.L.; Sciavolino, S.T.; Hailey, P.A. & Aldridge, P.K., On-Line Monitoring of Powder Blend Homogeneity by Near-Infrared Spectroscopy; *Anal. Chem.* 1996, 68, 509–513.

8. Hailey, P.A.; Doherty, P.; Tapsell, P.; Oliver, T. & Aldridge, P.K., Automated System for the On-Line Monitoring of Powder Blending Processes Using Near-Infrared Spectroscopy. Part I: System Development and Control; *J. Pharm. Biomed. Anal.* 1996, 14, 551–559.

9. Warman, M., Using Near Infrared Spectroscopy to Unlock the Pharmaceutical Blending Process; *American Pharmaceutical Review* 2004, 7(2), 54–57.

10. Trafford, A.D.; Jee, R.D.; Moffat, A.C. & Graham, P., A Rapid Quantitative Assay of Intact Paracetamol Tablets by Reflectance Near-Infrared Spectroscopy; *Analyst* 1999, 124, 163–168.

11. Rhodes, C.T. & Morissau, K., Tablet Evaluation Using Near Infrared Spectroscopy. In Swarbrick, J. & Boylan, J.C. (eds); Encyclopedia Pharmaceutical Technology, 2nd Edition; Marcel Dekker; New York, 2002; pp. 2689–2700.

12. Attributed to: Fred R. Barnard.

13. Clarke, F.C.; Jamieson, M.J.; Clark, D.A.; Hammond, S.V.; Jee, R.D. & Moffat, A.C., Chemical Image Fusion. The Synergy of FT-NIR and Raman Mapping Microscopy to Enable a More Complete Visualization of Pharmaceutical Formulations; *Anal. Chem.* 2001, 73(10), 2213–2220.

14. Lewis, E.N.; Carroll, J.E. & Clarke, F.C., A Near Infrared View of Pharmaceutical Formulation Analysis; *NIR News* 2001, 12(3), 16–18.

15. Lyon, R.C.; Jefferson, E.H.; Ellison, C.D.; Buhse, L.F.; Spencer, J.A.; Nasr, M.M. & Hussain, A.S., Exploring Pharmaceutical Applications of Near-Infrared Technology; *Amer. Pharma. Rev.* 2003, 6(3), 62–70.

16. Lyon, R.C.; Lester, D.S.; Lewis, E.N.; Lee, E.; Yu, L.X.; Jefferson, E.H. & Hussain, A.S., Near-Infrared Spectral Imaging for Quality Assurance of Pharmaceutical Products: Analysis of Tablets to Assess Powder Blend Homogeneity; *AAPS PharmSciTech.* 2002, 3(3), article 17, 1–15.

17. Clarke, F., Extracting Process-Related Information from Pharmaceutical Dosage Forms Using Near Infrared Microscopy; *Vib. Spectrosc.* 2004, 34, 25–35.

18. Clarke, F. & Hammond, S., NIR Microscopy of Pharmaceutical Dosage Forms; *European Pharmaceutical Review* 2003, 8(1), 41–50.

19. Barer, R.; Cole, A.R.H. & Thompson, H.W., Infra-Red Spectroscopy with the Reflecting Microscope in Physics, Chemistry and Biology; *Nature* 1949, 163, 198–200.

20. Harthcock, M.A. & Atkin, S.C., Imaging with Functional Group Maps Using Infrared Microspectroscopy; *Appl. Spectrosc.* 1988, 42(3), 449–455.

21. Treado, P.J.; Levin, I.W. & Lewis, E.N., Indium Antimonide (InSb) Focal Plane Array (FPA) Detection for Near-Infrared Imaging Microscopy; *Appl. Spectrosc.* 1994, 48(5), 607–615.

22. Tezcan, D.S.; Eminoglu, S. & Akin, T., A Low-Cost Uncooled Infrared Micorbolometer Detector in Standard CMOS Technology; *IEEE Trans. Electron. Devices* 2003, 50(2), 494–502.

23. Treado, P.J.; Levin, I.W. & Lewis, E.N., Near-Infrared Acousto-Optic Spectroscopic Microscopy: A Solid-State Approach to Chemical Imaging; *Appl. Spectrosc.* 1992, 46(4), 553–559.

24. Lewis, E.N. & Levin, I.W., Real-Time, Mid-Infrared Spectroscopic Imaging Microscopy Using Indium Antimonide Focal-Plane Array Detection; *Appl. Spectrosc.* 1995, 49(5), 672–678.

25. Wienke, D.; van den Broek, W. & Buydens, L., Identification of Plastics among Nonplastics in Mixed Waste by Remote Sensing Near-Infrared Imaging Spectroscopy. 2. Multivariate Rank Analysis for Rapid Classification; *Anal. Chem.* 1995, 67, 3760–3766.

26. Bellamy, M.K.; Mortensen, A.N.; Hammaker, R.M. & Fateley, W.G., Chemical Mapping in the Mid- and Near-IR Spectral Regions by Hadamard Transform/FT-IR Spectrometry; *Appl. Spectrosc.* 1997, 51(4), 477–486.

27. Lewis, E.N.; Treado, P.J.; Reeder, R.C.; Story, G.M.; Dowrey, A.E.; Marcott, C. & Levin, I.W., Fourier Transform Step-Scan Imaging Interferometry: High-Definition Chemical Imaging in the Infrared Spectral Region; *Anal. Chem.* 1995, 67, 3377–3381.

28. Reffner, J.A. & Martoglio, P.A., Uniting Microscopy and Spectroscopy. In Humecki, H.J. (ed.); *Practical Guide to Infrared Microscopy*; Practical Spectroscopy Series, Volume 19; Marcel Dekker; New York, 1995; pp. 41–84.

29. Claybourn, M., External Reflection Spectroscopy. In Chalmers, J.M. & Griffiths, P.R. (eds); *Handbook of Vibrational Spectroscopy, Sampling Techniques*, Volume 2; John Wiley & Sons; Chichester, 2002; pp. 969–981.

30. Griffiths, P.R. & Olinger, J.M., Continuum Theories in Diffuse Reflection. In Chalmers, J.M. & Griffiths, P.R. (eds); *Handbook of Vibrational Spectroscopy, Sampling Techniques*, Volume 2; John Wiley & Sons; Chichester, 2002; pp. 1125–1139.

31. Eilert, A.J. & Wetzel, D.J., Optics and Sample Handling for Near-Infrared Diffuse Reflection. In Chalmers, J.M. & Griffiths, P.R. (eds); *Handbook of Vibrational Spectroscopy, Sampling Techniques*, Volume 2; John Wiley & Sons; Chichester, 2002; pp. 1162–1174.

32. Kubelka, P. & Munk, F., Ein Beitrag zur Optik der Farbastriche; *Zeits. f. tech. Physik* 1931, 12, 593–601.

33. Savitzky, A. & Golay, M.J.E., Smoothing and Differentiation of Data by Simplified Least Squares Procedures; *Anal. Chem.* 1964, 36, 1627–1639.

34. Geladi, P.; MacDougall, D. & Martens, H., Linearization and Scatter-Correction for NIR Reflectance Spectra of Meat; *Appl. Spectrosc.* 1985, 39, 491–500.

35. Massart, D.L.; Vandeginste, B.G.M.; Buydens, L.M.C.; DeJong, S.; Lewi, P.J. & Smeyers-Verbeke, J., *Handbook of Chemometrics and Qualimetrics*; Part A: Data Handling in Science and Technology, 20A; Elsevier; Amsterdam, 1997.

36. Massart, D.L.; Vandeginste, B.G.M.; Buydens, L.M.C.; DeJong, S.; Lewi, P.J. & Smeyers-Verbeke, J., *Handbook of Chemometrics and Qualimetrics*; Part B: Data Handling in Science and Technology, 20B; Elsevier; Amsterdam, 1998.

37. Martens, H. & Næs, T., *Multivariate Calibration*; John Wiley & Sons; Chichester, 1989.

38. Mark, H., *Principles and Practice of Spectroscopic Calibration*; John Wiley & Sons; New York, 1991.

39. Mark, H. & Workman, J., Jr, *Statistics in Spectroscopy*; Academic Press; New York, 1991.

40. Burns, D.A. & Ciurczak, E.W. (eds), *Handbook of Near-Infrared Analysis*; Practical Spectroscopy Series, Volume 13; Marcel Dekker; New York, 1992.

41. Geladi, P. & Grahn, H., *Multivariate Image Analysis*; John Wiley & Sons; Chichester, 1996.

42. Carda-Broch, S.; Esteve-Romero, J. & Garca-Alvarez-Coque, M.C., Furosemide Assay in Pharmaceuticals by Micellar Liquid Chromatography: Study of the Stability of the Drug; *J. Pharm. Biomed. Anal.* 2000, 23(5), 803–817.

Chapter 8

Chemometrics in Process Analytical Chemistry

Charles E. Miller

8.1 Introduction

8.1.1 What is chemometrics?

The term 'chemometrics' was coined several decades ago to describe a new way of analyzing chemical data, in which elements of both **statistical** and **chemical** thinking are combined.[1] Since then, chemometrics has developed into a legitimate technical field of its own, and is rapidly growing in popularity within a wide range of chemical disciplines.

There are probably as many definitions for chemometrics as there are those who claim to practice it. However, there appear to be three elements that are consistently used in historical applications of chemometrics:

(1) empirical modeling
(2) multivariate modeling
(3) chemical data.

With this in mind, I ask the reader to accept my humble definition of 'chemometrics': *the application of multivariate, empirical modeling methods to chemical data.*[2]

The empirical modeling element indicates an increased emphasis on **data-driven** rather than **theory-driven** modeling of data. This is not to say that appropriate theories and prior chemical knowledge are ignored in chemometrics, but that they are not relied upon completely to model the data. In fact, when one builds a 'chemometric' calibration model for a process analyzer, one is likely to use prior knowledge or theoretical relations of some sort regarding the chemistry of the sample or the physics of the analyzer. For example, in process analytical chemistry (PAC) applications involving absorption spectroscopy, the Beer's Law relation of absorbance vs. concentration is often assumed to be true; and in reflectance spectroscopy, the Kubelka-Munk or $\log(1/R)$ relations are assumed to be true.

The multivariate element of chemometrics indicates that more than one response variable of the analyzer is used to build a model. This is often done out of necessity because no single response variable from the analyzer has sufficient selectivity to monitor a specific property without experiencing interferences from other properties.

The combination of empirical and multivariate modeling elements makes chemometrics **both very powerful and very dangerous**. The power of chemometrics is that it can be used to model systems that are both largely **unknown and complex**. Furthermore, these models are not restricted by theoretical constraints – which can be a big advantage if large deviations from theoretical behavior are known to be present in your system.

However, there are prices to pay for the advantages above. Most empirical modeling techniques need to be fed large amounts of **good** data. Furthermore, empirical models can be safely applied only to conditions that were represented in the data used to build the model (i.e. extrapolation of the model usually results in large errors). The use of multiple response variables to build models results in the temptation to overfit models, and obtain artificially optimistic results. Finally, multivariate models are usually much more difficult to explain to others, especially those not well versed in math and statistics.

8.1.2 *What does it do for analytical chemistry?*

Chemometrics tools can be used for a wide variety of tasks, including experimental design, exploratory data analysis, and the development of predictive models. In the context of analytical chemistry, however, chemometrics has been shown to be most effective for two general functions:

(1) Instrument specialization: Multivariate calibration models are built in order **to provide selectivity** for a multivariate analytical instrument, or
(2) Information extraction: Chemometrics tools are used to **'unlock' hidden information** already present in information-rich multivariate analytical instruments, to enable improved understanding of chemistry and analytical technology.

I venture to say that the majority of practical chemometrics applications in analytical chemistry are in the area of instrument specialization. The need to improve specificity of an analyzer depends on both the analytical technology and the application. For example, chemometrics is often applied to near-infrared (NIR) spectroscopy, due to the fact that the information in NIR spectra is generally non-specific for most applications. Chemometrics may not be critical for most ICP atomic emission or mass spectrometry applications because these techniques provide sufficient selectivity for most applications. On the other hand, there are some NIR applications that do not require chemometrics (e.g. many water analysis applications), and some ICP and mass spectrometry applications are likely where chemometrics is needed to provide sufficient selectivity.

The information extraction function of chemometrics is a very valuable one that is often overlooked, especially in the industrial world. It will be shown that this function can be used **concurrently** with the instrument specialization function, rather than relying upon additional exploratory data analysis.

8.1.3   *What about process analytical chemistry?*

Process analytical applications differ from laboratory analytical applications in that the analyzer hardware and software are in a more 'hostile' environment, the maintenance and operations personnel in the analyzer area are typically not trained like lab technicians, and the purpose of the analyzer is most often for rapid process control rather than trouble-shooting. As a result, process analytical applications tend to focus more on automation, minimization of maintenance, and long-term reliability. In this more pragmatic context, the instrument specialization function of chemometrics is used even more exclusively than the information extraction function.

An additional feature of chemometrics that is appealing to **process** analytical applications, is the use of **qualitative** models to detect (and possibly even characterize!) faults in the analyzer's calibration, the analyzer system itself, and even sample chemistry and process dynamics. Such faults can be used to trigger preventive maintenance and trouble-shoot specific analyzer or process problems – thus supporting the long-term reliability of the analyzer. Specific examples of such fault detection are given by Miller.[3,4]

8.1.4   *Some history*

The use of multivariate statistics to decipher multivariate data dates back to the 1930s,[5] and most early applications of these tools were in the fields of psychology, sociology, and economics – fields where large amounts of highly unspecific data are the rule, rather than the exception.

Despite its successes in other fields, chemometrics was not introduced in analytical chemistry until the late 1960s and 1970s. Somewhat surprisingly, the first appearance of chemometrics for analytical chemistry in the literature came from the food industry, rather than from academia. The work of Norris and Hruschka at the USDA, starting in the 1960s,[6–9] demonstrated the practical feasibility of multiple-wavelength NIR calibrations for rapid non-destructive analyses of foods and agricultural materials. With these practical successes as a background, a few intrepid researchers in the academic realm[10–16] pushed the integration of existing tools with this new field, as well as developed new tools that would become somewhat unique for chemometrics.

Through the 1980s and 1990s, the development of chemometrics for analytical chemistry was largely driven by the increasing popularity and utility of non-specific NIR spectroscopy technology. After the work of Norris, it was quickly realized that the great potential of NIR technology for a wide range of industrial applications (even outside of food and agriculture) could not be 'unlocked' without the instrument specialization abilities of chemometrics. At the same time, increased efficiency and accessibility of computing power allowed the power of chemometrics to be used not only by mathematicians and statisticians with mainframe computers but also by any scientist or engineer with access to a personal computer. These forces combined to generate a boom in the publication of chemometrics applied to analytical chemistry starting in the 1990s (NIR and others).

Unfortunately, it is this writer's opinion that, considering the voluminous publications on chemometrics applications, the number of **actual** effective process analytical chemometrics applications in the field is much less than expected. Part of this is due to the 'overselling' of chemometrics during its boom period, when personal computers (PCs) made these tools available to anyone who could purchase the software, even those who did not understand the methods. This resulted in misuse, failed applications, and a 'bad taste' with many project managers (who tend to have long memories...). Part of the problem might also be due to the lack of adequate software tools to develop and **safely** implement chemometric models in a process analytical environment. Finally, some of the shortfall might simply be due to lack of qualified resources to develop and maintain chemometrics-based analytical methods in the field.

At the same time, there are many effective chemometrics applications in PAC across a wide range of industries, although relatively few of them are published.[3,4,17,18] Nonetheless, they provide the driving force for analytical scientists and engineers to push for new applications in industry.

8.1.5 Some philosophy

Chemometrics is a rather unique technology – for two main reasons. First, most technologies are invented and developed at academic or R&D institutions, 'handed over' to industry, and then adapted for practical use by engineers and scientists in industry. Chemometrics technology is very different, in that industry played a very large role in its development, dating back to its infancy. This resulted because many chemometric tools are a product of the desire for **practical** applicability, rather than theoretical rigor. This often leads many in academic institutions to shun the technology as lacking sufficient fundamental basis or ignoring established rules of statistical analysis, even though it has proven itself in many industrial applications.

Secondly, as a result of the emphasis on empirical modeling and practical applicability, chemometrics must be a highly 'interfacial' discipline, in that specific tools are often developed with specific applications already in mind. For example, specific chemometric tools have been developed to align retention time axes in chromatograms, thus enabling the extraction of pure component information in second-order chromatographic techniques.[19] Such chemometric tools **require** association with another scientific discipline and/or sub-discipline (in this example, chromatography and second-order chromatography) in order to be relevant, and are essentially meaningless for most other applications. In contrast, other disciplines, such as statistics, are associated with well-defined 'stand-alone' tools (ANOVA, *t*-test, etc.) that can be applied to a wide array of different applications.

One consequence of this interfacial property of chemometrics is that one must often sift through a very large 'toolbox' of application-specific tools in order to find one that suits a particular application. This chapter will focus on tools that are more global in their relevance. Although these tools are widely considered to be within the realm of chemometrics, it is interesting to note that many of them were originally developed for **non**chemical applications.

Martens and Næs[1] provide an interesting perspective on chemometrics as 'a combination of chemical thinking and statistical thinking.' In this context, it is suggested that statisticians are well aware of the limitations and capabilities of statistical tools, but confined to a 'rigid' structure of these tools – not allowing chemical knowledge to aid in the experimental design, data modeling, or the interpretation of results. Conversely, chemists are often not aware of the theoretical limitations on experimental design and statistical analysis techniques, but have the chemical knowledge to optimize experimental designs and detect anomalies in the analyses. Both chemical thinking and statistical thinking are required to generate optimal results.

Finally, the highly empirical and practical nature of chemometrics leads many outside of the field to conclude that chemometrics practitioners do not understand the modeling tools they use, or the models that they develop – thus potentially leading to disastrous results. Although this perception might be accurate for many users, it need not be true for all users. This leads to the mission of this chapter: to enable users to better understand, and thus more effectively implement, chemometrics in process analytical applications.

8.2 The building blocks of chemometrics

8.2.1 Notation

In the upcoming discussions, I will use the following convention in expressing numerical values:

Scalars, which consist of single values, are expressed as ***lower-case, normal-face*** letters (e.g. a = [0.5]).

Vectors, which consist of a one-dimensional series of values, are expressed as ***lower-case, bold-face*** letters (e.g. **a** = [5 4 1 2]).

Matrices, which consist of a two-dimensional array of values, are expressed as ***upper-case, bold-face*** letters. For example,

$$\mathbf{A} = \begin{bmatrix} 1 & 2 & 3 & 4 \\ 5 & 6 & 7 & 8 \\ 9 & 10 & 11 & 12 \\ 13 & 14 & 15 & 16 \end{bmatrix}$$

In addition, specific letters are used to denote specific quantities that are common for the chemometrics tools discussed in this chapter:

(1) The letter 'X' is reserved for the ***independent variables*** in a regression problem, also called the '***X-variables***.' In PAC applications, these variables most often represent spectral intensities at different wavelengths, chromatographic intensities at different retention times, or similar analytical profiles.

(2) The letter 'Y' is reserved for the ***dependent variables*** in a regression problem, also called the '**Y-variables**.' In PAC applications, these variables most often represent constituent concentrations or properties of interest of the samples.

(3) The letter 'B' is reserved for the ***regression equation coefficients*** in a regression problem.

(4) The letter 'E' is reserved for the residuals of the independent variables in a modeling problem, also called the 'X-residuals.' These contain the independent variable information that is not explained by the model.

(5) The letter 'F' is reserved for the residuals of the dependent variables in a modeling problem, also called the 'Y-residuals.' These contain the dependent variable information that is not explained by the model.

Finally, the placement of a carrot symbol '^' on top of a letter indicates that the letter represents an ***estimated*** quantity, rather than a measured or theoretical quantity. For example: $\hat{X}$ indicates a matrix of independent variables that have been ***estimated by a model***, where X indicates a matrix of independent variables that have been ***actually measured*** on an analytical device.

Finally, some special notation for matrix operations are used throughout the text:

- Superscript '**t**' indicates the transpose of a matrix:

$$\text{if } \mathbf{D} = \begin{bmatrix} 1 & 5 \\ 2 & 6 \\ 3 & 7 \\ 4 & 8 \end{bmatrix}, \text{ then } \mathbf{D}^t = \begin{bmatrix} 1 & 2 & 3 & 4 \\ 5 & 6 & 7 & 8 \end{bmatrix}$$

- Superscript '−1' indicates the inverse of a matrix (note: the matrix must be square and invertible!):

$$\text{if } \mathbf{M} = \begin{bmatrix} 2 & 0 \\ 0 & 4 \end{bmatrix}, \text{ than } \mathbf{M}^{-1} = \begin{bmatrix} .5 & 0 \\ 0 & .25 \end{bmatrix}$$

- Two matrices or vectors placed adjacent to one another indicates matrix multiplication (note: the dimensionality of the matrices or vectors must match appropriately for multiplication to be possible):

$$\mathbf{DM} = \begin{bmatrix} 1 & 5 \\ 2 & 6 \\ 3 & 7 \\ 4 & 8 \end{bmatrix} \times \begin{bmatrix} 2 & 0 \\ 0 & 4 \end{bmatrix} \text{ (a 4 × 2 matrix multiplied by a 2 × 2 matrix)}$$

8.2.2 A bit of statistics

Despite the conflicting philosophies between statisticians and chemometricians, it is an unavoidable fact that statistics provides a strong foundation for chemometrics. One could easily argue that without classical statistics there could be no chemometrics.

One underlying principle of classical statistics is that any observation in nature has an **uncertainty** associated with it. One extension of this principle is that multiple observations will result in a **distribution** of values. One useful graphical representation of a distribution of values is a histogram (Figure 8.1), where the frequency of occurrence of a value is plotted against the value. The values represented in the histogram could be multiple observations of an NIR spectral analyzer at a single wavelength, multiple results obtained from a Karl-Fisher titration, or multiple grades obtained from a class examination.

Many statistical tools are based on a specific type of distribution, namely the Gaussian, or Normal distribution, which is shown in Figure 8.1 and has the following mathematical form:

$$f(x) = De^{-\frac{(x-\bar{x})^2}{2\sigma^2}} \tag{8.1}$$

where $\bar{x}$ is the mean of the values, and σ is the standard deviation of the values. The mean is the best single-value approximation of all of the values, and the variance is a measure of the extent to which the values vary. The mean and standard deviation as defined by the Gaussian distribution can be directly calculated from the multiple values:

$$\bar{x} = \frac{\sum_{i=1}^{n} x}{N} \tag{8.2}$$

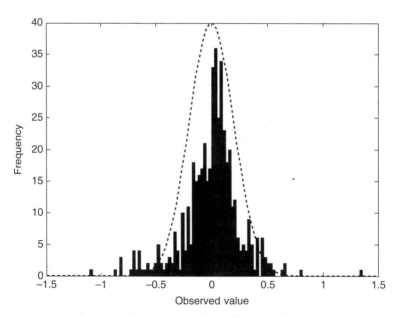

Figure 8.1 Histogram showing a distribution of observed values, along with a Normal (Gaussian) distribution function.

$$\sigma = \sqrt{\frac{\sum\limits_{i=1}^{n} (x - \bar{x})^2}{N}} \qquad (8.3)$$

In analytical chemistry, σ often represents an approximation of the noise of a particular measurement.

Although there are many useful statistical tools, there are two that have particular relevance to chemometrics: the t-test and the f-test.[20,21] The t-test is used to determine whether a single value is statistically different from the rest of the values in a series. Given a series of values, and the number of values in the series, the t-value for a specific value is given by the following equation:

$$t = \frac{|\bar{x} - x|}{\left(\dfrac{\sigma}{\sqrt{N}}\right)} \qquad (8.4)$$

A good way of remembering the t-value is that it is the number of standard deviations that the single value differs from the mean value. This t-value is then compared to the critical t-value obtained from a t-table, given a desired statistical confidence (i.e. 90, 95, or 99% confidence) and the number of degrees of freedom (typically $N - 1$), to assess whether the value is statistically different from the other values in the series. In chemometrics, the t-test can be useful for evaluating **outliers** in data sets.

The f-test is similar to the t-test, but is used to determine whether two different **standard deviations** are statistically different. In the context of chemometrics, the f-test is often used to compare distributions in regression model errors in order to assess whether one model is significantly different than another. The f-statistic is simply the ratio of the squares of two standard deviations obtained from two different distributions:

$$F = \frac{\sigma_1^2}{\sigma_2^2} \qquad (8.5)$$

As the f-statistic deviates further from one, there is a greater probability that the two standard deviations are different. To quantitate this probability, one needs to consult an f-table that provides the critical values of F as a function of the degrees of freedom for the two distributions (typically $N - 1$ for each of the data series), and the desired statistical confidence (90%, 95%, 99%, etc.).

8.2.3 Linear regression

Another critical building block for chemometrics is the technique of linear regression.[1,20,21] In chemometrics, this technique is typically used to build a linear model that relates an independent variable (X) to a dependent variable (Y). For example, in PAC, one could make a set of observations of the integrated area of a specific peak in an on-line chromatograph for a set of N samples, and a corresponding set of observations of an analyte concentration obtained from an off-line wet chemistry method, for the same set of N samples (Table 8.1). With this data, one can then use linear regression to develop a predictive model that can be used to estimate the analyte concentration from the integrated peak area of an unknown sample.

The model for linear regression is given below.

$$\mathbf{y} = b\mathbf{x} + b_0\mathbf{1} + \mathbf{f} \tag{8.6}$$

where $\mathbf{y}$ is a vector of measured independent variables, $\mathbf{x}$ is a matching vector of measured dependent variables, $\mathbf{1}$ is a vector of ones, and $\mathbf{f}$ is a vector containing the residuals of the linear regression model. All of the vectors mentioned above have N elements. This equation can be simplified to the following form:

$$\mathbf{y} = \mathbf{X}_{avg}\mathbf{b}_{avg} + \mathbf{f} \tag{8.7}$$

where $\mathbf{b}_{avg} = \begin{bmatrix} b \\ b_0 \end{bmatrix}$, and $\mathbf{X}_{avg} = \begin{bmatrix} x_1 & 1 \\ x_2 & 1 \\ \cdots & \cdots \\ x_N & 1 \end{bmatrix}$

As the name suggests, the residuals $(\mathbf{f})$ contain the variation in $\mathbf{y}$ that cannot be explained by the model.

Given the matching sets of measured data, $\mathbf{x}$ and $\mathbf{y}$, it is now possible to estimate the model regression coefficients $\mathbf{b}_{avg}$. Assuming that the model errors (values in $\mathbf{f}$) are Gaussian-distributed, it can be proven that the value of $\mathbf{b}_{avg}$ that minimizes the sum of squares of the model errors is determined using the least squares method:

$$\hat{\mathbf{b}}_{avg} = \left(\mathbf{X}_{avg}{}^t\mathbf{X}_{avg}\right)^{-1}\mathbf{X}_{avg}{}^t\mathbf{y} \tag{8.8}$$

Graphically, the elements of $\mathbf{b}_{avg}$ refer to the slope (b) and intercept (b_0) of the line of best fit through the observed data points. For our example data in Table 8.1, a graph of the linear regression fit is illustrated in Figure 8.2. Once the coefficients are estimated, the model error $(\mathbf{f})$ can also be estimated:

Table 8.1 Data values used for the linear regression example

Analyzer measurement	Analyte concentration
0.56	0.00
0.87	0.98
1.05	2.10
1.31	3.02
1.70	4.99

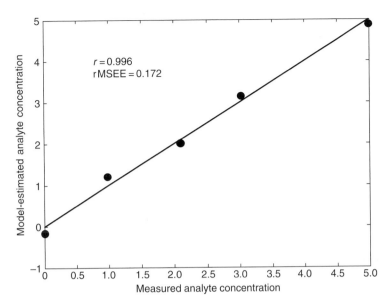

Figure 8.2 Graph of the data points represented in Table 8.1, along with the line of best linear fit. The correlation coefficient (*r*) and the RMSEE for the model fit are listed on the figure.

$$\hat{\mathbf{f}} = (\mathbf{y} - \hat{\mathbf{y}}) = \left(\mathbf{y} - \mathbf{X}_{avg}\hat{\mathbf{b}}_{avg}\right) \tag{8.9}$$

There are several properties of linear regression that should be noted. First, it is assumed that the model errors are normally distributed. Secondly, the relationship between X- and Y-variables is assumed to be linear. In analytical chemistry, the first assumption is generally a reasonable one. However, the second assumption might not be sufficiently accurate in many situations, especially if a strong non-linear relationship is suspected between X and Y. There are some non-linear remedies to deal with such situations, and these will be discussed later.

Another assumption, which becomes apparent when one carefully examines the model (Equation 8.7), is that all of the model error (**f**) is in the dependent variable (**y**). There is no provision in the model for errors in the independent variable (**x**). In PAC, this is equivalent to saying that there is *error only in the reference method, and no error in the on-line analyzer responses*. Although this is obviously not true, practical experience over the years has shown that linear regression can be very effective in analytical chemistry applications.

There are several figures of merit that can be used to describe the quality of a linear regression model. One very common figure of merit is the correlation coefficient, r, which is defined as:

$$r = \frac{\left(\sum_{i=1}^{N} \left(\hat{y}_i - \hat{\bar{y}}\right) * (y_i - \bar{y})\right)}{\sqrt{\left(\sum_{i=1}^{N} \left(\hat{y}_i - \hat{\bar{y}}\right) * \sum_{i=1}^{N} (y_i - \bar{y})\right)}} \tag{8.10}$$

where $\bar{y}$ is the mean of the **known** Y-values and $\hat{\bar{y}}$ is the mean of the **model-estimated** Y-values. A model with a perfect fit will yield a correlation coefficient of 1.0.

Another figure of merit for the fit of a linear regression model is the root mean square error of estimate (RMSEE), defined as:

$$\text{RMSEE} = \sqrt{\frac{\sum_{i=1}^{N}(y_i - \hat{y}_i)^2}{N-1}} \tag{8.11}$$

A model with a perfect fit will yield an RMSEE of zero. The r and RMSEE of the linear regression model built from the example data are listed in Figure 8.2.

Comparing the two figures of merit, the correlation coefficient has the advantage of taking into account the **range** of Y-values used in the regression. It has the disadvantage of being a unitless quantity, which might not provide sufficient meaning to the customer. An advantage of the RMSEE is that it is in the same units as the Y-variable, and therefore provides a good absolute assessment of the model fit error. However, if one is interested in the model error relative to the range, it would be more relevant to consider the RMSEE as a percentage of the range of Y-values, or just the correlation coefficient.

8.2.4 Multiple linear regression (MLR)

An extension of linear regression, MLR involves the use of **more than one independent variable**. Such a technique can be very effective if it is suspected that the information contained in a single independent variable (X) is insufficient to explain the variation in the dependent variable (Y). For example, it is suspected that a single integrated absorbance of the NIR water band at 1920 nm is insufficient to provide accurate concentrations of water contents in process samples. Such a situation can occur for several reasons, such as the presence of other varying chemical components in the sample that interfere with the 1920-nm band. In such cases, it is necessary to use more than one band in the spectrum to build an effective calibration model, so that the effects of such interferences can be compensated.

The MLR model is simply an extension of the linear regression model (Equation 8.6), and is given below:

$$y = Xb + b \cdot 1 + f \tag{8.12}$$

The difference here is that X is a matrix that contains responses from M (>1) different X-variables, and **b** contains regression coefficients for each of the M X-variables. If X and **b** are augmented to include an offset term (as in Eqn. 8.7) the coefficients for MLR (b_{avg}) are determined using the least squares method:

$$\hat{b}_{avg} = \left(X_{avg}^t X_{avg}\right)^{-1} X_{avg}^t y \tag{8.13}$$

At this point, it is important to note two limitations of the MLR method. First, note that the X-variable matrix (X) has the dimensionality of N by M. As a result, **the number of**

X-variables (M) cannot exceed the number of samples (N), otherwise the matrix inversion operation $(\mathbf{X^tX})^{-1}$ in Equation 8.13 cannot be done. Secondly, if any two of the X-variables are correlated to one another, then the same matrix inversion cannot be done. In real applications, where there is noise in the data, it is rare to have two X-variables exactly correlated to one another. However, a high degree of correlation between any two X-variables leads to an unstable matrix inversion, which results in a large amount of noise being introduced to the regression coefficients. Therefore, one must be wary of *inter-correlation* between X-variables when using the MLR method.

8.2.5 Data pre-treatment

It is most often the case that raw data from analytical instruments needs to be treated by one or more operations before optimal results can be obtained from chemometric modeling methods. Although such pre-treatments often result in improved model performance, it is critically important to understand the inherent assumptions of these pre-treatment methods in order to use them optimally.

Some of the pre-treatment methods discussed below will be illustrated using a set of NIR diffuse reflectance spectra. For reference, the uncorrected NIR spectra are shown in Figure 8.3.

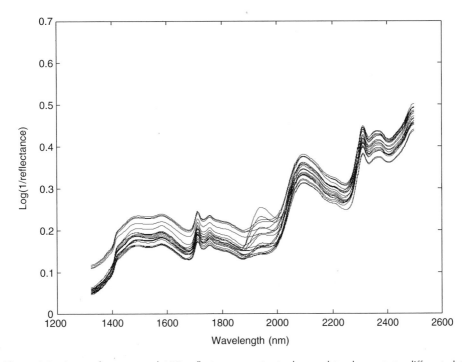

Figure 8.3 A set of uncorrected NIR reflectance spectra to be used to demonstrate different data pre-treatment methods.

8.2.5.1   Mean-centering

Mathematically, this operation involves the subtraction of each variable's response from the mean response of that variable over all of the samples in the data set. In the case where one wants to mean-center all of the independent variables in a data set, mean-centering is represented by the following equation:

$$\mathbf{X}_{mc} = \mathbf{X} - \mathbf{1}_N * \bar{\mathbf{x}} \qquad (8.14)$$

where $\mathbf{X}_{mc}$ is the mean-centered data, $\bar{\mathbf{x}}$ is a vector that contains the mean response values for each of the M-variables, and $\mathbf{1}_N$ is a vector of ones that is N-elements long.

The mean-centering operation effectively removes the absolute intensity information from each of the variables, thus enabling one to focus on the response **variations**. This can effectively reduce the 'burden' on chemometric modeling techniques by allowing them to focus on explaining variability in the data. For those who are still interested in absolute intensity information for model interpretation, this information is stored in the mean vector $(\bar{\mathbf{x}})$, and can be retrieved after modeling is done.

There are some applications where mean-centering is not necessarily optimal, and it might be advantageous to keep this mean information in the data during the modeling work. For example, if one is attempting to determine the amount of inorganic filler in polymers by NIR diffuse reflectance spectroscopy, the variation in filler concentration typically results in a shift in the absolute baseline of the reflectance spectrum. As a result, removal of the absolute reflectance baseline information by mean-centering could constitute the removal of critically relevant information for determining the filler concentration.

8.2.5.2   Autoscaling

This procedure is an extension of mean-centering. It is also a variable-wise pre-treatment that consists of mean-centering followed by division of the resulting intensities by the variable's standard deviation:

$$\mathbf{X}_{as} = \left(\mathbf{X} - \mathbf{1}_N * \bar{\mathbf{x}}\right)\left(diag(\mathbf{S}_M)\right)^{-1} \qquad (8.15)$$

where $\mathbf{X}_{as}$ is the autoscaled data, and $\mathbf{S}_M$ is a square diagonal matrix containing the standard deviations for each of the M-variables in the data.

Autoscaled data have the unique characteristic that each of the variables has a zero mean and a standard deviation of one. Like mean-centering, autoscaling removes absolute intensity information. However, unlike mean-centering, it also removes total variance information in each of the variables. It effectively puts each of the variables on 'equal footing' before modeling is done.

Autoscaling is often necessary in cases when the X-variables come from different types of instruments (e.g. FTIR, GC, ISEs), or when the units of measurement are not the same for all of the variables (e.g. ppm, %, °C, pH units). In such cases, if autoscaling is not done, then those variables with the largest absolute range will tend to dominate in the modeling process, and those with the lowest absolute range will tend to be ignored.

Autoscaling can also be used when all of the variables have the same units and come from the same instrument. However, it can be detrimental if the total variance information is relevant to the problem being solved. For example, if one wants to do an exploratory chemometric analysis of a series of FTIR (Fourier transform infrared) spectra in order to determine the relative sensitivities of different wavenumbers (*X*-variables) to a property of interest, then it would be wise to avoid autoscaling and retain the total variance information because this information is relevant for assessing the sensitivities of different *X*-variables.

8.2.5.3   Derivatives

For data in which the variables are expressed as a continuous physical property (e.g. spectroscopy data, where the property is wavelength or wavenumber), derivatives can be used to remove offset and background slope variations between samples. As the name suggests, this type of correction involves the mathematical derivation of a 'function,' where this function is simply the spectrum of a single sample over a range of wavelengths or wavenumbers. However, in analytical chemistry applications, where data are in a digitized form, a discrete form of derivation functions, called Savitsky-Golay filters,[22] can be used to calculate derivatives. These filters are essentially local functions that are applied to each spectrum in a moving-window manner across the wavelength/wavenumber axis, in order to evaluate the derivative at each wavelength/wavenumber. The use of these derivative filters requires the specification of a few parameters, which define the width and resolution of the local function.

In spectroscopy applications, a first derivative effectively removes baseline offset variations in the spectral profiles. As a result, first derivatives can be very effective in many spectroscopy applications, where spectral baseline offset shifts between samples are rather common.

Second derivative pre-treatment is rather common in NIR diffuse reflectance spectroscopy. This treatment results in the removal of both baseline offset differences between spectra **and differences in baseline slopes between spectra**. Its historical effectiveness in NIR reflectance applications suggests that baseline slope changes are common in these applications, although there is no theoretical basis for such variations.

Unlike the mean-centering and autoscaling pre-treatments, derivatives can be applied to a single sample's variable profile (i.e. spectrum) at a time. However, it is important to note that **derivative corrections do not mean-center or autoscale the data**. As a result, if one suspects that mean-centering or autoscaling is necessary, it should be done in addition to derivative pre-treatment.

Figures 8.4 and 8.5 show the effect of first derivative and second derivative pre-treatments, respectively, on the NIR reflectance spectra previously shown in Figure 8.3.

8.2.5.4   Application-specific scaling

Prior knowledge regarding the variables used in an application can be used to provide 'custom scaling.' For example, if the nominal signal-to-noise ratio (S/N) for each of the

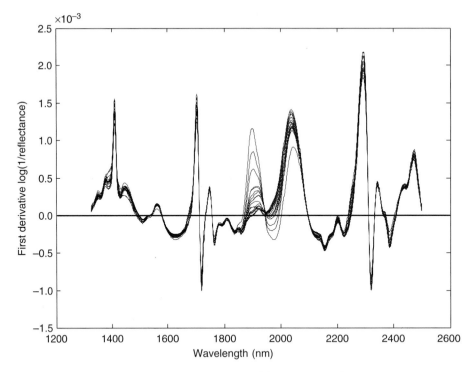

Figure 8.4   The result of first derivative pre-processing of the NIR spectra shown in Figure 8.3.

variables is known beforehand, then the variables could be scaled to their S/N before modeling, so that the variables with the best S/N can be given more influence during the modeling process.

Other knowledge that can be used for such custom scaling includes prior perception or theoretical estimation of each variable's importance in the analysis, or information regarding the financial cost for obtaining data for each specific variable.

8.2.5.5   Multiplicative signal correction (MSC)

This pre-treatment method[1,23] has been effectively used in many NIR diffuse reflectance applications, and in other applications where there are multiplicative variations between sample response profiles. In spectroscopy, such variations can be caused by differences in sample pathlength (or 'effective' pathlength, in the case of reflectance spectroscopy). It is important to note that *multiplicative variations cannot be removed by derivatives, mean-centering, or variable-wise scaling*.

The MSC model is given by the following equation:

$$\mathbf{x} = a\mathbf{x}_{\text{ref}} + b\mathbf{1}_n \tag{8.16}$$

where $\mathbf{x}_{\text{ref}}$ is a reference spectrum, a is a multiplicative correction factor, b is an additive correction factor, and $\mathbf{1}_M$ is a vector of ones that is M elements long. For most applications,

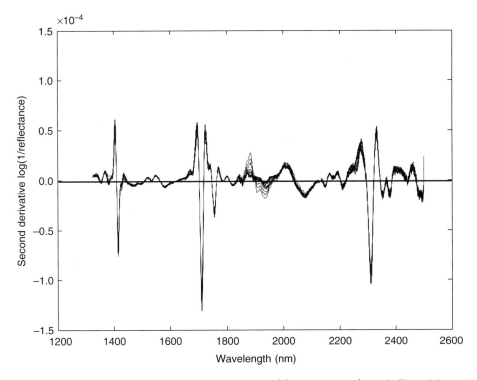

Figure 8.5 The result of second derivative pre-processing of the NIR spectra shown in Figure 8.3.

x_{ref} is simply the mean spectrum of a calibration data set, although this need not always be the case. Given a sample's spectrum and the reference spectrum, the MSC correction factors a and b are estimated using the linear regression method (Section 8.2.3). Once they are estimated, the corrected spectrum can be calculated:

$$\mathbf{x}_{corr} = \frac{\left(\mathbf{x} - \hat{b}\mathbf{1}_M\right)}{\hat{a}} \qquad (8.17)$$

When one carefully examines the MSC model (Equation 8.16), it becomes apparent that this model assumes that any sample spectrum can be simply estimated as a multiple of the reference spectrum, plus an offset. This underscores a very important limitation of the MSC method: *it assumes that offset and multiplicative spectral effects are much larger than effects from changes in sample chemistry*. As a result, uses of this method in applications where chemical-based variations in the data are much greater than the additive and multiplicative variations can lead to poor modeling results.

It should be noted that extensions of the MSC method that have been reported are designed to provide improved pre-treatment results in cases where spectral effects of sample chemistry are relatively large.[24,25]

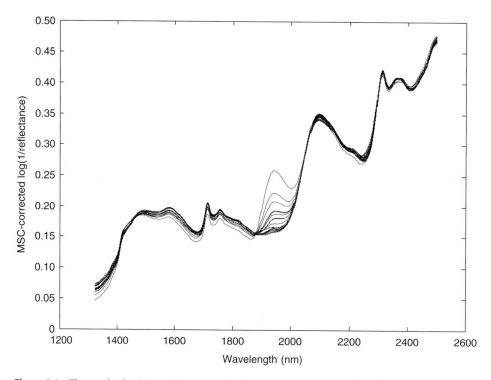

Figure 8.6 The result of MSC pre-processing of the NIR spectra shown in Figure 8.3.

Figure 8.6 shows the results obtained when MSC pre-processing is applied to the NIR reflectance spectra that were shown in Figure 8.3.

8.2.5.6 Standard normal variate (SNV)

This pre-treatment method is also focused on applications in spectroscopy, although it can be applied elsewhere. Like the MSC method, the SNV method performs both an additive and a multiplicative adjustment. However, the correction factors are determined differently. For each sample's spectrum, the offset adjustment is simply the mean of the values over all of the variables, and the multiplicative adjustment is simply the standard deviation of the values over all of the variables:

$$\mathbf{x}_{\mathrm{corr}} = \frac{\mathbf{x} - \bar{x}\mathbf{1}_M}{\sqrt{\dfrac{\displaystyle\sum_{j=1}^{M}\left(x_j - \bar{x}\right)^2}{M-1}}} \tag{8.18}$$

where $\bar{x}$ is the mean of the spectrum's intensities over all M-variables. The SNV method is performed on one spectrum at a time, and does not require the use of a reference spectrum.

8.2.6   Data compression

It was mentioned earlier that empirical multivariate modeling often requires a very large amount of data. These data can contain a very large number of samples (N), a very large number of variables (M), or both. In the case of PAC, where the analytical method is often a form of spectroscopy or chromatography, the number of variables collected per process sample can range from the hundreds to the thousands!

The presence of such a large number of variables presents both logistical and mathematical issues when working with multivariate data. From a logistical standpoint, a compressed representation of the data takes up less data storage space, and can be more quickly moved via hard-wired or wireless communication. A mathematical advantage of data compression is that it can be used to reduce unwanted, redundant or irrelevant information in the data, thus enabling subsequent modeling techniques to perform more efficiently.

Data compression is the process of reducing data into a representation that uses fewer variables, yet still expresses most of its information. There are many different types of data compression that are applied to a wide range of technical fields, but only those that are most relevant to process analytical applications are discussed here.

8.2.6.1   Variable selection: Compression by selection

The most straightforward means of data compression is to simply select a subset of variables that is determined to be relevant to the problem (or, conversely, to remove a subset of variables determined to be irrelevant to the problem). Assessment of relevance can be done manually (using a priori knowledge of the system being analyzed) or empirically (using statistical methods on the data itself).

Some of the earliest applications of chemometrics in PAC involved the use of an empirical variable selection technique commonly known as *stepwise multiple linear regression* (SMLR).[8,26,27] As the name suggests, this is a technique in which the relevant variables are selected sequentially. This method works as follows:

Step 1 A series of **linear regressions** of each X-variable to the property of interest is done.
Step 2 The single variable that has the best linear regression fit to the property of interest is selected (x_1).
Step 3 A series of **MLRs** of x_1 and each of the remaining variables to the property of interest is done.
Step 4 The variable that results in the best MLR fit to the property of interest *when combined with x_1* is selected.
Step 5 As the user desires, steps 3 and 4 can be repeated to select a third variable, fourth variable, etc.

The SMLR technique was shown to be rather effective during the early years of analytical NIR spectroscopy. However, the challenge of this method is to determine when to stop selecting variables. As more variables are selected, it is absolutely certain that the MLR fit (expressed as 'r' or 'RMSEE') for the set of variables to the property of interest will improve. However, the selection of too many variables can result in **overfitting**, where a large amount of noise is incorporated into the regression model, and the model is extremely sensitive to any uncalibrated condition. A more detailed discussion of over-fitting is presented in Section 8.3.7.

In a variant of the SMLR method, called the 'All Possible Combinations' (APC) method, the user simply inputs the number of variables to select, M_{sel}. The method then performs a series of MLRs using all possible subsets of M_{sel}-variables, and chooses the subset that provides the best fit to the property of interest. Unlike the SMLR method, this method is ensured of finding the variable subset with the best MLR fit for the data. However, it can take a rather long time to complete the search. The total number of regressions that must be done equals $M^{M_{sel}}$. In addition, one can overfit the model by selecting too many X-variables.

The above MLR-based variable selection methods, and other variants thereof, are capable of producing simple regression models that are relatively easy to explain. As I will explain later, simplicity of calibration models is of very high value in process analytical applications. However, it is important to recall the danger of highly correlated X-variables in MLR, discussed earlier. Although the nature of these variable selection methods should minimize incidences of highly correlated X-variables being chosen for MLR, they do not **explicitly** address such correlations. In process analytical spectroscopy, where spectral intensities at different wavelengths/wavenumbers are often correlated with one another, more explicit handing of these correlations can be advantageous.

8.2.6.2 *Principal Components Analysis (PCA): Compression by explained variance*

Principal Components Analysis is a data compression method that reduces a set of data collected on M-variables over N samples to a simpler representation that uses a much fewer number ($A \ll M$) of 'compressed variables,' called principal components (or **PCs**). The mathematical model for the PCA method is provided below:

$$\mathbf{X} = \mathbf{TP^t} + \mathbf{E} \tag{8.19}$$

where $\mathbf{T}$ is an N-by-A matrix containing the **scores** of the A PCs of $\mathbf{P}$ is an M-by-A matrix containing the **loadings** of the PCs, and $\mathbf{E}$ is an N-by-M matrix containing the PCA model residuals. The scores are the intensities of each of the A new 'compressed' variables for all of the N samples. The loadings are the 'definitions' of the A new variables in terms of the M-original variables. In the most commonly used algorithm for PCA,[1,28] the loadings for each PC are both normalized and orthogonal to one another, and the scores for each PC are orthogonal to one another. Conceptually, **orthogonality** can be described as two vectors being 'completely uncorrelated' with one another. Mathematically speaking, this translates to the following equations:

$$\mathbf{P^t P} = \mathbf{I}, \text{ and} \tag{8.20}$$

$$\mathbf{T^tT} = \text{diag}(t_a) \tag{8.21}$$

where **I** is the identity matrix (a diagonal matrix where each element on the diagonal is equal to one), and $\text{diag}(t_a)$ is a square diagonal matrix of dimensionality A, where each element is simply the sum of squares of the scores for each of the A PCs. An important consequence of orthogonality of the PCs obtained from PCA is that ***the issue of correlation between X-variables is completely eliminated if one chooses to use PCs instead of original X-variables.***

The most commonly used PCA algorithm involves sequential determination of each principal component (or each matched pair of score and loading vectors) via an iterative least squares process, followed by subtraction of that component's contribution to the data. Each sequential PC is determined such that it ***explains the most remaining variance in the X-data***. This process continues until the number of PCs (A) equals the number of original variables (M), at which time 100% of the variance in the data is explained. However, data compression does not really occur unless the user chooses a number of PCs that is much lower than the number of original variables ($A \ll M$). This necessarily involves ignoring a small fraction of the variation in the original X-data which is contained in the PCA model residual matrix **E**.

In practice, the choice of an optimal number of PCs to retain in the PCA model (A) is a rather subjective process, which balances the need to explain as much of the original data as possible with the need to avoid incorporating too much noise into the PCA data representation (overfitting). The issue of overfitting is discussed later in Section 8.3.7.

An example of PCA compression is made using the classic Fisher's Irises data set.[29] Table 8.2 lists ***part*** of a data set containing four descriptors (X-variables) for each of 150 different iris samples. Note that these iris samples fall into three known classes: Setosa, Verginica, and Versicolor. From Table 8.2, it is rather difficult to determine whether the four X-variables can be useful for discriminating between the three known classes.

The raw data in Table 8.2 are first autoscaled, and then the PCA method is applied. When this is done, it is found that the first two PCs explain almost 96% of the variation in

Table 8.2 The first ten samples from Fischer's iris data set, used to demonstrate PCA compression

Sample	Species	Petal width	Petal length	Sepal width	Sepal length
1	Setosa	2	14	33	50
2	Verginica	24	56	31	67
3	Verginica	23	51	31	69
4	Setosa	2	10	36	46
5	Verginica	20	52	30	65
6	Verginica	19	51	27	58
7	Versicolor	13	45	28	57
8	Versicolor	16	47	33	63
9	Verginica	17	45	25	49
10	Versicolor	14	47	32	70

Table 8.3 The explained variance in the *X*-data, for each PC, for the iris data set

PC number	% *X*-variance explained by PC	Cumulative % *X*-variance explained
1	72.8	72.8
2	22.9	95.7
3	3.7	99.4
4	0.6	100

the original data (Table 8.3). This result suggests that the four original *X*-variables, due to correlations with one another, really explain only two truly independent effects in the data. These two effects can be expressed more efficiently using the first two PC scores (see the scatter plot in Figure 8.7). Note that this PC scores scatter plot enables one to make a better assessment regarding the ability of the four original measurements to discriminate between the three classes.

Figure 8.8 provides a scatter plot of the first two PC loadings, which can be used to roughly interpret the two new 'compressed variables' in terms of the original four variables. In this case, it appears that the first PC is a descriptor between the sepal width and the other three *X*-variables, while the second PC is a descriptor of the two sepal measurements only. The plot also shows that the petal width and petal length are highly correlated with one another for the 150 irises used in this study. Further discussion on the interpretation of PC scores and loadings is found in Section 8.6.1.

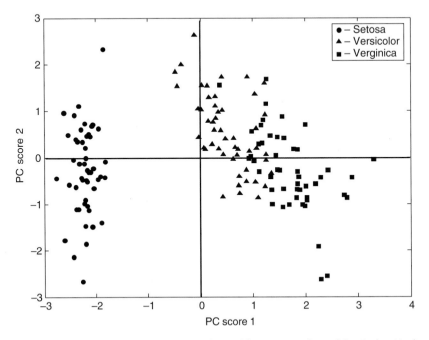

Figure 8.7 Scatter plot of the first two PC scores obtained from PCA analysis of the Fischer iris data set.

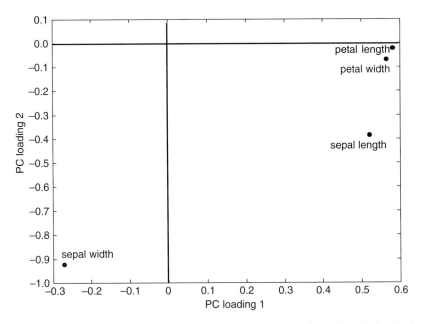

Figure 8.8 Scatter plot of the first two PC loadings obtained from PCA analysis of the Fischer iris data set.

8.2.6.3 *Fourier compression: Compression by frequency*

For data in which the variables are expressed as a continuous physical property (e.g. spectroscopy data, where the property is wavelength or wavenumber), the Fourier transform can provide a compressed representation of the data. The Fourier compression method can be applied to one analyzer profile (i.e. spectrum) at a time.

Like PCA compression, Fourier compression involves the reduction of an analyzer response profile into a simpler representation that uses basis functions that are linear combinations of the original variables. However, in the case of Fourier compression, these basis functions are pre-defined trigonometric functions of the original variables, whereas in PCA they (the PC loadings) are application-specific and must be determined through an analysis of the entire data set.

In Fourier compression, each profile $(\mathbf{x}_n)$ is essentially decomposed into a linear combination of sine and cosine functions of different frequency. If the spectrum $\mathbf{x}$ is considered to be a continuous function of the variable number m, then this decomposition can be expressed as:

$$\mathbf{x} = f(m) = a_\mathrm{o} + \sum_{k=1}^{\mathrm{NF}} a_k \sin\left(\frac{2\pi mk}{M}\right) + \sum_{k=1}^{\mathrm{NF}} b_k \cos\left(\frac{2\pi mk}{M}\right) \tag{8.22}$$

Where NF is the total number of Fourier factors to use in the decomposition, and k is the number of the Fourier factor. In practice, the Fourier coefficients (a's and b's) for all possible frequencies (corresponding to NF up to $M/2 - 1$) are determined using the discrete Fourier transform. These coefficients indicate the contribution of different frequency components to the profile. Data compression is effectively achieved by ignoring specific frequency components of the spectrum. For example, the highest-frequency components tend to describe spectral noise, and the lowest-frequency components tend to describe baseline offset and spectral background changes. In spectroscopy, the dominant frequencies in a spectrum depend on the width of the bands in the spectrum. As the bands get sharper, the higher frequencies tend to dominate, and as the bands get broader, the lower frequencies tend to dominate.

Like the scores in PCA, the Fourier coefficients can be used as a new compressed representation of the data, and thus can be directly used with regression techniques to provide quantitative prediction models.[30]

8.2.6.4   Wavelet transform: Compression by frequency and position

Although the Fourier compression method can be effective for reducing data into frequency components, it cannot effectively handle situations where the dominant frequency components vary as a function of position in the spectrum. For example, in Fourier transform near-infrared (FTNIR) spectroscopy, where wavenumber (cm^{-1}) is used as the x-axis, the bandwidths of the combination bands at the lower wavenumbers can be much smaller than the bandwidths of the overtone bands at the higher wavenumbers.[31,32] In any such case where relevant spectral information can exist at different frequencies for different positions, it can be advantageous to use a compression technique that compresses based on frequency but still preserves some position information. The Wavelet transform is one such technique.[33]

Only a general overview of the Wavelet transform will be presented here. Unlike PCA and Fourier compression, Wavelet compression involves the use of basis functions that are localized on the variable axis. The 'core' of the Wavelet transform is the 'mother wavelet,' which is an oscillatory-like function (similar to the sine or cosine) that defines the wavelet. Some examples of these functions are the Haar Wavelet and the Dabauchies Wavelet functions. This function is then represented on different x-axis scales, the largest scale being from 1 to M, the next largest scale from 1 to $M/2$, the next largest from 1 to $M/4$, etc. These different scale wavelets can then be fit to localized segments of the spectrum (using linear regression) in order to obtain the wavelet coefficients. In the Wavelet transform, the segmentation of the spectrum (or X-variable profile) must be done in a specific manner: the largest scale wavelet can only be fit to the entire spectrum, the next largest scale wavelets can only be fit to the first half of the spectrum and the second half of the spectrum, the next largest scale wavelets can only be fit to the four quarters of the spectrum, etc. Once this is done, the wavelet coefficients represent spectral intensities of different frequency (scale) components at different positions in the spectrum.

In general, wavelet functions are chosen such that they and their compressed representations are orthogonal to one another. As a result, the basis functions in Wavelet compression, like those in PCA and Fourier compression, are completely independent of one another. Several researchers have found that representation of spectral data in terms of Wavelet coefficients can result in effective multivariate calibrations.[34–36] They have also been used to efficiently store NIR spectral data.[37]

8.2.7 Spatial sample representation

It was mentioned earlier that data compression enables the X-data to be expressed in much simpler terms without losing too much information from the data. For example, most of the information in a FTIR spectrum containing 1000 data points (intensities at each of 1000 different wavenumbers) could be conveyed using only five or six PCA scores, Fourier coefficients, or Wavelet coefficients. Due to generally high covariance in spectral data, chromatographic data, and similar multivariate analytical data, this is the case for most process analytical applications. As a result, it becomes feasible to represent a single analytical profile in terms of a ***single data point in an A-dimensional space*** (where A is the number of compressed variables used to represent the data).

Such spatial representation of single-data profiles provides a very powerful view of the data, where all samples in a data set can be readily compared to one another. Figure 8.9 shows a more conventional 'profile' display of analytical data obtained from five different

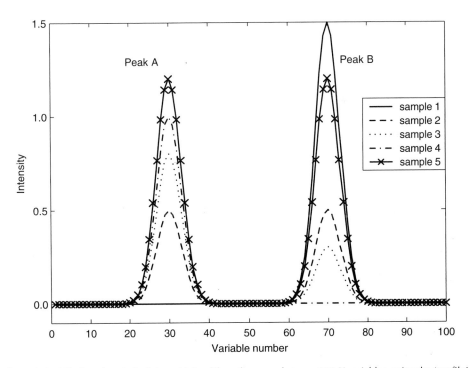

Figure 8.9 Display of analytical data obtained from five samples over 100 X-variables, using the 'profile' display format.

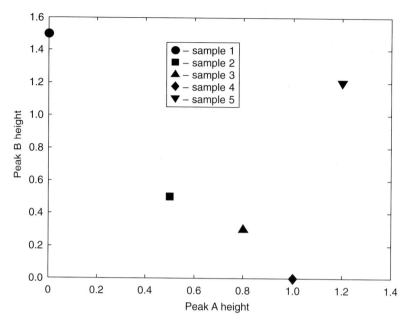

Figure 8.10 Display of the same data as in Figure 8.9, but in a 'spatial' display format.

samples over 100 X-variables. Such display formats are very common in a wide range of analytical techniques, including spectroscopy and chromatography. Figure 8.10 shows an alternative 'spatial' display of the same data, in which the samples are represented as points in space. Both graphs represent the same set of analytical data obtained on the five different samples. The 'spatial' plot is obtained by simply compressing the data into a two-value representation: namely the height of peak A and the height of peak B.

In these synthetic data, there are four samples that follow a general trend and a single sample that does not follow the trend. Using the spatial representation, it is much easier to identify the sample that does not follow the trend (sample 5), as well as to characterize the trend itself (as peak A increases, peak B decreases). Such interpretation of spatial plots, however, requires an understanding of the definition of the axes in the plot. In this example, these definitions are rather straightforward: the x-axis is the peak A intensity, and the y-axis is the peak B intensity. However, in PCA, where these definitions are determined using the loadings of the PCs, they can be more difficult to define. Nonetheless, it will be shown that, even in PCA and similar cases, very useful information can be obtained using a spatial sample representation.

8.2.8 *Experimental design*

Experimental design tools are widely considered to be some of the most important tools in the development of chemometric models.[20] At the same time, however, it is important to note that these tools can be effectively applied only in cases where sufficiently relevant calibration samples can be synthetically prepared. Consequently, in PAC, where the process samples are typically rather complex in chemical composition or at conditions that are difficult (or costly!) to prepare synthetically, there are many applications where these tools cannot be used. Nonetheless, for those applications where they can be used, experimental designs are very useful for balancing the need to provide sufficiently representative calibration samples with the need to avoid high costs associated with the preparation and analysis of calibration samples.

There are a wide variety of experimental designs which can be used for different experimental objectives. Therefore, the first step in choosing an experimental design is to **specify the experimental objective**. In the application of chemometrics to PAC, the experimental objective most often consists of two parts:

(1) Provide calibration data that sufficiently cover the range of expected analyzer responses during real-time operation.
(2) Provide calibration data that can be used to detect or assess non-linear effects in the analyzer data.

Depending on the application, there could be additional, or different, objectives to the experiment. For example, one might not want to build a quantitative regression model, but rather identify which design variables affect the analyzer response the most. In this case, a set of tools called screening designs[20] can be used to efficiently obtain the objective.

The next step in experimental design is to **identify appropriate design variables**. This step strongly depends on prior knowledge of the chemistry, physics, and process dynamics of the system, as well as the measurement principles of the analyzer. In the case of multivariate calibration, design variables are often the concentrations of process sample constituents and the values of physical properties (e.g. temperature, pressure), which are known to vary in the process samples. They can perhaps include even process parameters. During this step, it is critical to include any variables that are thought to be relevant to the ability of the process analyzer to measure the property of interest. These include other chemical constituents and physical properties that are expected to interfere with the analyzer's ability to measure the property of interest. At the same time, one should be careful not to include too many design variables because this can lead to an unacceptably large number of calibration samples.

After the design variables have been identified, the ***number* of target levels** for each design variable must be determined. Normally, one starts with two levels for each design variable. However, in chemometrics, it is often the case that one wants to use more than two levels of one or more design variables (especially the variable corresponding to the property of interest) to more thoroughly characterize any non-linear relationships between these design variables and the analyzer response. This is particularly important if non-linear modeling methods will be applied to the experimental data. Unfortunately, the total number of samples required in the experiment increases dramatically as the number of levels increases. Fortunately, it is possible to increase the number of levels used for each variable without greatly increasing the number of samples, and this will be discussed later.

Then, the ***values* of the target levels** for each design variable must be defined. Once again, this step depends greatly on prior knowledge of the system, as well as the objective of the experiment. For a typical calibration problem, one wants to assign values of the variables such that they cover the ranges of values (in a multivariate sense!) that are expected to occur during real-time operation of the analyzer. At the same time, however, one wants to avoid using ranges that are too wide because this might result in a degradation of model precision and accuracy due to inherent non-linear effects in the analyzer responses. For process spectrometers, where excessively high concentrations of some components can result in saturated absorption peaks, one is particularly suscep-tible to this trap. It is also worth mentioning that, if three or more target levels are specified for a design variable, it is not absolutely necessary to specify levels that are equidistant.

Once the target levels for each design variable are specified, one can choose from several different experimental **design types**. It is worth noting that the concept of spatial object representation, discussed earlier, is particularly helpful when discussing these different types of experimental designs. Figure 8.11 provides spatial representations of four different types of experimental designs that are useful for process analyzer calibra-tion. All of these representations involve only three design variables, each of which has two levels. Although all of these design types are very useful, they vary in terms of their ability to characterize interaction effects between design variables, as well as the total number of samples required. It is also important to note that all four of these designs as represented in Figure 8.11 include a ***centerpoint***, which represents a sample that has the average value of all of the design variables. In PAC, the centerpoint can be very useful for two reasons:

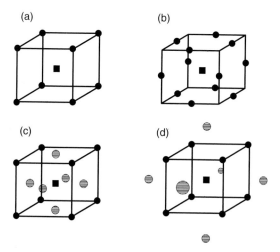

Figure 8.11 The spatial representations of four different types of experimental designs that are useful for process analyzer calibration: (a) full-factorial, (b) Box-Behnken, (c) face-centered cube, and (d) central composite.

(1) It enables an additional level of each design variable to be tested, so that the resulting data can provide *some* assessment of non-linear effects between design variables and analyzer responses (even if only two levels are specified for each design variable), and
(2) If replicate centerpoint samples are prepared (as is usually the case), they can be used to assess the repeatability of the analyzer response and the sample preparation method.

Table 8.4 provides a comparison of these four design types with regard to the number of samples that are required. Note that the inclusion of a single centerpoint to the full-factorial design results in an increase in the number of levels actually used by the design from two to three, for all design variables. The Box-Behnken, face-centered cube, and central composite design types also result in the use of three different levels for each design variable, and they also require the use of more samples than the full-factorial design. However, when one observes the spatial representation of these designs (Figure 8.11), it is clear that they do a better job of covering the design variable space than the full-factorial design, and are better able to detect non-linear relationships between variables. The central composite design has the added benefit of generating experiments that use four different levels for each variable, instead of only two or three.

Once the type of design is specified, sufficient information is available to generate a table containing a list of calibration samples, along with the values of each design variable for each sample. However, it is important to note that there are additional considerations which can further alter this experimental plan:

(1) *Randomization*: If one is concerned about time-dependent effects (e.g. analyzer response drift or chemical aging effects in the calibration standards), then the order of calibration sample analysis in the design should be randomized.
(2) *Blocking*: If there is a particular design variable that is difficult or costly to change, one might consider a sampling order in which this variable is blocked, so that it does not have to be changed frequently.
(3) *Truncation*: If prior knowledge indicates that certain samples in the original design refer to combinations of design variables that are not physically attainable, then it would be appropriate to remove these samples from the design.

Table 8.4 Comparison of four different types of experimental design with regard to the number of samples required

Design type	Number of samples required (2 levels)			Number of different levels *actually used* for each variable
	Three design variables	Four design variables	Five design variables	
Full factorial, no centerpoint	8	16	25	2
Full factorial, with centerpoint	9	17	26	3
Box-Behnken	13	25	41	3
Face-centered cube	15	25	43	3
Central composite	15	25	43	4

8.3 Quantitative model building

This section discusses common methods for building quantitative chemometric models in PAC. In this field, the user most often desires to build a model that converts values generated by an analytical instrument into values of properties or concentrations of interest for use in process control, quality control, industrial hygiene, safety, or other value-adding purposes. There are several chemometric techniques that can be used to build quantitative models, each of which has distinct advantages and disadvantages.

It should be noted that the purpose of quantitative model building discussed above constitutes a more pragmatic use of chemometric tools in PAC, which is generally a very pragmatic field. However, such modeling techniques can also be used for the more 'academic' exercise of improving the understanding of the process dynamics, measurement technology, or process chemistry. Such exploratory applications will be briefly discussed in Section 8.6.1.

8.3.1 'Inverse' multiple linear regression

When MLR is used to provide a predictive model that uses multiple analyzer signals (e.g. wavelengths) as inputs and one property of interest (e.g. constituent concentration) as output, it can be referred to as an *inverse* linear regression method. The word 'inverse' arises from the spectroscopic application of MLR because the inverse MLR model represents an inverted form of Beer's Law, where *concentration is expressed as a function of absorbance* rather than absorbance as a function of concentration:[1,38]

$$\mathbf{y} = \mathbf{X}_{sel}\mathbf{b} + \mathbf{f} \tag{8.23}$$

where $\mathbf{X}_{sel}$ represents a subset of the X-data, where only a selected subset of the M X-variables is used. The commonly used techniques of SMLR and APC linear regression, discussed earlier, can be classified as inverse linear regression because they typically use the same model function as Equation 8.23. The inverse MLR regression coefficients are determined using least squares method:

$$\hat{\mathbf{b}}_{MLR} = (\mathbf{X}_{sel}{}^{t}\mathbf{X}_{sel})^{-1}\mathbf{X}_{sel}{}^{t}\mathbf{y} \tag{8.24}$$

Once the regression coefficients of the inverse MLR model are determined (according to Equation 8.24), the property (Y-value) of an unknown sample can be estimated from the values of its selected X-variables ($\mathbf{X}_{sel,p}$):

$$\hat{y}_{p,MLR} = \mathbf{X}_{sel,p}\hat{\mathbf{b}}_{MLR} \tag{8.25}$$

The fit of an MLR model can be assessed using both the RMSEE and the correlation coefficient, discussed earlier (Equations 8.10 and 8.11). The correlation coefficient has the advantage that it takes into account the range of the Y-data, and the RMSEE has the advantage that it is in the same units as the property of interest.

The only significant implication of the inverse model is that **_all of the model error is assumed to be present in the Y-data_**, rather than in the *X*-data. Although this is not strictly true in any real applications (where the *X*-variables have noise!), this limitation has not prevented the effective use of inverse MLR in many process analytical applications.

Aside from univariate linear regression models, inverse MLR models are probably the simplest types of models to construct for a process analytical application. Simplicity is of very high value in PAC, where ease of automation and long-term reliability are critical. Another advantage of MLR models is that they are rather easy to communicate to the customers of the process analytical technology, since each individual *X*-variable used in the equation refers to a single wavelength (in the case of NIR) that can often be related to a specific chemical or physical property of the process sample.

One particular challenge in the effective use of MLR is the selection of appropriate *X*-variables to use in the model. The stepwise and APC methods are some of the most common empirical methods for variable selection. Prior knowledge of process chemistry and dynamics, as well as the process analytical measurement technology itself, can be used to enable a priori selection of variables or to provide some degree of added confidence in variables that are selected empirically. If a priori selection is done, one must be careful to select variables that are not highly correlated with one other, or else the matrix inversion that is done to calculate the MLR regression coefficients (Equation 8.24) can become unstable, and introduce noise into the model.

Another particularly dangerous property of inverse MLR is that the quality of the model fit (expressed as RMSEE or *r*) **must improve** as the number of variables used in the model increases. A mathematical proof of this property will not be presented here, but it makes intuitive sense that the ability to explain changes in the *Y*-variable is improved as one has more *X*-variables to work with. This leads to the temptation to **_overfit_** the model, through the use of too many variables. If the number of variables is already sufficient for determining the *Y*-property in the presence of interfering effects, then the addition of more unnecessary variables only presents the opportunity to add more noise to the model and make the model more sensitive to unforeseen disturbances. A discussion on over-fitting, as well as techniques for avoiding it, is provided in Section 8.3.7.

A specific example is used to illustrate the MLR methods, as well as the other quantitative calibration methods discussed in this section. In this example, a total of 70 different styrene–butadiene copolymers were analyzed by NIR transmission spectroscopy.[39] For each of these samples, four different properties of interest were measured: the concentrations of *cis*-butadiene units, *trans*-butadiene units, 1,2-butadiene units, and styrene units in the polymer. The NIR spectra of these samples are overlayed in Figure 8.12. These spectra contain 141 *X*-variables each, which cover the wavelength range of 1570–1850 nm.

To illustrate the MLR method, the SMLR calibration method is used to build a model for the *cis*-butadiene content in the polymers. In this case, four variables are specified for selection, based on prior knowledge that there are four major chemical components that are varying independently in the calibration samples. The SMLR method chooses the four *X*-variables 1706, 1824, 1670, and 1570 nm, in that order. These four selected variables are then used to build an MLR regression model for *cis*-butadiene content, the fit of which is shown in Figure 8.13. Table 8.5 lists the variables that were chosen by the SMLR method,

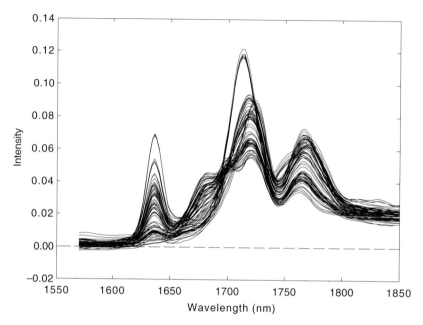

Figure 8.12 The NIR transmission spectra of 70 different styrene–butadiene copolymers, for use in demonstrating the quantitative regression model building methods.

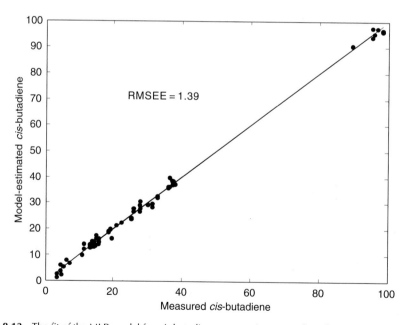

Figure 8.13 The fit of the MLR model for *cis*-butadiene content in styrene–butadiene copolymers, which uses four selected X-variables. The four variables were chosen by the stepwise method.

Table 8.5 Results obtained from building a SMLR predictive model for the *cis*-butadiene content in styrene–butadiene copolymers

Wavelength(s) chosen for MLR model (nm)	RMSEE (in % *cis*-butadiene)
1706	4.69
1706, 1824	2.19
1706, 1824, 1670	1.90
1706, 1824, 1670, 1570	1.39

as well as the calibration fit errors (in RMSEE) obtained for the MLR models that use one, two, three, and four different variables. Note that the fit error continually improves as the number of variables used in the model increases.

8.3.2 Classical least squares (CLS)

This method, which was designed for use with spectroscopic data, derives its name from the formulation of the model, which is a direct, classical expression of Beer's Law for a system with multiple analytes:

$$\mathbf{X} = \mathbf{CK} + \mathbf{E} \tag{8.26}$$

where $\mathbf{X}$ contains the spectra of N samples, $\mathbf{C}$ contains the pure concentrations of P analytes for the N samples, $\mathbf{K}$ contains the pure component spectra of the P analytes, and $\mathbf{E}$ contains the model error.[38] This model indicates that any sample's spectrum is simply a linear combination of the spectra of the pure components. In order to implement the CLS method for quantitative modeling, **the total number of analytes present in the samples must be known before-hand**. Also, please note that this model, unlike the inverse MLR model discussed earlier, assumes that all of the model error ($\mathbf{E}$) is in the spectral data, rather than in the concentration data.

A CLS model can be implemented using either **measured** pure component spectra ($\mathbf{K}$) or **estimated** pure component spectra ($\hat{\mathbf{K}}$). For most process analytical applications, the second option is most recommended because it is rare that high-quality pure component spectra of all analytes can be physically obtained. Furthermore, even if they could be obtained, they might not sufficiently represent the spectra of the pure components **in the process mixture**. However, the use of estimated pure component spectra requires that a series of standards of known concentration ($\mathbf{C}$) for **all analytes** be analyzed by a reference analytical method, thus generating a series of standard spectra ($\mathbf{X}_{std}$). If one can manage to do this, the estimated pure component spectra ($\hat{\mathbf{K}}$) can be calculated:

$$\hat{\mathbf{K}} = (\mathbf{C}^t\mathbf{C})^{-1}\mathbf{C}^t\mathbf{X}_{std} \tag{8.27}$$

The estimated pure component spectra $\hat{K}$ (or the measured pure component spectra, K, if they are sufficiently relevant) are the parameters for the CLS model. Once they are determined, the concentrations *for all analytes* in a future sample (c_p) can be estimated from the spectrum of that sample (x_p):

$$\hat{c}_p = x_p \hat{K}^t \left(\hat{K} \hat{K}^t \right)^{-1} \tag{8.28}$$

This equation represents a simple curve fit of the pure component spectra to the unknown's spectrum, where $\hat{c}_p$ contains the coefficients of the curve fit.

Unlike PCA compression or Fourier compression, where spectra are described in terms of abstract PCs or trigonometric functions respectively, CLS uses estimated pure component spectra as the basis for explaining each spectrum.

The same calibration fit metrics, RMSEE and r, can be used to express the fit of a CLS model. However, one must first calculate the model-estimated analyte concentrations for the calibration standards:

$$\hat{C} = X\hat{K} \left(\hat{K} \hat{K}^t \right)^{-1} \tag{8.29}$$

Then, separate RMSEE and r statistics can be derived for each analyte, using each column of the measured and estimated concentration matrices (C and $\hat{C}$, respectively).

The CLS method has several appealing advantages for PAC. First of all, a single model can be used to estimate multiple analytes. Secondly, because the responses of all analytes are being modeled simultaneously, it can be shown that the CLS model predictions (c_p) are invariant to pathlength or multiplicative variations in the sample spectra. This advantage can be particularly useful in spectroscopy applications, where sample thickness, pathlength, or 'effective pathlength' are known to vary with time. In addition, the use of estimated pure analyte spectra ($\hat{K}$) as a basis leads to the enhanced ability to extract qualitative information regarding the spectroscopy of the analytes, or even chemical interactions between analytes, in the process sample. The pure analyte spectrum basis is also much easier to understand, and much easier to communicate to the customers of the process analytical technology.

Unfortunately, for quantitative applications, the CLS method has some practical and technical disadvantages. From the practical viewpoint, it is only applicable for calibration to concentration properties, rather than non-concentration properties (such as viscosity, hydroxyl number, etc.). In addition, one must be sure that *all* of the analytes that could be present in a process sample have been identified. Furthermore, if one wants to use estimated pure component spectra as a basis for the CLS method, one must know the concentrations of *all* analytes in *all* of the calibration standards in order to build the model.

From the technical viewpoint, the matrix inversion $(C^tC)^{-1}$ in Equation 8.27 can be very unstable if any two of the analyte concentrations in the calibration data are highly correlated with one another. This translates into the need for careful experimental design in the preparation of calibration standards for CLS modeling, which is particularly challenging because multiple constituents must be considered. In addition, the CLS model assumes perfect linearity and additivity of the spectral contributions from each of the analytes in the sample. In many practical cases in PAC, there can be significant non-linearity of analyzer responses and strong spectral interaction effects between analytes, which can make the CLS method perform poorly. Finally, the CLS model assumes that all of the noise is in the spectral ($\mathbf{X}$) data, and there is no noise in the concentration ($\mathbf{C}$) data. This is not a particularly good assumption in most process analyzer applications, where the reference analytical data are often noisier than the spectral data.

Although the CLS model can be considered rather 'rigid' and limited in its scope of application, its advantages may be considerable in cases where it is applicable. In addition, it serves as a useful background for discussion of other regression methods.

In the styrene–butadiene polymer example, it is not physically possible to obtain spectra of pure *cis*-, pure *trans*-, and pure-1,2-butadiene because they cannot be synthesized. As a result, only the CLS method that uses estimated pure component spectra basis can be used. Fortunately, the concentrations of all four analytes (including styrene) are known for all of the samples, making it possible to estimate the pure component spectra (Figure 8.14). When these estimated pure component spectra are used to estimate the concentrations of the four constituents in the samples (Equation 8.29), it is found that the RMSEE for *cis*-butadiene is 1.53%.

8.3.3   Principal component regression (PCR)

Principal component regression is simply an extension of the PCA data compression method described earlier (Section 8.2.6.2), where a regression step is added after the data

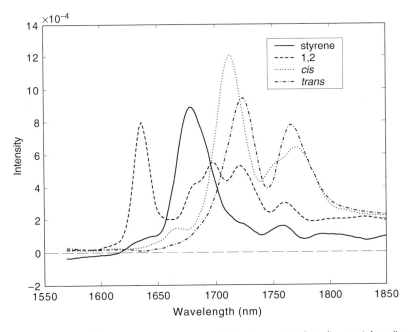

Figure 8.14   Estimates of the pure component spectra for styrene, 1,2-butadiene, *cis*-butadiene, and *trans*-butadiene units in styrene–butadiene polymers, obtained using the CLS method.

compression step.[1,40] It is very similar to inverse MLR, with the notable exception that *compressed variables* (namely PCs) are used as variables in the MLR model, rather than selected original X-variables. PCR involves a rather simple two-step process. Once the estimated PCA scores ($\hat{\mathbf{T}}$) and loadings ($\hat{\mathbf{P}}$) are obtained using the PCA modeling procedure (see Section 8.2.6.2), an MLR is carried out according to the following model:

$$\mathbf{y} = \hat{\mathbf{T}}\mathbf{q} + \mathbf{f}$$ (8.30)

where **y** denotes the known values of the property of interest, **q** denotes the Y-loadings for the PCR model, and **f** denotes the error in the Y-values. The estimated Y-loadings ($\hat{\mathbf{q}}$) are then obtained using the least squares procedure:

$$\hat{\mathbf{q}} = \left(\hat{\mathbf{T}}^t\hat{\mathbf{T}}\right)^{-1}\hat{\mathbf{T}}^t\mathbf{y}$$ (8.31)

Unlike MLR and CLS, where the matrix inversions $(\mathbf{X}^t\mathbf{X})^{-1}$ and $(\mathbf{C}^t\mathbf{C})^{-1}$ can be very unstable in some situations, PCR involves a very stable matrix inversion of $(\mathbf{T}^t\mathbf{T})^{-1}$ because the PCA scores are orthogonal to one another (recall Equation 8.21).

The PCR model can be easily condensed into a set of regression coefficients ($\hat{\mathbf{b}}_{PCR}$) using the following equation:

$$\hat{\mathbf{b}}_{PCR} = \hat{\mathbf{P}}\hat{\mathbf{q}}$$ (8.32)

The model can then be applied to provide the Y-value of an unknown sample from its array of X-values ($\mathbf{x}_p$):

$$\hat{y}_{p,PCR} = \mathbf{x}_p\hat{\mathbf{b}}_{PCR}$$ (8.33)

The calibration fit metrics previously discussed, RMSEE and *r*, can be used to express the fit of a PCR model. In this case, the RMSEE can be calculated directly from the estimated Y-residuals ($\hat{\mathbf{f}}$), which are calculated according to the following equation:

$$\hat{\mathbf{f}} = \mathbf{y} - \hat{\mathbf{T}}\hat{\mathbf{q}}$$ (8.34)

The main advantage of PCR over inverse MLR is that it accounts for covariance between different X-variables, thus avoiding any potential problems in the model computation mathematics, and removing the 'burden' on the user to choose variables that are sufficiently independent of one another. From a practical viewpoint, it is much more flexible than CLS because it can be used to develop regression models for any property, whether it is a concentration property or otherwise, and it only requires knowledge of the single property of interest for the calibration samples.

Like MLR, however, one must be careful to avoid the temptation of overfitting the PCR model. In this case, overfitting can occur through the use of too many PCs, thus adding unwanted noise to the model and making the model more sensitive to unforeseen disturbances. Model validation techniques (discussed in Section 8.3.7) can be used to avoid overfitting of PCR models.

For the styrene–butadiene example, the results of the use of the PCR method to develop a calibration for *cis*-butadiene are shown in Table 8.6. It should be mentioned that the data were mean-centered before application of the PCR method. Figure 8.15 shows the percentage of explained variance in both X (the spectral data) and Y (the *cis*-butadiene concentration data) after the use of each PC. After using four PCs, it does not appear that

Table 8.6 Results obtained from building a PCR predictive model for the *cis*-butadiene content in styrene–butadiene copolymers

Number of PCs in the PCR model	RMSEE (in % *cis*-butadiene)
1	17.00
2	5.80
3	3.18
4	1.26
5	1.25
6	1.25
7	1.23
8	1.23
9	1.15
10	1.15

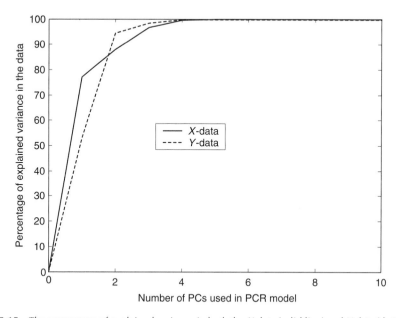

Figure 8.15 The percentage of explained variance in both the *X*-data (solid line) and *Y*-data (dotted line), as a function of the number of PCs in the **PCR** regression model for *cis*-butadiene content in styrene–butadiene copolymers.

the use of any additional PCs results in a large increase in the explained variance of X or Y. If a PCR regression model using four PCs is built and applied to the calibration data, a fit RMSEE of 1.26 is obtained.

It should be mentioned that, although the PCR method provides a more easily used and stable solution to the quantitative regression problem in PAC, the manner in which the PCR model parameters ($\mathbf{P}$, $\mathbf{q}$ and, subsequently, $\mathbf{b}$) are generated might not be optimal for building a quantitative prediction model. At this time, it is worth recalling that the PCA and PCR methods compress the X-data such that each sequentially determined compressed variable (PC) explains the most remaining variance in the X-data. In other words, the Y-data are not a factor in the data compression step. Consequently, in some applications, the variance in the X-data *that is relevant for predicting y* could be a rather small contributor to the total variance in X, and much of it could be lost through the elimination of 'weaker' PCs. This deficiency leads to the next topic of discussion.

8.3.4 Projection to latent structures (PLS) regression

The method of PLS, also known as 'Partial Least Squares,' is a highly utilized regression tool in the chemometrics toolbox,[1] and has been successfully used for many process analytical applications. Like the PCR method, PLS uses the exact same mathematical models for the compression of the X-data and the compression of the Y-data:

$$\mathbf{X} = \mathbf{T}_{PLS}\mathbf{P}_{PLS}{}^t + \mathbf{E} \tag{8.35}$$

$$\mathbf{y} = \mathbf{T}_{PLS}\mathbf{q} + \mathbf{f} \tag{8.36}$$

Like PCR, the compressed variables in PLS have the mathematical property of orthogonality, and the technical and practical advantages thereof. PLS models can also be built knowing only the property of interest for the calibration samples.

The difference between PLS and PCR is the manner in which the X-data are compressed. Unlike the PCR method, where X-data compression is done solely on the basis of explained variance in $\mathbf{X}$ followed by subsequent regression of the compressed variables (PCs) to $\mathbf{y}$ (a simple 'two-step' process), PLS data compression is done such that *the most variance in both X and Y is explained*. Because the compressed variables obtained in PLS are different from those obtained in PCA and PCR, they are *not* PCs! Instead, they are often referred to as *latent variables*.

There are actually several different PLS algorithms, the most common of which are the NIPALS algorithm[1,41] and the SIMPLS algorithm.[42] Both of these algorithms are somewhat more complex than the PCR algorithm, so the reader is referred to the above references for details. In the case of the NIPALS algorithm, several PLS model parameters are estimated: the X-scores and loadings ($\hat{\mathbf{T}}_{PLS}$ and $\hat{\mathbf{P}}_{PLS}$), the Y-loadings ($\mathbf{q}$) and the *loading weights* $\hat{\mathbf{W}}_{PLS}$. The loading weights have the same dimensionality as the X-loadings $\hat{\mathbf{P}}_{PLS}$, but, unlike the X-loadings, they are orthogonal to one another. The presence of two sets of 'loadings' ($\hat{\mathbf{P}}_{PLS}$ and $\hat{\mathbf{W}}_{PLS}$) for the NIPALS PLS algorithm is a consequence of the algorithm's constraint that the PLS scores ($\mathbf{T}$) be orthogonal to one another.

Like PCR, a PLS model can be condensed into a set of regression coefficients $\hat{\mathbf{b}}_{PLS}$. For the NIPALS algorithm discussed above, the regression coefficients can be calculated by the following equation:

$$\hat{\mathbf{b}}_{PLS} = \hat{\mathbf{W}}_{PLS}\left(\hat{\mathbf{P}}_{PLS}{}^{t}\hat{\mathbf{W}}_{PLS}\right)^{-1}\hat{\mathbf{q}} \tag{8.37}$$

The model can then be applied to provide the Y-value of an unknown sample from its array of X-values ($\mathbf{x}_p$):

$$\hat{y}_{p,PLS} = \mathbf{x}_p\hat{\mathbf{b}}_{PLS} \tag{8.38}$$

There are some distinct advantages of the PLS regression method over the PCR method. Because Y-data are used in the data compression step, it is often possible to build PLS models that are simpler (i.e. require fewer compressed variables), yet just as effective as more complex PCR models built from the same calibration data. In the process analytical world, simpler models are more stable over time and easier to maintain. There is also a small advantage of PLS for qualitative interpretative purposes. Even though the latent variables in PLS are still abstract, and rarely express pure chemical or physical phenomena, they are at least more relevant to the problem than the PCs obtained from PCR.

Unfortunately, the need to simultaneously use both X-data and Y-data for data compression results in increased complexity of PLS model determination and expression. As Martens and Næs mention, this need results in some necessary concessions in the expression of the PLS model itself: it must either be expressed using two different sets of X-loadings ($\mathbf{P}$ and $\mathbf{W}$), one which is orthogonal and another that is not, or be expressed using scores ($\mathbf{T}_{PLS}$) that are not orthogonal. Although the NIPALS algorithm was used as an example in this section, these concessions lead to the possibility of several different algorithms for calculating PLS models, each of which is more complicated than the simple two-step PCR algorithm. Details on the various PLS algorithms that have been used to date can be found in Martens and Næs.[1] For the purpose of this discussion, it will suffice to say that all PLS models can be condensed into a set of regression coefficients ($\hat{\mathbf{b}}_{PLS}$), which can be used to estimate the property of interest of an unknown sample from its analytical profile (as in Equation 8.30, for PCR).

Another potential disadvantage of PLS over PCR is that there is a higher potential to overfit the model through the use of too many PLS factors, especially if the Y-data are rather noisy. In such situations, one could easily run into a situation where the addition of a PLS factor helps to explain noise in the Y-data, thus improving the model fit without an improvement in real predictive ability. As for all other quantitative regression methods, the use of validation techniques is critical to avoid model overfitting (see Section 8.3.7).

In the styrene–butadiene copolymer example, Figure 8.16 shows the explained variance in both X and Y as a function of the number of latent variables. When this explained variance graph is compared with the same graph for the PCR model (Figure 8.15), it becomes clear that more of the Y-data are explained in the first few PLS latent variables

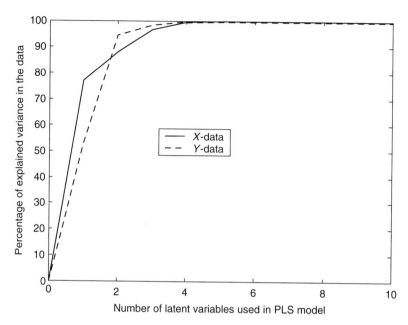

Figure 8.16 The percentage of explained variance in both the *X*-data (solid line) and *Y*-data (dotted line), as a function of the number of latent variables in a ***PLS*** regression model for *cis*-butadiene content in styrene–butadiene copolymers.

than in the first few PCR PCs. This result reflects the fact that the PLS method chooses its compressed variables with explained *Y*-data variance in mind, whereas the PCR method does not. The fit (RMSEE) of the PLS model that uses four latent variables is 1.25% cis-butadiene.

8.3.5 Artificial neural networks (ANN)

Developed several decades ago, ANNs are being increasingly applied to the development and application of quantitative prediction models.[43–45] ANNs simulate the parallel processing capabilities of the human brain, where a series of processing units (aptly called 'neurons') are used to convert input variable responses into a concentration (or property) output. Neural networks cover a very wide range of techniques that are used for a wide range of applications.

As a chemometric quantitative modeling technique, ANN stands far apart from all of the regression methods mentioned previously, for several reasons. First of all, the model structure cannot be easily shown using a simple mathematical expression, but rather requires a 'map' of the network 'architecture.' A simplified example of a feed-forward neural network architecture is shown in Figure 8.17. Such a network structure basically consists of three 'layers,' each of which represent a set of data values and possibly data processing instructions. The **input layer** contains the inputs to the model (I1–I4).

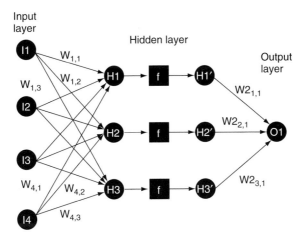

Figure 8.17 The structure of a typical feed-forward neural network.

Different linear combinations of the inputs (defined by the input weights, W) are then calculated to produce intermediate values (H1–H3), which are located in the **hidden layer** of the network. Within the hidden layer, these intermediate values are operated on by a transfer function (f) to produce processed intermediate values (H1′–H3′). Then, a linear combination of these processed intermediate values (defined by the output weights W2) is calculated to produce an output value (O1), which resides in the **output layer** of the network. In the context of analytical chemistry, the output (O1) refers to the property of interest.

It should be mentioned that there are other types of network architectures that can be used, but the feed-forward structure is shown here due to its relative simplicity and high relevance to quantitative method building.

Figure 8.17 shows a very specific case of a feed-forward network with four inputs, three hidden nodes, and one output. However, such networks can vary widely in their design. First of all, one can choose any number of inputs, hidden nodes, and number of outputs in the network. In addition, one can even choose to have more than one hidden layer in the network. Furthermore, it is common to perform scaling operations on both the inputs and the outputs, as this can enable more efficient training of the network. Finally, the transfer function used in the hidden layer (f) can vary widely as well. Many feed-forward networks use a non-linear function called the sigmoid function, defined as:

$$f(x) = \frac{1}{1 + e^{-x}} \tag{8.39}$$

However, one could also use a simple linear function or a different non-linear function.

It is interesting to note that the data processing that occurs during the PCR method is simply a special case of the ANN feed-forward data processing structure, where the input weights (W) are the PC loadings (**P**, in Equation 8.19), the output weights are the Y-loadings (**q**, in Equation 8.30), and there is no transfer function (f) in the hidden layer.[43] However, it is the use of **non-linear** transfer functions in the hidden layer (such as the common sigmoid function discussed earlier) that further distinguish the ANN method from the PCR method and all of the other quantitative methods discussed earlier. An important consequence of using non-linear transfer functions is that the ***ANN method has the flexibility to model linear or non-linear relationships*** between the X-variables and the Y-variable.

Of all the possible non-linear transfer functions used in ANNs, the sigmoid function (Equation 8.39) has the desirable property that it is approximately linear for small deviations of X around zero, and non-linear for larger deviations of X around zero. As a result, the degree of non-linearity of the transfer function can actually be altered through scaling of the inputs to the transfer function.

A final distinction of ANNs from other regression methods is the manner in which the model parameters (the two sets of weights, W and W2, and any scaling parameters) are determined. Unlike MLR, PCR, and PLS, which have well-defined (albeit sometimes complex) algorithms to determine the model parameters, ANNs use a ***learning process***. One common learning process, called 'Back-Propagation,'[46] involves the presentation of one calibration sample at a time in random order, and adjustment of the model parameters each time to optimize the current prediction error. This process is repeated until convergence of the model parameters. Næs *et al.*[43] provide a good comparison of back-propagation parameter estimation versus the more straightforward least squares parameter estimation method that is used in other regression techniques.

Although the ANN method is a very powerful tool for developing quantitative models, it is also probably the most susceptible to overfitting. For feed-forward networks, overfitting most often occurs through the use of too many nodes in the hidden layer. Although cross-validation techniques can be used to optimize the number of hidden nodes, this process is more cumbersome than for PCR or PLS because separate ANN models with different numbers of hidden nodes must be developed separately. In addition, from a practical point of view, the ANN model cannot be easily reduced into a series of regression coefficients (**b**), as in MLR, PCR, and PLS, and the real-time data processing instructions involve a series of steps, rather than a single vector dot product (as in Equations 8.25, 8.33, and 8.38). This might explain why there are very few software packages available for one to develop and implement an ANN model in real time. Finally, there is very little, or no, interpretive value in the parameters of an ANN model, which eliminates one useful means for improving the confidence of a predictive model.

With these limitations in mind, however, ANNs can be very effective at producing quantitative models in cases where unknown non-linear effects are present in the data.

In the styrene–butadiene copolymer application, a series of quantitative ANN models for the *cis*-butadiene content was developed. For each of these models, all of the 141 X-variables were used as inputs, and the sigmoid function (Equation 8.39) was used as the transfer function in the hidden layer. The X-data and Y-data were both mean-centered before being used to train the networks. A total of six different models were built, using one to six nodes in the hidden layer. The model fit results are shown in Table 8.7. Based on these results, it appears that only three, or perhaps four, hidden nodes are required in the model, and the addition of more hidden nodes does not greatly improve the fit of the model. Also, note that the model fit (RMSEE) is slightly less for the ANN model that uses three hidden nodes (1.13) than for the PLS model that uses four latent variables (1.25).

Table 8.7 Results obtained from building a neural network predictive model for the *cis*-butadiene content in styrene–butadiene copolymers

Number of nodes in the hidden layer	RMSEE (in % *cis*-butadiene)
1	2.70
2	1.52
3	1.13
4	1.05
5	0.96
6	1.07

This result could be an indicator of the improved ability of the ANN method to model non-linear relationships between the X-data and the Y-data. It could be the case that one of the four PLS latent variables is used primarily to account for such non-linearities, whereas the ANN method can more efficiently account for these non-linearities through the non-linear transfer function in its hidden layer.

Finally, it is interesting to note that the ANN model fit error actually increases as one goes from five to six nodes in the hidden layer. This increase is rather slight, but it indicates yet another distinction of the ANN method versus the PLS, PCR, and MLR methods discussed earlier. Unlike the other methods, where the model parameters are estimated using well-defined mathematical equations, the ANN method determines its parameters using an iterative learning process that starts with a random selection of parameters. As a result, if it runs the ANN model learning process on the same set of data more than once, it is likely that there will be slightly different results. Each of the models represented in Table 8.7 was built from only a single execution of the ANN learning process.

8.3.6 Other quantitative model-building tools

Although the above-mentioned methods cover most of the quantitative model building tools that are available in current commercial software, there are several other useful methods that were not mentioned. Many of these are simply extensions of the methods already discussed. A more complete list of such tools is also available.[1,44,47–49]

8.3.7 Overfitting and model validation

8.3.7.1 Overfitting and under-fitting

Throughout the above discussion of quantitative modeling tools, a recurrent theme is the danger of **overfitting** a model, through the use of too many variables in MLR, too many estimated pure components in CLS, too many factors in PCR and PLS, or too many hidden nodes in ANN. This danger cannot be understated, not only because it is so tempting to overfit, due to the improved model fit statistics that are obtained from doing so, but also because an overfit model carries with it a series of problems that are detrimental to any process analytical application. An overfit model has two distinct disadvantages over a properly fit model:

(1) It contains more of the noise from the analyzer and reference data.
(2) It is more specific to the exact data used to build it.

As a result, when the model is used in practice, it is much more sensitive to any condition that deviates from the conditions used to build the model. In process analytical applications, where there is significant error in the analyzer and reference data anyway, the second disadvantage is usually the most visible one.

A less tempting, but nonetheless dangerous alternative, is to **under-fit** a model. In this case, the model is not sufficiently complex to account for interfering effects in the analyzer data. As a result, the model can provide inaccurate results even in cases where it is applied to conditions that were used to build it! Under-fitting can occur if one is particularly concerned about overfitting, and zealously avoids any added complexity to the model, even if it results in the model explaining more useful information.

Figure 8.18 provides a graphical explanation of the phenomenon of overfitting and under-fitting, based on the explanation provided by Martens and Næs.[1] It shows that the overall prediction error of a model has contributions from two sources: (1) the interference error and (2) the estimation error. The interference error continually decreases as the complexity of the calibration model increases, as the added complexity enables the

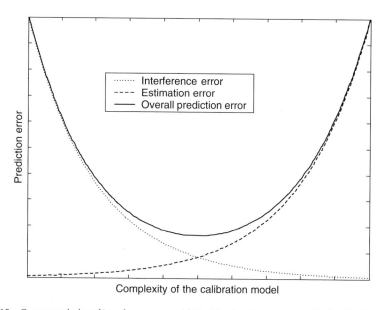

Figure 8.18 Conceptual plot of interference error (dotted line), estimation error (dashed line), and overall prediction error (solid line) as a function of model complexity.

model to explain more interferences in the analyzer data. At the same time, however, the estimation error of the model increases with the model complexity because there are more independent model parameters that need to be estimated from the same limited set of data. This results in a conceptual minimum in the overall prediction error of a model, where the interference error and estimation error are balanced. It should be noted that this explanation of a model's prediction error assumes that the calibration data are sufficiently representative of the data that will be obtained when the model is applied.

If overfitting and under-fitting are such big problems, then how can one avoid them? The most commonly used tools for combating them are called ***model validation techniques***. There are several tools that fall under this category, but they all operate with the same objective: ***attempt to assess the performance of the model when it is applied to data that were not used to build it!***

8.3.7.2 External validation

In external validation, a model is tested using data that were not used to build the model. This type of validation is the most intuitively straightforward of the validation techniques. ***If the external samples are sufficiently representative of the samples that will be applied to the model during its operation***, then this technique can be used to provide a reasonable assessment of the model's prediction performance on future samples, as well as to provide a good assessment of the optimal complexity of the model.

The main figure of merit in external validation is the root mean square error of prediction, or RMSEP:

$$\text{RMSEP} = \sqrt{\frac{\sum_{i=1}^{N} (\hat{y}_i - y_i)^2}{N}} \tag{8.40}$$

where $\hat{y}_i$ is the model-estimated property value for external validation sample i, y_i is the measured property value of external validation sample i, and N is the number of external validation samples. This equation is very similar in form to the equation for RMSEE (Equation 8.11), with one notable exception: the N samples represented in the equation are ***external validation samples***, rather than calibration samples. As a result, the RMSEP, which is in units of the property of interest, is an estimate of the model's prediction performance on future samples.

In order to assess the optimal complexity of a model, the RMSEP statistics for a series of different models with different complexity can be compared. In the case of PLS models, it is most common to plot the RMSEP as a function of the number of latent variables in the PLS model. In the styrene–butadiene copolymer example, an external validation set of 7 samples was extracted from the data set, and the remaining 63 samples were used to build a series of PLS models for *cis*-butadiene with 1 to 10 latent variables. These models were then used to predict the *cis*-butadiene of the seven samples in the external validation set. Figure 8.19 shows both the calibration fit error (in RMSEE) and the validation prediction error (RMSEP) as a function of the number of

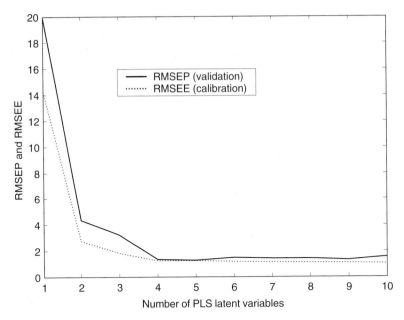

Figure 8.19 Plot of the calibration error (RMSEE) and the validation error (RMSEP) as a function of the number of latent variables, for the case where 63 of the styrene–butadiene copolymer samples were selected for calibration, and the remaining seven samples were used for validation.

latent variables in the PLS model. Note that the validation error in this example actually increases after four latent variables are added to the PLS model, whereas the calibration fit error continually decreases as expected. As a result, it appears that the use of four latent variables in this PLS model is optimal, and that the addition of more latent variables only adds noise to the model, thus damaging the model's prediction ability. The model fit error (RMSEC) at four latent variables is 1.26, and the validation error (RMSEP) is 1.34 at four latent variables. If one needed to express the estimated prediction performance of the PLS model with four latent variables, it would be more realistic to use the validation error because this reflects the performance of a model when it is applied to samples that were **not** used to build the model. This example illustrates the fact that the use of validation procedures can provide a more definitive estimate of the optimal complexity of a calibration model, as well as a more realistic estimate of the prediction error of this model.

The validation results shown in this specific example might lead one to make a generalized rule that the optimal complexity of a model corresponds to the level at which the RMSEP is at a minimum. However, it is not always the case that RMSEP-versus-complexity graph shows such a distinct minimum, and therefore such a generalized rule can result in overfit models. Alternatively, it might be more appropriate to choose the model complexity at which an increase in complexity does not significantly decrease the prediction error (RMSEP). This choice can be based on rough visual inspection of the prediction error-versus-complexity plot, or from statistical tools such as the f-test.[50,51]

Although external validation is probably the most rigorous of model validation techniques, it has several disadvantages. First of all, the external samples must be sufficiently representative of samples that the model will be applied to in the future. Otherwise, external validation can provide misleading results – in either the optimistic or pessimistic direction. This often means that a large number of external samples must be used, so that they can cover a sufficiently wide range of sample compositions that the model will experience during its operation. Under-representing these sample states in the validation set could result in overly optimistic validation results, and the use of sample states that are not represented in the calibration data can result in overly pessimistic validation results.

There is also a practical disadvantage of the external validation method. It requires that the reference analytical method be performed on an additional set of samples, namely the external validation samples. Considering the possible high cost of the reference analytical method, and the possibility of requiring a large number of external samples to provide sufficient representation of the calibration samples, this disadvantage can be rather costly.

8.3.7.3   Internal validation

In contrast to external validation, internal validation involves the use of the calibration data only, and does not require the collection and preparation of additional validation samples. There are several different techniques that are considered internal validation techniques, and these are mentioned in the following sections.

8.3.7.3.1   Model fit evaluation

As discussed earlier, the two figures of merit for a linear regression model, the RMSEE and the correlation coefficient (Equations 8.11 and 8.10), can also be used to evaluate the fit of any quantitative model. The RMSEE, which is in the units of the property of interest, can be used to provide a rough estimate of the anticipated prediction error of the model. However, such estimates are often rather optimistic because the exact same data are used to build and test the model. Furthermore, they cannot be used effectively to determine the optimal complexity of a model because increased model complexity will always result in an improved model fit. As a result, it is *very dangerous* to rely on this method for model validation.

8.3.7.3.2   Cross-validation

Probably the most common internal validation method, cross-validation, involves the execution of one or more internal validation procedures (hereby called *sub-validations*), where each procedure involves the removal of a part of the calibration data, use of the remaining calibration data to build a subset calibration model, and subsequent application of the removed data to the subset calibration model. Unlike the Model fit evaluation method discussed earlier, the same data are not used for model building and model testing for each of the sub-validations. As a result, they can provide more realistic estimates of a model's prediction performance, as well as better assessments of the optimal complexity of a model.

It is important to note that the main figure of merit for internal validation is also the RMSEP (Equation 8.40), although the model-estimated property values ($\hat{y}_i$s) are determined from different sub-validation experiments.

There are several types of cross-validation that are typically encountered in chemometrics software packages:

- *Selected subset cross-validation*: a single sub-validation is done, where a manually selected subset of the calibration data is removed.
- *Leave-one-out cross-validation*: a series of N sub-validations are done, where each sample in the calibration data is removed.
- *Random cross-validation*: a pre-specified number of sub-validations are done, where a random selection of a pre-specified number of samples are removed from the calibration data.
- *Block-wise cross-validation*: a pre-specified number of sub-validations are done, for each of which a contiguous block of a pre-specified number of calibration samples are removed.
- *Alternating sample cross-validation*: a pre-specified number (A, which is less than N) of internal validation procedures are done, by removing every Ath sample in the calibration data; the A different procedures result from starting at sample one to sample A.

Although all of these cross-validation methods can be used effectively, there could be an optimal method for a given application. The factors that most often influence the optimal cross-validation method are the design of the calibration experiment, the order of the samples in the calibration data set, and the total number of calibration samples (N).

In all cases, there are two traps that one needs to be aware of when setting up a cross-validation experiment:

(1) *The 'ill-conditioned sub-validation trap'*: This occurs when the selected subset of validation samples for a particular sub-validation is not representative of the samples in the rest of the data (which are used for modeling).
(2) *The 'replicate sample trap'*: This occurs when replicates of the same physical sample are present in both the selected subset of validation samples and the rest of the samples.

The **selected subset** cross-validation method is probably the closest internal validation method to external validation in that a single validation procedure is executed using a single split of subset calibration and validation data. Properly implemented, it can provide the least optimistic assessment of a model's prediction error. Its disadvantages are that it can be rather difficult and cumbersome to set it up so that it *is* properly implemented, and it is difficult to use effectively for a small number of calibration samples. It requires very careful selection of the validation samples such that not only are they sufficiently representative of the samples to be applied to the model during implementation, but also the remaining samples used for subset calibration are sufficiently representative as well. This is the case because there is only one chance given to test a model that is built from the data.

The **leave-one-out** method is among the simplest ones to use because it requires no input parameters. Because it runs N cross-validation procedures, it can take a rather long time if N is large, depending on the processing speed of the computer. The concern about representative validation samples is usually not an issue with this method because it tests models that use all but one of the calibration samples. However, one must accept the fact that overly pessimistic results will be obtained for those sub-validations where 'edge' samples in the calibration set are left out. These results are even more pessimistic if any of these 'edge' samples are very unique in their responses. Fortunately, the effects of such ill-conditioned sub-validations are usually 'diluted' by the results of the majority of better-conditioned sub-validations.

The **random selection** method is simply a quicker alternative to the leave-one-out method, where the user specifies both the number of samples to extract for each sub-validation and the total number of sub-validations to perform. The main advantages of this method are its speed and relative simplicity. However, there is no control over the selection of sample subsets for each sub-validation, and there is still a good chance that ill-conditioned sub-validations (such as those discussed above) will occur, through the random selection of a sample subset that contains many extreme samples. Furthermore, if this does occur, the user has no way of knowing, unless information regarding the sub-validation sample subsets is saved (which is not common in most chemometrics software packages). Therefore, if random selection is chosen to save time over the leave-one-out method, it is still a good idea to choose a large number of sub-validations.

Block-wise validation involves the selection of contiguous sample blocks for the validation subsets. This type of validation takes on special meaning only if the data profiles in the calibration data are arranged in a specific order (e.g. by sample collection time). Otherwise, it is essentially the same as random validation. If the samples are arranged by time, then block-wise validation can be used to provide a good estimate of model prediction performance over time. This is particularly appealing in PAC, where one is often concerned with model performance over long periods of time. However, the ill-conditioned sub-validation trap and replicate sample traps still apply, and require the user to carefully select the validation parameters (especially the number of samples per block). If replicates are blocked together in the calibration data, then this method can be used to avoid the replicate sample trap.

Like block-wise validation, **alternating sample** validation differs from random validation only if the calibration samples are arranged in a specific order. In this case, however, the potential for falling into the replicate sample trap is much greater, especially if replicate samples are blocked together in the calibration data (which they often are). In PAC, it is often the case that calibration samples are obtained from actual on-line measurements, and are consequently arranged by time. In this situation, alternating sample validation would not provide a good assessment of how the model will perform over time. This type of validation is suited for a specific case: where calibration samples from different classes are blocked in equal-sized blocks. In such cases, alternating sample validation (with careful choice of parameters!) can be used to avoid the ill-conditioned sub-validation trap by ensuring that each class is represented in both model and test sets for each sub-validation. Since this type of calibration sample order is rarely encountered in analytical applications, this type of validation is often not available in common chemometrics software packages.

8.3.7.3.3 *Other internal validation methods*

Although cross-validation is by far the most frequently used validation method in practice, it should be noted that there are other internal validation methods that have also been shown to be effective. A technique called leverage correction[1] involves an altered calculation of the RMSEE fit error of a model, where the contribution of the error from each calibration sample is weighted by that sample's leverage in the calibration set (a concept that will be discussed later).

8.3.8 *Improving quantitative model performance*

Once one builds a quantitative model and assesses its performance using either the validation methods discussed above or actual on-line implementation, the unavoidable question is: 'Can I do better?' In many cases, the answer is: 'quite possibly.' There are several different actions that one could take to attempt to improve model performance. The following is a list of such actions, which is an expansion of a previously published guide by Martens and Næs.[1]

8.3.8.1 *More calibration data*

The use of more calibration data can result in improved model performance, especially if the additional calibration data result in an improved representation of the sample states that need to be covered for calibration. Referring to the earlier plot of interference error and estimation error versus model complexity (Figure 8.18), an improved representation of the calibration space corresponds to a general drop in the level of the estimation error curve. This results in lower overall prediction error.

8.3.8.2 *Improved sampling protocol*

This means for improvement concerns the experimental procedures that are used to collect and analyze the calibration samples. In PAC, sample collection can involve either a highly automated sampling system, or a manual sampling process that requires manual sample extraction, preparation, and introduction. Even for an automated data collection system, errors due to fast process dynamics, analyzer sampling system dynamics, non-representative sample extraction, or sample instability can contribute large errors to the calibration data. For manual data collection, there are even more error sources to be considered, such as non-reproducibility of sample preparation and sample introduction to the analyzer.

With regard to analysis, there could be a different set of operating parameters for the analyzer instrument that would result in improved calibration data. For example, for an FTIR spectrometer, one could decrease the spectral resolution to obtain a better spectral S/N for the same total scan time. Such a strategy could be advantageous if the spectral features being used for calibration have relatively low resolution. There could also be specific sets of instrument-operating parameters where the instrument response is noisy or unstable over time, in which case they can be changed to improve precision and stability of the analyzer data.

There are several strategies and tools that can be used to assess, and subsequently minimize, such sampling-based errors in the calibration data. The use of replicate samples in the calibration experiment can help to detect noise issues around instrument repeatability as well as reference method repeatability. For many of the other error sources mentioned, one can consult the operating manual for the analyzer, or perform specialized experiments that involve an experimental design. For experiments that involve many variables and a large amount of data, there are several optimization tools that can be used to efficiently converge to an optimal solution.[52]

8.3.8.3   X-data pre-treatment

This improvement strategy involves the reduction of irrelevant information in the X-data, thus reducing the burden on the modeling method to define the correlation with the Y-data. Various types of X-data pre-processing, discussed earlier (Section 8.2.5), can be used to reduce such irrelevant information. Improvement can also be obtained through elimination of X-variables that are determined to be irrelevant. Specific techniques for the selection of relevant X-variables are discussed in a later section.

8.3.8.4   Improved reference analysis

In many process analytical applications, the largest source of error in the calibration model is often the reference analysis. Random error in the reference analysis causes noise in the reference data ($\mathbf{y}$), which is propagated into the model parameters (see Equations 8.24, 8.27, and 8.31, for MLR, CLS and PCR, respectively). As a result, any reduction in the error of the reference method can lead to highly improved calibration models. In cases where the cost of a reference method is not too great, one might run replicate reference analyses of each calibration sample. If the reference method error is truly random, then the average of the replicate analyses has less noise than the individual values by a factor of the square root of the number of replicates. Alternatively, one could treat each replicate as an independent calibration sample, in which case the signal averaging benefit is still obtained, but it is also possible to detect the presence of reference analysis outliers (or Y-outliers, discussed later) in the individual replicates.

Reduction of reference method error can also be achieved through modification of the analysis procedure itself. Because this strategy is very application-specific, it will not be discussed in detail here.

8.3.8.5   Subset modeling

As discussed earlier, chemometric models are very specific to the data that were used to build them. It is often the case that one attempts to build a single model that is relevant for a wide range of different sample states. In such cases, the modeling method might struggle to account for all of the possible interferences that can occur over this wide range of sample states. This often results in rather high model complexity, which leads to large amounts of noise in the model, as well as very high sensitivity of the model to uncalibrated sample states.

An alternative to the 'global' modeling approach discussed above is the 'subset' modeling approach, where a limited population of the calibration data is used to build a model that is specific to that population. This approach can lead to simpler, less noisy models. However, ***it is critical to communicate the limitation of the model relevance*** that was 'paid' in order to gain these improvements. Otherwise, there is a great potential for misuse of the model in its application. One means of avoiding the misuse of such subset models is the application of real-time outlier detection models, which will be discussed below in Section 8.4.3.

8.3.8.6 More appropriate model structure

It could be the case that the specific method that was used to develop the model required assumptions about the calibration data that were highly inaccurate. An example of such a case would be that of the CLS method applied to a set of data where all of the chemical constituents in the samples were ***not*** known. Or, the MLR, PCR, or PLS methods were used on a set of data where strong non-linearities between the property of interest and many of the X-variables exist. In such cases, the modeling method itself is ill suited to provide an optimal model.

In order to achieve model improvement by this means, it is first necessary to detect the inappropriateness of the current model structure. The means of detection depends on the nature of the inaccuracies of the model structure. For the CLS method, which has the most stringent assumptions, deviations from these assumptions will often produce strange, physically inconsistent artifacts in the estimated pure component spectra ($\hat{\mathbf{K}}$), such as negative absorptivities. For the other regression methods (MLR, PLS, PCR), the most common model inaccuracy comes from the assumption of linearity between the X-variables and the Y-variable. Although these methods are somewhat tolerant to weak non-linearities, they can provide poor results when the non-linearities are strong. Such non-linearities can often be detected through observation of the Y-residuals ($\mathbf{f}$), as a function of sample number or as a function of Y-value. For PLS and PCR, additional clues to non-linearity can be obtained by viewing each of the PC scores ($\mathbf{T}$) vs. the Y-value.[53]

If model inaccuracies are detected, there are several methods that can be used to improve model accuracy. For inaccuracies around the stringent requirements of the CLS method, it is often necessary to 'retreat' to one of the other regression methods which has less stringent requirements. For inaccuracies around linearity, one could try non-linear pre-processing of the X- and/or Y-data before applying the modeling method. One could also try one of the non-linear extensions of the PCR and PLS methods, such as non-linear PLS.[44,54] Alternatively, there is the ANN method, which has the inherent capability of handling non-linearities between the X-variables and the Y-variable. Finally, the subset modeling approach mentioned above can be used to deal with non-linearities. For this solution, several local models can be built over different sub-ranges of the Y-values, such that the X–Y relationship in these sub-ranges is approximately linear.

It should be noted, however, that the factor-based methods (PCR and PLS) have the ability to model weak non-linearities rather well. Consequently, the model errors due to non-linearity might not be large enough to justify a considerable increase in model complexity through the use of non-linear modeling and subset modeling methods. Therefore, the decision to change the model structure is very application-dependent.

8.4 Outliers

8.4.1 What are they, and why should we care?

Outliers can be defined as 'any observation that does not fit a pattern.' In analytical chemistry, an outlier could be a pure analyte concentration that is 'very different' from the rest of the analyte concentrations obtained from a set of calibration samples. In a typical quantitative calibration problem in PAC, one can encounter three different types of outliers:

(1) An *X-sample* outlier – a sample that has an extreme analytical (spectral) profile.
(2) A *Y-sample* outlier – a sample that has an extreme value of the property of interest.
(3) An *X-variable* outlier – a predictor variable that behaves quite differently than the rest of the predictor variables.

It is very important to note that the term *'outlier' does not imply 'incorrect.'* An outlier could be caused by an error or an incorrect action, but it could just as easily be caused by a real phenomenon that is relevant to the problem.

Outliers demand special attention in chemometrics for several different reasons. In calibration data, their extremeness often gives them an unduly high influence in the calculation of the calibration model. Therefore, if they represent erroneous readings, then they will add disproportionately more error to the calibration model. Even if they represent informative data, it might be determined that this specific information does not need to be included in the model.

Outliers are also very important when one is applying a model because they can be used to indicate whether the model is being applied to an inappropriate sample. Details regarding such on-line outlier detection are provided in a later section (Section 8.4.3).

8.4.2 Outliers in calibration

The fact that not all outliers are erroneous leads to the following suggested practice of handling outliers in calibration data: (1) detect, (2) assess, and (3) remove (if appropriate). In principle, this is the appropriate way to handle all outliers in a data set. In practice, however, there could be hundreds of calibration samples and thousands of X-variables. In such a case, individual detection and assessment of all outliers could be a rather time-consuming process. In fact, I would venture to say that this process is one of the most time-consuming processes in model development. However, it is one of the most important processes in model development, as well. The tools described below enable one to accomplish this process in the most efficient and effective manner possible.

8.4.2.1 Visual detection

The most obvious outliers in a calibration data set can be detected by simply plotting the data in various formats. Assessment of the outlier can be based on prior knowledge of the process, sample chemistry, or analyzer hardware. For PAC, one can detect X-sample and X-variable outliers by simply overlaying a series of analytical profiles. Figure 8.20 shows such a plot for a process spectrometer calibration data set. In this example, it is obvious that the profile for calibration sample 72 is very different from the profiles of the other calibration samples. Prior knowledge of the analyzer and its application suggests that the spectrum obtained for sample 72 represents an empty spectrometer sample cell. As a result, it was determined that removal of this sample from the data set was indeed appropriate. Furthermore, note that the intensities of the other four spectra in the range of variable numbers 300–310 are rather high and noisy. Once again, prior knowledge of the analyzer and its application tell us that such responses are likely to be caused by saturation effects in the spectrometer stemming from a highly absorbing sample at these wavelengths. This knowledge leads us to believe that these X-variables are also outliers, and should be removed from the calibration data.

In a similar manner, Y-sample outliers can also be detected by simply plotting the Y-values for all samples in the data set as a function of sample number, or as a histogram. The histogram format allows one to detect samples that have Y-values that are very different from those of the rest of the samples. The Y-value-versus-sample number plot is meaningful only if the samples are arranged in a specific order (e.g. by time). If the samples are arranged by time, then one can check for Y-values that do not follow an expected time trend. Assessment of such Y-sample outliers can involve use of production

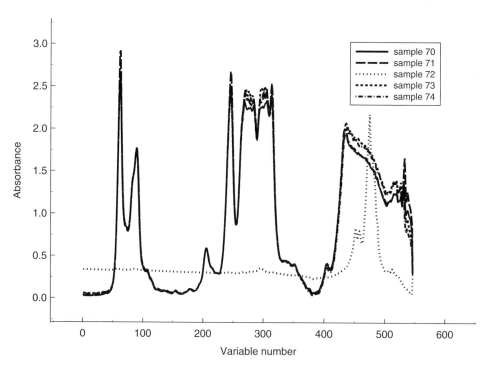

Figure 8.20 Detection of a single X-sample outlier by plotting the analytical profiles of five of the calibration samples in a process analytical calibration data set.

records for the samples, the reference analytical records for the samples, and prior knowledge of the standard deviation of the reference method.

It is advantageous to first screen the data for such strong outliers before using it for any chemometric modeling method. If such outliers are not removed, they will strongly influence the modeling procedure, thus producing strongly skewed or confusing results. They will ultimately need to be addressed at some point anyway, so it is best to get to them as early as possible.

8.4.2.2 Model-based detection

In today's process analytical instruments, where response noise and reproducibility have been greatly improved, it is quite possible to encounter outliers that are not easily visible by plotting the raw data. These outliers could involve single variables or samples that have relatively small deviations from the rest of the data, or they could involve sets of variables or sets of samples that have a unique multivariate pattern. In either case, these outliers, if they represent unwanted or erroneous phenomena, can have a negative impact on the calibration model.

For such outliers, detection and assessment can actually be accomplished using some of the modeling tools themselves.[1,3] In this work, the use of PCA and PLS for outlier detection is discussed. Since the PCA method only operates on the X-data, it can be used to detect X-sample and X-variable outliers. The three entities in the PCA model that are most commonly used to detect such outliers are the estimated PCA scores ($\hat{\mathbf{T}}$), the estimated PCA loadings ($\hat{\mathbf{P}}$), and the estimated PCA residuals ($\hat{\mathbf{E}}$), which are calculated from the estimated PCA scores and loadings:

$$\hat{\mathbf{E}} = \mathbf{X} - \hat{\mathbf{T}}\hat{\mathbf{P}}^{\mathbf{t}} \tag{8.41}$$

The scores and loadings are used to detect samples and X-variables (respectively) that have a very high importance in the PCA model. This detection is accomplished through *leverage statistics*, which are derived from the scores and loadings matrices. Sample leverage is calculated from the scores matrix by the following equation:

$$\mathrm{LEV}_i = \frac{1}{N} + \hat{\mathbf{t}}_i^{\mathbf{t}}\left(\hat{\mathbf{T}}^{\mathbf{t}}\hat{\mathbf{T}}\right)^{-1}\hat{\mathbf{t}}_i^{\mathbf{t}} \tag{8.42}$$

where $\hat{\mathbf{t}}_i$ is the vector containing the scores for sample i for each of the A PCs. The average value of the sample leverage over a data set is equal to $(1 + A)/N$. Samples for which the leverage greatly exceeds this value can be flagged as potential outliers.

Similarly, the leverages for each of the M X-variables can be calculated using the PCA loadings:

$$\mathrm{LEV}_j = \hat{\mathbf{p}}_j^{\mathbf{t}}\hat{\mathbf{p}}_j \tag{8.43}$$

where $\hat{\mathbf{p}}_j$ is the vector containing the loadings for X-variable j for each of the A PCs. The average variable leverage in a data set is equal to A/M. Variables for which the leverage greatly exceeds this value can be flagged as potential outliers.

In general, the **leverage** statistics mentioned above reflect the extremeness of samples or variables **within** the PCA model, once the optimal number of PCs (A) is determined. In contrast, statistics based on the model **residual** can be used to detect samples and variables that have a high amount of variance **outside** of the PCA model.

The average residual for the entire data set is a 'standard' by which other residuals can be compared. It is calculated as:

$$\text{RES}_{\text{TOT}} = \frac{\sum\limits_{i=1}^{N} \sum\limits_{j=1}^{M} \hat{e}_{i,j}^2}{df} \tag{8.44}$$

where $\hat{e}_{i,j}$ is the value in the estimated PCA residual matrix $\hat{\mathbf{E}}$ for the ith sample and jth variable, and df is the degrees of freedom for each of the $\hat{e}_{i,j}$s. The numerator in this expression is simply the sum of squares of all the elements in the estimated PCA residual matrix $\hat{\mathbf{E}}$. The degrees of freedom (df) is roughly equal to the number of samples times the number of variables ($N*M$). However, it is actually somewhat less than this value because several degrees of freedom were 'used up' to estimate the parameters in the PCA model. For PLS regression, Martens and Jensen[10] approximated df as follows:

$$df = (N*M) - K - A * \max(N, M) \tag{8.45}$$

where $\max(N, M)$ is the maximum value from N and M. A **sample-wise residual** (RES_i) is calculated as follows:

$$\text{RES}_i = \frac{\sum\limits_{j=1}^{M} \hat{e}_{i,j}^2}{\left(\frac{df}{N}\right)} \tag{8.46}$$

If a RES_i is much greater than the overall residual (RES_{TOT}), then the sample can be considered an outlier. A more rigorous outlier assessment can be made by applying the f-test to the residual ratio $\text{RES}_i/\text{RES}_{\text{TOT}}$, where $df/1$ degrees of freedom are used in the numerator and df degrees of freedom are used in the denominator. This way, a specific confidence level can be assigned for outlier detection (e.g. 95, 99, or 99.9%). Martens and Næs[1] propose a rough, practical guideline of $\text{RES}_i/\text{RES}_{\text{TOT}} > 2$ or 3 for outlier flagging, but this threshold can be higher if less sensitive outlier detection is desired. Samples with high residuals are those for which there is a large amount of information that is **not** explained by the PCA model containing A PCs.

In a similar manner, a **variable-wise residual** (RES_j) is calculated by the following equation:

$$\text{RES}_j = \frac{\sum\limits_{i=1}^{N} \hat{e}_{i,j}^2}{\left(\frac{df}{M}\right)} \tag{8.47}$$

In this case, if a RES_j is much greater than the RES_{TOT}, then the variable can be considered a potential outlier. Once again, the f-test, or the rough criterion of $RES_j/RES_{TOT} > 2$ or 3, can be used to assess the outlier status of different variables. Variables with high residuals tend to explain phenomena that are not included in the PCA model containing A PCs. This can be due to abnormal behavior, in that the response of such a variable does not follow any of the patterns observed for the majority of variables over all of the calibration samples.

Figure 8.21 illustrates the detection of X-sample outliers by PCA modeling, using the same analyzer data that were discussed in the previous section, with visual outliers removed. Each of the remaining calibration samples is represented as a single point in a scatter plot of the RES_i vs. the LEV_i. This plot indicates that calibration samples 469, 785, and 786 have both residuals and leverages that are much greater than those of the rest of the calibration samples. From a statistical point of view, there is rather strong evidence that these three samples are indeed X-sample outliers. However, it is still prudent to confirm their outlier status by observing their individual analytical profiles. This is especially the case if the leverage-versus-residual scatter plot shows that the suspected outliers are not as obvious as they are in Figure 8.21.

It should be noted that X-variable outlier candidates can be detected using a similar leverage-versus-residual scatter plot. However, in this case, the LEV_j is plotted versus the RES_j.

Of the three types of outliers listed earlier, there is one type that cannot be detected using the PCA method: **the Y-sample outlier.** This is simply because the PCA method does not use any Y-variable information to build the model. In this case, outlier detection can be done using the PLS regression method. Once a PLS model is built, the Y-residuals ($\hat{\mathbf{f}}$) can be estimated from the PLS model parameters $\hat{\mathbf{T}}_{PLS}$ and $\hat{\mathbf{q}}$:

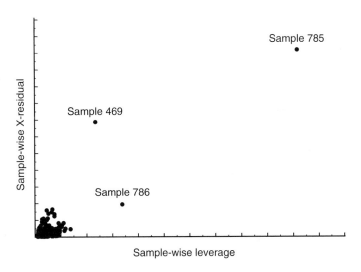

Figure 8.21 Plot of the RES_i vs. the LEV_i, obtained from PCA modeling of a calibration data set after the visual outliers were removed.

$$\hat{\mathbf{f}} = \mathbf{y} - \hat{\mathbf{T}}_{PLS}\hat{\mathbf{q}} \tag{8.48}$$

The individual N elements of the Y-residual vector ($\hat{\mathbf{f}}$) can then be observed to detect the presence of Y-sample outliers. The overall Y-residual (YRES$_{TOT}$) can be calculated as:

$$\text{YRES}_{TOT} = \frac{\sum_{i=1}^{N} \hat{f}_i^2}{df_y} \tag{8.49}$$

where $\hat{f}_i$ is the value in the Y-residual vector ($\hat{\mathbf{f}}$) for sample i, and df_y is the degrees of freedom that are 'left over' after estimating the PLS model parameters. An estimate of $df_y = N - 1 - A$ is usually appropriate.[1]

The sample-wise Y-residual (YRES$_i$) is simply the squared value of the sample's Y-residual divided by the appropriate degrees of freedom:

$$\text{YRES}_i = \frac{\hat{f}_i^2}{\left(\frac{df_y}{N}\right)} \tag{8.50}$$

If a YRES$_i$ is much greater than the YRES$_{TOT}$, then the sample can be considered an outlier. Such Y-sample outliers are usually caused by a large systematic error in the reference analytical method.

Figure 8.22 shows a scatter plot of the YRES$_i$s vs. the RES$_i$s that were obtained from a PLS model that was built using the same analyzer data, but with visual and PCA-detected outliers removed. The four samples that are circled in the figure are suspected Y-sample outliers due to their abnormally high Y-residuals. In this case, the calibration samples were arranged in chronological order, which made it possible to further investigate these samples by plotting the Y-values of all the samples versus the sample number.

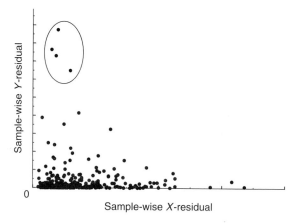

Figure 8.22 Plot of the YRES$_i$ vs. the RES$_i$, obtained from PLS modeling of a calibration data set after the visual and PCA-detected outliers were removed.

When this was done, it was found that all four of these samples were taken during transitions between product grades, when the sample composition is known to change quickly with time, and the concentration values (Y) are prone to large errors. With this prior knowledge in mind, it was decided to remove these samples from the data set.

8.4.3 Outliers in prediction

Once a chemometric model is built, and it is used to produce concentration or property values in real time from on-line analyzer profiles, the detection of outliers is a particularly critical task. This is the case for two reasons:

(1) Empirical models must not be applied to samples that were not represented in the calibration data.
(2) Application of empirical models to inappropriate samples *can* produce plausible, yet highly inaccurate results.

As a result, it is very important to evaluate process samples in real time for their appropriateness of use with the empirical model. Historically, this task has been often overlooked. This is very unfortunate not only because it is relatively easy to do, but also because it can effectively prevent the misuse of quantitative results obtained from a multivariate model. I would go so far as to say that *it is irresponsible to implement a chemometric model without prediction outlier detection*.

Such real-time evaluation of process samples can be done by developing a PCA model of the calibration data, and then using this model in real time to generate *prediction* residuals (RES_p) and leverages for each sample.[3] Given a PCA model of the analytical profiles in the calibration data (conveyed by $\hat{T}$ and $\hat{P}$), and the analytical profile of the prediction sample (x_p), the scores of the prediction sample can be calculated:

$$\hat{t}_p = x_p \hat{P} \tag{8.51}$$

The prediction leverage (LEV_p) can then be calculated using the scores of the prediction sample, along with the scores of the PCA model:

$$LEV_p = \frac{1}{N} + \hat{t}_p \left(\hat{T}^t \hat{T} \right)^{-1} \hat{t}_p^t \tag{8.52}$$

The ratio of this LEV_p to the average sample-wise leverage in the calibration data set, which can be called the *leverage ratio*, can then be used for outlier testing. The average sample-wise leverage of a data set is equal to $(1 + A)/N$, where A is the number of PCs used in the PCA model. Prediction samples for which this leverage ratio is much greater than, say, three or four could be flagged as potential outliers.

The calculation of the RES_p requires that the model estimate of the analytical profile of the prediction sample first be computed:

$$\hat{x}_p = \hat{t}_p \hat{P}^t \tag{8.53}$$

From this, the residual prediction profile ($\hat{\mathbf{e}}_p$) can be calculated:

$$\hat{\mathbf{e}}_p = \mathbf{x}_p - \hat{\mathbf{x}}_p \tag{8.54}$$

The RES_p is then calculated as the sum of squares of the elements of this $\hat{\mathbf{e}}_p$, divided by the appropriate degrees of freedom:

$$\text{RES}_p = \frac{\sum_{j=1}^{M} \hat{\mathbf{e}}_{p,j}^2}{(M - A)} \tag{8.55}$$

The ratio of this RES_p to the RES_{TOT} *obtained from the calibration data* (RES_{TOT}, Equation 8.44), which can be called the *residual ratio*, can then be used to determine whether the sample is a potential outlier. Prediction samples for which this ratio is much greater than, say, three or four could be flagged as potential outliers.

Figure 8.23 shows an example of a prediction outlier PCA model in operation for an on-line process spectrometer. In this example, both the residual ratio and the leverage ratio are tracked before and after a calibration model upgrade is performed. In this case, the calibration upgrade involves the addition of more recently obtained process data to the calibration data set, followed by rebuilding both the quantitative PLS calibration model and the PCA outlier model using the updated data set. Before the upgrade, the residual and leverage ratios were somewhat higher than desired, possibly because there

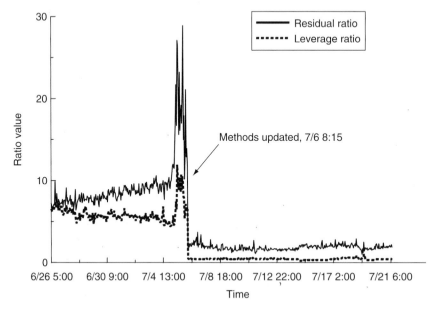

Figure 8.23 Example of the use of an on-line PCA model application to identify outliers during prediction. Solid line: ratio of prediction residual to average sample residual of the calibration samples; dotted line: ratio of prediction leverage to the average sample leverage of the calibration samples.

were some gradual long-term changes in the spectrometer response and/or the process sample composition that made the process spectra less relevant for the original methods. After the upgrade, the methods became more relevant for the process spectra, thus resulting in residual and leverage ratios that are within the desired range from one to four. Although it is not illustrated in Figure 8.23, such on-line outlier detection can also be used to detect disturbances in either the process analyzer or the process sample, which cause the process spectra to be invalid for use with the current calibration models.

It should be noted that there is no 'global' PCA prediction outlier model for a given application. In fact, several different prediction outlier models can be developed, using different subsets of calibration samples and/or X-variables, to provide alarms for several different outlier conditions.

8.5 Qualitative model building

At first thought, one might be skeptical about the need for qualitative model building in the field of PAC, where the customer most often wants a meaningful number from the analyzer. However, there are several circumstances where qualitative models can be very useful in PAC:

(1) Poor precision of reference concentration data causes unacceptably large errors in quantitative models.
(2) The inherent quantitative relationship between the X-variables and the Y-variable is very non-linear and very complex.
(3) One wants to detect unacceptable extrapolation of a quantitative model.
(4) One wants to provide an alarm for a dangerous condition in the process, which the analyzer is capable of detecting.

Cases 1 and 2 reflect an inherent advantage of qualitative models over quantitative ones: that it is not necessary to assume a quantitative relationship between the X-variables and the Y-variable. This results in less 'burden' on the modeling procedure because it is only necessary to detect and characterize patterns in the X-data. Consequently, there is a greater possibility to generate an effective model. The disadvantages of qualitative models are that they generate much less-specific results (pass/fail, or high/medium/low), and that they tend to be somewhat more complicated to implement in a real-time environment. Nonetheless, it is important to note that qualitative models can provide a viable alternative in cases 1 and 2, if the customer is willing to accept a non-quantitative result.

Case 3 indicates a means by which a qualitative model can be used to support a quantitative model, and was discussed in Section 8.4.3. Case 4 is a special application for PAC, in which the on-line analyzer can be set up to serve as 'health monitor' for the process.

This section will focus on classification methods that are in the category of ***supervised learning***, where a method is developed using a set of calibration data and complete prior knowledge about the classification of calibration samples. The other category,

unsupervised learning, is more appropriate for the exploratory analysis section which will be presented later.

In order to better understand supervised learning methods, it is useful to recall the concept of spatial sample representation which was mentioned earlier in Section 8.2.7. The development of any supervised learning method involves three steps:

(1) space definition
(2) specification of a distance measure in the space
(3) classification rule development.

8.5.1 Space definition

There are many different ways in which the classification space can be defined. The simplest space definition involves the use of individual selected *X*-variables to define each dimension in the space. One could also define the dimensions of the space using the first *A* PCs obtained from a PCA analysis of the calibration data.

The selected *X*-variable space is relatively easy to understand, and easy to implement in real time, because the original *X*-variables are used. However, one must be careful to avoid selecting redundant variables, or too many variables, which could lead to overfitting of the classification model. In practice, it is common to use empirical methods to select either original *X*-variables or compressed *X*-variables that provide the greatest ability to discriminate between classes. Some classification methods use empirical modeling tools to avoid selecting redundant *X*-variables. However, it is usually up to the user to limit the number of original or compressed *X*-variables that are used to define the space.

At this point, it is worth noting that the same validation methods that are used to avoid overfitting of quantitative models (Section 8.3.7) can also be used to avoid overfitting in qualitative models. The only difference is that the figure of merit in this case is ***not*** the Root Mean Squared Error of Prediction (RMSEP), but rather the percent correct classification or %CC:

$$\%\text{CC} = \left(\frac{N_{\text{correct}}}{N_{\text{total}}}\right) \times 100 \tag{8.56}$$

where N_{correct} is the number of correct classifications and N_{total} is the total number of samples to be classified. In a cross-validation experiment, N_{total} refers to the number of samples in the validation (or sub-validation) data set.

The principal component space has several advantages over selected *X*-variable spaces. Concerns about redundancy are eliminated because the PCs are orthogonal to one another. In addition, because each PC explains the most remaining variance in the *X*-data, it is often the case that fewer PCs than original *X*-variables are needed to capture the same information in the *X*-data. This leads to simpler classification models, less susceptibility to overfitting through the use of too many dimensions in the model space, and less noise in the model. Nonetheless, it is still a good idea to use validation techniques even when PCs are used to define the space.

The disadvantages of the principal component space are that it is more difficult to understand and explain to others, and it is more complex to implement in prediction. Instead of simply using individual X-variable intensities from a prediction sample's response profile ($\mathbf{x_p}$), one must first project these intensities onto the A significant PCs in order to obtain the prediction sample's PCA scores ($\hat{\mathbf{t_p}}$):

$$\hat{\mathbf{t_p}} = \mathbf{x_p}\hat{\mathbf{P}}_A \tag{8.57}$$

where the truncated loadings matrix ($\hat{\mathbf{P}}_A$) denotes the loadings for only the first A PCs.

8.5.2 Measures of distance in the space

Once the classification space has been defined, it is then necessary to define the distance in the space that will be used to assess the similarity of prediction samples and calibration samples. The most straightforward distance that can be used for this purpose is the **Euclidean distance** between two vectors (D_{ab}), which is defined as:

$$D_{ab} = \sqrt{(\mathbf{x}_a - \mathbf{x}_b)(\mathbf{x}_a - \mathbf{x}_b)^{\mathrm{t}}} \tag{8.58}$$

in the case where original X-variables are used to define the space, and

$$D_{ab} = \sqrt{(\mathbf{t}_a - \mathbf{t}_b)(\mathbf{t}_a - \mathbf{t}_b)^{\mathrm{t}}} \tag{8.59}$$

in the case where PCs are used to define the space. In classification applications, it is common to measure distances between vectors representing single calibration and prediction sample responses, or between vectors representing a single prediction sample and a mean of calibration samples that belong to a single class.

At first glance, one might think that the Euclidean distance is the only distance that one should consider when doing classification. However, a disadvantage of this distance measure is that it does not take into account the **variation** of each of the X-variables or PCs in the calibration data. This can be particularly important in many analytical chemistry applications (such as spectroscopy), where it is often the case that samples from individual classes have different variations in the different dimensions of the space. Geometrically speaking, these classes tend to form elliptical clusters, rather than spherical clusters, in the space. To illustrate this, Figure 8.24 shows the representation of a cluster of calibration samples (that belong to a single class) and two unknown prediction samples, in a two-dimensional space. It is important to note that both of the axes in this plot use an identical scale ranging from -1 to 1. In this situation, it is apparent that the Euclidean distances of the two unknown samples from the mean of the calibration samples are exactly the same. However, doesn't unknown 2 appear to be more likely to belong to the class of samples represented in the calibration data than unknown 1?

To address this issue, another type of distance measure, called the **Mahalanobis distance**, has been proposed. This distance is defined as:

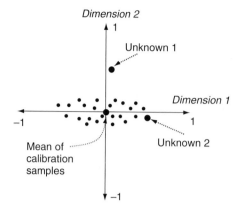

Figure 8.24 Spatial representation of a set of calibration samples (denoted by small circles) and two unknown samples (denoted by large circles). The mean of the calibration samples is at the origin.

$$M_{ab} = \sqrt{(\mathbf{x}_a - \mathbf{x}_b)\mathbf{S}^{-1}(\mathbf{x}_a - \mathbf{x}_b)^t} \tag{8.60}$$

in the case where original X-variables are used to define the space, and

$$M_{ab} = \sqrt{(\mathbf{t}_a - \mathbf{t}_b)\mathbf{S}^{-1}(\mathbf{t}_a - \mathbf{t}_b)^t} \tag{8.61}$$

in the case where PCs are used to define the space. $\mathbf{S}$ is the **covariance matrix** of the selected X-variables or PCs for the calibration data, which essentially describes the total variance described by each dimension as well as interactions between the different dimensions. In the case where PCs are used to define the space, there are no interactions between the dimensions, and $\mathbf{S}$ is simply a diagonal matrix containing the squared scores for each of the PCs.

The Mahalanobis distance is simply a **weighted** Euclidean distance, where each of the dimensions is **inversely** weighted by its overall variance in the calibration data. As a result, deviations in a dimension that has high variance in the calibration data will not be weighted as much as deviations in a dimension that has low variance in the calibration data. With this in mind, it becomes clear that the **Mahalanobis** distances of the two unknowns shown in Figure 8.24 are **not** equal. Furthermore, it becomes qualitatively apparent that unknown 2 becomes closer to the class mean than unknown 1 if Mahalanobis distance is used instead of Euclidean distance.

Although Euclidean and Mahalanobis distances are the ones most commonly used in analytical chemistry applications, there are other distance measures that might be more appropriate for specific applications. For example, there are 'standardized' Euclidean distances, where each of the dimensions is inversely weighted by the **standard deviation** of that dimension in the calibration data (standard deviation-standardized), or the **range** of that dimension in the calibration data (range-standardized).

8.5.3   Classification rule development

Once the classification space is defined and a distance measure selected, a classification rule can be developed. At this time, the calibration data, which contain analytical profiles for samples of known class, are used to define the classification rule. Classification rules vary widely depending on the specific classification method chosen, but they essentially contain two components:

(1) classification parameters
(2) classification logic.

Some common classification parameters are the mean values of the classes in the classification space, the variance of the class's calibration samples around the class mean, and the unmodeled variance in the calibration samples. Classification logic varies widely between classification methods. The following section provides details on some commonly encountered classification methods.

8.5.4   Commonly encountered classification methods

To illustrate some commonly encountered classification methods, a data set obtained from a series of polyurethane rigid foams will be used.[55] In this example, a series of 26 polyurethane foam samples were analyzed by NIR diffuse reflectance spectroscopy. The spectra of these foams are shown in Figure 8.25. Each of these foam samples belongs to one of four known classes, where each class is distinguished by different chemistry in the hard block parts of the polymer chain. Of the 26 samples, 24 are selected as calibration samples and 2 samples are selected as prediction samples. Prediction sample A is known to belong to class number 2, and prediction sample B is known to belong to class number 4. Table 8.8 provides a summary of the samples used to produce this data set.

8.5.4.1   K-nearest neighbor (KNN)

The KNN method is probably the simplest classification method to understand. It is most commonly applied to a principal component space. In this case, calibration is achieved by simply constructing a PCA model using the calibration data, and choosing the optimal number of PCs (A) to use in the model. Prediction of an unknown sample is then done by calculating the PC scores for that sample (Equation 8.57), followed by application of the classification rule.

The KNN classification rule involves rather simple logic:

- A parameter representing the number of 'nearest neighbors,' K ($\ll N$), is selected.
- The distances of the N calibration samples from the prediction sample in the classification space are ranked from least to greatest.
- Each of the K closest calibration samples submits a 'vote' for the class to which it belongs.

- The class containing the greatest number of 'votes' is chosen as the unknown sample's class.
- If there is a tie between two or more classes, the class whose samples have the lowest combined distance from the prediction sample is selected.

The KNN method has several advantages aside from its relative simplicity. It can be used in cases where few calibration data are available, and can even be used if only a single calibration sample is available for some classes. In addition, it does **not** assume that the classes are separated by linear partitions in the space. As a result, it can be rather effective at handling highly non-linear separation structures.

One disadvantage of the KNN method is that it does not provide an assessment of confidence in the class assignment result. In addition, it does not sufficiently handle the

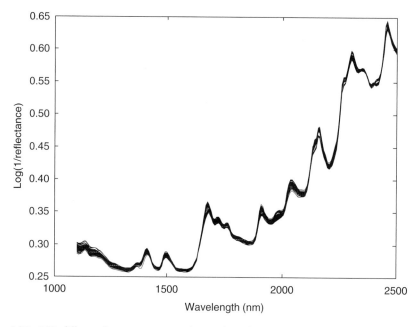

Figure 8.25 NIR diffuse reflectance spectra of 26 polyurethane foams, used to demonstrate different classification methods.

Table 8.8 A summary of the polyurethane foam samples used to demonstrate classification methods

Sample class number	Number of samples in class	Description of class
1	9	Aliphatic-based polyol
2	5	PET-based polyol
3	4	PET-based and amine-based polyols
4	8	PET-based and aliphatic-based polyols

cases where an unknown belongs to *none* of the classes in the calibration data, or to *more than one* class. A practical disadvantage is that the user must input the number of nearest neighbors (K) to use in the classifier. K is an adjustable parameter that should be influenced by the total number of calibration samples (N), the distribution of calibration samples between classes, and the degree of 'natural' separation of the classes in the sample space.

The KNN method is illustrated using the polyurethane rigid foams data set discussed earlier. First, the PCA method is applied to the NIR spectra in Figure 8.25, in order to define the principal component space that will be used to perform the KNN method. When this is done, it is found that three PCs are sufficient to use in the model, and that PCs one, two, and three explain 52.6, 20.5, and 13.6% of the variation in the spectral data, respectively. Figure 8.26 shows a scatter plot of the *first two* of the three significant PC scores of all 26 foam samples. Note that all of the calibration samples naturally group into four different clusters in these first two dimensions of the space. Furthermore, it can be seen that each cluster corresponds to a single known class. This result shows that the NIR method is generally able to discriminate effectively between the four types of polyurethane foams.

When only the first two of three PCs are considered (Figure 8.26), prediction sample A appears to be placed well within the cluster of class 2 samples, and prediction sample B appears to be somewhat between the group of class 3 samples and the group of class 4 samples (although it appears to be closer to class 4). If the KNN classification rule is applied to the two prediction samples using values of K from 1 to 10, correct classification of both

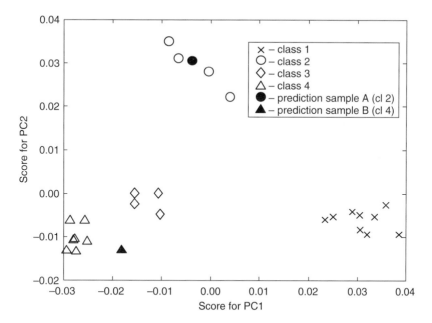

Figure 8.26 Scatter plot of the first two PC scores obtained from PCA analysis of the polyurethane foam spectra shown in Figure 8.25. Different symbols are used to denote samples belonging to the four known classes. The designated prediction samples are denoted as a solid circle (sample A) and a solid triangle (sample B).

samples is achieved in all cases. However, it should be noted that a misclassification of prediction sample A to class 4 would have been obtained if a K of 22 or more was chosen. This result indicates that one must be careful to avoid using a value of K that is much greater than the number of samples of the **least-represented class** in the calibration data.

8.5.4.2   PLS: Discriminant analysis (PLS-DA)

This classification method actually uses the **quantitative** regression method of PLS (described earlier) to perform **qualitative** analysis. In this case, however, the Y-variable is a 'binary' variable, in that it contains either zeros or ones, depending on the known class of the calibration samples. For example, if there are a total of four calibration samples and two different classes (A and B), and the first and last samples belong to class A, and the second and third samples belong to class B, then the y-vector can be expressed as the following:

$$\mathbf{y} = \begin{bmatrix} 0 \\ 1 \\ 1 \\ 0 \end{bmatrix}$$

Furthermore, **if more than two classes are involved**, it is necessary to have **several different Y-variables** in order to express the class membership of all the calibration samples. This results in the use of a Y-**matrix** (**Y**), which has the dimensionality of N by the number of Y-variables. For example, if there are three classes represented in the calibration data, then the Y-matrix might look something like this:

$$\mathbf{Y} = \begin{bmatrix} 0 & 1 & 0 \\ 0 & 1 & 0 \\ 0 & 0 & 1 \\ 1 & 0 & 0 \end{bmatrix}$$

In order to handle multiple Y-variables, an extension of the PLS regression method discussed earlier, called **PLS-2**, must be used.[1] The algorithm for the PLS-2 method is quite similar to the PLS algorithms discussed earlier. Just like the PLS method, this method determines each compressed variable (latent variable) based on the maximum variance explained in both **X** and **Y**. The only difference is that **Y** is now a **matrix** that contains several Y-variables. For PLS-2, the second equation in the PLS model (Equation 8.36) can be replaced with the following:

$$\mathbf{Y} = \mathbf{T}_{PLS}\mathbf{Q} + \mathbf{F} \tag{8.62}$$

For PLS-DA, calibration is achieved by applying the PLS-2 algorithm to the X-data and an appropriately constructed Y-matrix, to obtain a **matrix** of regression coefficients (**B**). Prediction of an unknown sample is then done by applying the regression coefficients to the analytical profile of the sample, which produces predicted values for each of the Y-variables. The classification rule can then be applied.

Although exact classification rules for PLS-DA can vary, they are all based on determining the membership of the unknown to a specific class by assessing the 'closeness' of each of the predicted Y-values to 1. For example, one could calculate the standard deviations for each of the estimated Y-variables in the calibration data to obtain uncertainties for each of the variables, and then use this uncertainty to determine whether an unknown sample belongs to a given class based on the proximity of the unknown's Y-values to 1.

There are several distinctions of the PLS-DA method versus other classification methods. First of all, the classification space is unique. It is not based on X-variables or PCs obtained from PCA analysis, but rather the latent variables obtained from PLS or PLS-2 regression. Because these compressed variables are determined using the known class membership information in the calibration data, they should be ***more relevant for separating the samples by their classes*** than the PCs obtained from PCA. Secondly, the classification rule is based on results obtained from quantitative PLS prediction. When this method is applied to an unknown sample, one obtains a predicted number for each of the Y-variables. Statistical tests, such as the f-test discussed earlier (Section 8.2.2), can then be used to determine whether these predicted numbers are sufficiently close to 1 or 0. Another advantage of the PLS-DA method is that it can, in principle, handle cases where an unknown sample belongs to more than one class, or to no class at all.

At the same time, there are some distinct disadvantages of the PLS-DA method. As for quantitative PLS, one must be very careful to avoid overfitting through the use of too many PLS factors. Also, the PLS-DA method does not explicitly account for response variations ***within*** classes. Although such variations in the calibration data can be useful information for assigning and determining uncertainty in class assignments during prediction, it will be treated essentially as 'noise' in the PLS model, thus being lost during prediction. Furthermore, this method assumes that the classes can be divided using linear partitions in the classification space. As a result, it might not perform well in cases where the X-data indicate strong natural clustering of samples in a manner that is not consistent with any of the known classes. In such cases, this natural but irrelevant clustering in the X-data can result in additional errors in the PLS-DA classifier, thus leading to less-certain results. In the polyurethane example, observation of the scores for the first two PCs (Figure 8.26) indicates that the different classes cannot be easily separated by linear partitions in the principal component space, and that the ***within-class*** variation is significant. As a result, it is not too surprising that the PLS-DA method is unable to build useful classifiers for this data.

8.5.4.3 Linear discriminant analysis (LDA)

This classification technique uses a space that is defined by a unique set of vectors called ***linear discriminants*** or LDs. Like the PCs obtained from PCA analysis, LDs are linear combinations of the original M-variables in the X-data that are completely independent of (or orthogonal to) one another. However, the criterion for determining LDs is quite different than the criterion for determining PCs: each LD is sequentially determined such that ***the ratio of between-class variability and within-class variability in the calibration data is maximized***. This has two consequences:

(1) The LDs provide a more relevant basis than PCs for discriminating between classes.
(2) In some cases, fewer LDs than PCs are needed to provide adequate discrimination on the same data.

Once the calibration data are expressed in terms of LDs, different types of models can be developed.[56] Common parameters that are used in LDA models are the mean of each known class in the space (to define the 'center' of each class) and the within-class standard deviations of each known class (to enable assessment of confidence of class assignment during prediction). Classification logic for an unknown sample usually involves calcula-tion of the distances of the unknown sample from each of the class centers, and subsequent assessment of confidence of the sample belonging to each class, based on the within-class standard deviations.

One advantage of the LDA method is that space definition is done using known class information, thus leading to a sample space that is relevant for classification. In fact, it is probably even more relevant than the space used for the PLS-DA method because the PLS-DA method does not consider the ratio of between-group variance and within-group variance. Consequently, it is often observed that fewer LDs than PCs or PLS factors are required to provide the same discriminating power on the same data. As mentioned earlier, such simplicity can be a distinct advantage for on-line applications. In addition, the LDA method explicitly accounts for variations ***within each sample class*** when assessing the confidence in a class assignment of an unknown sample, through the use of within-class variance in the classification rule. As a result, this method can handle cases where different classes have very different magnitudes and directions of within-class variations in the space.

Like the PLS-DA method, and many of the quantitative modeling methods discussed above, the LDA method is susceptible to overfitting through the use of too many compressed variables (or LDs, in this case). Furthermore, as in PLS-DA, it assumes that the classes can be linearly separated in the classification space. As a result, it can also be hindered by strong natural separation of samples that is irrelevant to any of the known classes.

8.5.4.4 Soft independent modeling of class analogies (SIMCA)

This classification method can be considered a qualitative equivalent to the 'subset regression' strategy for improving quantitative models, discussed earlier in Section 8.3.8.5. Instead of defining a common sample space for all classes and then determining class membership based on the location of the unknown sample in the common space, SIMCA actually defines ***a different space for each class***. For this discussion, the parameter Z will be used to denote the number of known classes in the calibration data.

For the SIMCA method, calibration is done by constructing a ***set*** of Z class-specific PCA models. Each PCA model is built using only the calibration samples from a single class. Consequently, one obtains a ***set*** of PCA model parameters for each of the Z classes (e.g. $\hat{T}_1, \hat{T}_2, \ldots \hat{T}_Z$). Then, when an unknown sample is encountered, its analytical profile is applied to each of the Z PCA models in order to generate Z sets of PCA prediction scores and leverages (see Equations 8.51 and 8.52) and Z sets of PCA RES_p values (see Equation 8.55). At this point, separate assessments of the unknown sample's membership to *each* class are made. These assessments involve both the distance between the unknown and the class mean in the PCA space, and comparison of the unknown's residual and leverage to the average values of these properties for the calibration data. Typically, a statistical *f*-test is applied to the prediction leverage and residual to determine whether the unknown belongs to the same class as the sample subset used to build the local PCA model, for a given level of confidence (say, 95 or 99%).

Several of the advantages of the SIMCA method result from the separate treatment of each of the classes, and the lack of a general classifier. First of all, it is more robust in cases where the different classes correspond to very different patterns, or highly non-linear differences, in the analytical profiles. The polyurethane foams example is particularly illustrative because the different classes of foams are not defined in terms of the concentration of a single chemical constituent, but rather in terms of the chemical structure of the hard-block segments in the polymer. In such cases, attempts to build a general classifier using PCA often require the use of an undesirably large number of PCs, thus unnecessarily adding to model complexity, and decreasing long-term stability when applied in real time. Secondly, the treatment of each class separately allows SIMCA to handle cases better where the within-class variance structure is quite different for each class. Using the Mahalanobis distance in a general PCA classifier forces one to assign a single covariance value to each PC, and does not allow for class-specific covariance values. Furthermore, the SIMCA method explicitly handles cases where the unknown sample belongs to more than one class, or to none of the known classes. In addition, its class-specific modeling approach is relatively easy to explain to customers. Finally, it can provide a quantitative assessment of the confidence of the class assignments for the unknown sample.

An important requirement of the SIMCA method is that one must obtain a relatively large number of calibration samples that are sufficiently representative of *each* of the Z classes. If any one of the classwise PCA models is poorly defined, then a large number of incorrect classifications can occur. Real-time implementation of a SIMCA model requires a relatively large number of parameters, but there are several packaged software products that facilitate such implementation. A theoretical limitation of SIMCA is that it does not use, or even calculate, *between-class* variability. As a result, there are some specific cases where the SIMCA method might not be optimal. Consider, for example, the data structure where there is strong natural clustering of samples into groups that do not correspond to the known classes, and there are representatives of each class in each of these 'natural' groups. In such a case, the inherent inter-class distance can be rather low compared to the intra-class variation, thus making it difficult to build an effective SIMCA classifier.

When the SIMCA method is applied to the polyurethane data, it is found that two PCs are optimal for each of the four local PCA class models. When this SIMCA model is applied to the prediction sample A, it correctly assigns it to class 2. When the model is applied to prediction sample B, it is stated that this sample does not belong to any class, but that it is closest to class 4. Although we know that this prediction sample belongs to class 4, the SIMCA result is actually consistent with our visual assessment of the placement of sample B in the PCA space (Figure 8.26): the sample is located between classes 3 and 4, but is closer to class 4.

8.5.4.5 Qualitative neural networks

As in quantitative analysis, ANNs can also provide an effective solution for qualitative analysis. There are many different ANN solutions for qualitative analysis, and these generally differ in terms of their network structures, processing functions (linear and non-linear), and parameter estimation ('learning') procedures.

Three commonly used ANN methods for classification are the perceptron network, the probabilistic neural network, and the learning vector quantization (LVQ) networks. Details on these methods can be found in several references.[57,58] Only an overview of them will be presented here. In all cases, one can use all available X-variables, a selected subset of X-variables, or a set of compressed variables (e.g. PCs from PCA) as inputs to the network. Like quantitative neural networks, the network parameters are estimated by applying a learning rule to a series of samples of known class, the details of which will not be discussed here.

The perceptron network is the simplest of these three methods, in that its execution typically involves the simple multiplication of class-specific weight vectors to the analytical profile, followed by a 'hard limit' function that assigns either 1 or 0 to the output (to indicate membership, or no membership, to a specific class). Such networks are best suited for applications where the classes are linearly separable in the classification space.

The probabilistic neural network does classification in a somewhat different manner. First, the distances of the input vector to each of a series of training vectors (which are each associated with a specific class) are calculated. Then, these distances, along with the known class identities of each of the weight vectors, are used to produce a set of probabilities for each class. Finally, a 'competitive layer' in the network is used to choose the class that has the highest probability of being correct. Probabilistic networks can produce very generalized classifiers, in that it is only necessary that the input vector be close to training vectors of its same class. As a result, they can provide useful results in cases where the classes are not linearly separable, or where there is strong natural clustering of samples that is irrelevant to the separation of classes.

The LVQ network is somewhat more complicated, but it directly addresses the issue where there can be strong natural clustering of calibration samples that is irrelevant to the separation of classes. When implementing an LVQ network, the input vector is passed to a competitive layer, which calculates the distances between the input vector and a series of S1 weight vectors (each of which should represent a *natural* subclass in the calibration data). Based on these distances, a winning subclass is chosen. This second layer maps the winning subclass to one of the S2 *known* classes. The number of subclasses must always be greater than the number of known classes (S1 > S2). The LVQ network can be very powerful for classification because it is possible that samples from very different regions of the input variable space can be mapped to the same known class. Rosenblatt[57] and Kohonen[58] give more detailed information on these qualitative neural networks.[57,58]

8.5.4.6   Other classification methods

The classification methods listed above are by no means an exhaustive list of tools that are available to the practicing chemometrician. There is a wide range of classification methods that are commonly used in non-chemical fields, such as machine learning and pattern recognition. Some notable methods include support vector machines (SVMs),[59] which attempt to draw ***non-linear*** boundaries between known classes in the space, and Bayesian classifiers,[60] which take advantage of prior knowledge regarding the probability of samples belonging to particular classes.

8.6   Exploratory analysis

Many of the chemometrics tools mentioned above can be used to extract valuable information about sample chemistry, the measurement system, as well as process dynamics. Such information can be very useful in a process analytical environment, where the emphasis is usually on developing and automating on-line quantitative analyses. Several different means for information extraction are discussed below.

8.6.1   Model parameters from PCA, PCR, and PLS

When one builds a quantitative model using PCR or PLS, one is often not aware that the model parameters that are generated present an opportunity to learn some useful information. Information extracted from these model parameters cannot only be used to better understand the process and measurement system, but also lead to improved confidence in the validity of the quantitative method itself.

In order to illustrate the ability to extract valuable information from such model parameters, a data set containing NIR transmission spectra of a series of polymer films will be used. In this example, films were extruded from seven different polymer blends, each of which was formulated using different ratios of high-density polyethylene (HDPE) and low-density polyethylene (LDPE). NIR spectra were then obtained for four or five replicate film samples at each blend composition. Table 8.9 lists the HDPE contents of the seven different formulations, and Figure 8.27 shows the NIR spectra that were obtained. Note that there is very little 'visible' separation between the spectra. However, this does not mean that there is little information in the spectra, as will be shown below.

8.6.1.1   Scores

The PC scores can be used to assess the relationships between different samples in the model. If this information is combined with known class information about the samples, an assessment of the effectiveness of the analytical method for distinguishing between classes can be made. Furthermore, the scores can be used to detect trends in the samples that might not be expected. It is common to plot PC scores using a two-dimensional scatter plot, where the axes represent different PC numbers.

Table 8.9 A summary of the HDPE/LDPE blend films used to demonstrate exploratory analysis methods

Blend composition, in % HDPE	Number of replicate samples analyzed
0	4
2.5	5
5	5
10	5
25	5
50	5
100	5

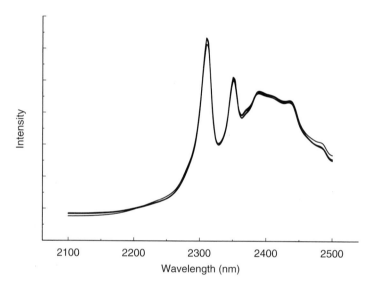

Figure 8.27 NIR spectra of a series of 34 films obtained from polymer blends with varying ratios of HDPE and LDPE.

Although the PCs and latent variables obtained from PCA, PCR, and PLS are useful for interpretive purposes, the basis for their calculation has nothing to do with interpretability. As a result, it is often the case that each PC or LV represents a mixture of two or more fundamental effects in the analytical data. Fortunately, however, the orthogonality of these compressed variables allows us to easily **rotate** them in order to improve their interpretability. Details regarding such uses of PC and LV rotation can be found in several references.[61,62] In these referenced examples, rotation of the PCs is done to varying degrees until a preferred structure of the PC scores scatter plot is obtained. It is important to note, however, that rotated PCs explain different amounts of the *X*-data variance than the original PCs, but that the total explained variance of all of the significant PCs remains the same.

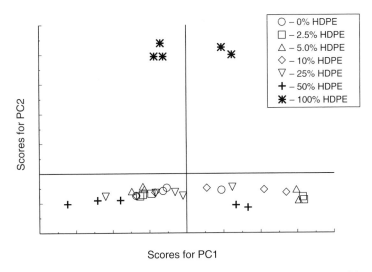

Figure 8.28　Scatter plot of the first two rotated PC scores obtained from PCA analysis of the NIR spectra of HDPE/LDPE blend films shown in Figure 8.27. Different symbols are used to denote different blend compositions.

Figure 8.28 is a scatter plot of the first two rotated PC scores obtained from a PCA analysis of the NIR spectra of polyethylene blend films, after rotation was done to improve interpretability. In this example, it was determined that three PCs were optimal for the PCA model, and the rotated PCs one, two, and three explain 46.67, 49.25, and 2.56% of the variation in the NIR spectra, respectively. Although the higher explained variance of the second **rotated** PC might seem anomalous based on the criterion that PCA uses explained variance to determine each PC, one must be reminded that these explained variances refer to rotated PCs, rather than the original PCs. There are two interesting things to note about this plot:

(1)　The replicate samples analyzed for each blend composition have large variability in the PC1 scores, thus indicating that the first rotated PC explains a phenomenon that causes replicate samples to be different.
(2)　The second rotated PC explains a phenomenon that causes the spectra of the 100% HDPE samples to be very different than the spectra of the other samples.

8.6.1.2　Loadings

The PC loadings can be used to assess relationships between different variables in the model, as well as help explain the fundamental basis of the phenomena explained by the PCs. More specifically, if one looks at the loadings for the **dominant PCs** in the model, they can be used to detect general correlations between X-variables. For this purpose, they are often plotted as two-dimensional scatter plots, where each axis corresponds to a different PC number. An example of such a loading scatter plot for the polyethylene data is shown in Figure 8.29, where the loadings for rotated PC1 are plotted against the loadings for rotated PC2. In this case, the first two PCs explain almost 96% of the *X*-data variation, and can therefore be considered the dominant PCs in the model. The *X*-variables that are highly correlated to one another appear on the same line through the origin. For example, note the dotted line in Figure 8.29 that runs through the origin. The variables in the loading plot that lie close to this line are highly correlated to one another. In this specific case, variables representing NIR wavelengths 2300, 2316, 2338, 2356, 2378, and 2434–2444 nm appear to be highly correlated. In addition, *if the X-variables were not autoscaled before modeling*, then the distance of each variable from the origin in the loading scatter plot reflects the relative magnitudes of their variability in the data set. In the polyethylene example, the spectra were not autoscaled before PCA analysis. In the loading scatter plot (Figure 8.29), the NIR wavelengths 2444 and 2304 nm appear furthest from the origin (upper left and lower right parts of the plot, respectively). Because these first two PCs explain about the same amount of variability in the data, one can conclude that these wavelengths have the greatest variability in this specific data set. If the *X*-variables are autoscaled before modeling, then one can only interpret directions (and not magnitudes) in the loading space.

In PAC, where variables are usually a continuous series of spectral intensities at different wavelengths, or a series of chromatographic intensities at different retention times, the loadings for a single PC plotted against the variable number can provide some useful interpretive information. In this view, such 'loadings spectra' provide analytical profiles of the dominant sources of variation in the *X*-data. Such plots can be combined with knowledge of the spectroscopy or chromatography to assign specific chemical phenomena to the PCs.

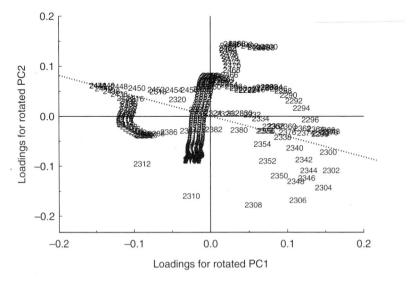

Figure 8.29 Scatter plot of the first two rotated PC loadings obtained from PCA analysis of the NIR spectra of HDPE/LDPE blend films shown in Figure 8.27. Labels on the scatter plot correspond to NIR wavelength.

In the polyethylene example, the first and second PC loadings spectra are shown in Figure 8.30. Note that these loadings 'spectra,' unlike conventional spectra, have both positive and negative peaks and seem rather cryptic to interpret. This is mainly due to two factors:

(1) The criterion for determining the PCs (maximize explained variance in the X-data) does not involve any consideration of interpretability.
(2) The loadings for each PC are mathematically constrained by orthogonality, which often results in artifacts that are not commonly associated with conventional spectra.

As a result, such loading spectra rarely represent pure chemical or physical phenomena in the samples. However, they can still provide very valuable information. For example, Figure 8.31 shows the result of subtracting a spectrum of polyethylene film before and after uniaxial stretching. Note the striking resemblance of this difference spectrum to the loading spectrum for rotated PC1 in Figure 8.30 (solid line). This result indicates that variation in the amount of molecular orientation in the polyethylene samples contributes a large amount of variability to the NIR data. Similar 'assignments' of the other PC loadings spectra are also possible.

8.6.1.3 X-residuals

When a PCA, PCR, or PLS model is constructed, one must choose the optimal number of PCs or latent variables, which is usually much less than the number of original

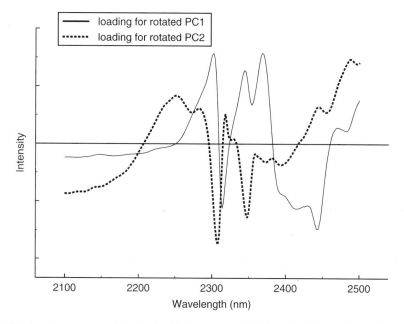

Figure 8.30 Loadings spectra of the first (solid line) and second (dotted line) rotated PCs obtained from PCA analysis of the NIR spectra of HDPE/LDPE blend films shown in Figure 8.27.

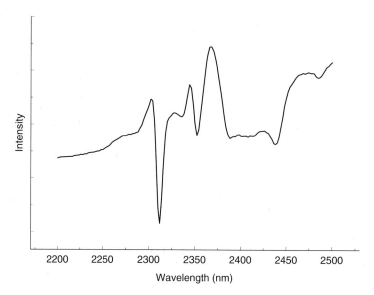

Figure 8.31   The result of ***subtracting*** a spectrum of polyethylene film obtained **after** uniaxial stretching from the spectrum of the same film **before** uniaxial stretching.

X-variables. This leads to the presence of model residuals (**E** in Equations 8.19 and 8.35). The residuals of the model can be used to indicate the ***nature*** of unmodeled information in the calibration data. For process analytical spectroscopy, plots of individual sample residuals versus wavelength ('residual spectra') can be used to provide some insight regarding chemical or physical effects that are not accounted for in the model. In cases where a sample or variable outlier is suspected in the calibration data, inspection of that sample or variable's residual can be used to help determine whether the sample or variable should be removed from the calibration data. When a model is operating on-line, the *X*-residuals of prediction (see Equation 8.55) can be used to determine whether the sample being analyzed is appropriate for application to a quantitative model (see Section 8.4.3). In addition, however, one could also view the prediction residual vector $\hat{e}_p$ as a profile (or 'residual spectrum') in order to provide some insight into the ***nature*** of the prediction sample's inappropriateness.

8.6.1.4   PLS regression vectors

The regression vector ($\hat{b}$) obtained from the PCR or PLS regression techniques can also provide useful qualitative information about the *X*-variables that are most important for determining the property of interest. In process analytical spectroscopy applications, the regression coefficients can also be plotted like a spectrum (e.g. regression coefficient vs. wavelength). However, like loadings 'spectra' discussed above, such a 'regression coefficient spectrum' can have both positive and negative peaks. Unlike the loadings, which reveal the 'spectral signatures' of the dominant sources of variability in the *X*-data, the coefficients indicate the ***relative importance of each X-variable for determining the property of interest***.

At the same time, however, one must be careful to avoid conclusions that the wavelengths (or variables) of strong positive or negative regression coefficients are strongly correlated with the property of interest. If they were, then there would be no need for a PLS model, and one could build an equally effective linear regression model using one or more of these wavelengths. The coefficient at a given wavelength (variable) must take into account spectral effects from phenomena that interfere with the property of interest, which can often be much stronger than spectral effects that are truly generated by the property of interest.

In the polyethylene example, Figure 8.32 shows the regression coefficient spectrum for the PLS calibration to the % HDPE content in the polyethylene blends. One particularly recognizable feature in this plot is the negative peak at approximately 2240 nm, where a prominent methyl ($-CH_3$) band is known to exist. Because the concentration of methyl groups in the polymer is expected to be inversely related to the % HDPE content, the negative methyl peak in the regression coefficient spectrum makes sense. However, please note that there are many other stronger and distinct features in the regression coefficient spectrum, suggesting that other effects in the polymers are important for predicting the % HDPE content.

8.6.2 Self-modeling curve resolution (SMCR)

The SMCR method can be particularly useful for extracting chemically interpretable information from sets of spectral data **without the use of reference concentration values**.

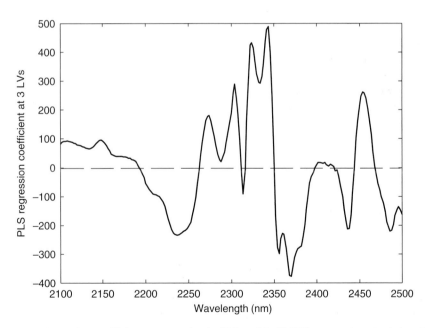

Figure 8.32 Regression coefficient spectrum for the PLS model of HDPE content in polyethylene blend films.

In the process analytical context, this method can be used to obtain reasonably good pure component spectra to aid in method development, as well as composition time profiles to enable assessment of reaction kinetics.[4,63]

The SMCR method uses the same 'Beer's Law' model as the CLS method, mentioned earlier:

$$\mathbf{X} = \mathbf{CK} + \mathbf{E} \tag{8.63}$$

However, unlike the CLS method, the SMLR method can be used without any knowledge of the concentrations of the different constituents in the calibration samples. Given only the spectral data ($\mathbf{X}$), an estimate of the number of ***chemically varying and spectrally active*** components represented in the data (P), and some ***constraints*** regarding the nature of the expected structures of the pure component profiles ($\mathbf{C}$) and pure component spectra ($\mathbf{K}$), this method can be used to provide estimates of the pure component spectra ($\hat{\mathbf{K}}$) and the component concentrations ($\hat{\mathbf{C}}$).

At first, one might intuitively question how the SMCR method can make such determinations from such little information. The answer is that the SMCR algorithm involves an iterative process, and that the constraints put on $\mathbf{C}$ and $\mathbf{K}$ during this iterative process are critical to enable the algorithm to converge on a useful solution. However, it should be noted that the presence of such constraints does not ***guarantee*** that the algorithm will converge on a meaningful solution, especially if the assumptions behind the constraints are not accurate. Common constraints that are put on the concentration profiles ($\mathbf{C}$) include non-negativity, closure (i.e. all constituent concentrations for a given sample must add up to 100%), and unimodality. The non-negativity constraint can also be applied to the pure component spectra, and in some cases the unimodality constraint can be applied to these as well.

In many cases, the 'standard' constraints mentioned above are sufficient for the SMCR algorithm to converge to a stable solution for $\mathbf{K}$ and $\mathbf{C}$, ***provided that a reasonable number of components are specified***. However, one must be careful to avoid interpreting $\hat{\mathbf{K}}$ and $\hat{\mathbf{C}}$ as 'absolute' pure component spectra and pure component time profiles, respectively. First of all, one rarely knows the true number of components in a given data set, and therefore it is quite tempting to overfit the SMCR model by specifying too many components. In addition, the SMCR model (Equation 8.63) assumes that the spectral contributions from each component are linearly additive. As a result, the presence of non-linear and inter-component interaction effects can lead to distortions in the SMLR-estimated profiles that do not make physical sense. Finally, the presence of correlations between pure chemical species must be also considered because this can result in a single 'SMCR component' representing more than one 'pure chemical component.'

To address the issue of inter-correlations between pure chemical species, it is possible to impose additional application-specific constraints to improve the interpretability of SMCR results. For example, *one might know* the location of a 'pure' component peak in the spectrum, or can obtain a sufficiently relevant pure component spectrum in the lab. Such additional constraints can be applied mathematically during the iterative SMCR process to improve the interpretability of the results.[4] Of course, if such 'extra' constraints are not truly accurate, then they will likely cause poor model fit results, or even prevent the SMCR algorithm from converging.

Provided that the algorithm converges, the residual of the SMCR model ($\hat{\mathbf{E}}$) can be calculated by the following equation:

$$\hat{\mathbf{E}} = \mathbf{X} - \hat{\mathbf{C}}\hat{\mathbf{K}} \tag{8.64}$$

The magnitude of these residuals, or the sum of squared elements in $\hat{\mathbf{E}}$, can be used to evaluate the fit of an SMCR model.

In a specific example, the SMCR method is applied to a set of 210 FTIR spectra obtained sequentially during a chemical reaction of the type $A + B \rightarrow C$. Figure 8.33 shows the original spectra of the samples obtained during the reaction. The SMCR method was implemented using non-negativity and closure constraints on the concentration profiles, and the non-negativity constraint on the spectral profiles. After running the SMCR algorithm several times on the data, it was found that four components were sufficient to explain the spectral data, and that the use of additional components did not significantly increase the fit of the model. The estimated concentration profiles ($\hat{\mathbf{C}}$) for these four components and the estimated pure spectral profiles ($\hat{\mathbf{K}}$) for these components are shown in Figures 8.34 and 8.35, respectively.

The concentration profiles (Figure 8.34) show a typical 'evolving' pattern, where the reactants are consumed early in the reaction, intermediates are formed and consumed in the middle of the reaction, and the product is formed at the end of the reaction. The interesting part of this result is that it suggests the production and consumption of more

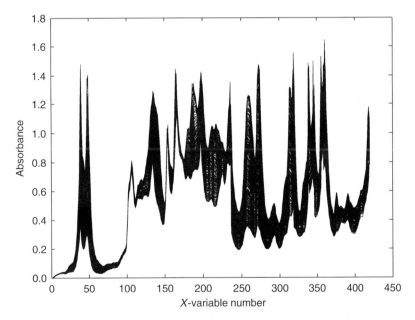

Figure 8.33   A series of 210 FTIR spectra obtained during the course of a chemical reaction.

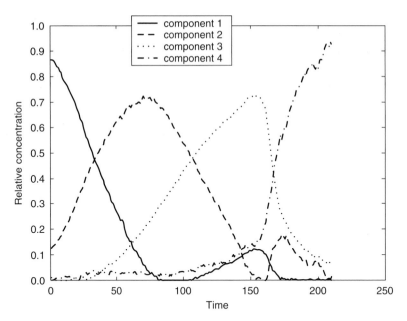

Figure 8.34 Estimated pure component concentration profiles extracted from the FTIR spectra obtained during the chemical reaction, using SMCR.

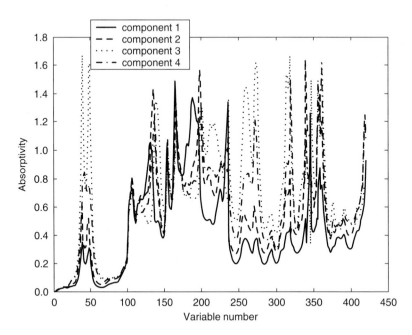

Figure 8.35 Estimated pure spectral profiles extracted from the FTIR spectra obtained during the chemical reaction, using SMCR.

than one intermediate species during the reaction. With prior knowledge regarding the chemistry of the reaction, as well as the timing of specific events during the reaction, one can use these results to extract valuable information regarding reaction chemistry and reaction kinetics.

The pure spectral profiles corresponding to each of the components (Figure 8.35) can be used to provide further information regarding the nature of the pure components that are represented in the concentration profiles. This is especially the case if one uses prior knowledge regarding the spectroscopic technique (more specifically, band assignments for specific functional groups in the process material).

There are several different iterative algorithms that have been used for SMCR, including alternating least squares (ALS)[63] and iterative target transformation factor analysis (ITTFA).[64] For more detailed information, the reader is referred to these references.

8.6.3 'Unsupervised' learning

In Section 8.5 it was shown that **supervised learning** methods can be used to build effective classification models, using calibration data in which the classes are known. However, if one has a set of data for which no class information is available, there are several methods that can be used to explore the data for 'natural' clustering of samples. These methods can also be called **unsupervised learning** methods. A few of these methods are discussed below.

8.6.3.1 Hierarchical cluster analysis (HCA)

Hierarchical cluster analysis is a common tool that is used to determine the 'natural' grouping of samples in a data set.[65] For many process analytical applications, where there are a large number of highly correlated X-variables, it is advantageous to use the PCs rather than the original X-variables as input to the HCA algorithm. The selection of the distance measure (Euclidean, Mahalanobis, etc.) depends on the user's prior knowledge about the expected nature of the separation between classes. If weaker PCs are expected to have large importance in determining class separation, then the Mahalanobis distance might provide better results.

Once the classification space and the distance measure are defined, one must also define a **linkage rule** to be used in the HCA algorithm. A linkage rule refers to the specific means by which the distance between different clusters is calculated. Some examples of these are provided below:

- *Single linkage, nearest neighbor* – uses the distance between the **two closest samples** in the different clusters.
- *Complete linkage, furthest neighbor* – uses the **greatest distance between any two samples** in the different clusters.
- *Pair-group average* – uses the **average distance between all possible pairs of samples** in the two clusters.

The HCA algorithm begins by assigning every sample to its own cluster. It then uses the pre-defined linkage rule to find the closest two clusters and merge them together into a single cluster. This process proceeds until either a predetermined final number of clusters is determined, or a single cluster is reached. The choice of the algorithm's termination condition depends on the objective of the analysis.

Comparing the different linkage rules listed above, the Single Linkage, Nearest Neighbor method tends to generate clusters that are elongated in the classification space. In contrast, the Complete Linkage, Furthest Neighbor method tends to generate more 'rounded' clusters in the space. Therefore, the appropriate linkage rule to use for a specific application can depend on the expected nature of the sample clustering. The Pair-Group Average linkage rule requires more computation time, but is equally appropriate to use in cases where natural sample clustering is 'elongated' or 'rounded.'

There are two common ways to display the results of an HCA: (1) a hierarchical tree plot or (2) a scatter plot of the samples in the classification space. The tree plot[65] is useful whether the algorithm proceeds to a single cluster or a predetermined number of clusters. It provides the user with information regarding both the subsets of samples that cluster together, and the threshold distances at which each sample joins a cluster. The scatter plot is of more limited use. Although it also provides information regarding the samples that are grouped together and the distances between groups, it is relevant only if the algorithm is terminated at a predetermined number of clusters. Furthermore, in cases where there are more than two input variables (i.e. more than two dimensions in the classification space), it can effectively show the clustering of samples only in two of the dimensions at a time.

To demonstrate the HCA method, the same polyurethane foam data, used earlier to demonstrate supervised learning methods, will be used (refer to Figures 8.25 and 8.26). In this case, HCA is done using the PCs obtained from a PCA analysis of *all 26 samples* (calibration and validation samples). Two different cluster analyses were done: one using the first two PCs (which explain 73.1% of the variation in the NIR spectral data) and one using the first three PCs (which explain 86.7% of the variation). The results of these analyses are shown in Table 8.10. Note that the HCA done using only two PCs results in 100% correct clustering of the four classes into their groups. However, the use of three PCs results in a clustering scheme where only the class 1 and class 4 samples are grouped in the same cluster, and many class 2 and class 3 samples are not grouped as expected. At this point, it is important to note that the PCA scatter plot in Figure 8.26 shows only the first two PCs, and not the third PC. This HCA result suggests that the third PC explains a phenomenon that is not useful for separating the samples into their classes.

8.6.3.2   *Self-organizing maps (SOMs)*

This tool, originally developed by Kohonen,[58,66] is really a data visualization technique that uses self-organizing neural networks to reduce a data set with a high level of dimensionality (i.e. large number of PCs used to define the space) into a one- or two-dimensional map that is much easier to interpret. In the map, not only are 'natural' sample groups identified according to their similar response profiles, but groups that are most similar to one another in terms of their responses also tend to lie next to one another on the map. As a result, one can obtain between-group information as well as sample

Table 8.10   Two sets of HCA results obtained on the polyurethane foam data, using 2 and 3 PCs

Sample number	Known class	Groupings obtained from HCA (four clusters specified)	
		2 PCs	3 PCs
1	1	1	1
2	1	1	1
3	1	1	1
4	1	1	1
5	1	1	1
6	1	1	1
7	1	1	1
8	1	1	1
9	1	1	1
10	2	2	2
11	2	2	3
12	2	2	2
13	2	2	2
14	2	2	2
15	3	3	4
16	3	3	4
17	3	3	4
18	3	3	4
19	4	4	4
20	4	4	4
21	4	4	4
22	4	4	4
23	4	4	4
24	4	4	4
25	4	4	4
26	4	4	4

grouping information. This is possible because the SOM algorithm explicitly considers the 'neighboring area' of a sample's point in the space, rather than just the point itself.

8.7   The 'calibration sampling paradox' of process analytical chemistry

In any chemometrics application, it is critical to obtain calibration samples that are both sufficiently *relevant* and sufficiently *representative* of the samples that the model will be applied to. Furthermore, for quantitative applications, the reference concentrations of the calibration samples must also be *accurate* and *precise* enough to result in a model that performs within desired specifications. In PAC, it is often difficult to obtain calibration samples that are both highly relevant and very accurate. For extracted process samples, which are the most relevant ones, reference lab analysis often suffers from biases and variability due to chemical instability, complicated sample pre-treatment, and sample handling variability. On the other hand, synthesized calibration standards, which are usually very accurate, might not be sufficiently representative of the actual process samples that the analyzer sees during real-time operation. This situation can be referred to as the ***calibration sampling paradox*** of PAC.

A useful means for discussing this paradox is a conceptual plot of relevance vs. accuracy, shown in Figure 8.36 (Martens, 1997, pers. comm.). Different strategies for calibration can be mapped onto this plot. If this is done, one finds that many of the common calibration strategies for process analyzers lie in either the high relevance/low accuracy region, or the low relevance/high accuracy region. For example, the strategy where calibration data are obtained from synthetically prepared standards analyzed on a lab analyzer should result in highly accurate reference concentrations, but the spectra might not be sufficiently representative of the spectra of actual process samples obtained by the process analyzer. Conversely, the strategy where calibration data are obtained from actual process spectra and the reference concentration values are estimated using process data should result in highly relevant spectra, but inaccurate reference concentrations. (By the way, if the process data were sufficiently accurate, there would be no need for a process analyzer at all!) The fact that many common calibration strategies in PAC fall into either of these two areas leads to the concept of a ***relevance/accuracy tradeoff***. This is unfortunate because the ideal calibration strategy lies in the upper right portion of this plot: the high relevance/high accuracy region.

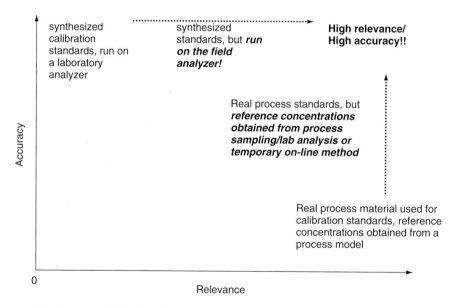

Figure 8.36 Conceptual plot of calibration relevance vs. calibration accuracy, used as a means to compare different analyzer calibration strategies.

The first step to addressing this problem is to be aware of, and fight against, the 'natural' forces that tend to push calibration strategies into the undesirable lower left part of the plot. These forces include chemical instability of synthesized standards, poor reference analysis protocol (which decrease accuracy), raw material variations, non-representative process sampling and instrument drift (which decrease relevance). At the same time, one can pursue calibration strategies that are closer to the desirable upper right region of the plot. For example, the strategy of using synthesized standards analyzed on a lab analyzer can be modified so that the standards are analyzed on the actual field analyzer (which increases relevance). Alternatively, the strategy of calibrating using actual process samples and process-model-estimated constituent concentrations can be modified so that the constituent concentrations are determined by either instituting a process sampling protocol coupled with off-line laboratory analysis, or installing a temporarily redundant on-line analyzer (which increases accuracy).

There are many different ways in which the calibration strategy can be moved toward the favorable upper right corner of the relevance vs. accuracy plot. In most cases, these rely heavily on prior knowledge of the specific process and measurement system involved in the problem. Such prior knowledge can involve the process dynamics, process chemistry, process physics, analytical objectives, analyzer hardware, analyzer sampling system, and other factors.

One example of a calibration strategy that is often practical for new analyzer installations is to start with synthesized standards only (low relevance, high accuracy), and then continually 'mix' in actual process samples (analyzed by a laboratory reference method) as the analyzer is operating on-line, in order to continuously improve method relevance. This method can be rather effective *if the standard samples are sufficiently representative of the process samples*. A specific example of this strategy is illustrated using the data obtained from an actual on-line spectrometer application. The spectral 'miscibility' of the synthetic calibration standards and the process samples is observed by doing a PCA analysis of a mixed data set containing both types of samples and observing scatter plots of the PC scores. The scatter plot of the first two PC scores is shown in Figure 8.37. Note that there is some 'mixing' of the process samples with the synthetic standards in the PC1/PC2 space, but that there are many process samples that lie in a region that is outside of the space covered by the synthetic standards. This result indicates that the spectra of the process samples contain some unique information that is not contained in the spectra of the synthetic standards. Fortunately, when a PLS regression model for the property of interest is built using the same mixed data set, a good model fit is obtained (Figure 8.38). This result indicates that, despite the fact that the process spectra contain some information that is missing from the synthetic standard spectra, these spectra can be effectively 'mixed' together to provide a good quantitative model.

8.8   Sample and variable selection in chemometrics

8.8.1   Sample selection

Earlier in this chapter, it was mentioned that a large amount of good calibration data is required to build good empirical models. However, it was also mentioned that these data

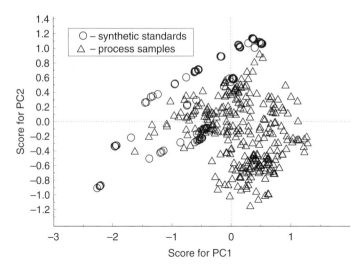

Figure 8.37   Scatter plot of the first two PCA scores obtained from a process analytical calibration data set containing both synthesized standards (circles) and actual process samples (triangles).

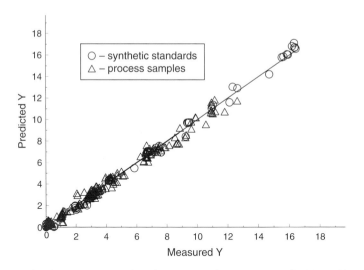

Figure 8.38   Results of a PLS regression fit to the property of interest, using all of the calibration samples represented in Figure 8.37.

must be sufficiently ***representative*** of the data that one expects to obtain during real-time operation of the analyzer. In this context, the concept of representative data means that not only must the calibration sample states cover the ***range*** of sample states expected during operation of the analyzer, but the ***distribution*** of the calibration sample states should also be sufficiently representative. This is particularly important for calibration data that is ***not*** collected according to a Design of Experiments (DOE, Section 8.2.8), such as data collected from the on-line analyzer during normal operation (otherwise known as happenstance data). In such cases, a particular collection of calibration samples might over-represent some states and under-represent others. As a result, any model built from this data will be somewhat 'skewed,' in that it will tend to perform better for the over-represented states and worse for the under-represented states. In addition, a more practical driving force for sample selection is that it decreases the time required to calculate a calibration model due to the fewer number of samples to be modeled.

There are several methods that can be used to select well-distributed calibration samples from a set of such happenstance data. One simple method, called ***leverage-based selection***, is to run a PCA analysis on the calibration data, and select a subset of calibration samples that have extreme values of the leverage for each of the significant PCs in the model. The selected samples will be those that have extreme responses in their analytical profiles. In order to cover the sample states better, it would also be wise to add samples that have low leverage values for each of the PCs, so that the 'center' samples with more 'normal' analytical responses are well represented as well. Otherwise, it would be very difficult for the predictive model to characterize any non-linear response effects in the analytical data. In PAC, where spectroscopy and chromatography methods are common, ***it is better to assume that non-linear effects in the analytical responses could be present than to assume that they are not***.

Another useful method for sample selection is ***cluster analysis-based selection***.[3,4,67] In this method, it is typical to start with a compressed PCA representation of the calibration data. An unsupervised cluster analysis (Section 8.6.3.1) is then performed, where the algorithm is terminated after a specific number of clusters are determined. Then, a single sample is selected from each of the clusters, as its 'representative' in the final calibration data set. This cluster-wise selection is often done on the basis of the maximum distance from the overall data mean, but it can also be done using each of the cluster means instead.

Figure 8.39 shows a three-dimensional scatter plot of the first three PC scores obtained from a PCA analysis of 987 calibration spectra that were collected for a specific on-line analyzer calibration project. In this case, cluster analysis was done using the first six PCs (all of which cannot be displayed in the plot!) in order to select a subset of 100 of these samples for calibration. The three-dimensional score plot shows that the selected samples are well distributed among the calibration samples, at least when the first three PCs are considered.

The cluster analysis-based method of sample selection is very useful when one wants to ensure that at least one sample from each of a known number of subclasses is selected for calibration. However, one must be careful to specify a sufficient number of clusters, otherwise all of the subgroups might not be represented. It is always better to err on the side of determining too many clusters. The specification of a number of clusters that is much greater than the number of 'natural' groups in the data should still result in well-distributed calibration samples.

8.8.2   *Variable selection*

The subject of *X*-variable selection was discussed earlier in this chapter, in the context of data compression methods (Section 8.2.6.1). It was mentioned that selection can be done

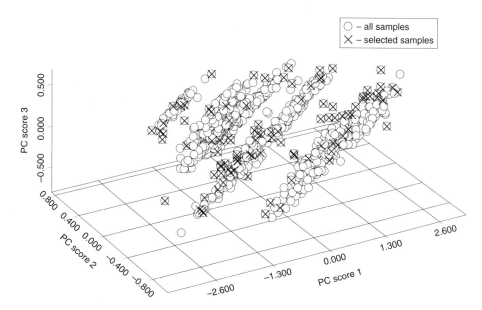

Figure 8.39 Three-dimensional scatter plot of the first three PCA scores obtained from a set of original calibration data. The calibration samples selected by the cluster analysis method are marked with an 'x.'

based on prior knowledge, and that there are some regression methods that actually incorporate empirical variable selection (e.g. SMLR). The purpose of this section is to supplement these selection methods with some empirical methods that can be used with non-MLR regression techniques, such as PLS, PCR, and ANN.

At first, one might wonder whether variable selection is really needed for PLS, PCR, and ANN at all. After all, these methods should 'naturally' avoid the use of *X*-variables that are irrelevant for solving the problem, right? The answer is: 'Not exactly.' Irrelevant *X*-variables in the data still present both random and systematic variability in the *X*-data, which could still find its way into the calibration model, simply because the reference concentration data (*Y*) are never noise-free in practice. As a result, such irrelevant variables can serve only to 'confuse' these modeling methods, and they can be considered an unneeded 'burden' on the methods. Furthermore, removing irrelevant variables is another means of *simplifying* the calibration model. It has already been stated that simpler models tend to be more stable, less sensitive to unforeseen disturbances, and easier to maintain. Finally, there is a practical advantage to reducing the number of *X*-variables: it decreases the computer run-time required to calculate a model (especially, if a large number of calibration samples is used!).

As for sample selection, I will submit two different methods for variable selection: one that is relatively simple and one that is more computationally intensive. The simpler method[68] involves a series of linear regressions of each *X*-variable to the property of interest. The relevance of an *X*-variable is then expressed by the ratio of the linear regression slope (b) to the variance of the elements in the linear regression model residual vector (**f**), called the 'S/N.' The higher the S/N value for a variable, the more likely that it is useful for determining the property of interest.

There are several ways in which the S/N results can be used to select variables. A statistical *f*-test can be used to determine the S/N threshold (at a given confidence level) where the boundary between useful and non-useful variables exists. Alternatively, one could use a more arbitrary threshold or a threshold **range**, based on prior experience on the application of this method to similar data. For an exact threshold, one simply selects all variables whose S/N values are above the threshold. For a threshold range, one can either select all variables above the range, or remove all variables below the range. This method has been used to provide improved results for PLS calibrations on process analyzers.[3] However, the main limitation of this method is that it only considers **univariate** relationships between a single *X*-variable and the property of interest. The fact that chemometrics is being considered for building the model in the first place is due to the suspicion (or knowledge!) that a **multivariate** relationship between several *X*-variables and the property of interest exists.

A more robust, yet time-consuming, alternative of the univariate approach is the multivariate 'trial and error' approach, where a large number of calibrations using different subsets of *X*-variables are developed and tested. However, for typical process analytical applications, where the number of *X*-variables is at least in the hundreds, the number of possible combinations of variables becomes an astronomical number (9.3×10^{157}, for 100 variables), and it is not practical to test each combination. Fortunately, there are 'searching' algorithms that allow one to make a more logical search of all of these possibilities, and converge upon a very good (if not mathematically optimal) variable subset. One algorithm that has been used for variable selection in chemometrics is called the **Genetic Algorithm** (GA). This algorithm operates in the following manner:

(1) The total number of variable subset candidates is specified (along with several other parameters), and an initial set of variable subset candidates are randomly generated.
(2) For each variable subset, a regression model is generated.
(3) Each of the regression models is evaluated for prediction ability, using cross-validation.
(4) Different pairs of variable subsets are 'mated' together, based on their prediction abilities assessed above. Those variable subsets that result in the best prediction abilities are most likely to 'mate' with one another. The mating process involves generating a new variable subset using some of the selected variables from one mate and some from the other. This process results in the generation of a new set of variable subset candidates.
(5) Steps (2) to (4) are repeated.

The algorithm can be terminated using several different termination conditions. For example, it can be terminated when a specified number of iterations has occurred, or when agreement between the different variable subsets reaches a certain level. Details on the GA method can be found in several references.[69,70]

The advantage of the GA variable selection approach over the univariate approach discussed earlier is that it is a true search for an optimal multivariate regression solution. One disadvantage of the GA method is that one must enter several parameters before it

can be executed. This leads to a significant learning curve because the user must determine the optimal parameter ranges for each different type of input data. The GA method
is also rather computationally intensive, especially when a large number of samples (N) or
variables (M) are present in the data. However, there are several means by which one can
address this issue, including sample selection (Section 8.8.1), and using averages of
adjacent variables in the GA (if the variables represent a continuous physical property
such as wavelength). With these issues in mind, the GA method can be a very effective
means of reducing the number of variables, as well as the number of compressed variables,
in a chemometric regression model.

Like the univariate method, there are several ways in which GA results can be used to
select variables. One could either select only those variables that appear in all of the best
models obtained at the end of the algorithm, or remove all variables that were not selected
in any of the best models. It is also important to note that the GA can provide different
results from the same input data and algorithm parameters. As a result, it is often useful to
run the GA several times on the same data in order to obtain a consensus on the selection
of useful variables.

It should be noted that there are other multivariate variable selection methods that one
could consider for their application. For example, the interactive variable selection (IVS)
method[71] is an actual modification of the PLS method itself, where different sets of
X-variables are removed from the PLS weights ($\hat{\mathbf{W}}$, see Equation 8.37) of each latent
variable in order to assess the usefulness at each X-variable in the final PLS model.

8.9   Calibration transfer

Consider a process spectroscopy application where ***all three*** of the following conditions
are present:

(1) There are several similar or duplicate processes at different geographical locations,
    and one would like to install the same type of analyzer at each location.
(2) Multivariate calibration (chemometrics) methods must be used to develop acceptable
    methods.
(3) Although the analyzers are all of the same make and model, their responses will not
    be identical to one another due to small variations in parts manufacturing and
    assembly.

In this situation, it would be ideal to produce a calibration on only one of the analyzers,
and simply ***transfer*** it to all of the other analyzers. There are certainly cases where this can
be done effectively, especially if response variability between different analyzers is low and
the calibration model is not very complex. However, the numerous examples illustrated
above show that multivariate (chemometric) calibrations could be particularly sensitive to
very small changes in the analyzer responses. Furthermore, it is known that, despite the
great progress in manufacturing reproducibility that process analyzer vendors have made
in the past decade, small response variabilities between analyzers of the same make and

model can still be present. For process spectrometers, there are typically two types of inter-instrument variability that are encountered: (1) response magnitude variability and (2) wavelength axis variability. Response magnitude variability refers to the differences in **absolute** response intensity of the instruments, and can be influenced by such factors as source intensity changes and detector response changes. Wavelength axis variability refers to shifting of the wavelength or wavenumber axes between instruments, and can be influenced by tiny differences in the alignment of the light-dispersing optics in the instruments. In cases where such instrument variabilities are present, a simple transfer of a calibration between analyzers is usually not effective.

In such cases, one can retreat to the 'brute force' approach of developing a separate calibration for each analyzer. However, for empirical multivariate calibrations, which require a large number of calibration samples each, this approach can become very laborious. Furthermore, if there is a significant cost associated with the collection, preparation, and analysis of calibration standards, the cost of this instrument-specific calibration approach can become prohibitive.

A more appealing approach would be to prepare and analyze calibration standards and develop a calibration on a single **master** analyzer, and then **standardize** all of the other **slave** analyzers so that the spectra they produce are the same as those that would have been produced on the master analyzer. This way, in principle, the same calibration could be effectively applied to all of the analyzers, and it would not be necessary to calibrate all of the analyzers separately.

Instrument standardization solutions can be split into two groups: hardware-based and software-based. Hardware-based standardization approaches, which have been increasingly used by many analyzer vendors,[72] include the use of internal intensity standards to standardize the response intensities of analyzer responses, and internal wavelength standards to standardize the wavelength/wavenumber axes of different analyzers. These hardware approaches go a long way toward standardizing analyzer responses, thus improving the transferability of calibrations. Nonetheless, there are some applications where inter-instrument response variability, even **after** hardware-based standardization, is still too high to enable the use of the appealing single-calibration approach. For these applications, some additional standardization help from chemometrics can lead to improved results. A few specific chemometric-based standardization methods are discussed below.

8.9.1 Slope/intercept adjustment

This method is probably the simplest of the software-based standardization approaches.[73,74] It is applied to each *X*-variable separately, and requires the analysis of a calibration set of samples on both master and slave instruments. A multivariate calibration model is built using the spectra obtained from the master instrument, and then this model is applied to the spectra of the same samples obtained from the slave instrument. Then, a linear regression of the predicted *Y*-values obtained from the slave instrument spectra and the known *Y*-values is performed, and the parameters obtained from this linear regression fit are used to calculate slope and intercept correction factors. In this method, instrument standardization is achieved by adjusting the predicted Y-values obtained from the slave instrument, rather than adjusting the spectra (X-data) obtained from the slave instrument.

The advantage of the slope/intercept standardization method is that it is rather simple to implement, in that it only requires two parameters for each slave instrument. In addition, it is particularly well suited for MLR calibration models, where only a few discrete wavelength variables are used in the model. However, there are several disadvantages that might limit its ability in some applications. First, it requires that all of the calibration samples be analyzed by all of the slave instruments, which can take a while if several slave instruments are involved. Secondly, it assumes that the sources of inter-instrument response variability are completely linear in nature. Therefore, it might not be optimal if significant non-linear differences, such as wavelength axis shifts between analyzers, are present. Finally, it assumes that the nature of the inter-instrument variability is constant over all wavelength ranges, and over all possible sample compositions.

During real-time operation, statistical f-tests can be used to determine whether the slope and the bias correction calculated for a particular unknown spectrum are within an expected range.[73] If not, then a warning can be issued indicating that the current sample is not appropriate to apply to the model.

8.9.2 Piecewise direct standardization (PDS)

The PDS method is a multivariate answer to the univariate slope/intercept method.[73,75] In this method, the responses of a set of transfer standards are obtained on both the master and the slave instruments (thus producing $\mathbf{X}_m$ and $\mathbf{X}_s$, respectively). It is then desired to obtain a transformation matrix $\mathbf{F}$ such that the spectra obtained on the slave instrument can be transformed into spectra that would have been obtained on the master instrument:

$$\mathbf{X}_m = \mathbf{X}_s \mathbf{F} \qquad (8.65)$$

Equation 8.65 indicates that the response that would have been obtained on the master instrument is simply a linear combination of the responses obtained on the slave instrument. In spectroscopy, it is very unlikely that the response at a specific wavelength on the master instrument should depend on any responses on the slave instrument that are at wavelengths far from the specified wavelength. As a result, the PDS method imposes a structure on the $\mathbf{F}$ matrix such that only wavelengths on the slave instrument that are within a specified range of the target wavelength on the master instrument can be used for the transformation.

An important advantage of the PDS method is its ability to address both intensity variations and wavelength axis shift variations between analyzers, although some might argue that the wavelength shift variations are not addressed explicitly in the PDS method. In addition, it can account for situations where the nature of the inter-instrument response variability is not constant over all wavelength regions. Another advantage that is not apparent from the above discussion is that PDS can be used effectively in cases where the master instrument is a high-quality (low-noise) instrument and the slave instrument is a lower-quality (high-noise) instrument. In this case, it is demonstrated that the use of the PDS method to enable the transfer of the master's calibration to the slave instrument results in better prediction performance of the slave analyzer than the 'brute force' method of developing a separate calibration for the slave analyzer.[76]

Some disadvantages of the PDS method are its relative complexity versus the slope/offset method, and the need to analyze the same set of standardization standards on all analyzers.

Another important issue that arises in the PDS method, as well as some other standardization methods, is the **selection** of the samples to use for standardization. It is critical that the standardization samples efficiently **convey the magnitude and nature of instrument-to-instrument variability artifacts** that are expected to be present in the analyzers while they are operating in the field. Note that this criterion is different than the criterion used for sample selection for calibration, which is to sufficiently **cover the compositions of the process samples** that the analyzer is expected to see during its operation. Sample selection strategies for instrument standardization have been given by many.[73,77–79]

8.9.3 Shenk-Westerhaus method

Shenk-Westerhaus method, one particular standardization method patented by Shenk and Westerhaus in 1989,[79,80] has seen high utility in NIR food and agricultural applications. This method involves a wavelength axis shift correction, followed by an intensity correction. The steps in this standardization method are as follows:

(1) For each wavelength on the master instrument

    (a) The correlations of the master wavelength with each of a series of wavelengths on the slave instrument that are within a predetermined window around each side of the master wavelength are calculated using linear regression.

    (b) Non-linear interpolation of the function of correlation versus slave instrument wavelength is done to estimate the exact slave wavelength of highest correlation.

(2) Once the exact slave wavelengths corresponding to all of the master wavelengths are determined, interpolation is done on the function of slave instrument response versus wavelength in order to obtain estimated slave responses at the previously specified slave wavelengths.

(3) Finally, linear regression of the master responses and corresponding estimated slave responses is done to compute slope and intercept correction factors for each slave wavelength.

More details on the operation of this method can be found in Bouveresse *et al.*[79]

Like other standardization methods, the effectiveness of this method depends greatly on the relevance of the standardization samples. In many cases, this method uses a set of sealed standardization samples that are relevant to the specific application. In one study,[79] it was found that the effectiveness of this method depended greatly on whether the standardization samples sufficiently cover the spectral intensity ranges that are to be experienced in the on-line process data.

The advantages of this method are that it explicitly addresses both intensity shifts and wavelength axis shifts. In addition, it can handle cases where the nature of the inter-instrument variability varies with wavelength. It is also relatively simple to explain. Like the PDS method mentioned above, it requires the analysis of all standardization samples on all analyzers. One must also be careful to choose standardization samples that sufficiently convey the magnitude and nature of inter-instrument effects in the particular application. In an ideal case, relatively few standardization samples will be required for this purpose. However, this can be determined only by testing the standardization method using standardization sample sets of varying size and membership.

8.9.4   Other standardization methods

There are several other chemometric approaches to calibration transfer that will only be mentioned in passing here. An approach based on finite impulse response (FIR) filters, which does not require the analysis of standardization samples on any of the analyzers, has been shown to provide good results in several different applications.[81] Furthermore, the effectiveness of 'three-way' chemometric modeling methods for calibration transfer has been recently discussed.[82] Three-way methods refer to those methods that apply to X-data that must be expressed as a third-order data array, rather than a matrix. Such data include excitation/emission fluorescence data (where the three orders are excitation wavelength, emission wavelength, and fluorescence intensity) and GC/MS data (where the three orders are retention time, mass/charge ratio, and mass spectrum intensity). It is important to note, however, that a series of spectral data that are continuously obtained on a process can be constructed as a third-order array, where the three orders are wavelength, intensity, and time.

8.10   Non-technical issues

8.10.1   Problem definition

It goes without saying that the first step to solving any problem is to define the problem itself. This is certainly the case for any problem that involves chemometrics in PAC. A large number of highly useful tools are at your disposal, but you are certainly not obligated to use all of them for every problem. Furthermore, the use of inappropriate tools can lead to disappointing results, wasted time, and overall confusion (leading to attrition!).

There is a wide range of tools in the Six-Sigma toolbox, used during the 'define' phase of a Six-Sigma project, which can be applied effectively for problem definition. Such tools include 'Voice of the Customer,' 'Project Charter,' and 'Quality Function Deployment' (QFD). In addition, given below is a non-exhaustive list of issues that are typically addressed during the problem definition phase of a process analytical project that involves chemometrics:

- What will the measurement be used for?
- What is the 'solved problem state?' (i.e. when am I finished?).
- What is the desired performance of the method?
- What are my available resources (time, money, personnel)?
- Are sampling system issues with the analyzer expected? (e.g. cell fouling or transfer-line plugging).
- Can process samples be taken and analyzed off-line? Are there issues with regard to sampling logistics (safety, environmental, costs)?
- How will I handle data storage and transfer logistics?
- Are synthetic standards possible? Are they sufficiently relevant?
- What are the available reference laboratory methods, and for each: what is the error of the method, and what is the cost per analysis?
- Can a calibration experiment be designed, or will I have to rely on happenstance data for calibration?

Answering all of the questions (and more!) can be a daunting task, but it can be argued that all of them are directly relevant to the strategy that one should choose for chemo-metric modeling development, testing, and deployment.

8.10.2   Where does all the time go?

Of course, I can only speak for myself with regard to the amount of time that is required to develop, test, deploy, and maintain chemometric models for on-line analytical applications. However, it should be apparent that there are several tasks involved in this process, and many of them have nothing to do with building models. I am tempted to provide my estimate of the fraction of time that is typically spent on each of these steps, but I feel obligated to retreat, so that I do not report too biased an assessment (even though I have received feedback from several colleagues on this matter). Instead, I will simply list the tasks and supply a general indicator of the relative level of time spent on each one.

- problem definition – *high*
- design of experimental protocol for calibration sample selection – *high*
- set-up of calibration data collection – *low/medium*
- organization of calibration data in preferred model development software – *medium/ high*
- exploratory modeling (trying different pre-treatments, sample/variable selections, modeling methods, etc.) – *medium/high*

- constructing the final model – *low*
- installing and testing the model – *medium*
- if needed, updating the model using new calibration samples – *medium/high*.

I also acknowledge that the relative time commitment levels can vary widely by application. However, it is important to note that a large fraction of the time is spent addressing issues that are more practical, rather than technical. One particularly illustrative example is the calibration data organization step. During this step, one must often review, edit, divide, merge, re-format, export, import, and validate very large blocks of data. The length of time required for this work scales up as the number of variables and the number of calibration samples increase. It also requires meticulous detail work. However, hopefully all of this work is rewarded by a highly effective method, or some newly discovered information about the system under study.

8.10.3 *The importance of documentation*

In the previous section, and earlier in this chapter, it was mentioned that an important part of chemometric method development involves the exploration of different data pre-treatments, modeling methods, sample selection/segregation, variable selection, and other factors in the model building process. With this in mind, one is likely to produce at least several dozen, and possibly more than 100 different 'test' models for a given project! Furthermore, today's industrial chemists and engineers are likely to have several different projects running at any given time, and must often resort to 'time-sharing' in order to make sufficient progress on each project.

The above facts underscore the critical need for documentation of chemometric model building. With sufficient documentation in place, the user is able to continue working on model development on an interrupted project in a time-efficient manner, and without duplicating prior effort. Documentation on the final model(s) that is currently being used on-line can also serve to provide necessary information for satisfying regulatory agencies such as ISO and the FDA, provided that they contain the required information about the models.

The above discussion leads to the following question: Exactly what information about a chemometric model should be documented? For cases where documentation is required by law, I refer the reader to the appropriate publications.[83,84] For those who would like to document to improve their efficiency during the exploratory modeling and final model building steps, I suggest the following as a minimum:

- Date of model;
- Model developer;
- X-data variables used for modeling (excluding outliers);
- Name of Y-variable;
- List of calibration samples used for modeling (excluding outliers);
- Pre-treatment procedures applied to the data before modeling, in order of application, with parameters for each procedure;

- Modeling method, with parameters, if necessary (e.g. for PLS model, the number of latent variables);
- Validation method, with validation parameters;
- Model quality attributes:
  - Fit error (RMSEE or $r2$)
  - Error of cross-validation.

Chemometrics software vendors have come a long way in facilitating good documentation of models.

8.10.4 People issues

Before a chemometrics practitioner can achieve any level of success in a process analytical environment, they must first realize that they do not operate in a vacuum. It has been mentioned that chemometrics is a highly interfacial discipline, which brings together folks from a wide range of different application venues and instrument types. Furthermore, in PAC, one must work effectively with folks from plant operations, plant maintenance, process engineering, plant operations management, technical management, laboratory management, safety, environmental and laboratory operations, and potentially other job functions. One must work effectively with people that have very different backgrounds, experience levels, and value systems. In any project, these differences can be the source of challenge and conflict, and tend to be the root of many project failures if they are not properly addressed.

Based on my experiences, and experiences from my colleagues, there are several different people issues that seem to recur in process analytical projects, and some that are unique to those that involve chemometrics. Instead of discussing each in detail, which could easily take up another book, I will simply list and provide a short explanation for each one.

- The familiarity level of statistics (and certainly chemometrics) among plant personnel is generally low. As a result, there is often skepticism and apprehension about process analytical methods that rely on chemometrics.
- The typical plant laboratory is seldom actively involved in the development and support of process analyzers. This is unfortunate because laboratory technical personnel are usually more likely to understand chemometrics, and thus allay the concerns of others regarding the technology.
- In process analytics, chemometrics can undergo a 'staling' process: during the early phases of implementation, the benefits of the technology are most apparent (e.g. interference rejection, multi-component analysis capabilities, and fault monitoring), but over time, as conditions change and new interferences appear, the models can lose their accuracy. This can lead to suspicions of 'bait and switch.'
- Plant maintenance organizations are most accustomed to simpler sensors that do not require chemometrics for calibration. Specialized training is required to understand and support more complex chemometrics-enhanced analyzers. To some, such specialization can be considered a career 'dead end.'

Most of these issues can be addressed through interpersonal skills, which are not the subject of this chapter. However, the 'staling' process of chemometric models can be addressed by performing three important tasks during the project: (1) publicize the limitations of your models as soon as you are aware of them, (2) do the 'hard work' early on (experimental designs for calibration, collection and analysis of many samples) in order to avoid embarrassments through model extrapolations, and (3) keep an eye on your methods as they are operating, and update/adjust them promptly.

8.11 The final word

Admittedly, navigating through a reference text on chemometrics can be a rather daunting task. There are many different tools in the toolbox, even though it is likely that you will only need a small subset of these to solve a specific problem. This chapter discussed many of the more commonly used and commonly available tools that are at your disposal. It is hoped that these discussions will improve your comfort level and confidence in chemometrics as an effective technology for solving your problems.

Regardless of your problem, and the specific chemometric tools that you choose to implement, there are three guiding principles that should always be kept in mind:

(1) When building a method for on-line use, **keep it simple**! Strive for simplicity, but be wary of complexity.
(2) Do your best to **cover the relevant analyzer response space** in your calibration data. If this cannot be achieved, then at least **know** the limitations in your calibration data.
(3) Regardless of your background, try to use both **chemical** and **statistical** thinking:

   (a) Use your prior knowledge to guide you (chemical thinking);
   (b) But still 'listen' to the data – it might tell you something new (statistical thinking)!

With these principles in mind, and the tools at your disposal, you are ready to put chemometrics to work for your process analytical problems.

References

1. Martens, H. and Næs, T., *Multivariate Calibration*; John Wiley & Sons; Chichester, 1989.
2. Miller, C.E., Chemometrics and NIR: A Match Made in Heaven?; *Am. Pharm. Rev.* 1999, 2(2), 41–48.
3. Miller, C.E., The Use of Chemometric Techniques in Process Analytical Method Development and Operation; *Chemometrics Intell. Lab Syst.* 1995, 30, 11–22.
4. Miller, C.E., Chemometrics for On-Line Spectroscopy Applications: Theory and Practice; *J. Chemom.* 2000, 14, 513–528.
5. Hotelling, H., Analysis of a Complex of Statistical Variables into Principal Components; *J. Educ. Psychol.* 1933, 24, 417–441 and 498–520.

6. Norris, K.H. and Hart, J.R., Direct Spectrophotometric Determination of Moisture Content of Grains and Seeds. In Wexler, A. (ed.); Proceedings 1963 International Symposium on Humidity and Moisture in Liquids and Solids, vol. 4; Reinhold; New York, 1965; p. 19.

7. Massie, D.R. and Norris, K.H., The Spectral Reflectance and Transmittance Properties of Grains in the Visible and Near-Infrared; *Trans. Am. Soc. Agric. Eng.* 1965, 8, 598.

8. Hruschka, W.R. and Norris, K., Least Squares Curve Fitting of Near-Infrared Spectra Predicts Protein and Moisture Content in Ground Wheat; *Appl. Spectrosc.* 1982, 36, 261–265.

9. Norris, K.H., Barnes, R.F., Moore, J.E. and Shenk, J.S., Predicting Forage Quality by Infrared Reflectance Spectroscopy; *J. Anim. Sci.* 1976, 43, 889–897.

10. Sjoestroem, M., Wold, S., Lindberg, W., Persson, J.A. and Martens, H., A Multivariate Calibration Problem in Analytical Chemistry Solved by Partial Least Squares Models in Latent Variables; *Anal. Chim. Acta* 1983, 150, 61–70.

11. Martens, H., Factor Analysis of Chemical Mixtures: Non-Negative Factor Solutions for Cereal Amino Acid Data; *Anal. Chim. Acta* 1979, 112, 423–441.

12. Wold, H., Soft Modeling: The Basic Design and Some Extensions. In Joreskog, K.G. and Wold, H. (eds); *Systems Under Indirect Observation, Causality-Structure-Prediction*, Part 2; North-Holland Publishing Co.; Amsterdam, 1982; pp. 1–54.

13. Gerlach, R.W., Kowalski, B.R. and Wold, H.O.A., Partial Least Squares Path Modeling With Latent Variables; *Anal. Chim. Acta* 1979, 112, 417–421.

14. Malinowski, E.R. and Howery, D.G., *Factor Analysis in Chemistry*, 1st Edition; John Wiley & Sons; New York, 1980.

15. Kowalski, B.R., *Chemometrics: Mathematics and Statistics in Chemistry*; D. Reidel; Dordrecht (Netherlands), 1980.

16. Wold, S., Cross-Validatory Estimation of the Number of Components in Factor Analysis and Principal Component Models; *Technometrics* 1978, 20, 397–406.

17. Kaltenbach, T.F., Colin, T.B. and Redden, N., Chemometric Modeling for Process Mass Spectrometry Applications; *AT-Process* 1997, 3, 43–52.

18. Heikka, R., Minkkinen, P. and Taavitsainen, V.-M., Comparison of Variable Selection and Regression Methods in Multivariate Calibration of a Process Analyzer; *Process Contr. Qual.* 1994, 6, 47–54.

19. Prazen, B.J., Synovec, R.E. and Kowalski, B.R., Standardization of Second-Order Chromatographic/Spectroscopy Data for Optimal Chemical Analysis; *Anal. Chem.* 1998, 70, 218–225.

20. Box, G.E.P., Hunter, W.G. and Hunter, J.S., *Statistics for Experimenters: An Introduction to Design, Data Analysis and Model Building*; John Wiley & Sons; New York, 1978.

21. Taylor, J.R., *An Introduction to Error Analysis*; University Science Books; Mill Valley, CA USA, 1982.

22. Mobley, P.R., Kowalski, B.R., Workman, J.J. and Bro, R., Review of Chemometrics Applied to Spectroscopy: 1985–1995, Part 2; *Appl. Spec. Rev.* 1996, 31, 347–368.

23. Geladi, P., MacDougall, D. and Martens, H., Linearization and Scatter Correction for Near-Infrared Reflectance Spectra of Meat; *Appl. Spectrosc.* 1985, 39, 491–500.

24. Miller, C.E. and Næs, T., A Pathlength Correction Method for Near-Infrared Spectroscopy; *Appl. Spectrosc.* 1990, 44, 895–898.

25. Martens, H. and Stark, E., Extended Multiplicative Signal Correction and Spectral Interference Subtraction: New Preprocessing Methods for Near Infrared Spectroscopy; *J. Pharm. Biomed. Anal.* 1991, 9, 625–635.

26. Tkachuk, R., Kuzina, F.D. and Reichert, R.D., Analysis of Protein in Ground and Whole Field Peas by Near-Infrared Reflectance; *Cereal Chem.* 1987, 64, 418–422.

27. Tenge, B. and Honigs, D.E., *The Effect of Wavelength Searches on NIR Calibrations*; Abstracts of Papers of the American Chemical Society 1987, 193(April), 112.

28. Golub, G. and Van Loan, C.F., *Matrix Computations*; Johns Hopkins University Press; Baltimore, MD, USA, 1983.

29. Fisher, R.A., The Use of Multiple Measurements in Taxonomic Problems; *Annals of Eugenics* 1936, 7, 179–188.

30. McClure, W.F., Hamid, A., Giesbrecht, F.G. and Weeks, W.W., Fourier Analysis Enhances NIR Diffuse Reflectance Spectroscopy; *Appl. Spectrosc.* 1984, 38, 322–329.

31. Miller, C.E., Chemical Principles of Near-Infrared Technology. In Williams, P. and Norris, K. (eds); *Near Infrared Technology for the Agricultural and Food Industries*; American Association of Cereal Chemists; St Paul, MN, USA, 2001; pp. 19–36.

32. Bonanno, A.S., Ollinger, J.M. and Griffiths, P.R., The Origin of Band Positions and Widths in Near Infrared Spectroscopy. In Hildrum, K.I., Isaksson, T., Næs, T. and Tandberg, A. (eds); *Near-Infrared Spectroscopy: Bridging the Gap Between Data Analysis and NIR Applications*; Ellis-Horwood; New York, 1992; 20–28.

33. Mallat, S.G., *A Wavelet Tour of Signal Processing*; Academic Press; New York, 1998.

34. Walczak, B., van den Bogaert, B. and Massart, D.L., Application of Wavelet Packet Transform in Pattern Recognition of Near-IR Data; *Anal. Chem.* 1996, 68, 1742–1747.

35. Chaoxiong, M. and Xueguang, S., Continuous Wavelet Transform Applied to Removing the Fluctuating Background in Near-Infrared Spectra; *J. Chem. Inf. Comput. Sci.* 2004, 44, 907–911.

36. Mittermayr, C.R., Tan, H.W. and Brown, S.D., Robust Calibration with Respect to Background Variation; *Appl. Spectrosc.* 2001, 55, 827–833.

37. Fearn, T. and Davies, A.M.C., A Comparison of Fourier and Wavelet Transforms in the Processing of Near-Infrared Spectroscopic Data: Part 1. Data Compression; *J. Near Infrared Spectrosc.* 2003, 11, 3–15.

38. Brown, C.W. and Obremski, C.W., Multicomponent Quantitative Analysis; *Appl. Spec. Rev.* 1984, 20, 373–418.

39. Miller, C.E., Eichinger, B.E., Gurley, T.W. and Hermiller, J.G., Determination of Microstructure and Composition in Butadiene and Styrene–Butadiene Polymers by Near-Infrared Spectroscopy; *Anal. Chem.* 1990, 62, 1778–1785.

40. Mandel, J., Use of the Singular Value Decomposition in Regression Analysis; *The American Statistician* 1982, 36, 15–24.

41. Wold, S., Albano, C., Dunn, W.J. *et al.*, Pattern Recognition: Finding and Using Patterns in Multivariate Data. In Martens, H. and Russwurm, H. Jr. (eds); *Food Research and Data Analysis*; Applied Science Publishers; London, 1983; pp. 147–188.

42. de Jong, S., SIMPLS: An Alternative Approach to Partial Least Squares Regression; *Chemometrics and Intelligent Laboratory Systems* 1993, 18, 251–263.

43. Næs, T., Kvaal, K., Isaksson, T. and Miller C., Artificial Neural Networks in Multivariate Calibration; *J. Near Infrared Spectrosc.* 1993, 1, 1–11.

44. Blanco, M., Coello, J., Iturriaga, H., Maspoch, S. and Pages, J., NIR Calibration in Non-linear Systems: Different PLS Approaches and Artificial Neural Networks; *Chemometrics Intell. Lab. Syst.* 2000, 50, 75–82.

45. Jagemann, K.U., Fischbacher, C., Muller, U.A. and Mertes, B., Application of Near-Infrared Spectroscopy for Non-Invasive Determination of Blood-Tissue Glucose using Neural Networks; *Zeitschrift Phys. Chemie- Int. J. of Research in Physical Chemistry and Chemical Physics* 1995, 191, 179–190.

46. Rumelhart, D.E., Hinton, G.E. and Williams, R.J., Learning Internal Representations by Error Propagation. In Rumelhart, D. and McClelland, J. (eds); *Parallel Data Processing*, 1; MIT Press; Cambridge, MA, USA, 1986; pp. 318–362.

47. Næs, T., Multivariate Calibration by Covariance Adjustment; *Biometrical J.* 1986, 28, 99–107.

48. Fearn, T., Misuse of Ridge Regression in the Calibration of Near-Infrared Reflectance Instruments; *Appl. Statistics* 1983, 32, 73–79.

49. Friedman, J.H. and Stuetzle, W., Projection Pursuit Regression; *J. Am. Stat. Assoc.* 1981, 76, 817–823.

50. Elbergali, A., Nygren, J. and Kubista, M., An Automated Procedure to Predict the Number of Components in Spectroscopic Data; *Anal. Chim. Acta* 1999, 379, 143–158.

51. Faber, K. and Kowalski, B.R., Critical Evaluation of Two f-tests for Selecting the Number of Factors in Abstract Factor Analysis; *Anal. Chim. Acta* 1997, 337, 57–71.

52. Bayne, C.K. and Rubin, I.B., *Practical Experimental Design and Optimization Methods for Chemists*; VCH Publishers; Deerfield Beach, FL, USA, 1986.

53. Oman, S.D., Næs, T. and Zube, A., Detecting and Adjusting for Non-Linearities in Calibration of Near-Infrared Data Using Principal Components; *J. Chemom.* 1993, 7, 195–212.

54. Berglund, A., Kettaneh, N., Uppgard, L.-L., Wold, S., Bendwell, N. and Cameron, D.R., The GIFI Approach to Non-Linear PLS Modeling; *J. Chemom.* 2001, 15, 321–336.

55. Miller, C.E. and Eichinger, B.E., Analysis of Rigid Polyurethane Foams by Near-Infrared Diffuse Reflectance Spectroscopy; *Appl. Spectrosc.* 1990, 44, 887–894.

56. Fukunaga, K. *Introduction to Statistical Pattern Recognition*; Academic Press; San Diego, CA, USA, 1990.

57. Rosenblatt, F., *Principles of Neurodynamics*; Spartan Press; Washington DC, USA, 1961.

58. Kohonen, T., *Self-Organization and Associative Memory*, 2nd Edition; Springer-Verlag; Berlin, 1987.

59. Burges, C.J.C., A Tutorial on Support Vector Machines for Pattern Recognition; *Data Min. Knowl. Disc.* 1998, 2, 121–167.

60. Woody, N.A. and Brown, S.D., Hybrid Bayesian Networks: Making the Hybrid Bayesian Classifier Robust to Missing Training Data; *J. Chemom.* 2003, 17, 266–273.

61. Miller, C.E. and Eichinger, B.E., Analysis of Reaction-Injection-Molded Polyurethanes by Near-Infrared Diffuse Reflectance Spectroscopy; *J. Appl. Polym. Sci.* 1991, 42, 2169–2190.

62. Windig, W. and Meuzelaar, H.L.C., Numerical Extraction of Components from Mixture Spectra by Multivariate Data Analysis. In Meuzelaar, H.L.C. and Isenhour, T.L. (eds); *Computer Enhanced Analytical Spectroscopy*; Plenum; New York, 1987; pp. 81–82.

63. Tauler, R., Kowalski, B. and Fleming, S., Multivariate Curve Resolution Applied to Spectral Data from Multiple Runs of an Industrial Process; *Anal. Chem.* 1993, 65, 2040–2047.

64. Gemperline, P., Zhu, M., Cash, E. and Walker, D.S., Chemometric Characterization of Batch Reactions; *ISA Transactions* 1999, 38, 211–216.

65. Mardia, K.V., Kent, J.T. and Bibby, J.M., *Multivariate Analysis*; Academic Press; London, 1979.

66. Kohonen, T., *Self-Organizing Maps*; Springer-Verlag; New York, 1997.

67. Isaksson, T. and Næs, T., Selection of Samples for Calibration in Near-Infrared Spectroscopy. Part II: Selection Based on Spectral Measurements; *Appl. Spectrosc.* 1990, 44, 1152–1158.

68. Brown, P.J., Wavelength Selection in Multicomponent Near-Infrared Calibration; *J. Chemom.* 1991, 6, 151–161.

69. Leardi, R., Genetic Algorithms in Chemometrics and Chemistry: A Review; *J. Chemom.* 2001, 15, 559–569.

70. Jouan-Rimbaud, D., Massart, D.-L., Leardi, R. and De Noord, O.E., Genetic Algorithms as a Tool for Wavelength Selection in Multivariate Calibration; *Anal. Chem.* 1995, 67, 4295–4301.

71. Lindgren, F., Geladi, P., Berglund, A., Sjostrom, M. and Wold, S., Interactive Variable Selection (IVS) for PLS. Part II: Chemical Applications; *J. Chemom.* 1995, 9, 331–342.

72. Workman, J. and Coates, J., Multivariate Calibration Transfer: The Importance of Standardizing Instrumentation; *Spectroscopy* 1993, 8(9), 36–42.

73. Bouveresse, E., Hartmann, C., Massart, D.L., Last, I.R. and Prebble, K.A., Standardization of Near-Infrared Spectrometric Instruments; *Anal. Chem.* 1996, 68, 982–990.

74. Osborne, B.G. and Fearn, T., Collaborative Evaluation of Universal Calibrations for the Measurement of Protein and Moisture in Flour by Near-Infrared Reflectance; *J. Food Technol.* 1983, 18, 453–460.

75. Wang, Y. and Kowalski, B.R., Calibration Transfer and Measurement Stability of Near-Infrared Spectrometers; *Appl. Spectrosc.* 1992, 46, 764–771.

76. Wang, Y., Lysaght, M.J. and Kowalski, B.R., Improvement of Multivariate Calibration through Instrument Standardization; *Anal. Chem.* 1992, 64, 562–564.

77. Wang, Y., Veltkamp, D.J. and Kowalski, B.R., Multivariate Instrument Standardization; *Anal. Chem.* 1991, 63, 2750–2756.

78. Kennard, R.W. and Stone, L.A., Computer Aided Design of Experiments; *Technometrics* 1969, 11, 137–149.

79. Bouveresse, E., Massart, D.L. and Dardenne, P., Calibration Transfer Across Near-Infrared Spectroscopic Instruments using Shenk's Algorithm: Effects of Different Standardisation Samples; *Anal. Chim. Acta* 1994, 297, 405–416.

80. Shenk, J.S. and Westerhaus, M.O., *Optical Instrument Calibration System*; U.S. Pat. 4866644, 12 September 1989.

81. Blank, T.B., Sum, S.T., Brown, S.D. and Monfre, S.L., Transfer of Near-Infrared Multivariate Calibrations without Standards; *Anal. Chem.* 1996, 68, 2987–2995.

82. Bro, R. and Ridder, C. (2002) *Notes on Calibration of Instruments (Spectral Transfer)*, http://www.models.kvl.dk/users/rasmus/presentations/calibrationtransfer/calibrationtransfer.pdf.

83. http://21cfrpart11.com/pages/library/.

84. http://www.iso.org/iso/en/iso9000-14000/iso9000/2000rev7.html.

Chapter 9
On-Line Applications in the Pharmaceutical Industry

Steven J. Doherty and Charles N. Kettler

9.1 Background

A significant portion of the world's population uses products manufactured and distributed by pharmaceutical companies. The processes used to discover, develop and manufacture drug products are regulated by different governmental agencies throughout the world. Through the efforts of these manufacturers and regulating agencies many life saving drugs have been provided to patients. Statutes in the Federal Food, Drug and Cosmetic Act regulate drug manufacturing in the United States. Adherence to the Current Good Manufacturing Practices (cGMPs) described in these statutes is mandated by these regulations. In addition to the regulations set forth by the United States Food and Drug Administration (US FDA), other regulatory agencies around the world have similar expectations of drug manufacturers.

In August 2002, the FDA launched a new program – Pharmaceutical cGMPs for the Twenty-First Century: A Risk-Based Approach. The Agency was concurrently advocating a related proposal, known as the Process Analytical Technology (PAT) Initiative.[1] A draft guidance[2] related to the PAT initiative was issued on 25 August 2003. The evolution of these two programs will have profound impact on the development of on-line spectroscopic tools in pharmaceutical development and manufacturing control for the next several decades.

In the past, careful scientific study was used to develop the processes for manufacturing drug products. The processes were typically carried out in stepwise fashion and utilized post-manufacture analysis of the product to ensure the quality of the material and its fitness for use in the next step or for packaging and distribution to the patient. Control of these manufacturing steps was typically through measurement and control of time, temperature, pressure and flow. The intention was to have a sequence of steps that could be validated as capable of producing product that met specifications and fulfilled the requirements of cGMP. After validation of the process and approval of the FDA, there was no impetus to change or improve the manufacturing process. This regulatory environment, for a variety of reasons, discouraged the implementation of new manufacturing technologies and measurements for processes that were already validated and approved.

Recognizing that the pharmaceutical industry was hesitant to introduce new technologies, the FDA launched the initiative to develop risk-based cGMPs. The goals of the program are:

- The most up-to-date concepts of risk management and quality systems approaches are incorporated into the manufacture of pharmaceuticals while maintaining product quality.
- Manufacturers are encouraged to use the latest scientific advances in pharmaceutical manufacturing and technology.
- The agency's submission review and inspection programs operate in a coordinated and synergistic manner.
- Regulations and manufacturing standards are applied consistently by the agency and the manufacturer, respectively.
- Management of the agency's risk-based approach encourages innovation in the pharmaceutical manufacturing sector.
- Agency resources are used effectively and efficiently to address the most significant health risks.

The intention of the agency is to encourage the pharmaceutical industry to develop their manufacturing processes with enhanced scientific and engineering rigor. The ability to use new information about a process and to apply advanced data analysis techniques to understand manufacturing performance must be encouraged throughout the life cycle of a drug. By utilizing the best and most appropriate measurement and control technologies, potential product and process quality risks can be identified and mitigated. The FDA characterizes the future of pharmaceutical manufacturing as exhibiting the following attributes:

- Product quality and performance are ensured through the design of effective and efficient manufacturing processes;
- Product and process specifications are based on a mechanistic understanding of how formulation and process factors affect product performance;
- Continuous *real-time* quality assurance;
- Relevant regulatory policies and procedures are tailored to accommodate the most current level of scientific knowledge.

Simply stated, in-depth scientific study of a pharmaceutical process will generate unambiguous knowledge leading to process understanding. Applying the required measurement techniques and engineering control concepts to the design of the manufacturing process will define the quality of the drug product. This product quality by design approach will have to be integrated into the research, development and manufacturing steps of pharmaceutical manufacturing. This evolutionary process will ultimately provide higher-quality and lower-cost medicine for the patient and a reduction in the regulatory burden borne by FDA-regulated manufacturing companies.

The FDA's guidance document for PAT divides the tools for process understanding and control into four different categories:

(1) multivariate data acquisition and analysis tools;
(2) process analyzers and sensors;

(3) process and endpoint monitoring and control;

(4) tools for information management for ongoing improvement.

This chapter will review spectroscopic tools that fall into categories (2) and (3) above, across the continuum from discovery through process development and on to final release. Where appropriate, illustrative examples from the literature will be cited. Since the spectroscopic tools themselves have been described in greater detail elsewhere in this text (Chapters 3–7), the focus will be on practical demonstration of process control (either open-loop or closed-loop control) using spectroscopic tools. Although the FDA PAT guidance includes off-line (or laboratory), at-line (plant floor) and near-line (manufacturing area) techniques, those topics will not be specifically addressed in this chapter, and such applications will be covered in greater detail in the following chapter. Multivariate tools (category 1) have been discussed in the preceding chapter.

One indication of the developing interest in PATs in the pharmaceutical area is the number of book chapters and review articles in this field that have appeared in the last few years. Several chapters in *The Handbook of Vibrational Spectroscopy*[3] are related to the use of various optical spectroscopies in pharmaceutical development and manufacturing. Warman and Hammond also cover spectroscopic techniques extensively in their chapter titled 'Process Analysis in the Pharmaceutical Industry' in the text *Pharmaceutical Analysis*.[4] Pharmaceutical applications are included in an exhaustive review of near-infrared (NIR) and mid-infrared (mid-IR) by Workman,[5] as well as the periodic applications reviews of Process Analytical Chemistry and Pharmaceutical Science in the journal *Analytical Chemistry*. The *Encyclopedia of Pharmaceutical Technology* has several chapters on spectroscopic methods of analysis, with the chapters on Diffuse Reflectance and Near-Infrared Spectrometry particularly highlighting on-line applications.[6] There are an ever-expanding number of recent reviews on pharmaceutical applications, and a few examples are cited for Raman,[7,8] NIR,[9–11] and mid-IR.[12]

The FDA Center for Drug Evaluation and Research (CDER) maintains an archival site with a significant amount of information on the PAT initiative.[1] Ronald Miller provided a synopsis of the initial meetings of the FDA's Process Analytical Technology Advisory Committee for Pharmaceutical Science in a series of articles in *The American Pharmaceutical Review*.[13] There are a number of journals with recurring coverage of developments in the FDA's PAT initiative, as well as related technological advances, including the molecular spectroscopy workbench column in *Spectroscopy*, in addition to regular publications in both *European Pharmaceutical Review* and *American Pharmaceutical Review*.

While there are a number of factors driving the implementation of PAT in pharmaceutical development and manufacture, it can also be said that there are a number of barriers. The magnitude of these barriers is profoundly affected by the organizational culture within each pharmaceutical manufacturer, and frequently across sites within a single company. Several of these cultural issues have been spelled out in the literature,[14,15] and there are additional issues related to the absence of infrastructure (both physical and human). A number of the potential impediments are cited here to provide context for those unfamiliar with the area.

Perhaps most significant is the regulatory uncertainty associated with a new approach to manufacturing control. As described in Chapter 2, validation requirements are different for on-line and in-line tools than for laboratory assays, and there is concern that any uncertainties during the regulatory review process will lead to delayed product launches.

There are related issues affecting computer system validation and computer software (21 CFR Part 11), particularly when a chemometric model that may require model maintenance and re-calibration is involved. For process control applications in manufacturing plants, sample taps are rarely available for interfacing an on-line tool, and many sites are reticent to take vessels out of service for modification. Historically, the regulatory implications for making alterations to validated processes provided such significant barriers to change, that it simply was not done. Consequently, many PAT applications will likely be incorporated with the process technology as it is transferred from the development organization to manufacturing.

There are numerous advantages to developing control technologies during the original development program, rather than after the process is in manufacturing. Systematically performing data-rich experiments throughout development increases the depth of process knowledge, increasing the likelihood of developing a robust, well-understood process. Collecting similar process data across several scales and in different processing equipment will highlight the scale-dependent factors, and perhaps identify problems earlier in development when there is greater opportunity to correct the issues. Furthermore, process understanding that arises early in development enables the identification and evaluation of the simplest control technology, which can be evaluated in the actual process during subsequent development.

While standardized tools and data across development and manufacturing might be desirable, one also needs to recognize the different requirements of PAT in these disparate environments. In a development environment, one would desire a flexible, capable analytical tool that would be expected to require a higher level of support (in quantity and sophistication) than might be expected in a manufacturing environment. In a manufacturing environment, the PAT tool should resemble more a robust sensor, with modular components and a low cost of ownership. While it might be ideal to bring the same instrument used in development to the plant, what may be more practical is to have similar data or an analogous approach to model the chemical system's performance across scales. This is related to the concept of a 'process signature,' and also touches on the FDA concept of information management.

In the following examples of in-line spectroscopic analysis, the applications will flow through a pharmaceutical development process, basically following the path of the molecular entity. Incoming material, frequently analyzed using PAT, will be discussed in Chapter 10, but the other applications will step through active pharmaceutical ingredient (API) reaction and crystallization, followed by drying, milling, blending, granulation, tabletting and release. The material will be presented in this order, with applications from both R&D and manufacturing being cited. Where appropriate, distinctions will be made with respect to the environment of the cited work.

9.2 Reaction monitoring and control

The majority of APIs are produced by conventional synthetic organic chemistry, though there are a significant number of therapeutic agents produced through biosynthetic means. For the former, the process of manufacture is generally a series of synthetic organic processing steps, interspersed with other unit operations, such as filtration and drying, that are directly analogous to commodity and specialty chemical manufacturing. As such, the tools used for process monitoring and control can frequently be the same equipment and approaches as described in Chapter 11 of this text, albeit with additional requirements arising from the regulatory environment in which they are used.

However, there are substantial incentives for the investment in developing process analytical tools during the development phase. First and foremost are the significant time pressures in the pharmaceutical area. Due to the limited amount of time for market exclusivity arising from the patent system, there are generally tremendous pressures to push viable products through development and into the market. In the interest of time to market, marginal manufacturing processes have been ushered into manufacturing, with the understanding that the process can be refined and optimized over time. However, this requires that the chemical manufacturing and controls (CMC) documents registered with the process allow sufficient latitude to make changes post-launch. This flexibility must be weighed against the concern that any ambiguities in the CMC document will lead to elevated scrutiny by the regulatory agencies. Although it is possible to make changes to registered processes, this is rarely done in practice due to the costs and logistic issues associated with filing and tracking changes with the numerous regulatory bodies. Process analytical tools can offer significant return on investment by producing robust manufacturing processes out of development, which do not have to be 'de-bugged' in commercial equipment. This is one component of a mantra cited in several pharmaceutical manufacturing organizations of 'right the first time.'

In addition, there is a need for maximizing process understanding while consuming the minimum amount of material due to the value (and frequently limited supply) of the materials themselves. Many of the synthetic processes involve numerous reaction steps and numerous manipulations, so it is common for the supply of material to be the limiting factor to the number of experiments that can be performed, making the information content of each experiment of paramount importance.

Automation of experiments in process development groups, particularly in combination with design of experiments (DOE), have been the focus of extensive work, including a special section in the journal *Organic Product Research and Development*,[16] but a preponderance of the published work in this field still cites near-line or off-line HPLC for analysis. A number of application notes and technical reports that describe on-line spectroscopic measurements can be found at some of the automation vendors' sites, such as Mettler Autochem and Argonaut.[17]

The molecular specificity of Fourier transform infrared (FTIR) lends itself quite well to applications in pharmaceutical development labs, as pointed out in a review article with some historical perspective.[10] One of the more common applications of mid-IR in development is a real-time assessment of reaction completion when used in conjunction with standard multivariate statistical tools, such as partial least squares (PLS) and principal component analysis (PCA).[18,19] Another clever use of FTIR is illustrated in Figure 9.1, where the real-time response of a probe-based spectroscopic analyzer afforded critical control in the charge of an activating agent (trifluoroacetic anhydride) to activate lactol. Due to stability and reactivity concerns, the in situ spectroscopic approach was

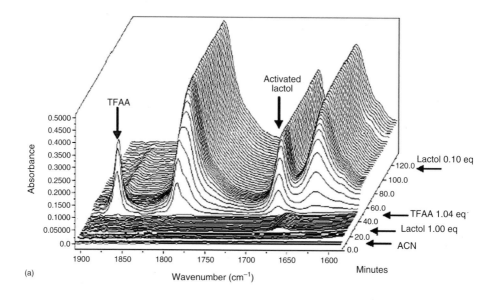

(a)

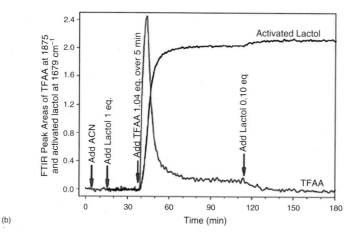

(b)

Figure 9.1 (a) Waterfall plot of FTIR spectra during a lactol/TFAA activation reaction in acetonitrile at −5°C; (b) Corresponding profiles of TFAA and activated lactol. Reprinted from Chen *et al.* (2003)[20] with permission from Elsevier.

found to be superior to sample extraction with off-line chromatographic analysis.[20] Similarly, a mid-IR application to track the progress of a deprotection reaction eliminated the need for extractive sampling of a highly potent compound, while providing a real-time measure of reaction conversion on a labile product.[21] Figure 9.2 shows the concentration profile over time for the on-line and extractive measurements, as well as the correlation plot.

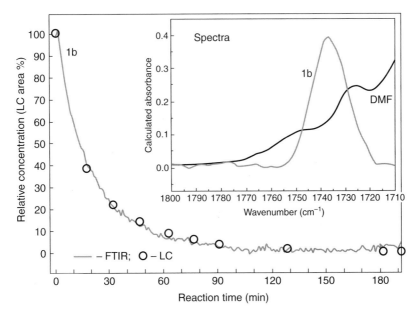

Figure 9.2 Concentration profile of TFAA and activated lactol obtained by predicting the prediction run with the calibration model built from five PLS factors in the spectral range of 1780–1703 and 1551–1441 cm^{-1}. Inside is the corresponding correlation plot of TFAA and activated lactol for prediction data set predicted by the same calibration model. Reprinted from Cameron *et al.* (2002)[21] with permission from Elsevier.

Another application where FTIR has been successfully deployed is to study hydrogenation reactions, as described in an article by Marziano and colleagues.[22] In that work the authors demonstrated real-time qualitative and quantitative monitoring of hydrogenations on three different functional groups. No chemometric analysis was required due to the molecular specificity of the technique. Individual species could be probed, which is not possible with the less-specific techniques of measuring hydrogen uptake or performing calorimetry.

Frequently, however, the lack of specificity in an analytical technique can be compensated for with sophisticated data processing, as described in the chemometrics chapter of this text (Chapter 8). Quinn and associates provide a demonstration of this approach, using fiber-optic UV-vis spectroscopy in combination with chemometrics to provide real-time determination of reactant and product concentrations.[23] Automatic window factor analysis was used to evaluate the spectra. This technique was able to detect evidence of a reactive intermediate that was not discernable by off-line HPLC, and control charting of residuals was shown to be diagnostic of process upsets. Similarly, fiber-optic NIR was demonstrated by some of the same authors to predict reaction endpoint with suitable precision using a single PLS factor.[24]

One way to circumvent issues with generating and maintaining multivariate calibrations is to use first-principle kinetic models that are fit to robust experimental data. This approach has been demonstrated using a fiber-optic NIR with a diode-array detector on a batch reaction in a microscale reactor.[25] Though off-line (reference) measurements and conventional calibration spectra are not required, making the process 'calibration free,' care must be taken to avoid overfitting of the spectral data. Significant up-front effort is necessary to identify the correct reaction mechanisms. The authors claim that these limitations can be overcome by good experimental design, and that this method is less susceptible to time-variant instrument response (e.g. baseline drift or source change) than traditional multivariate methods.

Am Ende and associates[26,27] provide examples of how the use of NIR and FTIR in development and scale-up can result in 'increased process knowledge, improved efficiency in the development process, and reduced worker exposure to hazardous materials.' They provide examples of reaction endpoint monitoring, Grignard reagent preparation, and a polymorph conversion, frequently in combination with reaction automation. In another example of Grignard reagent synthesis, scientists at Merck monitored a series of Grignard reactions using in-line FTIR to characterize a number of reaction species. The spectroscopic technique served to provide reference data during the validation of a potentiometric measurement.[28]

An article by McConnell *et al.*[29] describes the use of multidimensional on-line analytical technologies as one approach to creating reaction process data more efficiently. Using reaction calorimetry, condensed phase measurement (with in situ IR spectrometry), and gas-phase analysis (using mass spectrometry) concurrently, the authors were able to demonstrate higher data density than was possible by traditional methods. Figure 9.3 illustrates the set-up for their experiments. An article by David *et al.*[30] describes an 'integrated development laboratory' approach with a highly instrumented miniplant. In their example, reaction, crystallization and drying were all serially examined with both

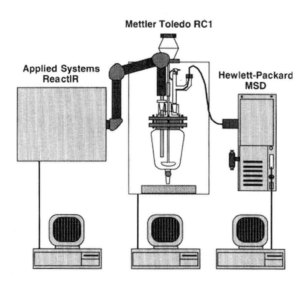

Mettler Toledo RC1

Applied Systems ReactIR

Hewlett-Packard MSD

Figure 9.3 Schematic illustration of the analyzer consisting of reaction calorimetry, IR spectroscopy, and mass spectrometry. Reprinted with permission from McConnell *et al.* (2002).[29]

Table 9.1 Analytical technology utilized for glycolization reaction/crystallization and drying during scale-up experiment. Adapted from David et al. (2004)[30]

Unit operation	Analytical tool	Manufacturer
Reaction	In situ FTIR	Mettler Toledo, ReactIR
Objective: determine reaction end point	Off-line LC	Agilent 1100 LC
Crystallization	In situ FTIR	Mettler Toledo, ReactIR
Objectives: determine desaturation rate and	In situ FBRM	Lasentec M200
polymorphic conversion	In situ PVM	Lasentec Model 900
	Off-line LC	Agilent 1100 LC
Drying	Mass	Agilent 5793 with MS Sensor
Objective: determine drying rate and end point	Spectrometry	Software (Diablo Analytical)
	Off-line GC	Agilent 6890 GC

on-line and off-line techniques, as compiled in Table 9.1. In their conclusions, the authors allude to the importance of integrating the functions of chemistry, engineering and analytics during API process development and scale-up. One of the steps frequently associated with reactions is the changing of solvents in a reactor via distillation. Authors from Merck describe the use of NIR in the distillate stream to monitor such a solvent switch process. They demonstrated stable operation over two months in a series of pilot plant runs, potentially eliminating the need for extractive sampling and time-consuming analysis by GC.[31]

Even a technique as complicated as direct liquid-introduction mass spectrometry has been coupled with reactor systems to provide real-time compositional analysis, as described in a series of articles by Dell'Orco and colleagues.[32-34] In their work, these authors used a dynamic dilution interface to provide samples in real time to un-modified commercial ionization sources (electrospray (ESI) and atmospheric pressure chemical ionization (APCI)). Complete speciation was demonstrated due to the unambiguous assignment of molecular weights to reactants, intermediates, and products.

In addition to traditional chemical-based synthetic processes, extensive work has been performed in support of fermentation monitoring. Early work using NIR looked at simply monitoring cell density non-invasively through the wall of a glass vessel[35] or tracking product and nutrient concentrations in real time,[36] while more recent work has looked at the details involved in developing robust methods and systems validation.[37]

Mid-IR has also been demonstrated for real-time concentration monitoring of a fermentation using a standard transmission cell, and the spectral response was similar, regardless of whether the broth was filtered or not.[38] A PLS regression was used to perform quantitation of the substrate, the lactic acid bacterium, and the major metabolites. In another article describing quantitative mid-IR for fermentation studies, a model system was investigated under various fermentation conditions.[39] The mid-IR provided insight into the relative concentrations of carbohydrates, nucleic acids, proteins and lipids in the host cells. Mid-IR has also been demonstrated for multi-component quantitation on pilot-scale (75-L and 280-L scale) reactors,[40] showing the real-time control of substrate feed. Instrument response was shown to be stable with respect to variations in operating conditions for the fermentation, with the exception of fermenter temperature, which was corrected for in the PLS model.

9.3 Crystallization monitoring and control

Crystallization is frequently associated with a chemical reaction, though it is commonly treated as a distinct unit operation. Crystallization is a critical process since it can dictate critical product-quality attributes, such as particle size distribution, crystal shape and solid-state form. As will be described later, this last attribute is significant enough to be incorporated in a distinct section on polymorphs. The FDA has demonstrated an appreciation for the importance of understanding and controlling crystallization processes, and highlighting the relevance of PAT technologies, in a review article with several case studies.[41] Another review article compares the type and quality of data provided by both on-line and off-line analyses, highlighting the benefits and relative value of each.[42]

One of the more commonly measured attributes during crystallization is the degree of product supersaturation, which is commonly done with mid-IR.[43] An example of such an experimental apparatus is illustrated in Figure 9.4. On-line supersaturation can be combined with turbidity measurement or chord length distribution measurement (as a surrogate for particle size distribution) on reactors that are temperature-programmed to automatically determine the metastable zone width (MSZW, which is the supersaturation a solution can tolerate without spontaneous nucleation).[44–46] In an elegant extension of this approach, a crystallization process with a well-characterized MSZW was systematically probed (with respect to cooling rates and seeding) and monitored to understand the conditions resulting in nucleation or crystal growth. A closed-loop control strategy, illustrated in Figure 9.5, was ultimately developed, which produced crystals of optimal size by maintaining the system in conditions favorable to crystal growth.[47]

A more comprehensive examination of crystallization processes is described in an article describing the engineering and design of an integrated laboratory for batch crystallization.[48] Capabilities for both on-line and off-line analyses are described, including ATR-FTIR for supersaturation, acoustic monitoring for crystal size, X-ray diffraction for polymorphic form, and UV-vis turbidity measurement for MSZW and onset of crystallization events. Directed at probing crystallizations in 2- and 20-L vessels, model systems would also be evaluated by laser Doppler anemometry to understand mixing effects, utilizing computational fluid dynamic modeling.

In situations where crystallization does not produce the desired form or there are issues with robust final form production, crystallization engineers frequently perform solvent-mediated or slurry transformations to the desired product. Several articles have described the use of in situ spectroscopic probes to evaluate, track and control such transformations post-crystallization. Both Raman[49–51] and NIR[52,53] have been demonstrated for such

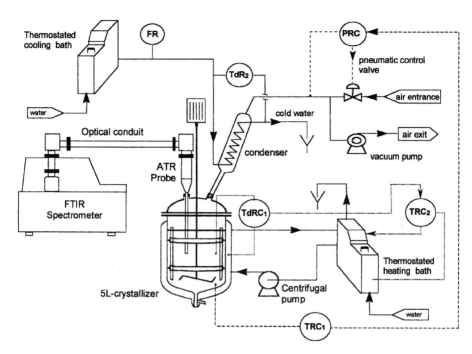

Figure 9.4   Experimental set-up: Bench-scale multi-purpose batch crystallizer equipped with an in situ ATR FTIR probe. Reprinted from Fevotte (2002)[43] with permission from Elsevier.

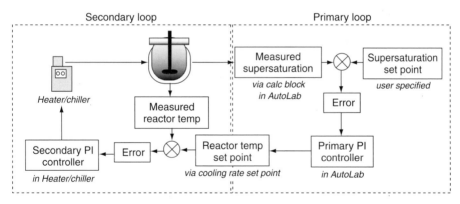

Figure 9.5   Cascaded feedback-control scheme for supersaturation control. Reprinted with permission from Liotta (2004).[47]

a purpose. Wang and colleagues at Merck[49] demonstrated the significance of particle size and shape on quantitative Raman in a slurry process. Wachter and coworkers[51] determined the temperature-dependent transformation rate of progesterone in situ using a quantitation procedure (Figure 9.6) that simply used the peak centroid for the mixture to deduce relative contribution of the two forms.

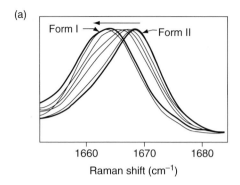

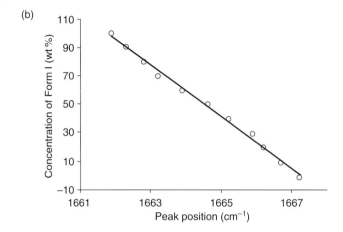

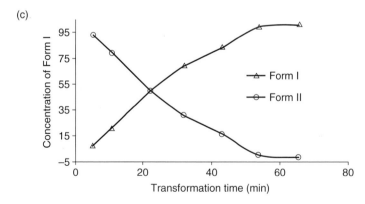

Figure 9.6 (a) In situ spectra showing peak shift during the phase transformation process at 45°C; (b) Calibration curve for the relative concentration of Form I in the mixture of Form I and Form II of progesterone slurry in water; (c) In situ concentration profiles of Form I and Form II of progesterone throughout the phase transformation process at 45°C. Reprinted with permission from Wachter *et al.* (2002).[51]

9.4 Drying

Drying is an important unit operation since consistent drying can have a profound effect on product quality as well as processability in subsequent handling and drug product operations. Drying typically occurs during both API manufacture (sometimes multiple times in a synthetic process) as well as during granulation of drug product. Drying operations generally have significant economic considerations as well. Drying occasionally represents the rate-limiting step in an API manufacturing process, limiting the overall throughput from a production train. Furthermore, operation of the drying equipment itself is typically expensive and complicated, involving energy-intensive hardware, elevated temperatures, vacuum systems and rotating equipment. Manual sampling and analysis have additional risks associated with operator and analyst exposure, as well as fundamental questions about maintaining sample integrity (particularly for the analysis of volatile components).

There are two basic approaches to dryer monitoring, distinguished by whether the measurement is made directly on the solid, or by correlation, typically with the vapor composition of a purge stream. Examples of the latter include an application using an acousto-optic tunable filter NIR (AOTF-NIR) coupled to a gas cell located on a vapor line between a dryer and the vacuum system,[54] and a demonstration of mass spectrometry directly coupled to the vacuum line on vacuum dryers.[55] In an interesting demonstration of a process-induced transformation, Doherty and colleagues describe a system where the final solid-state form of an API could be directed by the drying conditions in the manufacturing equipment.[56] In that experiment, mass spectrometry was used to monitor the vapor composition of the vacuum during the course of drying, while in situ Raman spectroscopy tracked the solid-state form of the material. As seen in Figure 9.7, the three forms that were tracked had distinct spectral features, and by plotting the ratio of characteristic bands (following single normal variant signal treatment), the trajectory of the drying could be easily deduced and trended in real time. In all three of the cases of vapor monitoring, multiple solvent species could be monitored, due to the flexibility and sensitivity of the analytical techniques.

In contrast, moisture is the focus of most applications of direct measurement, historically performed principally using NIR. The preponderance of applications tracking moisture during drying reflects the high sensitivity of NIR to spectral bands associated with water. An in-line NIR has been used to predict the endpoint of a microwave drying process following granulation.[57] In another application involving a cytotoxic compound, the non-invasive nature of NIR monitoring minimized the risk of exposure to operators and analysts.[58] This issue was pointed out to be particularly important in a development environment due to the number of experiments required to characterize the process. Some of the same authors described in a subsequent article how PCA methods could be employed to spectrally distinguish between surface (unbound) water and water of hydration incorporated in the crystal lattice of the substrate.[59] Figure 9.8 shows different spectra with varying degrees of bound and unbound moisture. Figure 9.9 illustrates the real-time measurement of composition during the course of a drying experiment. Using this approach, they were able to demonstrate control of the drying process to consistently and predictably generate product with the desired degree of hydration.

(a)

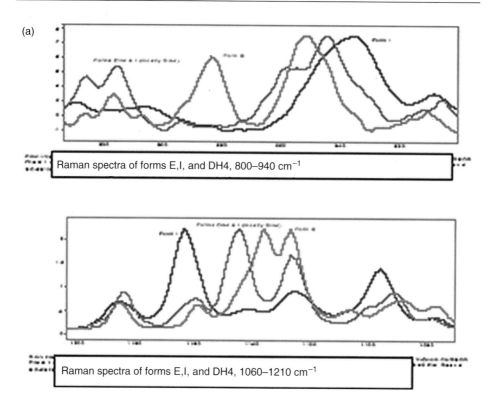

Raman spectra of forms E,I, and DH4, 800–940 cm^{-1}

Raman spectra of forms E,I, and DH4, 1060–1210 cm^{-1}

(b)

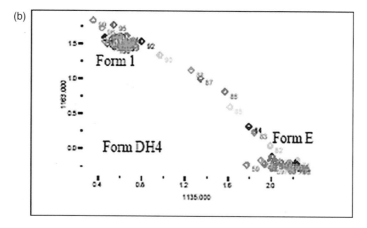

Figure 9.7   (a) Raman spectra of pure distinct forms; (b) Form trajectory tracked in real time with in situ Raman. Reprinted with permission from Doherty *et al.* (2001).[56]

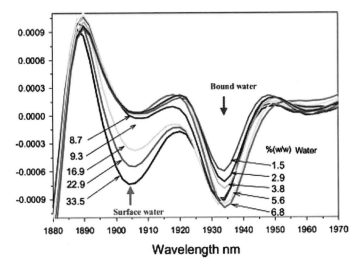

Figure 9.8 Second-derivative spectra of M1 samples with different levels of water with eight data points averaging. Reprinted with permission from Zhou *et al.* (2003).[59]

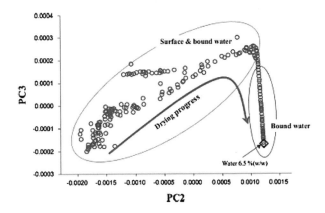

Figure 9.9 Principle component analysis plot of PC2 vs. PC3 during the drying process of M1 wetcake by NIRS in-line monitoring. Reprinted with permission from Zhou *et al.* (2003).[59]

Freeze-drying is another unit operation that has been interrogated with in situ NIR spectroscopy. A group in AstraZeneca R&D has demonstrated how the process insight provided by NIR was complementary to and corroborated by the information obtained from temperature monitoring.[60] Their work was demonstrated in a pilot-scale facility and involved the construction of special equipment to ensure a representative sample.

In an extension of simply monitoring and controlling dryers, Morris and colleagues have demonstrated how in situ probes can provide real-time analysis that enables optimization of a drying process.[61,62] In these articles, the authors demonstrate how evaporative cooling can be used to expedite drying in formulations where heat transfer dominates the drying mechanism, and propose a model for predicting whether or not a process is amenable to this approach.

9.5 Milling

In order to obtain a drug product with the desired bio-availability attributes, it is common for the API to require particle size reduction via milling. The principal requirement for control of this unit operation is a reliable measure of particle size distribution. Laboratory analyses are conventionally performed by laser diffraction, and there are now commercial on-line implementations of this technology. However, optical techniques have been known almost since their inception to exhibit features related to the sample's particle size. The difficulty is in extracting size data encoded in the raw signal without also extracting unrelated information. Early work correlating NIR absorbance with particle size demonstrated correlations that were either linear[63] or non-linear but correctable.[64] Subsequent work used more sophisticated data treatments to demonstrate improved quantitative measures of particle size of API and excipients.[65] Each of these evaluations was performed in laboratory settings, but demonstrated the potential for the technique. More recently, workers at Merck demonstrated an on-line application of NIR with suitable accuracy for control of a slurry mill used to produce nanoparticles.[66] In this paper, the authors also propose a phenomenological model (Figure 9.10) to explain both the change in slope and offset of the observed spectra, and the increase in overtone/ combination band absorbance.

Frequency domain photon migration (FDPM) has been investigated as an optical technique with potential application to particle size analysis,[67] albeit in a laboratory environment. The approach could be readily implemented in situ, with appropriate

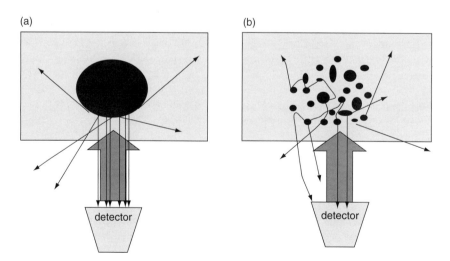

(a)

(b)

detector

detector

Figure 9.10 Schematic of light scattering by particles near the (a) beginning; (b) endpoint of the media milling process. Reprinted with permission from Higgins *et al.* (2003).[66]

interfacing with the process. FDPM uses intensity-modulated light to probe slurries or powders, and the detected signal contains information allowing the independent determination of both scattering and absorption coefficients. The scattering coefficient is relevant to particle size determination.

9.6   Cleaning validation

Many pharmaceutical manufacturing facilities are multi-purpose batch facilities that run different chemical processes in multi-lot campaigns. Significant processing time can be lost during inter-batch or intra-batch turnover while waiting for cleaning validation results. This testing commonly requires swabbing of reactor and equipment surfaces, with subsequent analysis of the swabs by HPLC. In situ analysis can lead to more rapid equipment turnaround times. One approach to addressing this measurement in situ is to track the signal arising from chemical residue in the cleaning solvent matrix. One implementation using UV spectroscopy with a photodiode array offers sub-ppm detection limits, provided the targets have active chromophores.[68] An alternative approach is to use a grazing angle sample probe coupled with a fiber-optic mid-IR spectrometer to measure analyte concentrations directly on working surfaces.[69]

9.7   Solid dosage form manufacture

From the perspective of the patient, the preferred route of receiving medicine is the oral ingestion of a tablet. The process of developing an oral dosage form for a new drug product is an effort that requires contributions from pharmaceutical formulators, analytical chemists, physical chemists and pharmaceutical engineers. There are many variables to consider and control so that the risk to the patient posed by this multivariate process can be minimized and the delivery of the designated dosage is assured. The API must be mixed with excipients so that when a tablet is formed from the blended materials the release of the API is predictable and the quantity delivered is correct. Figure 9.11 provides a block diagram of many of the tools available to formulators for the development of drug products. The following sections will discuss the use of spectroscopic techniques that are employed to monitor the various steps in the development and manufacture of oral, solid dosage forms.

9.8   Granulation and blending of API and excipients

The first step in manufacturing the formula to be pressed into a solid dosage form is the mixing of the selected materials. Formulators often specify the use of different binders and excipients that require the use of added liquid to assist in the process of blending the ingredients. This process, commonly known as wet granulation, takes place in a high-speed mixer granulator or fluid-bed granulator. The process of granulation provides

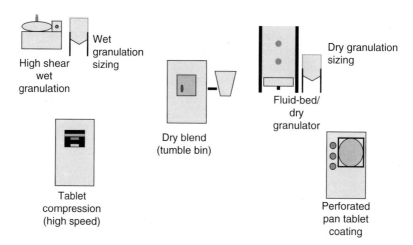

Figure 9.11   Unit operations often found in a dry-product manufacturing process.

a mechanism to uniformly distribute trace components throughout the mixture while inducing an increase in the particle size distribution of the mixture. This imparts uniformity to the mix, which reduces the chances of size segregation and facilitates material flow. There are formulations that do not require granulation. For these mixtures the ingredients are often dry-blended in a V-type or ribbon blender. The mixture resulting from the granulating or dry blending process requires sampling to determine the progress of the process. This assessment of the blend provides guidance as to the length of time required for blending or granulation of the ingredients to ensure a homogeneous mixture. The validation of the blending and granulation processes has received additional attention as the result of a court ruling.[70]

NIR spectroscopy offers sampling techniques and data acquisition and analysis approaches that allow it to be used as a quality improvement technique that is being implemented in the pharmaceutical industry. Plugge and coworkers[71] found that the use of NIR with a polar qualification system could be used to distinguish between identical substances that only differed slightly in physical or chemical character. Although NIR was not a technique that was new to the pharmaceutical industry, it was the beginning of an era defined by the incorporation of on-line and at-line spectroscopic techniques coupled with novel signal analysis algorithms for characterizing the processes used to make medicine. Implementation of NIR for the purpose of process control and monitoring should have proven economic advantages. The opportunity to decrease sample time, enhance unit operation throughput and minimize risks to the patient must be present in order to ensure that implementation efforts make good business sense and offer a return on the necessary investment. Axon and colleagues[11] reported examples where NIR, used as a qualitative assessment of a manufacturing process, can provide opportunities for productivity and quality improvements. The following sections will provide an overview of the use of spectroscopic techniques in the characterization and control of granulation and blending processes. A separate section will discuss the use of spectroscopy to assess the impact of granulation and blending on the physical state of the API.

9.8.1 Granulation

The focus of most of the spectroscopic investigations of granulation involves the determination of water during wet granulation or post-granulation drying. Control of the rate of moisture introduction and loss of moisture through the air vent of a fluid-bed granulator are critical to the proper distribution of trace components in the granulation as well as the rate at which particles grow in size. Watano and coworkers[72] utilized an IR moisture sensor to elucidate the relationship between the IR absorbance spectrum and the moisture content of granules. They investigated the water-absorbing potential of different lactose/corn starch ratios and found that the water-absorbing potential was correlated to the relationship between granule moisture and IR absorbance spectra. The particle size of the granules showed no affect. For fluid-bed drying of the granulation, the particle size of the granules had a demonstrated effect on the drying. Rantanen and coworkers[73] calibrated a four wavelength NIR sensor using an IR gravimetric dryer as the reference method. They were able to measure the moisture of granules with a standard error of prediction of 0.2%. Differences in moisture profiles of the granules were readily distinguished for different granulations. In a subsequent publication[74] the researchers investigated the effects of particle size and binder on the NIR measurement of water during granulation. For this work, a full spectrum FTNIR instrument was used to make off-line measurements of formulations at different stages of granulation. The full NIR spectrum information provided the means to determine which wavelengths would be reliable for the determination of water in the presence of different binders and particles sizes. Frake and associates[75] utilized a fiber-optic probe in a fluid-bed granulator (Figure 9.12) to monitor moisture during granulation. Utilizing the NIR spectra with loss on drying and Karl

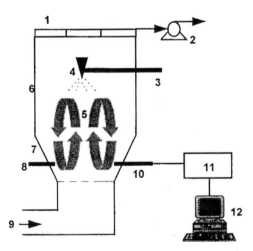

Figure 9.12 Equipment Schematic: 1 – Filters; 2 – Air fan; 3 – Spray arm; 4 – Spray gun/nozzle, 5 – Direction of movement of product bed; 6 – Expansion chamber; 7 – Product bowl; 8 – Sample probe; 9 – Heated/dehumidified inlet air-flow; 10 – NIR probe; 11 – NIR monochromator; 12 – PC. Reprinted from Frake *et al.* (1997)[75] with permission from Elsevier.

Fischer titration as the reference methods, calibration for water yielded a standard error of calibration of 0.5% which was deemed acceptable to achieve the level of control needed for the process. Models to predict the particle size from the NIR data were comparable to sieve analysis of the granulation. Given the complex nature of modeling NIR data to predict particle size, they were unable to generate calibration models that were useful for quantitative determinations of particle size. The concept of granulation process control using NIR was carried to a higher level by Rantanen;[76] by coupling the four wavelength NIR data with process variables such as air flow, air temperature, mass flow of granulation liquid, process state indicators and differential pressures they were able to generate a 20-element vector of the process. They used the self-organizing map technique for data dimension reduction and process-state monitoring. The technique was able to present the different states of granulation (mixing, spraying, and drying) and subtle differences between batches could be observed. Miwa and coworkers[77] used an IR sensor to examine the behavior of water on the surface of different excipients. Based on measurements and observations, the range of water addition suitable for granulation could be determined for each excipient. The work suggests that the amount of water to be added for a granulation trial could be determined a priori using their method.

9.8.2   Blending

Some formulations only require the simple mixing of API and excipients. Like granulation, the use of blending devices will yield a mixture that requires analysis according to prescribed protocols and methods. Formulators have always wanted a method that could indicate to them that a blend had achieved homogeneity. NIR was utilized to determine blend homogeneity by analysis of samples that were thief-sampled from a blend in progress with subsequent analysis using an UV-vis assay.[78] The investigators utilized NIR reflectance measurements of samples contained in vials. The reflectance measurements were made through the bottom of the glass vial. Excellent correlation of the NIR data with subsequent API analysis proved the capability of NIR to follow a blending process with fast analysis and minimal sample preparation. Scientists from Pfizer, in a series of papers,[79–81] examined on-line monitoring of the powder-blending process using NIR. Figure 9.13 shows how the fiber-optic NIR probe was installed on two of the three types of blenders studied. These works not only addressed the hardware requirements for effective monitoring of blends, but also did a thorough examination of signal acquisition, processing and evaluation. The use of various forms of NIR signal pre-processing steps such as standard normal variate (SNV), second derivative (2D) and detrending were examined to find the optimal method for isolating the physical and chemical information of the spectra. The data pretreatments were coupled with techniques such as dissimilarity calculations, PCA and spectral standard deviation to qualitatively evaluate the blending of materials. The importance of these initial works is the recognition by the authors that additional issues must also be addressed, including: number of probes required to fully characterize a blend, effects of variability in particle size and morphology of excipients, capability of these techniques in wet granulators and the extent of correlation of NIR

(a) Bin blender (b) Y-cone blender configuration

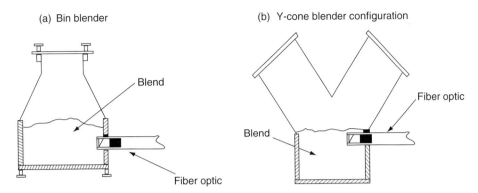

Figure 9.13 On-line blender configurations (a) Bin blender; (b) Y-cone blender configuration. Reprinted from Hailey *et al.* (1996)[79] with permission from Elsevier.

methods with current HPLC-based blend analysis protocols. Some of the sampling and signal processing techniques have resulted in an approved US patent.[82] NIR has also been used to track magnesium stearate during blending in a Bohle bin blender.[83] A model system of lactose, Avicel and magnesium stearate was used, and the resulting data showed that NIR was capable of tracking the variability of all the components of the mixture. The method was capable of determining magnesium stearate as low as 0.5%. Prediction confidence intervals of above 98% were reported. Investigators at MIT[84] used light-induced fluorescence (LIF) to interrogate blends in a micro-scale blender. The LIF signal was found to come to equilibrium when a blend was homogeneous as determined by analysis of thief samples. Void volumes and bulk density variations within the blend were sources of signal variability. To minimize these sources of variability, measurement location has to be carefully determined.

9.9 Detection of drug polymorphism

The majority of all APIs manufactured demonstrate structural polymorphism. In order to formulate a drug product that is physically and chemically stable, the formulation team must identify the most thermodynamically stable polymorph of the API. By identifying the proper polymorph, the patient's need for a drug product with reproducible bioavailability during the course of typical and atypical shelf-life conditions will be met. This section will review the use of spectroscopic techniques for identifying polymorphism of APIs during API crystallization and formulation. For a more comprehensive discussion of polymorphism the reader is directed to a work by Singhal[85] and references cited therein.

NIR spectroscopy was utilized by Aldridge and coworkers[86] to determine, in a rapid manner, the polymorphic quality of a solid drug substance. Two computational methods, Mahalonobis distance and soft independent modeling of class analogy (SIMCA) residual variance, were used to distinguish between acceptable and unacceptable samples. The authors not only determined that the Mahalonobis distance classification yielded the best results, they addressed one of the key implementation issues regarding NIR as a PAT tool.

The method was successfully transferred to six other NIR instruments without the use of multivariate calibration transfer protocols. Fevotte and coworkers[53] present an important discussion of the utility of NIR spectroscopy in monitoring the solid state during API crystallization. In this work, they focused on exploiting the advantages of fiber-optic probes for rapid in situ data acquisition and developing calibrations with a minimum of rigor to return the maximum value. These calibrations were sensitive to the polymorphic composition of the solid. NIR provided important information on the kinetics of polymorph transition during filtration where it was determined that the amount of water in the solvent was the rate-determining variable. The reader is directed to DeBraekeleer and coworkers[87] for additional detail and references on the utility of IR/NIR for crystallization control. They investigated temperature influence on a NIR data set to understand the effect on polymorph conversion. Chemometric techniques were used to model temperature effects on the NIR spectra and correct these effects in the resulting concentration profiles.

Drug substance formulators use the pharmaceutical unit operations identified in Figure 9.11 to develop and manufacture drug product. Each of the operations is capable of inducing transformations of the API through process-induced stress. These physical transformations are known to exist, have been reported in specific instances[88] and are the subject of a theoretical overview.[89] PAT tools, using spectroscopic techniques, are very capable of making at-line and in-line determinations of polymorph changes in drug products.

Patel and coworkers[90] quantified polymorphs of sulfamethoxazole (SMZ) in binary and multi-component mixtures using diffuse reflectance NIR. Figure 9.14 illustrates the capability of NIR to quantify the two forms of SMZ at a single wavelength using 2D spectra. Figure 9.15 shows the excellent correlation of predicted polymorph composition with the theoretical composition. The authors compared three different calibration approaches for polymorph composition: univariate, multiple linear regression and PLS. The results of the calibrations are summarized in Table 9.2. The authors undertook a comprehensive examination of limit of detection, limit of quantitation, precision, method

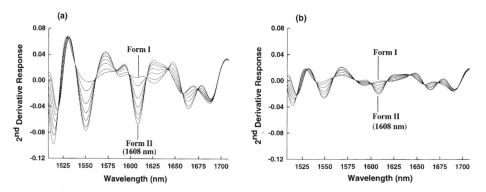

Figure 9.14 The 2D spectra of SMZ Form I and Form II: (a) binary mixture; (b) diluted with 60% lactose. Each spectrum represents decreasing percent of Form I in Form II from top to bottom at the indicated wavelength. Reprinted from Patel *et al.* (2000)[90] with permission from Elsevier.

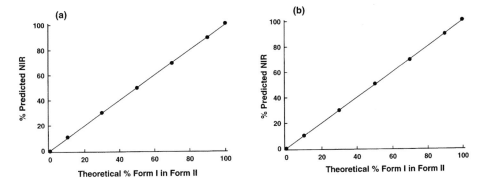

Figure 9.15 Percent theoretical versus percent predicted by NIRS for SMZ Form I in Form II: (a) binary mixture ($R^2 = 0.9998$, SEC = 0.63 at 1608/1784 nm); (b) diluted with 60% lactose ($R^2 = 0.9999$, SEC = 0.41 at 1608/2060 nm). Solid lines represent the data fit with a linear regression model. Reprinted from Patel *et al.* (2000)[90] with permission from Elsevier.

Table 9.2 Near-infrared calibration results for SMZ Form I and Form II in binary and multi-component mixtures. Adapted from Patel *et al.* (2000)[90]

Composition	Univariate		MLR[a]		PLS[b]	
	R^2	SEC[c]	R^2	SEC	R^2	SEC
Binary mixture	0.9999	0.48	1.0000	0.12	1.0000	0.19
Binary mixture diluted with:						
95% Sodium chloride	0.9998	0.54	1.0000	0.14	0.9999	0.52
90% Avicel	0.9959	2.70	0.9974	2.81	0.9991	2.02
90% Lactose	0.9997	0.79	0.9999	0.43	1.0000	0.22
80% Avicel:Lactose (1:1)	0.9999	0.31	0.9999	0.67	0.9999	0.71

Notes:
[a] Multiple linear regression with three wavelengths.
[b] Partial least squares regression with four factors.
[c] Standard error of the calibration (as % polymorph composition).

error and instrument reproducibility. Their conclusion was the overall error for the method was in the range of 1–2% for the binary and multicomponent mixtures. Their validated limits of detection and quantitation demonstrated that NIR is capable of determining polymorphs in the range of 2–5% even at significant levels of dilution.

Terahertz pulse spectroscopy was used to observe the polymorphs of ranitidine hydrochloride.[91] Sample preparation for this technique is the same as for Raman and FTIR spectroscopy and the data generated is complementary to Raman. Terahertz pulse spectroscopy provides information on the low-frequency intermolecular modes that are difficult to study with Raman due to the proximity of the laser excitation line. The authors concluded that this technique has many applications in pharmaceutical science including formulation, screening and stability studies.

Räsänen and coworkers[92] demonstrated the ability of NIR to detect different hydration states of the molecule during the wet granulation of theophylline. NIR results were compared with differential scanning calorimetry and wide-angle X-ray scattering patterns. NIR was found to be capable of observing the formation of theophylline monohydrate and subsequent free water molecules as greater amounts of water were added to the granulation. This work was followed by an assessment of NIR and Raman spectroscopy as tools for following the wet granulation of caffeine and theophylline.[93] NIR and Raman were found to be complementary methods in studying hydrate formation. Raman allows for the observation of the drug molecule itself during the formation of hydrates. For the purpose of observing granulation processes, NIR is superior due to its ability to differentiate water of hydration from free water. The utility of Raman spectroscopy to evaluate solid-state forms in tablets was investigated by Taylor and Langkilde.[94] Using an FT-Raman instrument they researched sampling and analysis techniques and studied five different pharmaceutical products. They found Raman spectroscopy to be capable of investigating the solid-state forms of the API and concluded that differentiation of different polymorphs should be possible in many instances.

9.10　Spectroscopic techniques for tablets

The process of compacting a powder, pressing the compaction into a tablet and then coating the tablet is a multivariate problem. Similar to the tablet-making process are the steps required to accurately fill capsules with a drug formulation so that the medicine is delivered to the patient with the proper efficacious dosage. This section will discuss the use of NIR and other spectroscopic techniques during the manufacture of solid dosage forms.

Miller[95] researched the value of NIR to map roller compaction steps: blend concentration, compactor roll pressure, vacuum de-aeration, granule size and variation in tablet concentrations. NIR provided understanding of the influence of compactor parameters on the product characteristics. Gustaffson and coworkers[96] utilized FTIR and FTNIR to investigate the characteristics of hydroxypropyl methylcellulose and the relationship to final tablet properties. Using multivariate data analysis it was possible to predict apparent particle density, particle shape factor, tablet tensile strength and drug release. Table 9.3 provides a summary of the properties that were studied and the variance explained for X, the spectral differences, and Y, the property variance. Even for the limited numbers of samples used in this study, the results indicated that high value information is available in the PLS models developed. The development and value of NIR as a method for analytical control of different steps during tablet production were investigated by Han and Faulkner.[97] NIR was utilized during fluid-bed drying to monitor moisture and as an end of granulation process assay. NIR was also used as an end of process assay for blending. During the compression of tablet cores NIR was used to sample the process at the beginning, in the middle and at the end of the compression run. NIR was used to analyze the thickness of tablet coating and as an assay of the coated tablet during the coating process. The tablets were identified by NIR in the blister packs during the packaging process. Blanco and

Table 9.3 Predictions obtained using PLS models. The number of components was two for all properties. Adapted from Gustafsson *et al.* (2003)[96]

Property	NIR		FTIR	
	Explained Variance in X (R^2X)	Explained Variance in Y (R^2Y)	Explained Variance in X (R^2X)	Explained Variance in Y (R^2Y)
Methoxy	0.95	0.99	0.91	0.99
Hydroxypropyl	0.92	0.99	0.83	0.88
Total degree of substitution	0.93	0.97	0.87	0.96
Apparent density of powders	0.95	0.99	0.90	0.99
Heywood's shape factor	0.97	0.99	0.91	0.97
Volume-specific surface area of powders	0.97	0.99	0.94	0.97
Radial tensile strength	0.97	0.99	0.91	0.97
Axial tensile strength	0.94	0.99	0.86	0.94
Volume-specific surface area of tablets	0.97	0.99	0.87	0.98
Porosity	0.81	0.86	0.80	0.80
Pore radius	0.96	0.99	0.92	0.93
Elastic recovery after ejection	0.88	0.90	0.81	0.83
Apparent yield pressure	0.67	0.79	0.66	0.69
Drug release after 1 hour	0.96	0.99	0.96	0.86

coworkers[98] investigated similar capabilities of NIR during tablet production. They found that all of the samples could be analyzed using a single calibration. The calibration was built using samples with a broad concentration range of APIs that were generated in the laboratory and samples from the production process. The production samples included variation in the raw materials, particle size distribution, compactness, granulation and other unit operation sources of variability. Product samples were chosen by PCA. A PLS model was built, which provided a relative standard deviation for production samples of less than 1%.

Tablet hardness is a property that, when measured, destroys the sample. The destructive nature of the test, coupled with the variability of the test itself does not contribute to an incentive to test a large number of samples. Morisseau and Rhodes[99] correlated the diffuse reflectance NIR spectra of tablets pressed at different pressures and subsequently tested the tablet hardness with an Erweka Hardness Tester. The tablet hardness, as predicted by the NIR method, was at least as precise as the laboratory test method. Kirsch and Drennen[100] evaluated NIR as a method to determine potency and tablet hardness of Cimetidine tablets over a range of 1–20% potency and 107-kPa compaction pressure. Hardness at different potency levels was used to build calibration models using PCA/ principal component regression and a new spectral best-fit algorithm. Both methods provided acceptable predictions of tablet hardness.

The rapid and non-destructive analysis features of NIR make it an ideal method to investigate film-coated tablets. Kirsch and Drennen[101] analyzed theophylline tablets coated with ethylcellulose. They were able to develop calibrations for the prediction of dissolution rate for tablets with different coating thickness, determine the coating thickness and predict the hardness of the tablet. The authors have also investigated the use of NIR to monitor a film-coating process.[102] They placed the NIR spectrometer at the location of the coating operation to facilitate the rapid acquisition of spectral data from samples withdrawn from the process with a sample thief. Ethylcellulose and hydroxypropyl methylcellulose coatings were investigated. In order to predict the weight gain due to coating over a range of 0–30% w/w, several calibration approaches were investigated. Multiplicative scatter correction (MSC) and 2D spectral pretreatments were used to remove spectral baseline shifts caused by variation in the positioning of tablets during NIR spectral acquisition. Calibrations were constructed using multiple wavelengths or multiple linear regression using principal component spectra following MSC or 2D spectral pre-processing steps. Andersson *et al.*[103] also used NIR to monitor a tablet-coating process. Sample presentation was researched and two different approaches (Figure 9.16) were investigated. Although differences in the spectral offset were observed for the two presentations, the calibrations were essentially the same. The bolt-style tablet holder was found to be easier to utilize. This work was extended from an at-line approach to an in-line measurement using a fiber-optic diffuse reflectance probe.[104] This was a very thorough and detailed research of the coating process using an in situ NIR technique. The authors used a batch calibration process that correlated the NIR observations with image analysis of dissected pellets and a non-linear coating thickness growth model. Their development and discussion of the multivariate batch calibration technique is particularly noteworthy.

Mowery and coworkers[105] utilized laser-induced breakdown spectroscopy as a means to rapidly discern the thickness and variability of an enteric tablet coating. The tablet core contained a high concentration of calcium, while the tablet coating contained magnesium, silicon and titanium. The emission spectrum of all four elements was monitored as a series of laser pulses was applied to a single location. The depth of penetration of each laser pulse was determined using profilometry. As each laser pulse was triggered the

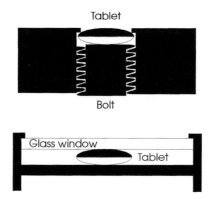

Figure 9.16 The two tablet holders, not drawn to scale. The upper is the bolt holder, where the tablet is fixed by a bolt, and the lower is the glass window holder, where the tablet is sandwiched between a glass window and a metal support. Reprinted from Andersson *et al.* (1999)[103] with permission from Elsevier.

thickness of the coating was monitored by the increase in the calcium signal and the decrease in the magnesium, silicon and titanium signals. The technique could detect a change in coating of 2 wt% on a 100-mg tablet. Ten tablets could be analyzed in less than 15 minutes making this a capable at-line analysis technique.

NIR has been used to determine the drug content in solid dosage forms. The relative merits of diffuse reflectance and transmission techniques were investigated by Thosar *et al.*[106]. While both methods yielded acceptable results, the best standard error of prediction models was obtained using transmission spectra and the PLS algorithm. Abrahamsson and coworkers[107] investigated variable selection methods for NIR transmission measurements of intact tablets. Four different methods were used for wavelength selection in building the PLS regression models: genetic algorithm, iterative PLS, uninformative variable elimination by PLS, and interactive variable selection for PLS. The predictive capability of models built using these techniques was compared to the results using models with manually selected wavelengths. The iterative methods improved the prediction capabilities; however, the authors offer the caveat that additional research is needed before any general conclusions can be made. The determination of paracetamol in intact tablets has been accomplished using reflectance NIR[108] and transmission NIR.[109] These two works were followed up with a method for validation of the techniques[110] in the analysis of two related formulations. Ekube and coworkers[111] investigated NIR as a technique to predict microcrystalline types in powdered and tablet form as well as prediction of lubricant concentration and tablet hardness.

The diffuse reflectance infrared FT (DRIFT) spectroscopy technique was coupled with artificial neural networks (ANN) to analyze ranitidine HCl.[112] A simple ANN calibration with two outputs was able to discern the percentage of each of the two polymorphs in the bulk drug. Another calibration successfully determined the two API polymorphs as well as three tablet excipients. Neubert and coworkers[113] used step-scan FTIR photoacoustic spectroscopy to determine the drug content of a semisolid drug formulation. This technique, with its minimal sample preparation, reported excellent correlation with the HPLC and capillary zone electrophoresis reference methods. Raman spectroscopy was utilized to analyze tablets containing the illegal drug 'ecstasy' and other related compounds.[114] They found that with 2-minute data acquisition periods the target analyte and very similar compounds could be easily identified, including different polymorphs and hydrates of the target and related compounds. Vergote and coworkers[115] realized similar utility of Raman spectroscopy in the analysis of diltiazem hydrochloride tablets. Raman analysis was compared to HPLC analysis and was essentially the same in capability. Raman offers the advantage of speed and absence of solvent disposal. In a similar study, Wang[116] compared Raman and HPLC as tools for monitoring the stability of aspirin tablets. The methods were found to be comparable for this application. Raman and transmission NIR were utilized to determine the API level of four different dosages of Escitalopram® tablets.[117] Utilizing PLS calibrations to quantify the active ingredient, the best results were obtained using transmission NIR analysis. This result reinforces the measurable difference between simple surface analysis by Raman and the volume interrogation of the NIR technique.

In the interest of patient safety, there exists a need for final identification testing of products as they are packaged for delivery to the patient. Herkert *et al.*[118] evaluated a reflectance NIR instrument and found it to be capable of on-line identity check of tablets at-line speeds of 12 000 tablets a minute.

References

1. http://www.fda.gov/cder/OPS/PAT.htm.
2. http://www.fda.gov/cder/guidance/5815dft.pdf.
3. Chalmers, J.M. & Griffiths, P.R. (eds), *The Handbook of Vibrational Spectroscopy*; John Wiley & Sons; Chichester, West Sussex, UK, 2002.
4. Warman, M. & Hammond, S., Process analysis in pharmaceutical industry, in Lee, D.C. & Webb, M.L. (eds), *Pharmaceutical Analysis*; Blackwell Publishing Ltd; Oxford, UK, 2003; pp. 324–356.
5. Workman, J.J., Jr, Review of process and non-invasive near infrared and infrared spectroscopy: 1993–1999; *Appl. Spec. Rev.* 1999, 34, 1–89.
6. Swarbrick, J. & Boylan, J.C. (eds), *Encyclopedia of Pharmaceutical Technology*, Second Edition; Marcel Dekker, Inc.; New York, 2002.
7. Vankeirsbilck, T.; Vercauteren, A.; Baeyens, W. *et al.*, Applications of Raman spectroscopy in pharmaceutical analysis; *TrAC, Trends Anal. Chem.* 2002, 21, 869–877.
8. Folestad, S. & Johansson, J., Raman spectroscopy: A technique for the process analytical technology toolbox; *Am. Pharm. Rev.* 2004, 7, 82–88.
9. Ciurczak, E.W., Growth of near-infrared spectroscopy in pharmaceutical and medical sciences; *Am. Pharm. Rev.* 2002, 5, 68–73.
10. Bakeev, K.A., Near-infrared spectroscopy as a process analytical tool; *Spectroscopy* 2003, 18, 32–35 and 2004, 19, 39–42.
11. Axon, T.G.; Brown, R.; Hammond, S.V. *et al.*, Focusing near infrared spectroscopy on the business objectives of modern pharmaceutical production; *J. Near Infrared Spectrosc.* 1998, 6, A13–A19.
12. Rein, A.J.; Donahue, S.M. & Pavlosky, M.A., In situ FTIR reaction analysis of pharmaceutical-related chemistry and processes; *Curr. Opin. Drug Disc. Dev.* 2000, 3, 734–742.
13. Miller, R.W., Process analytical technology (PAT) – Parts 1&2; *Am. Pharm. Rev.* 2002, 5, 25–29 and 2003, 6, 52–61.
14. Cooley, R.E. & Egan, J.C., The impact of process analytical technology (PAT) on pharmaceutical manufacturing; *Am. Pharm. Rev.* 2004, 7, 62–68.
15. Arrivo, S.M., The role of PAT in pharmaceutical research and development; *Am. Pharm. Rev.* 2003, 6, 46–53.
16. Laird, T. (Ed.), Laboratory automation in process R&D; *Org. Process Res. Dev.* 2001, 5, 272–339.
17. http://www.rxeforum.com & http://www.argotech.com/documentation/index.html.
18. De Braekeleer, K.; De Juan, A.; Sanchez, F. *et al.*, Determination of the end point of a chemical synthesis process using on-line measured mid-infrared spectra; *Appl. Spectrosc.* 2000, 54, 601–607.
19. Ge, Z.; Thompson, R.; Cooper, S. *et al.*, Quantitative monitoring of an epoxidation process by Fourier transform infrared spectroscopy; *Process Contr. Qual.* 1995, 7, 3–12.
20. Chen, Y.; Zhou, G.X.; Brown, N. *et al.*, Study of lactol activation by trifluoroacetic anhydride via in situ Fourier transform infrared spectroscopy; *Anal. Chim. Acta* 2003, 497, 155–164.

21. Cameron, M.; Zhou, G.X.; Hicks, M.B. *et al.*, Employment of on-line FT-IR spectroscopy to monitor the deprotection of a 9-fluorenylmethyl protected carboxylic acid peptide conjugate of doxorubicin; *J. Pharm. Biomed. Anal.* 2002, 28, 137–144.

22. Marziano, I.; Sharp, D.C.A.; Dunn, P.J. *et al.*, On-line mid-IR spectroscopy as a real-time approach in monitoring hydrogenation reactions; *Org. Process Res. Dev.* 2000, 45, 357–361.

23. Quinn, A.C.; Gemperline, P.J.; Baker, B. *et al.*, Fiber-optic UV/visible composition monitoring for process control of batch reactions; *Chemometrics Intell. Lab. Syst.* 1999, 451, 199–214.

24. Coffey, C.; Cooley, B.E., Jr & Walker, D.S., Real time quantitation of a chemical reaction by fiber optic near-infrared spectroscopy; *Anal. Chim. Acta* 1999, 395, 335–341.

25. Gemperline, P.; Puxty, G.; Maeder, M. *et al.*, Calibration-free estimates of batch process yields and detection of process upsets using in situ spectroscopic measurements and nonisothermal kinetic models: 4-(dimethylamino)pyridine-catalyzed esterification of butanol; *Anal. Chem.* 2004, 76, 2575–2582.

26. am Ende, D.J. & Preigh, M.J., Process optimization with in situ technologies; *Curr. Opin. Drug Disc. Dev.* 2000, 3, 699–706.

27. am Ende, D.J.; Norris, T. & Preigh, M.J., In-situ FTIR and NIR spectroscopy in the development and scale-up of API synthesis; *Am. Pharm. Rev.* 2000, 3, 42–46.

28. Chen, Y.; Wang, T.; Helmy, R.; Zhou, G.X. *et al.*, Concentration determination of methyl magnesium chloride and other Grignard reagents by potentiometric titration with in-line characterization of reaction species by FTIR spectroscopy; *J. Pharm. Biomed. Anal.* 2002, 29, 393–404.

29. McConnell, J.R.; Barton, K.P.; LaPack, M.A. *et al.*, Streamlining process R&D using multidimensional analytical technology; *Org. Process Res. Dev.* 2002, 6, 700–705.

30. David, P.A.; Roginski, R.; Doherty, S. *et al.*, The impact of process analytical technology in pharmaceutical chemical process development; *JPAC* 2004, 9, 1–5.

31. Ge, Z.; Buchanan, B.; Timmermans, J. *et al.*, On-line monitoring of the distillates of a solvent switch process by near-infrared spectroscopy; *Process Contr. Qual.* 1999, 11, 277–287.

32. Brum, J. & Dell'Orco, P., Online mass spectrometry: Real-time monitoring and kinetics analysis for the photolysis of idoxifene; *Rapid Commun. Mass Spectrom.* 1998, 12, 741–745.

33. Dell'Orco, P.; Brum, J.; Matsuoka, R.; Badlani, M. *et al.*, Monitoring process-scale reactions using API mass spectrometry; *Anal. Chem.* 1999, 71, 5165–5170.

34. Brum, J.; Dell'Orco, P.; Lapka, S. *et al.*, Monitoring organic reactions with on-line atmospheric pressure ionization mass spectrometry: The hydrolysis of isatin; *Rapid Commun. Mass Spectrom.* 2001, 15, 1548–1553.

35. Ge, Z.; Cavinato, A.G. & Callis, J.B., Noninvasive spectroscopy for monitoring cell density in a fermentation process; *Anal. Chem.* 1994, 66, 1354–1362.

36. Hammond, S.V. & Brookes, I.K., Near-infrared spectroscopy: A powerful technique for at-line and on-line analysis of fermentations, in Ladisch, M.R. and Bose, A. (Eds); Harnessing Biotechnol. 21st Century, *Proc. Int. Biotechnol. Symp. Expo.*, 9th; ACS, Washington, DC, 1992; pp. 325–333.

37. Arnold, S.A.; Harvey, L.M.; McNeil, B. *et al.*, Employing near-infrared spectroscopic methods of analysis for fermentation monitoring and control. Part 1: Method development; *BioPharm Int.* 2002, 15, 26–34, and Arnold, S.A.; Harvey, L.M.; McNeil, B. *et al.*, Employing near-infrared spectroscopic methods of analysis for fermentation monitoring and control.: Part 2: Implementation strategies; *BioPharm Int.* 2003, 16, 47–49, 70.

38. Fayolle, P.; Picque, D. & Corrieu, G., Monitoring of fermentation processes producing lactic acid bacteria by mid-infrared spectroscopy; *Vib. Spectrosc.* 1997, 14, 247–252.

39. Grube, M.; Gapes, J.R. & Schuster, K.C., Application of quantitative IR spectral analysis of bacterial cells to acetone-butanol-ethanol fermentation monitoring; *Anal. Chim. Acta* 2002, 471, 127–133.

40. Pollard, D.J.; Buccino, R.; Connors, N.C. *et al.*, Real-time analyte monitoring of a fungal fermentation, at pilot scale, using in situ mid-infrared spectroscopy; *Bioprocess Biosystem Eng.* 2001, 24, 13–14.

41. Yu, L.X.; Lionberger, R.A.; Raw, A.S. *et al.*, Applications of process analytical technology to crystallization processes; *Adv. Drug Deliv. Rev.* 2004, 56, 349–369.

42. Dell'Orco, P.; Diederich, A.M. & Rydzak, J.W., Designing for crystalline form and crystalline physical properties: The roles of in-situ and ex-situ monitoring techniques; *Am. Pharm. Rev.* 2002, 5, 46–54.

43. Fevotte, G., New perspectives for the on-line monitoring of pharmaceutical crystallization processes using in situ infrared spectroscopy; *Int. J. Pharm.* 2002, 241, 263–278.

44. Fujiwara, M.; Chow, P.S.; Ma, D.L. *et al.*, Paracetamol crystallization using laser backscattering and ATR-FTIR spectroscopy: Metastability, agglomeration, and control; *Crys. Growth Des.* 2002, 2, 363–370.

45. http://www.lasentec.com/M-2-138_abstract.html.

46. http://www.lasentec.com/M-2-155.pdf.

47. Liotta, V. & Sabesan, V., Monitoring and feedback control of supersaturation using ATR-FTIR to produce an active pharmaceutical ingredient of a desired crystal size; *Org. Process Res. Dev.* 2004, 8, 488–494.

48. Cao, Z.; Groen, H.; Hammond, R.B. *et al.*, Monitoring the crystallization of organic specialty chemical products via on-line analytical techniques; *International Symposium on Industrial Crystallization*, 14th, Cambridge, UK, September 12–16, 1999, 2267–2278.

49. Zhou, G.; Wang, J.; Ge, Z. *et al.*, Ensuring robust polymorph isolation using in-situ Raman spectroscopy; *Am. Pharm. Rev.* 2002, 5, 74, 76–80.

50. Starbuck, C.; Spartalis, A.; Wai, L. *et al.*, Process optimization of a complex pharmaceutical polymorphic system via in situ Raman spectroscopy; *Crys. Growth Des.* 2002, 2, 515–522.

51. Wang, F.; Wachter, J.A.; Antosz, F.J. *et al.*, An investigation of solvent-mediated polymorphic transformation of progesterone using in situ Raman spectroscopy; *Org. Process Res. Dev.* 2000, 4, 391–395.

52. Norris, T.; Aldridge, P.K. & Sekulic, S.S., Determination of end-points for polymorph conversions of crystalline organic compounds using online near-infrared spectroscopy; *Analyst* 1997, 122, 549–552.

53. Fevotte, G.; Calas, J.; Puel, F. & Hoff, C., Applications of NIR to monitoring and analyzing the solid state during industrial crystallization processes; *Int. J. Pharm.* 2004, 273, 159–169.

54. Coffey, C.; Predoehl, A. & Walker, D.S., Dryer effluent monitoring in a chemical pilot plant via fiber-optic near-infrared spectroscopy; *Appl. Spectrosc.* 1998, 52, 717–724.

55. am Ende, D.J.; Preigh, M.J.; Hettenbach, K. *et al.*, On-line monitoring of vacuum dryers using mass spectrometry; *Org. Process Res. Dev.* 2000, 4, 587–593.

56. Doherty, S.; LaPack, M.; David, P. *et al.*, Integrated analytical technology: Instrumental in achieving efficient pharmaceutical process development; *National ACS Meeting*, October 2001, Chicago, IL.

57. White, Jackie G., Online moisture detection for a microwave vacuum dryer; *Pharm. Res.* 1994, 11, 728–732.

58. Hicks, M.B.; Zhou, G.X.; Lieberman, D.R. *et al.*, In situ moisture determination of a cytotoxic compound during process optimization; *J. Pharm. Sci.* 2003, 92, 529–535.

59. Zhou, G.X.; Ge, Z.; Dorwart, J. *et al.*, Determination and differentiation of surface and bound water in drug substances by near infrared spectroscopy; *J. Pharm. Sci.* 2003, 92, 1058–1065.

60. Bruells, M.; Folestad, S.; Sparen, A. *et al.*, In-situ near-infrared spectroscopy monitoring of the lyophilization process; *Pharm. Res.* 2003, 20, 494–499.

61. Morris, K.R.; Stowell, J.G.; Byrn, S.R. *et al.*, Accelerated fluid bed drying using NIR monitoring and phenomenological modeling; *Drug Dev. Ind. Pharm.* 2000, 26, 985–988.

62. Wildfong, P.L.D.; Samy, A.-S.; Corfa, J. *et al.*, Accelerated fluid bed drying using NIR monitoring and phenomenological modeling: Method assessment and formulation suitability; *J. Pharm. Sci.* 2002, 91, 631–639.

63. Ciurczak, E.W.; Torlini, R.P. & Demkowicz, M.P., Determination of particle size of pharmaceutical raw materials using near infrared reflectance spectroscopy; *Spectroscopy* 1986, 1, 36–39.

64. Ilari, J.L.; Martens, H. & Isaksson, T., Determination of particle size in powders by scatter correction in diffuse near infrared reflectance; *Appl. Spectrosc.* 1988, 42, 722–728.

65. O'Neil, A.J.; Jee, R. & Moffatt, A.C., The application of multiple linear regression to the measurement of the median particle size of drugs and pharmaceutical excipients by near-infrared spectroscopy; *Analyst* 1998, 123, 2297.

66. Higgins, J.P.; Arrivo, S.M.; Thurau, G. *et al.*, Spectroscopic approach for on-line monitoring of particle size during the processing of pharmaceutical nanoparticles; *Anal. Chem.* 2003, 75, 1777–1785.

67. Sun, Z.; Torrance, S.; McNeil-Watson, F.K. *et al.*, Application of frequency domain photon migration to particle size analysis and monitoring of pharmaceutical powders; *Anal. Chem.* 2003, 75, 1720–1725.

68. http://www.paa.co.uk/process/products/cleanscan.asp.

69. Mehta, N.K.; Goenaga-Polo, J.; Hernandez-Rivera, S.P. *et al.*, Development of an in situ spectroscopic method for cleaning validation using mid-IR fiber optics; *Spectroscopy* 2003, 18, 14–19.

70. Berman, J. & Planchard, J.A., Blend uniformity and unit dose sampling; *Drug Dev. Ind. Pharm.* 1995, 21, 1257–1283.

71. Plugge, W. & van der Vlies, C., Near-infrared spectroscopy as a tool to improve quality; *J. Pharm. Biomed. Anal.* 1996, 14, 891–898.

72. Watano, S.; Takashima, H.; Sato, Y. *et al.*, IR absorption characteristics of an IR moisture sensor and mechanism of water transfer in fluidized bed granulation; *Advanced Powder Tech.* 1996, 7, 279–289.

73. Rantanen, J.; Lehtola, S. & Rämet, P., On-line monitoring of moisture content in an instrumented fluidized bed granulator with a multi-channel NIR moisture sensor; *Powder Technol.* 1998, 99, 163–170.

74. Rantanen, J.; Räsänen, E.; Tenhunen, J. *et al.*, In-line moisture measurement during granulation with a four-wavelength near infrared sensor: An evaluation of particle size and binder effects; *Eur. J. Pharm. Biopharm.* 2000, 50, 271–276.

75. Frake, P.; Greenhalgh, S.M.; Grierson, J.M. *et al.*, Process control and end-point determination of a fluid bed granulation by application of near infra-red spectroscopy; *Int. J. Pharm.* 1997, 151, 75–80.

76. Rantanen, J.T.; Laine, S.J.; Antikainen, O.K. *et al.*, Visualization of fluid-bed granulation with self-organizing maps; *J. Pharm. Biomed. Anal.* 2001, 24, 343–352.

77. Miwa, A.; Yajima, T. & Itai, S., Prediction of suitable amount of water addition for wet granulation; *Int. J. Pharm.* 2000, 195, 81–92.

78. Wargo, D.J. & Drennen, J.K., Near-infrared spectroscopic characterization of pharmaceutical powder blends; *J. Pharm. Biomed. Anal.* 1996, 14, 1415–1423.

79. Hailey, P.A.; Doherty, P.; Tapsell, P. *et al.*, Automated system for the on-line monitoring of powder blending processes using near-infrared spectroscopy: Part I, system development and control; *J. Pharm. Biomed. Anal.* 1996, 14, 551–559.

80. Sekulic, S.S.; Wakeman, J.; Doherty, P. *et al.*, Automated system for the on-line monitoring of powder blending processes using near-infrared spectroscopy: Part II, qualitative approaches to blend evaluation; *J. Pharm. Biomed. Anal.* 1998, 17, 1285–1309.

81. Sekulic, S.J.; Ward II, H.W.; Brannegan, D.R. *et al.*, On-line monitoring of powder blend homogeneity by near-infrared spectroscopy; *Anal. Chem.* 1996, 68, 509–513.

82. Aldridge, P.K., Apparatus for mixing and detecting on-line homogeneity, US Patent 5946088; Assigned to Pfizer, Inc.; 1999.

83. Duong, N.-H.; Arratia, P.; Muzzio, F. *et al.*, A homogeneity study using NIR spectroscopy: Tracking magnesium stearate in Bohle bin-blender; *Drug Dev. Ind. Pharm.* 2003, 29, 679–687.

84. Lai, C.-K.; Holt, D.; Leung, J.C. *et al.*, Real time and noninvasive monitoring of dry powder blend homogeneity; *AIChE J.* 2001, 47, 2618–2622.

85. Singhal, D. & Curatolo, W., Drug polymorphism and dosage form design: A practical perspective; *Adv. Drug Deliv. Rev.* 2004, 56, 335–347.

86. Aldridge, P.K.; Evans, C.L.; Ward II, H.W. *et al.*, Near-IR detection of polymorphism and process-related substances; *Anal. Chem.* 1996, 68, 997–1002.

87. DeBraekeleer, K.; Cuesta Sánchez, F.; Hailey, P.A. *et al.*, Influence and correction of temperature perturbations on NIR spectra during the monitoring of a polymorph conversion process prior to self-modelling mixture analysis; *J. Pharm. Biomed. Anal.* 1997, 17, 141–152.

88. Otsuka, M.; Hasegawa, H. & Matsuda, Y., Effect of polymorphic transformation during the extrusion-granulation process on the pharmaceutical properties of carbamazepine granules; *Chem. Pharm. Bull.* 1997, 45, 894–898.

89. Morris, K.R.; Griesser, U.J.; Eckhardt, C.J. *et al.*, Theoretical approaches to physical transformations of active pharmaceutical ingredients during manufacturing processes; *Adv. Drug Deliv. Rev.* 2001, 48, 91–114.

90. Patel, A.D.; Luner, P.E. & Kemper, M.S., Quantitative analysis of polymorphs in binary and multi-component powder mixtures by near-infrared reflectance spectroscopy; *Int. J. Pharm.* 2000, 206, 63–74.

91. Taday, P.F.; Bradley, I.V.; Arnone, D.D. *et al.*, Using terahertz pulse spectroscopy to study the crystalline structure of a drug: A case study of the polymorphs of ranitidine hydrochloride; *J. Pharm. Sci.* 2003, 92, 831–838.

92. Räsänen, E.; Rantanen, J.; Jørgensen, A. *et al.*, Novel identification of pseudopolymorphic changes of theophylline during wet granulation using near infrared spectroscopy; *J. Pharm. Sci.* 2001, 90, 389–396.

93. Jørgensen, A.; Rantanen, J.; Karljalainen, M. *et al.*, Hydrate formation during wet granulation studied by spectroscopic methods and multivariate analysis; *Pharm. Res.* 2002, 19, 1285–1291.

94. Taylor, L.S. & Langkilde, F.W., Evaluation of solid-state forms present in tablets by Raman spectroscopy; *J. Pharm. Sci.* 2000, 89, 1342–1353.

95. Miller, R.W., The use of near infrared technology to map roller compaction processing applications, Proceedings Institute for Briquetting and Agglomeration, Biennial Conference (2000), 1999, 26, 17–26.

96. Gustafsson, C.; Nyström, C.; Lennholm, H. *et al.*, Characteristics of hydroxypropyl methylcellulose influencing compactability and prediction of particle and tablet properties by infrared spectroscopy; *J. Pharm. Sci.* 2003, 92, 460–470.

97. Han, S.M. & Faulkner, P.G., Determination of SB 216469-S during tablet production using near-infrared reflectance spectroscopy; *J. Pharm. Biomed. Anal.* 1996, 14, 1681–1689.

98. Blanco, M.; Coello, J.; Eustaquio, A. *et al.*, Analytical control of pharmaceutical production steps by near infrared reflectance spectroscopy; *Anal. Chim. Acta* 1999, 392, 237–246.

99. Morisseau, K.M. & Rhodes, C.T., Near-infrared spectroscopy as a nondestructive alternative to conventional tablet hardness testing; *Pharm. Res.* 1997, 14, 108–111.

100. Kirsch, J.D. & Drennen, J.K., Nondestructive tablet hardness testing by near-infrared spectroscopy: A new and robust spectral best-fit algorithm; *J. Pharm. Biomed. Anal.* 1999, 19, 351–362.

101. Kirsch, J.D. & Drennen, J.K., Determination of film-coated tablet parameters by near-infrared spectroscopy; *J. Pharm. Biomed. Anal.* 1995, 13, 1273–1281.

102. Kirsch, J.D. & Drennen, J.K., Near-infrared spectroscopic monitoring of the film coating process; *Pharm. Res.* 1996, 13, 234–237.

103. Andersson, M.; Josefson, M.; Lankilde, F.W. *et al.*, Monitoring of a film coating process for tablets using near infrared reflectance spectroscopy; *J. Pharm. Biomed. Anal.* 1999, 20, 27–37.

104. Andersson, M.; Folestad, S.; Gottfries, J. *et al.*, Quantitative analysis of film coating in a fluidized bed process by in-line NIR spectrometry and multivariate batch calibration; *Anal. Chem.* 2000, 72, 2099–2108.

105. Mowery, M.D.; Sing, R.; Kirsch, J. *et al.*, Rapid at-line analysis of coating thickness and uniformity on tablets using laser induced breakdown spectroscopy; *J. Pharm. Biomed. Anal.* 2002, 28, 935–943.

106. Thosar, S.S.; Forbes, R.A.; Ebube, Y.C. *et al.*, A comparison of reflectance and transmittance near-infrared spectroscopic techniques in determining drug content in intact tablets; *Pharm. Dev. Tech.* 2001, 6, 19–29.

107. Abrahamsson, C.; Johansson, J.; Sparén, A. *et al.*, Comparison of different variable selection methods conducted on NIR transmission measurements on intact tablets; *Chemometrics Intell. Lab. Syst.* 2003, 69, 3–12.

108. Trafford, A.D.; Jee, R.D.; Moffat, A.C. *et al.*, A rapid quantitative assay of intact paracetamol tablets by reflectance near-infrared spectroscopy; *Analyst* 1999, 124, 163–167.

109. Eutaquio, A.; Blanco, M.; Jee, R.D. *et al.*, Determination of paracetamol in intact tablets by use of near infrared transmittance spectroscopy; *Anal. Chim. Acta* 1999, 383, 283–290.

110. Blanco, M.; Eustaquio, A.; González, J.M. *et al.*, Identification and quantitation assays for intact tablets of two related pharmaceutical preparations by reflectance near-infrared spectroscopy: Validation of the procedure; *J. Pharm. Biomed. Anal.* 2000, 22, 139–148.

111. Ekube, N.K.; Thosar, S.S.; Roberts, R.A. *et al.*, Application of near-infrared spectroscopy for nondestructive analysis of Avicel powders and tablets; *Pharm. Dev. Tech.* 1999, 4, 19–26.

112. Agatonovic-Kustrin, S.; Tucker, I.G. & Schmierer, D., Solid state assay of ranitidine HCl as a bulk drug and as active ingredient in tablets using DRIFT spectroscopy with artificial neural networks; *Pharm. Res.* 1999, 16, 1477–1482.

113. Neubert, R.; Collin, B. & Wartewig, S., Direct determination of drug content in semisolid formulations using step-scan FT-IR photoacoustic spectroscopy; *Pharm. Res.* 1997, 14, 946–948.

114. Bell, S.E.J.; Burns, D.T.; Dennis, A.C. *et al.*, Rapid analysis of ecstasy and related phenethylamines in seized tablets by Raman spectroscopy; *Analyst* 2000, 125, 541–544.

115. Vergote, G.J.; Vervaet, C.; Remon, J.P. *et al.*, Near-infrared FT-Raman spectroscopy as a rapid analytical tool for the determination of diltiazem hydrochloride in tablets; *Eur. J. Pharm. Sci.* 2002, 16, 63–67.

116. Wang, C.; Vickers, T.J. & Mann, C.K., Direct assay and shelf-life monitoring of aspirin tablets using Raman spectroscopy; *J. Pharm. Biomed. Anal.* 1997, 16, 87–94.

117. Dyrby, M.; Engelsen, S.B.; Nørgaard, L. *et al.*, Chemometric quantitation of the active substance (containing C−N) in a pharmaceutical tablet using near-infrared (NIR) transmittance and NIR FT-Raman spectra; *Appl. Spectrosc.* 2002, 56, 579–585.

118. Herkert, T.; Prinz, H. & Kovar, K.-A., One hundred percent online identity check of pharmaceutical products by near-infrared spectroscopy on the packaging line; *Eur. J. Pharm. Biopharm.* 2001, 51, 9–16.

Chapter 10

Use of Near-Infrared Spectroscopy for Off-Line Measurements in the Pharmaceutical Industry

Marcelo Blanco and Manel Alcalá

10.1   Introduction

Drug manufacturing is an industrial activity comprehensively regulated by both national institutions (pharmacopoeias, FDA, drug regulatory agencies) and international bodies (ICH, EMEA). Applicable regulations require strict control of not only raw materials, but also production processes. Such control is currently performed mainly by using chromatographic techniques. While chromatography is suitable for most of the analytical processes involved, it has some disadvantages including sluggishness, the need for careful sample preparation and the production of waste with a potential environmental impact, all of which have promoted a search for alternative techniques avoiding or minimizing these problems. Spectroscopic techniques are highly suitable candidates in this respect.

Molecular spectroscopic techniques have been widely used in pharmaceutical analysis for both qualitative (identification of chemical species) and quantitative purposes (determination of concentration of species in pharmaceutical preparations). In many cases, they constitute effective alternatives to chromatographic techniques as they provide results of comparable quality in a more simple and expeditious manner. The differential sensitivity and selectivity of spectroscopic techniques have so far dictated their specific uses. While UV-vis spectroscopy has typically been used for quantitative analysis by virtue of its high sensitivity, infrared (IR) spectrometry has been employed mainly for the identification of chemical compounds on account of its high selectivity. The development and consolidation of spectroscopic techniques have been strongly influenced by additional factors such as the ease of sample preparation and the reproducibility of measurements, which have often dictated their use in quality control analyses of both raw materials and finished products.

Spectra differ markedly in appearance depending on the particular technique. UV-vis and near-infrared (NIR) spectra typically consist of a few broad bands that make unambiguous identification difficult. By contrast, IR and Raman spectra contain a number of sharp bands at well-defined positions that facilitate the identification of specific compounds. Recent improvements in sensitivity and reproducibility of spectroscopic equipment have substantially expanded the scope of these techniques. The introduction of

Fourier transform IR spectrometry (FTIR) has facilitated the development of spectral libraries of a high value for the identification of active pharmaceutical ingredients (APIs) and excipients. The ability of some techniques (e.g. IR, NIR and Raman spectroscopies) to provide spectra for undissolved solid samples has considerably expedited analyses and facilitated the determination of additional physical properties such as average particle size, polymorphism, etc. IR and Raman spectra also provide structural information that allows not only the identification of characteristic functional groups facilitating the establishment of chemical structures, but also the unequivocal identification of some compounds. NIR spectra can be obtained without sample preparation and due to this fact NIR has become one of the most promising spectroscopic tools for process monitoring and manufacturing control in the pharmaceutical industry. Also, NIR chemical imaging has enabled the identification of critical properties of both raw materials and finished products.

For the above-described reasons, molecular spectroscopic techniques have become the most common choices for pharmaceutical analysis in addition to chromatography. The latter, however, are being gradually superseded by the former in some industrial pharmaceutical processes. Recent technological advances have led to the development of more reliable and robust equipment. The ubiquity of computers has enabled the implementation of highly powerful chemometric methods. All this has brought about radical, widespread changes in the way pharmaceutical analyses are conducted.

Chemometric techniques have gained enormous significance in the treatment of spectral information by virtue of their ability to process the vast amount of data produced by modern instruments over short periods of time with a view to extracting the information of interest they contain and improving the quality of the results. In some cases, the operator is unacquainted with the chemometric techniques (spectral smoothing, baseline drift correction) embedded in the software used by the instrument; in others, the chemometric tools involved are inherent in the application of the spectroscopic technique concerned (e.g. in NIR spectroscopy) and thus indispensable to obtaining meaningful results.

Because NIR spectroscopy has played the most prominent role in this context, the discussion that follows mainly revolves around it. However, the most salient contributions of other spectroscopic techniques (viz. UV-vis, IR and Raman spectroscopies) are also commented on.

10.1.1   Operational procedures

The procedures used to record NIR spectra for samples are much less labor-intensive than those involved in other spectroscopic analytical techniques. NIR spectral information can be obtained from transmission, reflectance and transflectance measurements; this allows the measurement process to be adapted to the physical characteristics of the sample and expedites analyses by avoiding the need for sample preparation.

One major difference between NIR spectroscopy and other spectroscopic techniques is its ability to provide spectra for untreated or minimally treated samples. As a result, the instrumentation is adjusted to the characteristics of the samples rather than the opposite.

The absorbance of a liquid or solution can be readily measured by using quartz or sapphire cuvettes and fiber-optic probes of variable pathlengths. The most suitable solvents for this purpose are those not containing O—H, N—H and C—H groups and therefore exhibit little or no absorption in this spectral region.

The NIR spectrum for a solid sample can be obtained by using various types of devices. The most frequent choices when the sample is in powder or grain form are reflectance cuvettes with a transparent window material (e.g. quartz or glass) and fiber-optic probes. The latter considerably facilitate recording of spectra; however, light losses resulting from transport along the fiber result in increased signal noise at higher wavelengths. The spectra for samples in tablet form require no pretreatment such as powdering, sieving or homogenization and can be recorded by using three different types of equipment, namely:

(1) Reflectance cuvettes for tablets, which provide spectra subject to strong scattering arising from dead spaces between tablets placed in the cuvette.
(2) Devices for recording reflectance spectra for individual tablets.
(3) Instruments that allow transmission spectra for individual tablets to be recorded.

It should be noted that the type of standard to be used for reflectance measurements remains a subject of debate. According to American Society for Testing and Materials (ASTM), the perfect standard for this purpose is a material absorbing no light at any wavelength and reflecting light at an angle identical with the incidence angle. Teflon is the material that fulfills these conditions best. Because no single material meets these requirements, the standards used in this context should be stable, homogeneous, non-transparent, non-fluorescent materials of high, fairly constant relative reflectance. Springsteen and Ricker[1] discussed the merits and pitfalls of materials such as barium sulfate, magnesium oxide, Teflon and ceramic plates as standards for reflectance measurements. Some manufacturers supply National Institute of Standards and Technology (NIST)-validated reference standards for standardizing spectroscopic equipment.[2]

10.1.2 Instrument qualification

Ensuring that the instrument to be used performs correctly is the first step in developing an instrumental methodology. The European Pharmacopoeia[3] and the Guidelines of the Pharmaceutical Analytical Sciences Group (PASG)[4] recommend that NIR instruments be qualified as per the manufacturer's instructions, which should include at least the following:

(a) Checking for wavelength accuracy by using one or more suitable wavelength standards exhibiting characteristic maxima at the wavelengths of interest.
(b) Checking for wavelength repeatability by using one or more suitable standards (e.g. polystyrene or rare-earth oxides). The repeatability of measurements should be consistent with the spectrophotometer specification.

(c) Checking for repeatability in the response by using one or more suitable standards (e.g. reflective thermoplastic resins doped with carbon black). The standard deviation of the maximum response should be consistent with the spectrophotometer specification.
(d) Checking for photometric linearity by the use of a set of transmittance or reflectance standards (e.g. Spectralon, carbon black).
(e) Checking for the absence of photometric noise by using a suitable reflectance standard (e.g. white reflective ceramic tiles or reflective thermoplastic resins, such as Teflon). The reflectance standard should be scanned in accordance with the recommendations of the spectrophotometer's manufacturer and peak-to-peak photometric noise at a given calculated wavelength. The amount of photometric noise should be consistent with the spectrophotometer specification.

However, two different instruments will inevitably exhibit small spectral differences that can be reduced by applying standardization corrections based on appropriate references or standards. Reference standardization uses NIST traceable photometric standards (80% reflectance) to measure the absolute reflectance of the ceramic tile reference at each wavelength. When a sample is then compared to the ceramic reference, because the absolute absorbance of the reference at each wavelength is known, the absolute absorbance of the sample can be accurately measured – no negative absorbance values are possible. Master standardization uses a sample of similar spectral pattern to that of the products being analyzed to be scanned first on the Master instrument and then on the Host instrument. A standardization file is then created for the Host instrument. When a sample is scanned in the Host instrument, the spectrum is adjusted using the standardization file to cause it to more closely resemble that measured on the Master. This is a 'cloning' algorithm. The above described corrections allow calibration models to be readily transferred between instruments.

10.2 Qualitative analysis

Identifying pharmaceuticals, whether APIs or excipients used to manufacture products, and the end products themselves is among the routine tests needed to control pharmaceutical manufacturing processes. Pharmacopoeias have compiled a wide range of analytical methods for the identification of pharmaceutical APIs and usually several tests for a product are recommended. The process can be labor-intensive and time-consuming with these conventional methods. This has raised the need for alternative, faster methods also ensuring reliable identification. Of the four spectroscopic techniques reviewed in this book, IR and Raman spectroscopy are suitable for the unequivocal identification of pharmaceuticals as their spectra are compound-specific: no two compounds other than pairs of enantiomers or oligomers possess the same IR spectrum. However, IR spectrometry is confronted with some practical constraints such as the need to pretreat the sample. The introduction of substantial instrumental improvements and the spread of attenuated total reflectance (ATR) and IR microscopy techniques have considerably expanded the scope of IR spectroscopy in the pharmaceutical field. Raman spectroscopy, which has been developed and incorporated by the pharmaceutical industry somewhat more recently, is similar to IR spectroscopy. Some of its more interesting and recent applications are the identification of impurities and adulterants.[5] On the other hand, UV-vis spectra are too uncharacteristic to allow the identification of specific compounds.

Near-infrared spectroscopy, which is subject to none of the previous shortcomings, has been used to monitor all the steps in the drug manufacturing process including chemical synthesis, analysis of raw materials, process intermediates (following blending, grinding, drying, tabletting) and preparations, and finally in quality control. The quality attributes studied using this technique include identity (authentication), characterization of polymorphs and measurement of mixture uniformity in finished products. Also, chemical composition can be determined without losing any physical information (e.g. particle size, density, hardness), which can be very useful in characterizing some products. The construction of spectral libraries avoids the need to develop a specific method for each individual product and enables its identification in a simple, direct, expeditious, economic manner. The whole process can be automated, computer-controlled and conducted by unskilled operators.

10.2.1   Foundation of identification (authentication)

NIR spectral bands are usually broad and strongly overlapped. This precludes the direct identification of compounds by matching spectra to an individual spectrum in a library, which is the usual procedure in IR spectroscopy. Rather, NIR identification relies on pattern recognition methods (PRMs). Essentially, the identification process involves two steps, namely: (1) recording a series of spectra for the product in order to construct a 'spectral library'; and (2) recording the sample spectrum and comparing it to those in the previously compiled library on the basis of mathematical criteria for parametrizing spectral similarity. If the similarity level exceeds a preset threshold, then the spectra are considered to be identical and the sample is identified with the corresponding product in the library. A wide variety of PRM techniques exist for use in specific fields. Figure 10.1 classifies the most commonly used PRMs.

Many PRMs are based on similarity measurements. Similarity here is taken to be the extent to which an object (a spectrum) is identical with another; very frequently, similarity is expressed in terms of correlation[6] or distance.[7] Depending on whether the objects are known to belong to specific classes, PRMs are called 'supervised' or 'unsupervised.' Unsupervised methods search for clustering in an N-dimensional space without knowing the class to which a sample belongs. Cluster analysis,[8] minimal spanning tree (MST)[9] and unsupervised (Kohonen) neural networks[10] are among the most common unsupervised PRMs.

Supervised methods rely on some prior training of the system with objects known to belong to the class they define. Such methods can be of the discriminant or modeling types.[11] Discriminant methods split the pattern space into as many regions as the classes encompassed by the training set and establish bounds that are shared by the spaces. These methods always classify an unknown sample as a specific class. The most common discriminant methods include discriminant analysis (DA),[12] the *K*-nearest neighbor

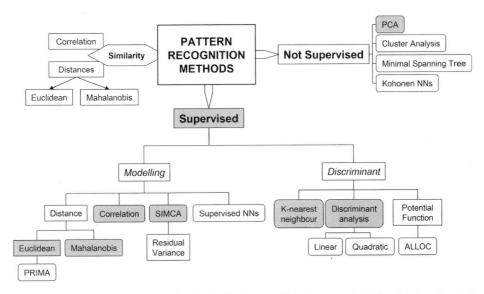

Figure 10.1 Pattern recognition methods classification used in pharmaceutical identification. The dark grey blocks refer the techniques more commonly used.

(KNN)[13,14] and potential function methods (PFMs).[15,16] Modeling methods establish volumes in the pattern space with different bounds for each class. The bounds can be based on correlation coefficients, distances (e.g. the Euclidian distance in the Pattern Recognition by Independent Multicategory Analysis methods [PRIMA][17] or the Mahalanobis distance in the Unequal [UNEQ] method[18]), the residual variance[19,20] or supervised artificial neural networks (e.g. in the Multi-layer Perceptron[21]).

Not every PRM is suitable for constructing spectral identification libraries. These are usually compiled by using supervised modeling methods, and unknown samples are identified with those classes they resemble most.

10.2.2   Construction of NIR libraries

A NIR identification library should encompass all the raw materials used by the manufacturer in order to be able to identify all possible substances and avoid or minimize errors. The method to be used should allow the unequivocal identification of each compound present in the library and the exclusion of those not present. It should also be able to distinguish between very similar compounds used in different applications (e.g. products with different particle sizes, polymorphs, products in different grades or from different suppliers).

Figure 10.2 schematizes the use of spectral information for constructing multivariate calibration models. The pretreatment of the sample spectrum and variable selection processes involve constructing both identification libraries and calibration models for quantifying APIs; however, the samples and spectra should suit the specific aim in each case.

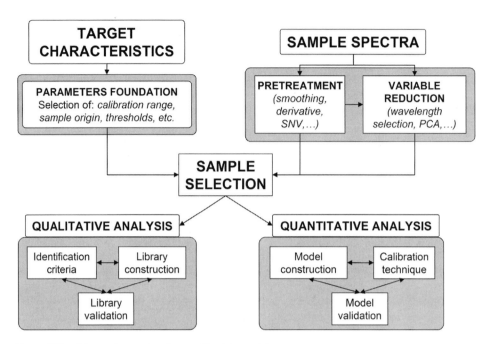

Figure 10.2 Schema for constructing multivariate models.

The correlation coefficient (in wavelength space) is especially suitable for constructing the general library as it has the advantage that it is independent of library size, uses only a few spectra to define each product and is not sensitive to slight instrumental oscillations. This parameter allows the library to be developed and validated more expeditiously than others. Correlation libraries can also be expanded with new products or additional spectra for an existing product in order to incorporate new sources of variability in an expeditious manner.

One of the crucial factors to ensure adequate selectivity in constructing a spectral library is the choice of an appropriate threshold. Such a threshold is the lowest value (for correlation) required to unequivocally assign a given spectrum to a specific class. Too low a threshold can lead to ambiguities between substances with similar spectra, whereas too high a threshold can result in spectra belonging to the same class being incorrectly classified. Choosing an appropriate threshold entails examining the spectra included in the library in an iterative manner. The threshold is successively changed until that resulting in the smallest number of identification errors and ambiguities is reached. In some cases, the threshold thus selected may not distinguish some compounds whose spectra are too similar. This problem can be overcome by identifying the compounds concerned in two steps, using a general library in the first and a smaller one (a subset of the general set with a higher discriminating power) in the second. This methodology has been termed 'cascading identification' as it involves identifying the unknown sample against the general library and, if the result is inconclusive, using a sub-library for qualification.

This procedure can consist of different stages and in each one a sub-library is constructed. The first one enables the identification of the product within an ample library, a second defines a characteristic that produces an important variability in the spectra (e.g. particle size) and a third defines a smaller spectral variability (difference in the content of impurities, origin of manufacture, etc.).

Once the general library has been constructed, those products requiring the second identification step are pinpointed and the most suitable method for constructing each sub-library required is chosen. Sub-libraries can be constructed using various algorithms including the Mahalanobis distance or the residual variance. The two are complementary,[22] so which is better for the intended purpose should be determined on an individual basis.

The procedure to be followed to construct the library (Figure 10.2) involves five steps, namely:

(1) *Recording the NIR spectra.* For a set of samples of known identity, either certified by the supplier or confirmed using alternative identification tests.

(2) *Choosing spectra.* For each substance, the spectra used to construct the calibration set should be from various batches so that physico-chemical variability can be effectively incorporated. Not every spectrum recorded should be used to construct the library as some should be set aside for its validation.

(3) *Constructing the library.* First, one must choose the pattern recognition method to be used (e.g. correlation, wavelength distance, Mahalanobis distance, residual variance, etc.), the choice being dictated by the specific purpose of the library. Then, one must choose construction parameters such as the math pretreatment (SNV, derivatives), wavelength range and threshold to be used. When a method involving a preliminary variable reduction procedure, such as principal component analysis (PCA) is used, one must also decide upon the values for other parameters such as the number of latent variables, the proportion of variance accounted for, etc. The next step involves validating the library internally in order to check for incorrectly labeled spectra – spectra yielding identification errors or unidentified (or ambiguously identified) substances. Based on the validation results, one must determine if some change in the threshold, spectral range, pretreatment, etc., must be introduced. This is an iterative process that must be repeated until the developer obtains the desired specificity level. Ambiguous identifications should be resolved by developing appropriate sub-libraries.

(4) *Constructing sub-cascading libraries.* Each sub-library should include all mutually related substances, which will have very similar spectra and may result in ambiguous identification with the general library. An adequate number of representative spectra should be included in the sub-libraries in order to ensure accurate description of each class. The number of spectra used should exceed that of spectra required to define an individual class in the general library. Once spectra have been chosen, the process is similar to that followed to construct the general library: an appropriate PRM is selected – the residual variance and the Mahalanobis distance are especially suitable here – the characteristics of the library (viz. threshold, spectral range and pretreatment, proportion of variance accounted for or number of PCs, etc.) are established and calculations performed. Based on the results, the procedure is repeated until all compounds can be unequivocally distinguished.

(5) *External validation.* The general library and its sub-libraries must be validated by checking that external spectra (for the validation step) are correctly, unambiguously identified. Likewise, samples not present in the library should not be identified with any of the compounds it contains.

10.2.3   Identification of raw materials and pharmaceutical preparations

Control analyses rely on the use of appropriate procedures or measurements allowing the identity of the materials involved in each step of the manufacturing process from receipt of raw materials to delivery of the finished products to be assured. NIR spectroscopy is an advantageous alternative to wet chemical methods and instrumental techniques such as IR, Raman and nuclear magnetic resonance (NMR) spectroscopies.

The earliest reported studies on the identification of substances by NIR focused on structural characterization, in the vein of many studies in the mid-IR region. However, it is usually difficult to identify a substance from the mere visual inspection of its NIR spectrum; rather, a pattern recognition method must be applied to a body of spectra (a library) for this purpose. The reliability of the results obtained in the qualitative analysis of a product depends on whether the library is constructed from appropriate spectra. For each product encompassed by the library, one should ensure that the spectra include all potential sources of variability associated with their recording and their manufacturing process. The former should be incorporated by including spectra for the same sample recorded by different operators on different days, and the latter by including spectra from different production batches. How many spectra are to be included in order to construct a 'proper' library is difficult to tell a priori. For a product manufactured in a highly reproducible manner, the manufacturing variability can usually be encompassed by samples from 5–10 batches and 20–40 spectra. On the other hand, if the manufacturing procedure is scarcely reproducible, the number of batches and spectra can easily double.

One other important consideration in constructing a library is checking that all spectra are correct. This can be done by visual inspection of the spectra and outlier detection tools. For various reasons (e.g. an inadequately filled cuvette, large temperature fluctuations), spectra may differ from others due to factors other than natural variability. Any such spectra should be excluded from the library. Spectral libraries are used for two different purposes, namely: identification (of chemical structure) and qualification (of chemical and physical attributes).

10.2.3.1   Identification

This involves the confirmation of a certain chemical entity from its spectrum by matching against the components of a spectral library using an appropriate measure of similarity such as the correlation coefficient,[6,23] also known as the spectral match value (SMV). SMV is the cosine of the angle formed by the vectors of the spectrum for the sample and the average spectrum for each product included in the library.

In theory, if the product spectrum coincides with one in the library, then its correlation coefficient should be unity. However, the random noise associated with all spectral measurements precludes an exact match. This parameter has the advantage that it is independent of the size of the library and of oscillations in concentration, so it enables correct identifications with libraries consisting of a small number of spectra. However, it has a relatively low discriminating power, which may require the use of near-unity thresholds in order to distinguish very similar products. It can also result in ambiguous identification of batches of the same substance with different physical characteristics (e.g. samples of the same product with different grain sizes can have very similar correlation coefficients and be confused) or different substances with high chemical similarity. In these cases, cascading libraries, as previously discussed, can be used.

10.2.3.2 Qualification

NIR spectroscopy not only advantageously replaces other techniques for identification purposes, but also goes one step further in qualitative analysis with its qualification capabilities. The pharmaceutical industry must assure correct dosing and manufacturing of stable products. This entails strictly controlling the raw materials and each of the steps involved in their processing by determining parameters such as potency, moisture, density, viscosity or particle size in order to identify and correct any deviations in the manufacturing process in a timely fashion. Such controls can be achieved by quantifying the different parameters required using appropriate analytical methods or by comparing the NIR spectrum of the sample with a body of spectra of the samples meeting the required specifications and encompassing all possible sources of manufacturing variability. The latter choice is known as 'qualification,' which can be defined as the fulfillment of the product specifications. If the parameter values for a given sample do not fall within the normal variability ranges, then the sample is subjected to thorough analyses. The qualification process can be considered the second stage of an identification process in cascade. Similarity in this context is expressed as a distance. Distance-based methods are subject to two main shortcomings. They use a large number of correlated spectral wavelengths, which increases the distance without providing additional information. They also use more samples than do correlation methods. The Euclidian distance in the wavelength space and the Mahalanobis distance – which corrects correlation between spectra – in principal component space are the two most widely used choices.

Distance-based methods possess a superior discriminating power and allow highly similar compounds (e.g. substances with different particle sizes or purity grades, products from different manufacturers) to be distinguished. One other choice for classification purposes is the residual variance, which is a variant of Soft Independent Modeling of Class Analogy (SIMCA).

Table 10.1 lists selected references to the identification of pharmaceutical raw materials. As can be seen, a variety of techniques have been used, which fall short of the discriminating power of the previous ones and have not yet been included by manufacturers in their equipment software. This makes them inconvenient to implement.

Table 10.1 Applications of NIR spectroscopy to API and excipient identification

Analyte	Pattern recognition method	Remarks	References
Oxytetracycline	PCA and SIMCA	Sample identity determination before multivariate quantitation	24
Pharmaceuticals	PCA and visual classification	Drug authenticity assurance, counterfeiting control	25
Excipients	PCA and cluster analysis	Transferability between different spectrometers	26
Raw materials	Mahalanobis distance and residual variance	Classification models development	21
Metronidazole	Hierarchical and stepwise classification, stepwise discriminant analysis and PCA	Assure the quality of the drug	27
Excipients	Polar qualification system	Sample identification and qualification	28
Raw materials	Correlation coefficient, Mahalanobis distance, residual variance	Libraries and sub-libraries	29
Corn starch	Correlation coefficient and distance	Spectral library as quality control system	30
Herbal medicines	Wavelength distance, residual variance, Mahalanobis distance and SIMCA	PRMs comparison	31
Acetaminophen	Maximum distance in wavelength space	Identification of tablets brand	32
Excipients	SIMCA	Classification models development	33
Pharmaceuticals	Correlation coefficient, wavelength distance, polar coordinates, gravity center, scanning probability windows	PRMs comparison	34

10.2.4 Determination of homogeneity

Ensuring homogeneous mixing of the components (APIs and excipients) of a pharmaceutical preparation is a crucial prerequisite for obtaining proper solid dosages (tablets and capsules). Pharmaceutical manufacturers invest much time, labor and material resources in this process. Blending is intended to ensure uniform distribution of all components in the end products,[35] so that each dose will contain the correct amount of the active.

As a rule, controlling the blending process entails stopping the blender, collecting samples at preset points within it – usually by hand or by using a sample thief – and analysing them by liquid chromatography (HPLC) or UV-vis spectroscopy.[36] Near-infrared reflectance spectroscopy (NIRRS) enables the analysis of complex matrices without the need to manipulate samples. This results in substantially decreased analysis time relative to wet chemical methods. The large number of qualitative and quantitative uses of NIRS reported in recent years testify to its high efficiency. Many such uses have focused on the determination of APIs in pharmaceutical preparations[37,38] or physical parameters of samples.[39,40] Others have enabled the identification and characterization of raw materials[41,42] or even the monitoring of product degradation.[43] There are, however, comparatively few references to the validation of blending processes,[44–46] all of which have involved a search for objective criteria to determine how long blending should be continued. In fact, the visual inspection of spectra recorded during the blending process usually helps, but may not allow one to precisely identify the time when the mixture becomes uniform. The need to arrange these objective criteria continues to be reflected in current studies in this context.

Some of the procedures used to assess mixture uniformity (e.g. the visual inspection of spectra) are subjective. Those based on mathematical parameters are objective. However, most require spectral information from homogeneous samples, so criteria can be established only after the process is finished.

The qualification concept was used by Wargo and Drennen[45] to verify the homogeneity of solid mixtures. Qualitative analytical algorithms based on Bootstrap Error-adjusted Single-sample Technique (BEST) proved sensitive to variations in sample homogeneity. Plugge and Van der Vlies converted NIR spectra into polar coordinates and used them to calculate the Mahalanobis distance[47] in order to identify non-homogeneous samples.

Hailey *et al.*[48] and Sekulic *et al.*[49] developed systems for monitoring the homogenization of solid mixtures based on measurements made with a fiber-optic probe fitted to the mixer. The most salient advantage of these systems is that they allow one to identify the end point of the homogenization process non-invasively in real time. In both cases, mixture homogeneity was determined by plotting the standard deviation for several replicates against the homogenization time. Cuesta[46] conducted a similar study using samples withdrawn from the blender. Sekulic[50] explored various parameters of the blending process and applied various spectral pretreatments including detrending, SNV and the second spectral derivative in conjunction with various criteria (running block standard deviation, dissimilarity and PCA) to identify the end point of the process.

Blanco[51] proposed the use of the mean square difference between two consecutive spectra plotted against the blending time in order to identify the time that mixture homogeneity was reached.

10.2.5   Characterization of polymorphs

The significance of polymorphism in the pharmaceutical industry lies in the fact that polymorphs can exhibit differential solubility, dissolution rate, chemical reactivity, melting point, chemical stability or bioavailability, among others. Such differences can have considerable impact on their effectiveness. Usually, only one polymorph is stable at a given temperature, the others being metastable and evolving to the stable phase with time. Regulatory authorities have established the need to control the polymorph selected for pharmaceutical use in each case and to ensure its integrity during formulation and storage. Polymorphs are usually identified, characterized and determined using X-ray diffraction (the reference method), IR, solid-state NMR or Raman spectroscopy. NIR spectroscopy[52,53] is an effective alternative as polymorphs usually differ in their NIR spectra because the vibrations of their atoms are influenced by the differential ordering in their crystal lattices (especially by the presence of hydrogen bonds). NIR spectroscopy enables the fast, reproducible recording of spectra without the need to pretreat or alter the sample in any way, thereby preventing polymorph interconversion. This makes NIR spectroscopy especially attractive for this purpose. NIR can be used to rapidly identify the different polymorphs. Applicable references include the discrimination of polymorphs by use of pattern recognition methods,[54] the on-line monitoring of polymorph conversions in reactors,[55] the monitoring of such conversions in active drug ingredients using multiple linear regression (MLR)[56] and the qualitative monitoring of polymorph conversion as a function of ambient moisture.[57] Polymorphs can be characterized in an expeditious, convenient manner by using the above-described spectral libraries; the spectral differences involved are almost always large enough to use the correlation coefficient as the discriminating criterion. By using sub-libraries one can define classes to the level of distinguishing pure polymorphs from contaminated forms containing relatively small amounts of other polymorphs.[29]

Some chiral compounds exhibit a different crystal structure for each pure enantiomer and their mixture; the difference can be used to distinguish them. Again, the use of cascading libraries allows pure compounds to be distinguished from compounds contaminated with relatively small amounts of the other chiral forms.

10.3   Quantitative analysis

Properly controlling the different steps of the production process requires knowledge of the API concentration throughout the process (from the raw materials to the end product). The characteristics of the NIR spectrum require the use of multivariate calibration techniques in order to establish a quantitative relationship between the spectrum and the chemical (composition) and/or physical parameters (particle size, viscosity and density).

Calibration is the process by which a mathematical model relating the response of the analytical instrument (a spectrophotometer in this case) to specific quantities of the samples is constructed. This can be done by using algorithms (usually based on least squares regression) capable of establishing an appropriate mathematical relation such as single absorbance vs. concentration (univariate calibration) or spectra vs. concentration (multivariate calibration).

The calibration methods most frequently used to relate the property to be measured to the analytical signals acquired in NIR spectroscopy are MLR,[59,60] principal component regression (PCR)[61] and partial least-squares regression (PLSR).[61] Most of the earliest quantitative applications of NIR spectroscopy were based on MLR because spectra were then recorded on filter instruments, which afforded measurements at a relatively small number of discrete wavelengths only. However, applications involving PCR and PLSR have become more widely used since the availability of commercial instruments allowing the whole NIR region to be scanned.

Essentially, the procedure for quantitation using multivariate calibration involves the following steps: (a) selecting a representative sample set; (b) acquiring the spectra; (c) obtaining the reference values; (d) mathematically processing the spectra; (e) selecting the model that relates the property to be determined and the signals; and (f) validating the model. Each step is described in detail below,[62] with special emphasis on the problems specific to NIR analyses of pharmaceuticals.

10.3.1 Selection of samples

The starting point for any calibration is collecting a set of representative samples encompassing all possible sources of variability for the manufacturing process and spanning a wide-enough range of values of the target parameter. These requirements are easily met for liquid samples, but not quite so for solid samples (e.g. grain, tablet or pill forms).

The NIR spectrum depends not only on the chemical composition of the sample, but also on some physical properties such as the size, form, particle distribution and degree of compression of the sample. This is useful in some cases as it enables the spectroscopic determination of some physical parameters of the sample. However, physical differences can lead to multiplicative effects in the spectrum, which, together with other additive effects such as baseline shift or chemical absorption, can complicate calibration models and detract from the quality of the results of quantitative analyses if not properly accounted for.

These effects can be modeled by ensuring that the calibration set encompasses all possible sources of variability in the samples, which in turn can be ensured by using samples from an adequate number of production batches.[63]

Different production batches of the same pharmaceutical preparation can have very similar concentrations of API ($\pm5\%$ around the nominal value) and excipients close to the nominal value. It is difficult to arrange a set of production samples of the target product spanning the concentration range required to obtain a calibration model that will allow the pharmacopoeial recommendations to be met and different-enough values ($\pm20\%$ around the nominal one) of potency, content uniformity, etc., accommodated. This can be accomplished by using various types of samples, namely:

(a) *Pilot plant samples.* A pilot plant mimicking the actual production plant is used to obtain a set of samples with physical properties similar to those of the production process and containing API at concentrations spanning the required range. This procedure is rather labor-intensive and not always affordable. However, it is the only one that ensures the incorporation of all sources of production variability into the calibration set. As such, it is usually employed to construct models for the analysis of solid samples.

(b) *Overdosed and underdosed production samples.* Properly powdered production samples are supplied with known amounts of the API (overdosing) or a mixture of the excipients (underdosing) in order to obtain the desired range of API concentrations. The amounts of API or excipients added are usually small, so the ensuing changes in the physical properties of the samples are also small and their matrix can be assumed to be essentially identical with that of the production samples. This is an effective, less labor-intensive method than the previous one and allows the required API concentration range to be uniformly spanned.

(c) *Laboratory samples.* These are prepared by weighing known amounts of the API and each excipient, and blending them to homogeneity. This procedure has some advantages. It is less laborious and expensive than the previous two; also, the required sample concentrations are easily obtained. However, laboratory-made samples can differ from actual production samples in some physical characteristics as a result of the differences between the two preparation procedures. This can lead to errors in predicting production samples. This shortcoming, however, can be circumvented by including an adequate number of production samples in the calibration set. This procedure has proved effective in work conducted by the authors' research group for determination of the content of active ingredient in some pharmaceutical preparations.[64,65] In some cases, however, the spectral differences between laboratory-made and production samples are so large that the ensuing models are rather complex and scarcely robust.

The body of samples selected is split into two subsets, namely the calibration set and the validation set. The former is used to construct the calibration model and the latter to assess its predictive capacity. A number of procedures for selecting the samples to be included in each subset have been reported. Most have been applied to situations of uncontrolled variability spanning much wider ranges than those typically encountered in the pharmaceutical field. One especially effective procedure is that involving the selection of as many samples as required to span the desired calibration range and encompassing the whole possible spectral variability (i.e. the contribution of physical properties). The choice relies on a plot of PCA scores obtained from all the samples.

Too small a calibration set may result in some source of variability in the product being excluded and hence in spurious results in analysing new samples. According to some authors, the optimum number of samples to be included depends on their complexity, the concentration range to be spanned and the particular calibration methods used.[66,67] Thus, when the aim is to quantify using 1–4 PLS factors and the samples exhibit relatively small differences in their physical and chemical properties, a calibration set comprising a minimum of 15–20 samples will be more than adequate.

10.3.2   Determination of reference values

In order to construct a calibration model, the values of the parameters to be determined must be obtained by using a reference method. The optimum choice of reference method will be that providing the highest possible accuracy and precision. The quality of the results obtained with a multivariate calibration model can never exceed that of the method used to obtain the reference values, so the choice should be carefully made as the quality of the model will affect every subsequent prediction. The averaging of random errors inherent in regression methods can help construct models with a higher precision than the reference method.

10.3.3 Acquisition of spectra

The spectra for the samples in the above-described set are typically recorded on the same equipment that will subsequently be used for the measurement of the unknown samples. Such spectra will include the variability in the instrument. The essential prerequisites for constructing the calibration model are that the spectrum should contain the information required to predict the property of interest and that the contribution of such a property to the spectrum should be much greater than that of the noise it may contain. Some authors recommend recording the spectra for the calibration samples in a more careful manner than those for the other samples in order to minimize their noise (e.g. by strictly controlling the environmental conditions, averaging an increased number of scans, recording spectra on different days) on the assumption that a better model will result in better predictions – even if the latter spectra are noisier.

The spectra will contain information about the physical properties of the samples that do not contribute to determining ingredient concentrations. Models can be simplified by subjecting the spectra to effective mathematical pretreatments in order to cancel or reduce such contributions. The spectral pretreatments most frequently used for this purpose are smoothing derivatives, multiplicative scatter correction (MSC) and standard normal variate (SNV), which can be implemented via algorithms included in the software bundled with many instruments. Choosing an appropriate pretreatment and the way it is applied are important as not all are equally effective in removing or reducing unwanted contributions. For example, spectral derivatives increase signal/noise ratio, so derivative spectra are frequently smoothed with a suitable algorithm such as that of Savitsky and Golay.

10.3.4 Construction of the calibration model

Although many calibration models use the whole spectrum, using a selected range of wavelengths can facilitate the construction of more simple, accurate and hence robust models. The selection can be made with the aid of various algorithms, the most straightforward and intuitive of which is probably that based on the range where the analyte exhibits its strongest response and hence its greatest contribution to the spectral signal for the sample.

The choice of the calibration method is dictated by the nature of the sample, the number of components to be simultaneously determined, the a priori knowledge about the system studied and available data on it. The most salient features of the different types of calibration methods are described below.

Multiple linear regression is the usual choice with filter instruments and is also used with those that record whole spectra. It is an effective calibration approach when the analytical signal is linearly related to the concentration, spectral noise is low and the analyte does not interact with other sample components. In addition, the MLR technique affords modeling some linear relations as it assumes modeling errors to arise from concentrations. However, it can only be used at a small number of wavelengths, which, if incorrectly selected, may result in overfitting (i.e. in the modeling of noise or random errors). Also, if spectral data are highly collinear, then the precision of the results will suffer appreciably with the use of multiple collinear wavelengths. A detailed description of available procedures for determining how many and which wavelengths should be used can be found elsewhere.[61,68,69]

Whole-spectrum methodologies (viz. PCR and PLSR) have the advantage that they use every single wavelength in a recorded spectrum with no prior selection. Also, they allow the simultaneous determination of several components in the same sample and avoid the problems arising from collinearity among spectral data and noise-related variability. One important decision regarding calibration is the choice of the number of factors to be used in order to avoid overfitting (viz. modeling noise) and the ensuing sizeable errors in the prediction of unknown samples.

Non-linearity in NIR signals is ascribed to non-linear detector responses that result in curved signal–concentration plots, as well as to physical and/or chemical factors giving rise to shifts and width changes in spectral bands.[70,71] In some cases, slight non-linearity can be corrected by including an additional factor in the PLS model; strong non-linearity, however, requires the use of a non-linear calibration methodology such as 'Neural networks,'[72–76] 'Locally weighted regression,'[77,78] 'Projection pursuit regression'[79,80] or quadratic versions of the PCR and PLSR algorithms.[81,82]

10.3.5   Validation of the model

NIR models are validated in order to ensure quality in the analytical results obtained in applying the method developed to samples independent of those used in the calibration process. Although constructing the model involves the use of validation techniques that allow some basic characteristics of the model to be established, a set of samples not employed in the calibration process is required for prediction in order to confirm the goodness of the model. Such samples can be selected from the initial set, and should possess the same properties as those in the calibration set. The quality of the results is assessed in terms of parameters such as the relative standard error of prediction (RSEP) or the root mean square error of prediction (RMSEP).

10.3.6   Prediction of new samples

Once its goodness has been checked, the model can be used to predict production samples. This entails applying it to new samples that are assumed to be similar in nature to those employed to construct the model. This, however, is not always the case as uncontrolled changes in the process can alter the spectra of the samples and lead to spurious values for the target parameter. Instrument-control software typically includes statistical methods for identifying outliers and excluding them prior to quantitation; the use of sample identification and qualification techniques prior to the quantitation solves this problem. One other important consideration is that the model retains its quality characteristics when interpolating, that is when the values to be predicted fall within the concentration range spanned by the calibration curve. In fact, extrapolation is unacceptable in routine analyses.

10.4 Method validation

Every analytical method should be validated in order to ensure that it is clearly and completely understood. This is accomplished by applying appropriate methodologies to determine properly selected parameters. The methodology usually employed to validate analytical methods has been designed for univariate techniques (e.g. HPLC, UV-vis spectrophotometry). Its adaptation to NIR spectroscopy is made difficult by the multivariate nature of calibration. Recently, EMEA,[83] the Pharmaceutical Analytical Sciences Group,[4] ICH[84] and pharmacopoeias have issued guidelines for their specific application to the validation of NIR methods. The quality parameters recommended for the determination of major compounds include specificity, linearity, range, accuracy, repeatability, intermediate precision and robustness. Other parameters such as the detection limit, quantitation limit and reproducibility are unnecessary. In a review, Moffat[85] described and critically compared the characteristics of different validation parameters. The special features of NIR warrant some comment. The greatest difference lies in specificity. As noted earlier, NIR spectrometry is unselective, so it does not necessarily allow the presence of a given analyte in a pharmaceutical preparation to be uniquely identified. The application of a quantitative method to a spectrum can yield spurious results if the samples examined are different from those used to construct the model. Probably the only reliable approach to the demonstration of specificity is the use of both identification and quantitation tests; a pharmaceutical preparation passing the identification test will be of the same type as those used for calibration and hence fit for application of the model. This makes a substantial difference to the methodology used to validate methods based on other techniques.

10.5 Matching models

Constructing a multivariate model is a time-consuming task that involves various operations including the preparation of a sizeable number of samples, recording their spectra and developing and validating the calibration equation. Obtaining an appropriate calibration model therefore involves a substantial investment of time – and of money as a result. In the process, the usual variability in laboratory work (e.g. in the recording of spectra by different operators and/or on different days, in the samples to be analyzed) is incorporated into the model. The incorporation of variability sources makes the model

more robust and avoids adverse effects of typical laboratory variations on the determination of the target parameter.

The model can lose its predictive capacity for reasons such as a dramatic change in
the nature of the samples. A new calibration model better adjusted to the new situation
must therefore be constructed. In other cases, however, the model can be re-adjusted
(re-standardized) in order to ensure a predictive capacity not significantly different from
the original one.

The calibration model may exhibit a decrease in the ability to predict new samples due
to changes in the instrumentation. The differences in instrumentation may introduce
variability in the spectra. The potential sources of variability include the following:

(a) Replacing the analytical equipment with one of similar, better or rather different
    characteristics. If both systems are available simultaneously for an adequate length of
    time, then spectra can be recorded on both in order to derive as much information as
    possible in order to solve the problem. The original equipment is known as the *Master*
    and that replacing it as the *Slave*.
(b) Transferring calibration to another work center. Because the ability to use the Master
    and Slave simultaneously may not exist, the amount of useful spectral information
    may be inadequate.
(c) Breakdown of the instrument or end of its service lifetime. Repairing or replacing an
    instrument may alter spectral responses if the instruments are not perfectly matched.
    Because this is an unexpected event, spectra recorded prior to the incident are rarely
    available.

One plausible solution involves recalibration with the new equipment. This drastic
choice entails ignoring the whole work previously done for this purpose, which is only
acceptable in extreme cases where no alternative solution exists.

A host of mathematical techniques for standardizing calibration models is available,
which focus on the coefficients of the model, the spectral response or the predicted values.[86]

Standardizing the coefficients of the model entails modifying the calibration equation.
This procedure is applicable when the original equipment is replaced (situation (a)
above). Forina *et al.*[87] developed a two-step calibration procedure by which a calibration
model is constructed for the Master (Y–X), its spectral response correlated with that of the
Slave (X–X') and, finally, a global model correlating variable Y with both X and X' is
obtained. The process is optimized in terms of SEP and SEC for both instruments as it
allows the number of PLS factors used to be changed. Smith *et al.*[88] propose a very simple
procedure to match two different spectral responses.

Standardizing the spectral response is mathematically more complex than standardizing
the calibration models but provides better results[89] as it allows slight spectral differences –
the most common between very similar instruments – to be corrected via simple
calculations. More marked differences can be accommodated with more complex and
specific algorithms. This approach compares spectra recorded on different instruments
which are used to derive a mathematical equation, allowing their spectral response to be
mutually correlated. The equation is then used to correct the new spectra recorded on the
Slave, which are thus made more similar to those obtained with the Master. The simplest

methods used in this context are of the univariate type, which correlate each wavelength in two spectra in a direct, simple manner. These methods, however, are only effective with very simple spectral differences. On the other hand, multivariate methods allow the construction of matrices correlating bodies of spectra recorded on different instruments for the above-described purpose. The most frequent choice in this context is piecewise direct standardization (PDS),[90] which is based on the use of moving windows that allow corrections to be applied in specific zones. Other methods[86] used for this purpose include simple wavelength standardization (SWS), multivariate wavelength standardization (MWS) and direct standardization (DS).

Standardizing the predicted values is a simple, useful choice that ensures smooth calibration transfer in situations (a) and (b) (p. 380). The procedure involves predicting samples for which spectra have been recorded on the Slave using the calibration model constructed for the Master. The predicted values, which may be subject to gross errors, are usually highly correlated with the reference values. The ensuing mathematical relation, which is almost always linear, is used to correct the values subsequently obtained with the Slave.

Alternative mathematical methods such as artificial neural networks (ANN), maximum likelihood PCA and positive matrix factorization have also proved effective for calibration transfer, but are much more complex than the previous ones and are beyond the scope of this chapter.

10.6 Pharmaceutical applications

10.6.1 Determination of physical parameters

The NIR spectrum contains information about the physical characteristics of samples, as well as chemical information, which are the basis for the previously discussed applications. This makes it suitable for determining physical parameters.

Particle size determinations are of paramount significance to the pharmaceutical industry and incorrect grain-size analyses can lead to altered properties such as coating power, hinder subsequent mixing of powders (for tablet formulations) or powder and liquids (suspensions), and result in defective pressing of solid mixtures in manufacturing tablets. Because particle size is one of the physical parameters most strongly influencing NIR spectra, it is an effective alternative to the traditional methods involving sieving, light scattering by suspensions, gas adsorption on solid surfaces or direct inspection under a microscope.

Ciurczak *et al.*[91] used the linear relationship of band intensity at a constant concentration on the average particle size at a preset wavelength to determine the average particle size of pure substances and granules. They found plots of absorbance vs. particle size at variable wavelengths to consist of two linear segments, and the effect of particle size on reflectance measurements to be significantly reduced below 80 µm.

Ilari *et al.*[92] explored the feasibility of improving the determination of the average particle size of two highly reflecting inorganic compounds (viz. crystalline and amorphous

NaCl) and an NIR-absorbing species (amorphous sorbitol), using the intercept and slope obtained by subjecting spectra to MSC treatment as input parameters for PLSR.

Blanco *et al.*[93] determined the average particle size of Piracetam® over the average size range of 175–325 μm with an error of ±15 μm. NIR was used for quantification of particle size with the aid of MLR and PLSR calibrations.

While particle size continues to be the physical property most frequently determined by NIR spectroscopy, several other parameters including the dissolution rate and the thickness and hardness of the ethylcellulose coating on theophylline tablets have also been determined using NIR, all with good errors of prediction.[94] Tablet hardness, which dictates mechanical stability and dissolution rate, has been determined in hydrochlorothiazide tablets.[95]

The presence of an unwanted polymorph, even in small proportions, in an API can significantly affect its response by inducing its polymorphic transformation. The determination of polymorphs and *pseudo*-polymorphs, which can be regarded as impurities, is gaining significance, thanks to the fact that NIR can advantageously replace traditional X-ray diffraction methods on account of its high sensitivity and expeditiousness. A number of API polymorphs have been determined using NIR[96–98] or Raman spectroscopy.[99] Blanco *et al.* applied this technique to the determination of the crystalline forms in amorphous antibiotics[100,101] and in tablets.[97]

Some authors have examined polymorphic transformations resulting from hydration of the product[102] or the passage of time.[47,55]

10.6.2 Determination of moisture

Water is one of the most strongly absorbing compounds in the NIR spectral region. This makes it an unsuitable solvent and considerably restricts the determination of species in water. Due to its strong absorbance, water can be determined as an analyte at very low concentrations. Direct inspection of a NIR spectrum allows one to distinguish differences in moisture content among batches of the same product. The presence of crystallization or adsorbed water in pure substances and pharmaceutical preparations, whether during treatment or storage of the sample, causes significant changes in those properties that influence chemical decay rates, crystal dimensions, solubility and compaction power, among others. NIR spectroscopy is an effective alternative to traditional analytical methods such as the Karl Fischer (KF) titration for quantifying moisture. The NIR spectrum for water exhibits five absorption maxima at 760, 970, 1190, 1450 and 1940 nm. The positions of these bands can be slightly shifted by temperature changes[103–105] or hydrogen bonding between the analyte and its matrix.[106,107] The bands at 760, 970 and 1450 nm correspond to the first three overtones of O–H stretching vibrations, whereas the other two are due to combinations of O–H oscillations and stretching. The specific band to be used to determine water will depend on the desired sensitivity and selectivity levels.[108] As a rule, the overtone bands are appropriate for this purpose when using solutions in solvents that do not contain O–H groups; on the other hand, the band at 1940 nm provides increased sensitivity.

Table 10.2 Applications of NIR spectroscopy in pharmaceutical determination

Analyte	Sample type	Calibration method	Remarks	References
Paracetamol	Tablet	PLS	Errors less than 1.5%	64
Rutin, Ascorbic acid	Tablet	MLR	Non-destructive quantitation method. Errors (SEP) less than 0.7%. Method validated using ICH-adapted guidelines	129
Acetylsalicylic acid	Tablet	PLS	Non-destructive quantitation method in three different formulations. Reflectance and transmittance modes compared. Errors less than 3%	130
Otilonium bromide	Powder and tablet	PLS	Non-destructive quantitation method. Errors less than 0.8%	131
Acetylsalicylic acid	Tablet	ANN	Errors less than 2.5%	132
Caffeine	Tablet	PLS	Identification and quantitation in mixtures (foodstuffs, sugar, others). Errors less than 2.3%	133
Analgin	Solid	PLS	Method validated following ICH-adapted guidelines. Errors less than 1.5%	134
Analgin	Solid	PCR, PLS, ANN	Different calibration strategies compared. Errors less than 0.7%	135
Cocaine, Heroin, MDMA	Solid	PLS	Non-destructive quantitation method	136
Miokamycin	Solid	PLS	Validation of the proposed method	137
Mannitol	Solid	PCR, PLS, ANN	Different routine parameters compared. Errors less than 0.15%	138
Sennosides	Senna pods	PLS	Errors less than 0.1%	139
Sennosides	Senna pods	PLS	Transflectance mode used. Method validated using ICH-adapted guidelines	140
Resorcinol	Water solution	MLR	Non-destructive quantitation method. Errors (SEP) less than 0.7%. Method validated using ICH-adapted guidelines	141
Dexketoprofen	Gel	PLS	Non-destructive quantitation method. Errors less than 0.8%	58
Ketoprofen	Gel	PLS	Non-destructive quantitation method in three different formulations. Reflectance and transmittance modes compared. Errors less than 3%	142

The earliest applications for quantitative analysis of liquid samples[109] and solid preparations[107] entailed sample dissolution in an appropriate solvent and using an MLR calibration. MLR calibration models can also be applied in methods based on reflectance measurement of spectra.

A number of moisture determinations in APIs and pharmaceutical preparations based on both reflectance and transmission measurements have been reported.[110–114] Their results are comparable to those of the KF method.

The high sensitivity provided by the NIR technique has fostered its use in the determination of moisture in freeze-dried pharmaceuticals.[115–120] The non-invasive nature of NIR has been exploited in determination of moisture in sealed glass vials.[121–123]

10.6.3   Determination of active pharmaceutical ingredients

The strict regulations of the pharmaceutical industry have a significant effect on the quality control of final products, demanding the use of reliable and fast analytical methods. The capacity that the technique has for the simultaneous determination of several APIs with no need of, or with minimum sample pretreatment, has considerably increased its application in pharmaceutical analytical control. The main limitation of NIR is the relative reduced sensitivity that limits the determination of APIs in preparations when its concentration is smaller than 1%. Nevertheless, instrumental improvements allow the determination below this limit depending on the nature of the analyte and the matrix, with comparable errors to the ones obtained with other instrumental techniques. The reference list presents an ample variety of analytical methodologies, types of samples, nature of analyte and calibration models. A detailed treatment of each one eludes to the extension of this chapter. Many applications have been gathered in recent reviews.[124–128] Table 10.2 summarizes the most recent reported uses of NIR in this context.

References

1. Springsteen, A. and Ricker, T. (1996) Standards for the measurement of diffuse reflectance. Part I; *NIR News* 7, 6.
2. Barnes, P.Y., Early, E.A. and Parr, A.C. (1988) NIST measurement services: Spectral reflectance. *NIST Special Publication*, SP250–48.
3. *European Pharmacopoeia*, 4th Edition (2002) Section 2.2.40: Near infrared spectrophotometry.
4. Broad, N., Graham, P., Hailey, P., Hardy, A., Holland, S., Hughes, S., Lee, D., Prebble, K., Salton, N. and Warren, P. (2002) Guidelines for the development and validation of near-infrared spectroscopic methods in the pharmaceutical industry. In Chalmers, J.M. and Griffiths, P.R. (eds); *Handbook of Vibrational Spectroscopy*, vol. 5; John Wiley & Sons Ltd; Chichester, UK, pp. 3590–3610.
5. Olinger, J.M., Griffiths, P.R. and Burger, T. (2001) Theory of diffuse reflectance in the NIR region. In Burns, D. and Ciurczak, E.W. (eds); *Handbook of Near-Infrared Analysis*, 2nd Edition; Marcel Dekker; New York, pp. 19–52.

6. Blanco, M., Coello, J., Iturriaga, H., Maspoch, S., Pezuela, C. and Russo, E. (1994) Control analysis of a pharmaceutical preparation by near-infrared reflectance spectroscopy: A comparative study of a spinning module and fibre optic probe; *Anal. Chim. Acta* **298**, 183–191.

7. Mark, H. (2001) Qualitative discriminant analysis. In Burns, D. and Ciurczak, E.W. (eds); *Handbook of Near-Infrared Analysis*, 2nd Edition; Marcel Dekker; New York, pp. 351–362.

8. Bratchell, N. (1989) Cluster analysis; *Chemom. Intell. Lab. Syst.* **6**, 105–125.

9. Strouf, O. (1986) *Chemical Pattern Recognition*; Research Studies Press; Letchworth.

10. Cáceres-Alonso, P. and García-Tejedor, A. (1995) Non-supervised neural categorisation of near infrared spectra. Application to pure compounds; *J. Near Infrared Spectrosc.* **3**, 97–110.

11. Derde, M.P. and Massart, D.L. (1986) Supervised pattern recognition: The ideal method? *Anal. Chim. Acta* **191**, 1–16.

12. Lachenbruch, P.A. (1975) *Discriminant Analysis*; Hafner Press; New York.

13. Coomans, D. and Massart, D.L. (1982) Alternative K-nearest neighbour rules in supervised pattern recognition. Part 1: K-nearest neighbour classification by using alternative voting rules; *Anal. Chim. Acta* **136**, 15–27.

14. Coomans, D. and Massart, D.L. (1982) Alternative K-nearest neighbour rules in supervised pattern recognition. Part 2: Probabilistic classification on the basis of the kNN method modified for direct density estimation; *Anal. Chim. Acta* **138**, 153–165.

15. Coomans, D. and Broeckaert, I. (1986) *Potential Pattern Recognition*; Wiley; New York.

16. Coomans, D., Derde, M.P., Massart, D.L. and Broeckaert, I. (1981) Potential methods in pattern recognition. Part 3: Feature selection with ALLOC; *Anal. Chim. Acta* **133**, 241–250.

17. Jurickskay, I. and Veress, G.E. (1981) PRIMA: A new pattern recognition method; *Anal. Chim. Acta* **171**, 61–76.

18. Derde, M.P. and Massart, D.L. (1986) UNEQ: A disjoint modelling technique for pattern recognition based on normal distribution; *Anal. Chim. Acta* **184**, 33–51.

19. VISION User Manual (1998) Foss NIRSystems, Silver Spring, Maryland, USA.

20. Gemperline, P.J. and Webber, L.D. (1989) Raw materials testing using soft independent modeling of class analogy analysis of near infrared reflectance spectra; *Anal. Chem.* **61**, 138–144.

21. Bertran, E., Blanco, M., Coello, J., Iturriaga, H., Maspoch, S. and Montoliu, I.J. (2000) Near infrared spectrometry and pattern recognition as screening methods for the authentication of virgin olive oils of very close geographical origins; *J. Near Infrared Spectrosc.* **8**, 45–52.

22. Shah, N.K. and Gemperline, P.J. (1990) Combination of the Mahalanobis distance and residual variance pattern recognition techniques for classification of near-infrared reflectance spectra; *Anal. Chem.* **62**, 465–470.

23. Workman, J.J., Jr, Mobley, P.R., Kowalski, B.R. and Bro, R. (1996) Review of chemometrics applied to spectroscopy: 1985–1995, Part I; *Appl. Spectrosc. Rev.* **31**, 73–124.

24. Smola, N. and Urleb, U. (1999) Near infrared spectroscopy applied to quality control of incoming material in the pharmaceutical industry; *Farmacevtski Vestnik* **50**, 296–297.

25. Scafi, S.H. and Pasquini, C. (2001) Identification of counterfeit drugs using near-infrared spectroscopy; *Analyst* **126**, 2218–2224.

26. Ulmschneider, M., Barth, G. and Trenka, E. (2000) Building transferable cluster calibrations for the identification of different solid excipients with near-infrared spectroscopy; *Drugs* **43**, 71–73.

27. Ren, Y., Li, W., Guo, Y., Ren, R., Zhang, L., Jin, D. and Hui, C. (1997) Study on quality control of metronidazole powder pharmaceuticals using near infrared reflectance first-derivative spectroscopy and multivariate statistical classification technique; *Jisuanji Yu Yingyong Huaxue* **14**, 105–109.

28. Yoon, W.L., North, N.C., Jee, R.D. and Moffat, A.C. (2000) Application of a polar qualification system in the near infrared identification and qualification of raw pharmaceutical excipients. In

Davies, A.M.C. and Giangiacomo, R. (eds); *Near Infrared Spectroscopy, Proceedings of the International Conference, 9th*, Verona, Italy, June 13–18, 1999; NIR Publications; Chichester, UK, pp. 547–550.

29. Blanco, M. and Romero, M.A. (2001) Near-infrared libraries in the pharmaceutical industry: A solution for identity confirmation; *Analyst* **126**, 2212–2217.

30. Hackmann, E.R.M.K., De Abreu, E.M.C. and Santoro, M.I.R.M. (1999) Corn starch identification by near infrared spectroscopy; *Revista Brasileira de Ciencias Farmaceuticas* **35**, 141–146.

31. Woo, Y., Kim, H. and Cho, J. (1999) Identification of herbal medicines using pattern recognition techniques with near-infrared reflectance spectra; *Microchem. J.* **63**, 61–70.

32. Hernandez Baltazar, E. and Rebollar, B.G. (2002) Development of identification method by near-infrared spectroscopy: Acetaminophen tablets; *Revista Mexicana de Ciencias Farmaceuticas* **33**, 42–47.

33. Candolfi, A., De Maesschalck, R., Massart, D.L., Hailey, P.A. and Harrington, A.C.E. (1999) Identification of pharmaceutical excipients using NIR spectroscopy and SIMCA; *J. Pharm. Biomed. Anal.* **19**, 923–935.

34. Khan, P.R., Jee, R.D., Watt, R.A. and Moffat, A.C. (1997) The identification of active drugs in tablets using near infrared spectroscopy; *J. Pharm. Sci.* **3**, 447–453.

35. Uhl, V.W. and Gray, J.B. (1967) *Mixing: Theory and Practice*; Academic Press; New York.

36. Schirmer, R.E. (1991) *Modern Methods of Pharmaceutical Analysis*, vol. 2, 2nd Edition; CRC Press; Boca Raton; pp. 31–126.

37. Jensen, R., Peuchant, E., Castagne, I., Boirac, A.M. and Roux, G. (1988) One-step quantification of active ingredient in pharmaceutical tablets using near-infrared spectroscopy; *Spectrosc.* **6(2)**, 63–72.

38. Lonardi, S., Viviani, R., Mosconi, L., Bernuzzi, M., Corti, P., Dreassi, P., Murrazzu, C. and Corbin, G.J. (1989) Drug analysis by near-infrared reflectance spectroscopy. Determination of the active ingredient and water content in antibiotic powders; *J. Pharm. Biomed. Anal.* **7**, 303–308.

39. Dreassi, E., Ceramelli, G., Corti, P., Lonardi, P.S. and Perruccio, P. (1995) Near-infrared reflectance spectrometry in the determination of the physical state of primary materials in pharmaceutical production; *Analyst* **120**, 1005–1008.

40. Reeves, J.B. (1995) Near- versus mid-infrared spectroscopy: Relationships between spectral changes induced by water and relative information content of the two spectral regions in regard to high-moisture samples; *Appl. Spectrosc.* **49**, 295–903.

41. Plugge, W. and Van der Vlies, C.J. (1993) Near-infrared spectroscopy as an alternative to assess compliance of ampicillin trihydrate with compendial specifications; *J. Pharm. Biomed. Anal.* **11**, 435–442.

42. Corti, P., Dreassi, E., Ceramelli, G., Lonardi, S., Viviani, R. and Gravina, S. (1991) Near Infrared Reflectance Spectroscopy applied to pharmaceutical quality control. Identification and assay of cephalosporins; *Analusis* **19**, 198–204.

43. Drennen, J.K. and Lodder, R.A. (1993) Pharmaceutical applications of near-infrared spectrometry. In Patonay, G. (ed.); *Advances in Near Infrared Measurements*, vol. 1; JAI Press; Greenwich, CT, pp. 93–112.

44. Ciurczak, E.W. (1991) Following the progress of a pharmaceutical mixing study via near-infrared spectroscopy; *Pharm. Tech.* **15**, 140.

45. Wargo, D.J. and Drennen, J.K. (1996) Near-infrared spectroscopic characterization of pharmaceutical powder blends; *J. Pharm. Biomed. Anal.* **14**, 1415–1423.

46. Cuesta, F., Toft, J., Van den Bogaert, B., Massart, D.L., Dive, S.S. and Hailey, P. (1995) Monitoring powder blending by NIR spectroscopy; *Fresenius J. Anal. Chem.* **352**, 771–778.

47. Van der Vlies, C., Kaffka, K.J. and Plugge, W. (1995) Qualifying pharmaceutical substances by fingerprinting with near-IR spectroscopy and the polar qualification system; *Pharm. Technol. Eur.* **7**, 46–49.

48. Hailey, P.A., Doherty, P., Tapsell, P., Oliver, T. and Aldridge, P.K. (1996) Automated system for the on-line monitoring of powder blending processes using near-infrared spectroscopy. Part I: System development and control; *J. Pharm. Biomed. Anal.* **14**, 551–559.

49. Sekulic, S.S., Ward, H.W., Brannegan, D.R., Stanley, E.D., Evans, C.L., Sciavolino, S.T., Hailey, P.A. and Aldridge, P.K. (1996) On-line monitoring of powder blend homogeneity by near-infrared spectroscopy; *Anal. Chem.* **68**, 509–513.

50. Sekulic, S.S., Wakeman, J., Doherty, P. and Hailey, P.A. (1998) Automated system for the online monitoring of powder blending processes using near-infrared spectroscopy. Part II: Qualitative approaches to blend evaluation; *J. Pharm. Biomed. Anal.* **17**, 1285–1309.

51. Blanco, M., Gonzalez, R. and Bertran, E. (2002) Monitoring powder blending in pharmaceutical processes by use of near infrared spectroscopy; *Talanta* **56**, 203–212.

52. Ciurczak, E.W. (1987) Use of near infrared in pharmaceutical analyses; *Appl. Spectrosc. Rev.* **23**, 147.

53. Osborne, B.G. and Fearn, T. (1986) *Near Infrared Spectroscopy in Food Analysis*; Wiley; New York, 1986.

54. Aldridge, P.K., Evans, C.L., Ward, H.W., Colgan, S.T., Boyer, N. and Gemperline, P.J. (1996) Near-IR detection of polymorphism and process-related substances; *Anal. Chem.* **68**, 997–1002.

55. Norris, T., Aldridge, P.K. and Sekulic, S.S. (1997) Determination of end-points for polymorph conversions of crystalline organic compounds using on-line near-infrared spectroscopy; *Analyst* **122**, 549–552.

56. Gimet, R. and Luong, A.T. (1987) Quantitative determination of polymorphic forms in a formulation matrix using the near infra-red reflectance analysis technique; *J. Pharm. Biomed. Anal.* **5**, 205–211.

57. Buckton, G., Yonemochi, E., Hammond, J. and Moffat, A. (1998) The use of near infra-red spectroscopy to detect changes in the form of amorphous and crystalline lactose; *Int. J. Pharm.* **168**, 231–241.

58. Blanco, M. and Romero, M.A. (2002) Near infrared transflectance spectroscopy. Determination of dexketoprofen in a hydrogel; *J. Pharm. Biomed. Anal.* **30**, 467–472.

59. Thomas, E.V. and Haaland, D.M. (1990) Comparison of multivariate calibration methods for quantitative spectral analysis; *Anal. Chem.* **62**, 1091–1099.

60. Draper, N. and Smith, H. (1981) *Applied Regression Analysis*, 2nd Edition; Wiley; New York.

61. Martens, H. and Næs, T. (1991) *Multivariate Calibration*; Wiley; New York.

62. ASTM E1655-00. Standard practices for infrared, multivariate, quantitative analysis; In: Annual Book of Standards, vol. 3.06, American Society for Testing and Materials; West Conshohocken, PA, 2000.

63. Blanco, M., Coello, J., Iturriaga, H., Maspoch, S. and Pezuela, C. (1997) Strategies for constructing the calibration set in the determination of active principles in pharmaceuticals by near infrared diffuse reflectance spectrometry; *Analyst* **122**, 761–765.

64. Trafford, D., Jee, R.D., Moffat, A.C. and Graham, P. (1999) A rapid quantitative assay of intact paracetamol tablets by reflectance near-infrared spectroscopy; *Analyst* **124**, 163–167.

65. Corti, P., Ceramelli, G., Dreassi, E. and Mattii, S. (1999) Near infrared transmittance analysis for the assay of solid pharmaceutical dosage forms; *Analyst* **124**, 755–758.

66. Corti, P., Dreassi, E., Murratzu, C., Corbini, G., Ballerini, L. and Gravina, S. (1989) Application of NIRS to the quality control of pharmaceuticals. Ketoprofen assay in different pharmaceutical formulae; *Pharmaceutica Acta Helvetiae* **64**, 140–145.

67. Buchanan, B. and Honigs, D. (1986) Trends in near-infrared analysis; *Trends Anal. Chem.* **5**, 154–157.

68. Honigs, D.E., Hieftje, G.M. and Hirschfeld, T. (1984) A new method for obtaining individual component spectra from those of complex mixtures; *Appl. Spectrosc.* **38**, 317–322.

69. Cowe, I.A., McNicol, J.W. and Cuthbertson, D.C. (1985) A designed experiment for the examination of techniques used in the analysis of near-infrared spectra. Part 2: Derivation and testing of regression models; *Analyst* **110**, 1233–1240.

70. Næs, T. and Isaksson, T. (1993) Non-linearity problems in NIR spectroscopy; *NIR News* **4**(3), 14.

71. Miller, C.E. (1993) Sources of non-linearity in near infrared methods; *NIR News* **4**(6), 3.

72. Long, R.L., Gregoriov, V.G. and Gemperline, P.J. (1990) Spectroscopic calibration and quantitation using artificial neural networks; *Anal. Chem.* **62**, 1791–1797.

73. Gemperline, P.J., Long, J.R. and Gregoriov, V.G. (1991) Nonlinear multivariate calibration using principal components regression and artificial neural networks; *Anal. Chem.* **63**, 2313–2323.

74. Næs, T., Kvaal, K., Isaksson, T. and Miller, C. (1993) Artificial neural networks in multivariate calibration; *J. Near Infrared Spectrosc.* **1**, 1–11.

75. Walczak, B. and Wegscheimer, W. (1993) Non-linear modelling of chemical data by combinations of linear and neural networks methods; *Anal. Chim. Acta* **283**, 508–517.

76. Liu, Y., Upadhyaya, B.R. and Naghedolfeizi, M. (1993) Chemometric data analysis using artificial neural networks; *Appl. Spectrosc.* **47**, 12–23.

77. Næs, T., Isaksson, T. and Kowalski, B.R. (1990) Locally weighted regression and scatter correction for near-infrared reflectance data; *Anal. Chem.* **62**, 664–673.

78. Isaksson, T., Miller, C.E. and Næs, T. (1992) Nondestructive NIR and NIT determination of protein, fat, and water in plastic-wrapped, homogenized meat; *Appl. Spectrosc.* **46**, 1685–1694.

79. Friedman, J.H. and Stuetzle, W. (1981) Projection pursuit regression; *J. Am. Stat. Assoc.* **76**, 817–823.

80. Beebe, K.R. and Kowalski, B.R. (1988) Nonlinear calibration using projection pursuit regression: Application to an array of ion-selective electrodes; *Anal. Chem.* **60**, 2273–2278.

81. Vogt, N.B. (1989) Polynomial principal component regression: An approach to analysis and interpretation of complex mixture relationships in multivariate environmental data; *Chemom. Intell. Lab. Syst.* **7**, 119–130.

82. Wold, S., Wold, N.K. and Skagerber, B. (1989) Nonlinear PLS modelling; *Chemom. Intell. Lab. Syst.* **7**, 53–65.

83. http://www.emea.eu.int/pdfs/human/qwp/330901en.pdf. Note for guidance. The European Agency for the Evaluation of Medicinal Products, 2003.

84. ICH Guidelines: Q2A Text on Validation of Analytical Procedures 1994, Q2B Validation of Analytical Procedures: Methodology 1996. International Conference on Harmonisation of Technical Requirements for Registration of Pharmaceuticals for Human Use.

85. Moffat, A.C., Trafford, A.D., Jee, R.D. and Graham, P. (2000) Meeting of the international conference on harmonisation's guidelines on validation of analytical procedures: Quantification as exemplified by a near-infrared reflectance assay of paracetamol in intact tablets; *Analyst* **125**, 1341–1351.

86. Feudale, R.N., Woody, H.T., Myles, A.J., Brown, S.D. and Ferre, J. (2002) Transfer of multivariate calibration models: A review; *Chem. Intell. Lab. Syst.* **64**, 181–192.

87. Forina, M., Drava, G., Armanino, C., Boggia, R., Lanteri, S., Leardi, R., Corti, P., Conti, P., Giangiacomo, R., Galliena, C., Bigoni, R., Quartari, I., Serra, C., Ferri, D., Leoni, O. and Lazzeri, L. (1995) Transfer of calibration function in near-infrared spectroscopy; *Chem. Intell. Lab. Syst.* **27**, 189–203.

88. Smith, M.R., Jee, R.D. and Moffat, A.C. (2002) The transfer between instruments of a reflectance near-infrared assay for paracetamol in intact tablets; *Analyst* **127**, 1682–1692.

89. Wang, Y., Veltkamp, D.J. and Kowalski, B.R. (1991) Multivariate instrument standardization; *Anal. Chem.* **63**, 2750–2756.

90. Bouveresse, E., Hartmann, C., Massart, D.L., Last, I.R. and Prebble, K.A. (1996) Standardization of near-infrared spectrometric instruments; *Anal. Chem.* **68**, 982–990.

91. Ciurczak, E.W., Torlini, R.P. and Demkowicz, M.P. (1986) Determination of particle size of pharmaceutical raw materials using near-infrared reflectance spectroscopy; *Spectroscopy* **1**(7), 36.

92. Ilari, J.L., Martens, H. and Isaksson, T. (1988) Determination of particle size in powders by scatter correction in diffuse near-infrared reflectance; *Appl. Spectrosc.* **42**, 722–728.

93. Blanco, M., Coello, J., Iturriaga, H., Maspoch, S., González, F. and Pous, R. Near infrared spectroscopy. In Hildrum, K.I., Isaksson, T., Næs, T. and Tandberg, A. (eds); *Bridging the Gap between Data Analysis and NIR Applications*; Ellis Horwood; Chichester, 1992; pp. 401–406.

94. Kirsch, J.D. and Drennen, J.K. (1995) Determination of film-coated tablet parameters by near-infrared spectroscopy; *J. Pharm. Biomed. Anal.* **13**, 1273–1281.

95. Morisseau, K.M. and Rhodes, C.T. (1997) Near-infrared spectroscopy as a nondestructive alternative to conventional tablet hardness testing; *Pharm. Res.* **14**, 108–111.

96. Luner, P.E., Majuru, S., Seyer, J.J. and Kemper, M.S. (2000) Quantifying crystalline form composition in binary powder mixtures using near-infrared reflectance spectroscopy; *Pharm. Dev. Tech.* **5**, 231–246.

97. Otsuka, M., Kato, F., Matsuda, Y. and Ozaki, Y. (2003) Comparative determination of polymorphs of indomethacin in powders and tablets by chemometrical near-infrared spectroscopy and X-ray powder diffractometry; *AAPS PharmSciTech.* **4**, 147–158.

98. Patel, A.D., Luner, P.E. and Kemper, M.S. (2000) Quantitative analysis of polymorphs in binary and multi-component powder mixtures by near-infrared reflectance spectroscopy; *Int. J. Pharm.* **206**, 63–74. Erratum in: *Int. J. Pharm.* **212**, 295.

99. Deeley, C.M., Spragg, R.A. and Threlfall, T.L.A. (1991) Comparison of Fourier transform infrared and near-infrared Fourier transform Raman spectroscopy for quantitative measurements: An application; *Spectrochim. Acta, Part A* **47A**, 1217–1223.

100. Blanco, M., Coello, J., Iturriaga, H., Maspoch, S. and Perez-Maseda, C. (2000) Determination of polymorphic purity by near infrared spectrometry; *Anal. Chim. Acta* **407**, 247–254.

101. Blanco, M., Valdes, D., Bayod, M.S., Fernandez-Mari, F. and Llorente, I. (2004) Characterization and analysis of polymorphs by near-infrared spectrometry; *Anal. Chim. Acta* **502**, 221–227.

102. Higgins, J.P., Arrivo, S.M. and Reed, R. (2003) Approach to the determination of hydrate form conversions of drug compounds and solid dosage forms by near-infrared spectroscopy; *J. Pharm. Sci.* **92**, 2303–2316.

103. Lin, L. and Brown, C.W. (1992) Near-IR spectroscopic determination of sodium chloride in aqueous solution; *Appl. Spectrosc.* **46**, 1809–1815.

104. Delwiche, S.R., Norris, K.H. and Pitt, R.E. (1992) Temperature sensitivity of near-infrared scattering transmittance spectra of water-adsorbed starch and cellulose; *Appl. Spectrosc.* **46**, 782–789.

105. Lin, J. and Brown, C.W. (1993) Near-IR fiber-optic temperature sensor; *Appl. Spectrosc.* **47**, 62–68.

106. Sinsheimer, J.E. and Poswalk, N.M. (1968) Pharmaceutical applications of the near infrared determination of water; *J. Pharm. Sci.* **57**, 2007–2010.

107. Issa, R.M., El-Marsafy, K.M. and Gohar, M.M. (1988) Application of the near infrared spectrophotometry as an analytical procedure for the determination of water in organic compounds and pharmaceutical products; *An. Quim.* **84**, 312–315.

108. Böhme, W., Liekmeier, W., Horn, K. and Wilhelm, C. (1990) Water determination by near infrared spectroscopy; *Labor Praxis* **14**, 86–89.

109. Ludvik, J., Hilgard, S. and Volke, J. (1988) Determination of water in acetonitrile, propionitrile, dimethylformamide and tetrahydrofuran by infrared and near-infrared spectrometry; *Analyst* **113**, 1729–1731.

110. Keutel, H., Gobel, J. and Mehta, V. (2003) NIR diode array spectrometer in pharmaceutical production control; *Chemie Technik* **32**, 48–50.

111. Leasure, R.M. and Gangwer, M.K. (2002) Near-infrared spectroscopy for in-process moisture determination of a potent active pharmaceutical ingredient; *Am. Pharm. Rev.* **5**(1), 103–104, 106, 108–109.

112. Dunko, A. and Dovletoglou, A. (2002) Moisture assay of an antifungal by near-infrared diffuse reflectance spectroscopy; *J. Pharm. Biomed. Anal.* **28**, 145–154.

113. David, A.Z., Antal, I., Acs, Z., Gal, L. and Greskovits, D. (2000) Investigation of water diffusion in piracetam by microwave moisture measurement and near-infrared spectroscopy; *Hung. J. Ind. Chem.* **28**, 267–270.

114. Zhou, X., Hines, P. and Borer, M.W. (1998) Moisture determination in hygroscopic drug substances by near infrared spectroscopy; *J. Pharm. Biomed. Anal.* **17**, 219–225.

115. Stokvold, A., Dyrstad, K. and Libnau, F.O. (2002) Sensitive NIRS measurement of increased moisture in stored hygroscopic freeze dried product; *J. Pharm. Biomed. Anal.* **28**, 867–873.

116. Birrer, G.A., Liu, J., Halas, J.M. and Nucera, G.G. (2000) Evaluation of a container closure integrity test model using visual inspection with confirmation by near infrared spectroscopic analysis; *J. Pharm. Sci. Tech.* **54**, 373–382.

117. Buhler, U., Maier, E. and Muller, M. (1998) Determination of the water content in lyophilizates with NIR; *J. Pharm. Tech.* **19**, 49–55.

118. Savage, M., Torres, J., Franks, L., Masecar, B. and Hotta, J. (1998) Determination of adequate moisture content for efficient dry-heat viral inactivation in lyophilized factor VIII by loss on drying and by near infrared; *Biologicals* **26**, 119–124.

119. Han, S.M. and Faulkner, P.G. (1996) Determination of SB 216469-S during tablet production using near-infrared reflectance spectroscopy; *J. Pharm. Biomed. Anal.* **14**, 1681–1689.

120. Derksen, M.W.J., Van De Oetelaar, P.J.M. and Maris, F.A. (1998) The use of near-infrared spectroscopy in the efficient prediction of a specification for the residual moisture content of a freeze-dried product; *J. Pharm. Biomed. Anal.* **17**, 473–480.

121. Moy, T., Calabria, A. and Hsu, C. (2001) Near infrared: A non-invasive method to determine residual moisture of lyophilized protein pharmaceuticals; *American Chemical Society 221st National Meeting*, BIOT-016. Meeting abstract.

122. Jones, J.A., Last, I.R., MacDonald, B.F. and Prebble, K.A. (1993) Development and transferability of near-infrared methods for determination of moisture in a freeze-dried injection product; *J. Pharm. Biomed. Anal.* **11**, 1227–1231.

123. Kamat, M.S., Lodder, R.A. and DeLuca, P.P. (1989) Near-infrared spectroscopic determination of residual moisture in lyophilized sucrose through intact glass vials; *Pharm. Res.* **6**, 961–965.

124. Romanach, R.J. and Santos, M.A. (2003) Content uniformity testing with near infrared spectroscopy; *Am. Pharm. Rev.* **6**(2), 62, 64–67.

125. Lyon, R.C., Jefferson, E.H., Ellison, C.D., Buhse, L.F., Spencer, J.A., Nasr, M. and Hussain, A.S. (2003) Exploring pharmaceutical applications of near-infrared technology; *Am. Pharm. Rev.* **6**(3), 62, 64–66, 68–70.

126. Ciurczak, E.W. and Drennen, J.K. Pharmaceutical applications of near-infrared-spectroscopy. In Raghavachari, R. (ed.); *Near-Infrared Applications in Biotechnology*; Marcel Dekker Inc.; New York, 2001; pp. 349–366.

127. Blanco, M., Coello, J., Iturriaga, H., Maspoch, S. and Pezuela, C. (1998) Near-infrared spectroscopy in the pharmaceutical industry; *Analyst* **123**, 135–150.

128. Ciurczak, E.W. and Drennen, J.K. Near-infrared spectroscopy in pharmaceutical and biomedical applications. In Burns, D. and Ciurczak, E.W. (eds); *Handbook of Near-Infrared Analysis*, 2nd Edition; Marcel Dekker, Inc.; New York; 2001; pp. 609–632.

129. Sun, S., Du, D., Zhou, Q., Leung, H.W. and Yeung, H.W. (2001) Quantitative analysis of rutin and ascorbic acid in compound rutin tablets by near-infrared spectroscopy; *Anal. Sci.* **17**, 455–458.

130. Tian, L., Tang, Z., Liu, F., Gou, Y., Guo, Y. and Ren, Y. (2001) Nondestructive quantitative analysis of aspirin by near infrared spectroscopy and artificial neural network; *Fenxi Shiyanshi* **20**(1), 79–81.

131. Blanco, M., Coello, J., Iturriaga, H., Maspoch, S. and Pou, N. (2000) Development and validation of a near infrared method for the analytical control of a pharmaceutical preparation in three steps of the manufacturing process; *Fres. J. Anal. Chem.* **368**, 534–539.

132. Merckle, P. and Kovar, K.A. (1998) Assay of effervescent tablets by near-infrared spectroscopy in transmittance and reflectance mode: Acetylsalicylic acid in mono and combination formulations; *J. Pharm. Biomed. Anal.* **17**, 365–374.

133. Laasonen, M., Harmia-Pulkkinen, T., Simard, C., Raesaenen, M. and Vuorela, H. (2003) Development and validation of a near-infrared method for the quantitation of caffeine in intact single tablets; *Anal. Chem.* **75**, 754–760.

134. Ren, Y., Li, W., Guo, Y., Zhang, L., Sun, Y. and Chen, B. (1997) Noninvasive analysis of analgin by PLS-near-infrared diffuse reflectance spectroscopy; *Jilin Daxue Ziran Kexue Xuebao* **3**, 99–102.

135. Ren, Y., Gou, Y., Tang, Z., Liu, P. and Guo, Y. (2000) Nondestructive quantitative analysis of analgin powder pharmaceutical by near-infrared spectroscopy and artificial neural network technique; *Anal. Lett.* **33**, 69–80.

136. Ryder, A.G., O'Connor, G.M. and Glynn, T.J. (1999) Identifications and quantitative measurements of narcotics in solid mixtures using near-IR Raman spectroscopy and multivariate analysis; *J. Forensic Sci.* **44**, 1013–1019.

137. Blanco, M., Coello, J., Eustaquio, A., Itturriaga, H. and Maspoch, S. (1999) Development and validation of methods for the determination of miokamycin in various pharmaceutical preparations by use of near infrared reflectance spectroscopy; *Analyst* **124**, 1089–1092.

138. Yang, N., Cheng, Y. and Qu, H. (2003) Quantitative determination of mannitol in Cordyceps sinensis using near infrared spectroscopy and artificial neural networks; *Fenxi Huaxue* **31**, 664–668.

139. Pudelko-Koerner, C., Fischer, A., Lentzen, H., Glombitza, K.W. and Madaus, A.G. (1996) Quantitative Fourier transform-near infrared reflectance spectroscopy of sennosides in Senna pods; *Pharm. Pharmacol. Lett.* **6**, 34–36.

140. Pudelko-Koerner, C. (1998) Quantitative near-infrared reflectance spectroscopy of sennosides from Sennae fructus angustifoliae in in-process and quality control including method validation; *Pharmazeutische Industrie* **60**, 1007–1012.

141. Cinier, R. and Guilment, J. (1996) Quantitative analysis of resorcinol in aqueous solution by near-infrared spectroscopy; *Vib. Spectrosc.* **11**, 51–59.

142. Kemper, M.S., Magnuson, E.J., Lowry, S.R., McCarthy, W.J., Aksornkoae, N., Watts, D.C., Johnson, J.R. and Shukla, A.J. (2001) Use of FT-NIR transmission spectroscopy for the quantitative analysis of an active ingredient in a translucent pharmaceutical topical gel formulation; *Pharm. Sci.* **3**(3), article 23.

Chapter 11
Applications of Near-Infrared Spectroscopy (NIR) in the Chemical Industry

Ann M. Brearley

11.1 Introduction

The drive to improve asset productivity, lower manufacturing costs, and improve product quality, which has been occurring throughout the chemical industry for the last several decades, has led to increased focus on process understanding, process improvement and the use of advanced process control. A key aspect of many of these efforts is improvement in the quality of the analytical data used to control the process or release product. Process analyzers are increasingly being used in order to improve the timeliness, frequency, or precision of analytical data. Multivariate spectroscopic analyzers (NIR, IR, Raman, etc.) can be used to monitor the chemical composition of a process stream non-invasively, in near real time, and are available commercially from numerous vendors. However, even after several decades of work, the technology is still relatively new to the chemical industry and there remains a steep learning curve for any given installation. A major reason for this is that process analytical chemistry and chemometrics are not generally part of the standard educational curriculum for an analytical chemist, or for a process chemist or process engineer, or even for an analyzer engineer (the latter traditionally focusing on univariate sensors for pressure, temperature, flow, level, pH, etc.).

The objective of this chapter is to reduce the learning curve for the application of near-infrared (NIR) spectroscopy, or indeed any process analytical technology, to the chemical industry. It attempts to communicate realistic expectations for process analyzers in order to minimize both unrealistically positive expectations (e.g. 'NIR can do everything and predict the stock market!') and unrealistically negative expectations (e.g. 'NIR never works; don't waste your money'). The themes of this chapter are value and challenge; specifically business value and technical challenge, in that order. Realistic, thorough assessment of the business value to be gained by a process analyzer installation is critical to the success of the project. This assessment should be the very first step of the project since it serves to guide, motivate and prioritize all subsequent work. Technical (and non-technical) challenges are part of any process analyzer installation. If they are not properly addressed, they can become a huge sink for time and money and even lead to failure of the project. The discussion and examples in this chapter are designed to help potential users of NIR analyzers anticipate and be prepared for the common challenges. NIR analyzers have enabled significant improvement of many industrial processes, and can do the same for many others, but only if they are implemented for the right (business) reasons by people with the knowledge and experience to properly address the challenges.

The first part of this chapter (Section 11.2) discusses the value of and challenges involved in implementing NIR technology in the chemical industry, and describes a process for successful process analyzer implementation. The second part of the chapter (Section 11.3) gives a number of examples of actual NIR applications in the chemical industry, which serve both to illustrate the points from the first part and to demonstrate the range of applications for NIR technology.

11.2   Successful process analyzer implementation

The focus of this part of the chapter is on how to 'do it right.' Implementation of a process analyzer (whether NIR or other technology) in a chemical manufacturing plant requires a large upfront investment of time, effort, and money. Analyzer implementations are complex projects, both technically and organizationally, requiring the coordination of resources and people across multiple organizational, functional, and geographic boundaries. If the implementation is done correctly, the analyzer saves the business very large amounts of money and enables the plant to improve quality to a previously unattainable degree. If the implementation is not done correctly, the business loses its investment and acquires a lasting bad opinion of process analyzers, and the people involved in the project get black marks on their records. This is clearly undesirable. There are many factors that increase the chances of failure of an analyzer implementation, and these are discussed in Section 11.2.5. However, it is useful to keep the primary emphasis on success rather than failure and focus on what the process analytical chemist needs to do to ensure a successful implementation.

Section 11.2.1 describes a process for 'doing it right.' It is written broadly enough to apply to process analyzers in general, not just NIR analyzers; it can also be applied to the development of new methods using existing analyzer hardware (e.g. a new method or calibration model to monitor an additional analyte in the same process stream). After the overview of the guidelines, there is more detailed discussion in Sections 11.2.2–11.2.7 on what needs to be considered at the various steps.

11.2.1   A process for successful process analyzer implementation

The first issue that should be addressed when a process analyzer project is proposed concerns roles: who will be doing what. There are four basic roles in an analyzer project: the customer(s), the sponsor, the project manager, and the receiver. These are described below.

(1) *Customer(s)*: The person(s) who will use the information from the proposed new method or analyzer. This could be a control engineer interested in improving control of the manufacturing process (tuning control loops, understanding dynamics, etc.), a process

engineer interested in improving process yield or up-time (shortening startups or transitions, shortening batches), an R&D scientist interested in improving understanding of the process, or a QA/QC (quality control) person interested in speeding up product testing and release. The customer must understand and be able to quantify the business opportunity or need.

(2) *Sponsor*: The person who approves and pays for the project. This is generally a management-level person such as a plant manager or technical director, often in the line management of the customer.

(3) *Project manager*: The person responsible for development, installation, and validation of the new method or analyzer. This person is often a process analytical chemist, an analytical chemist, or an analyzer engineer. Sometimes the technical aspects of the role may be contracted out to a consultant or vendor: in this case the project manager may be a process engineer or lab chemist; nevertheless s/he still has the responsibility for the success of the project. The project manager coordinates the efforts of people across organizational, functional, and often geographic lines, including vendors, analytical personnel, R&D people, operations people (engineers, chemists, process technicians), crafts people (electricians, etc.), QA/QC personnel, etc. S/he is responsible for communicating effectively so that all people involved in the project (including the sponsor and the customer) are kept informed of the project plan and their part in it.

(4) *Receiver*: The person responsible for ensuring that the new method or analyzer is used and maintained correctly on site, once it is installed and validated. This is nearly always an operations person, often a dedicated analyzer technician or a lab analytical technician.

Once the roles are established the implementation process can formally begin. The ten-step process described below can greatly increase the probability that the analyzer will be a success. Note that while the process is described as a linear series of ten steps, in reality there may be loops: for example, if feasibility testing (Step 3) shows that the technical requirements cannot be met with existing technology, one will need to loop back and revisit the technical requirements (Step 2). This process is not intended to be burdensome, but rather to keep the process analytical chemist out of trouble. The length of time each step takes will depend on the magnitude of the proposed project. For example, a business value assessment may involve as little as a 30-minute discussion between two people, or as much as a month-long exercise involving dozens of people from multiple plants on multiple continents. The project manager does not personally carry out every step – some of the tasks can or must be delegated – but s/he is responsible for ensuring that each of the steps is done properly, documented, and the results communicated to all concerned.

Step 1 *Business value assessment*: Identify and quantify the business value for the proposed new method or analyzer. This is done jointly by the customer and the project manager. Ideally the value of the technology can be expressed concretely in monetary terms (e.g. dollars) as, for example, a net present value or an internal rate of return. It is critical to include a realistic estimate of the costs of implementing and maintaining the analyzer, as well as the benefits to be realized from it. This assessment is used to prioritize effort and expenses and to justify any capital purchases needed. (Ways in which process analyzers can contribute to business value are discussed in Section 11.2.2.)

Step 2 *Technical requirements*: Determine the specific technical requirements for the new method or analyzer that will ensure that it captures the previously determined business value. This is also done jointly by the customer and the project manager. Avoid 'would be nice ifs' and focus on what is necessary. Technical requirements may include accuracy, precision, specificity, turnaround time, response time, robustness, maintenance needs, etc. They should be focused on *what* the analyzer needs to do (e.g. determine the concentration of component X with 1.5% relative precision, once a minute, in a corrosive process stream at 80°C), not *how* it will do it. (Issues that need to be considered in setting technical requirements are discussed in Section 11.2.3.)

Step 3 *Feasibility assessment*: Determine (via literature search, consultation, etc.) suitable candidate methods and assess the feasibility of the candidates. The purpose of this step is to collect enough information to accurately define (or 'scope') and prioritize the project. The amount of information required may vary considerably, depending on the magnitude of the proposed project. One needs a lot less information to spend a week developing a new method on an existing off-line analyzer than is needed to justify spending several million dollars on a major installation of on-line analyzers in a large manufacturing complex. (The capabilities and limitations of NIR process analyzers are discussed in Section 11.2.4.)

Step 4 *Project initiation and approval*: Define a project to develop and implement the proposed new method or analyzer and get it approved by the sponsor. The documentation required by the sponsor is usually company-specific, but will typically include: goal, business value, critical technical issues to be addressed, ramifications, proposed approach, resources needed (including people's time), deliverables, and timing.

Step 5 *Off-line method or analyzer development and validation*: This step is simply standard analytical chemistry method development. For an analyzer that is to be used off-line, the method development work is generally done in an R&D or analytical lab and then the analyzer is moved to where it will be used (QA/QC lab, at-line manufacturing lab, etc.). For an analyzer that is to be used on-line, it may be possible to calibrate the analyzer off-line in a lab, or pseudo-on-line in a lab reactor or a semi-works unit, and then move the analyzer to its on-line process location. Often, however, the on-line analyzer will need to be calibrated (or re-calibrated) once it is in place (see Step 7). Off-line method development and validation generally includes: method development and optimization, identification of appropriate check samples, method validation, and written documentation. Again, the form of the documentation (often called 'the method' or 'the procedure') is company-specific, but it typically includes: principles behind the method, equipment needed, safety precautions, procedure steps, and validation results (method accuracy, precision, etc.). It is also useful to document here what approaches did *not* work, for the benefit of future workers.

Step 6 *Installation*: This is where the actual analyzer hardware is physically installed in its final location and started up. For both off-line and on-line analyzers, this may involve obtaining equipment (analyzer, sampling system, analyzer shelter, etc.), modifying facilities (piping, electrical, etc.), installing equipment, writing procedures, and training people. The project manager is responsible for ensuring that the method or analyzer is used correctly on site; s/he may choose to personally train all technicians and operators, to delegate the training but periodically audit their performance, or some combination.

Step 7 *On-line analyzer calibration*: Calibrating an analyzer entirely on-line is a last resort, for reasons discussed in Section 11.2.6. It is preferable to somehow calibrate the analyzer off-line, or to transfer to the on-line analyzer a method developed on an off-line analyzer or on another on-line analyzer. However, sometimes this is not possible. On-line analyzer calibration is again simply standard analytical chemistry method development, except that getting sufficient variation in composition to build a robust calibration model may be difficult or may take a long time. (Ways to address the challenges involved in on-line calibration are discussed in Section 11.2.6.)

Step 8 *Method or analyzer qualification*: Qualify the method on-site: that is, assess the ability of the method *as practiced* to meet the original technical requirements such as speed, accuracy, or precision. This is best done jointly by the project manager and the receiver.

Note: It is assumed that if the analyzer meets the previously agreed-upon technical requirements, it will then deliver the agreed-upon business value. Some companies may require that this be verified, for example, by tracking the savings over a specified period of time, and the results included in the project closure documentation.

Step 9 *Method maintenance*: Incorporate the new method or analyzer into the existing method maintenance systems for the site, to ensure that the method as practiced continues to meet the technical requirements for as long as it is in use. This is done by the receiver. Method maintenance systems may include check sample control-charting, intra- and/or inter-lab uniformity testing, on-site auditing, instrument preventive maintenance (PM) scheduling, control-charting the method precision and/or accuracy, etc.

Step 10 *Project closure*: Document the results of the project in a technology transfer package or project closure report. The form of the documentation will again be company-specific, but typically includes the documentation from the previous steps (business value assessment, technical requirements, etc.). It may also (especially for on-line analyzers) include engineering drawings, vendor manuals, standard operating procedures (SOPs), training materials, maintenance schedules, etc. Formal sign-off on the project closure package by the customer(s), the receiver, and the sponsor may be required to formally end the project manager's responsibilities.

That's it! All done! Throw a party and celebrate success!

11.2.2   How NIR process analyzers contribute to business value

Process analyzer technology can contribute value to the business in two basic ways: by reducing risks or by reducing costs.

Risks can be related to safety, such as the risk of a runaway reaction, or to money, such as the risk of ruining a batch of an expensive product by mistakenly adding a wrong ingredient. Process analyzers can reduce safety risks by monitoring safety-critical process parameters in real time and automatically shutting down the process if a parameter goes outside of safe limits. On-line process analyzers can reduce safety risks by reducing the need for routine process sampling, which reduces the exposure of process operators to potentially hazardous conditions. Process analyzers can reduce monetary risks by rapidly inspecting raw materials or in-process materials to assure that they are the correct material.

Cost reductions usually arise out of improvements to the control of the process for both continuous processes and batch processes. Process analyzers enable chemical composition to be monitored in essentially real time. This in turn allows control of the process to be improved by shortening startup times and transition times (for continuous processes) or batch cycle times (for batch processes), by improving the ability to respond to process disturbances, by enabling process oscillations to be detected and corrected, and by reducing product variability. Real-time monitoring of chemical composition in a process allows a manufacturing plant to:

- increase product yield by decreasing off-spec production;
- increase product yield by reducing startup times, transition times, or batch cycle times;
- improve product quality (consistency);
- reduce product 'giveaway' (e.g. run closer to aim);
- reduce in-process inventory;
- reduce consumption of expensive ingredients;
- reduce production of waste streams that are expensive to treat or dispose.

It is sometimes attempted to justify a process analyzer on the basis that it will reduce the number of grab samples that need to be taken from the process and analyzed, and therefore reduce the sample load on the QC lab. This is not usually a good justification for an on-line analyzer for two reasons: (1) The reduction in lab analyses rarely offsets the cost of analyzer installation and operation; (2) An on-line analyzer relies on high-quality lab data for calibration, validation, and long-term model maintenance, so the reference method can never be eliminated entirely.

Business value example
A fictitious example illustrates the large potential value of even small improvements in control of a manufacturing process. Suppose one has a continuous process in which the final product (a polymer) is sampled and analyzed to be sure the copolymer composition is within specifications. A sample is taken from the process once every 2 hours, and it takes about 2 hours for the lab to dissolve the polymer and measure its composition. This process produces a number of different copolymer compositions, and it transitions from one product to another about twice a month on average. The 2-hour wait for lab results means that during a transition the new product has been within specification limits for 2 hours before the lab confirms it and the operators are able to send the product

to the 'in-spec' silo. Consequently, on every transition, 2 hours' worth of in-spec polymers are sent to the off-spec silo.

Suppose that implementation of NIR technology in this process (either off-line or on-line) allows the lab delay to be reduced from 2 hours to a few minutes (effectively zero). This means that, during a transition, the operators know immediately when the new product has reached specifications and can immediately send it to the in-spec silo. This reduces the transition time by 2 hours for every transition. Suppose this plant makes 15 000 pounds of polymer per hour, and that off-spec polymer sells for US$0.70 less per pound than in-spec polymer. The 2-hour reduction in transition time, then, saves US$504 000 or half a million dollars a year. This is a conservative estimate because it does not include other contributions that real-time monitoring will make, such as shortening the *beginning* of product transitions (by not sending the previous product to the 'off-spec' silo until it actually goes out of specification limits), shortening startups, and enabling quicker recoveries from process upsets.

Note: The costs of implementing this fictitious analyzer have not been included in the value calculation, though it is a critical aspect. The costs would include hardware, software, facility modifications, people's time, travel, and ongoing maintenance costs (including the analyzer technician's time) over the expected analyzer lifetime.

11.2.3   Issues to consider in setting technical requirements for a process analyzer

This section describes the issues that need to be considered during Step 2 (technical requirements) of a process analyzer implementation. As stated earlier, the technical requirements should be focused on *what* the analyzer needs to do (e.g. determine the concentration of component X with 1.5% relative precision, once a minute, in a corrosive process stream at 80°C), not *how* it will do it (e.g. with on-line GC or off-line reflectance NIR), although the issues do overlap somewhat. It is critical at this stage to separate the 'would be nice' requirements from those that are truly necessary to gain the already-determined business value. For example, if the business value is gained by shortening product transition times by 2 hours (as in the fictitious example above), the new analyzer only needs to be fast; it does not need to be any more accurate or precise than the existing laboratory method. If the business value is to be gained by detecting and responding earlier to process upsets, then the analyzer needs to be fast, sensitive, and non-noisy; but it does not need to be quantitatively accurate.

The issues are organized in the form of questions to be asked. The questions should be asked more than once to people in different functions and levels, since they may have different levels of knowledge about the business and the process, and different perspectives on the relative importance of various issues. The first answer is not necessarily the right answer.

What type of measurement needs to be done? Do you need quantitative information (concentrations) or qualitative information (identification, qualification (ok/not ok), or trends)? If quantitative, what accuracy and precision are required? If qualitative, what level of errors can be tolerated?

What are the analytes of interest? What are their typical concentration ranges? Do the concentrations of any of the analytes covary? What is the accuracy and precision of the reference method for each analyte?

What other components are present in the process stream of interest: (a) under normal conditions? and (b) under process upset conditions? What are their concentration ranges? Do the concentrations of any components covary with those of any analytes?

What type of process is the measurement being done in (batch or continuous, single product or multiple products, long campaigns or short ones)?

What are the measurement results going to be used for? Are they going to be used for process control (closed loop, open loop, feed-forward, feedback)? Are they going to be used as a safety interlock? (This puts very strict requirements on the analyzer reliability, and may even require duplicate analyzers.) Are they going to be used to accept raw materials or to release product? Are they going to be used to sort or segregate materials? How frequently does the measurement need to be done? How rapid does the measurement need to be? How accurate does it have to be? How precise? How much analyzer downtime is acceptable?

Are there external regulations (such as FDA, EPA, or USDA regulations) that apply to this process and any measurements made on it?

Is the measurement going to be done on-line or off-line? For on-line analyzers, is there a grab sample port nearby? Is the sample stable once withdrawn from the process? (Only an on-line analyzer can enable real-time process control. However, an on-line analyzer implementation will generally be more expensive, more complex, and involve more people in more functions than an off-line one.)

What is the nature of the material to be measured? Is it a gas, a liquid, a slurry or a solid? If solid, is it a powder, pelletized resin, fiber, film, or web? How large is it? Is it clear, cloudy, or opaque? Is it colored? Viscous? Bubbly? Hot? Cold? Under high pressure? Under vacuum? Corrosive? Highly toxic? Is it moving, and if so, how rapidly? Does it need to remain sterile?

These questions will undoubtedly lead to more questions specific to the business and process at hand. They are not intended to be exhaustive, but to remind the process analytical chemist of the kinds of issues that need to be discussed during Step 2.

11.2.4 Capabilities and limitations of NIR

Before selecting a process analytical technology to implement, it is helpful to understand the capabilities and limitations of the technology. Good introductions can be found elsewhere in this book to NIR spectroscopy and instrumentation (Chapter 3), and to chemometric methods (Chapter 8). Williams,[1] Siesler,[2] and Burns[3] provide additional information on NIR techniques and applications.

Note: The focus of this book is on multivariate analyzers, but it is worth keeping in mind that there are still a lot of process analyzer applications in the chemical industry that are successfully using older filter NIR and IR technology. In some cases, business value could be gained by replacing the older technology with newer multivariate analyzers, but in other cases the cost of doing so exceeds the benefit. It is not necessarily true that the latest technology has the greatest value to the business, a fact which it is sometimes easy for technical people to forget. What is important about a technology is not how new it is, but whether it has the specific capabilities needed to capture business value (e.g. shorten transition times, increase first-quality yields, etc.).

In general, NIR can be used to monitor organic or inorganic compounds that contain C–H, O–H, or N–H bonds. NIR is generally used to monitor bulk composition (components at levels of 0.1% and higher), rather than trace-level composition. However, NIR is very sensitive to water, and to alcohols and amines, and under the right circumstances can be used to monitor them at levels well below 0.1%. NIR can be used for quantitative analyses. It can also be used for qualitative analysis, including identification or qualification of incoming raw materials, verification of product identity or quality, monitoring the trajectory of a reaction, and rapid sorting. The results of NIR monitoring (whether quantitative or qualitative) can be used to control a process (feed-forward or feedback control, closed or open loop control), to decide when a batch reaction is complete, to accept raw materials, or to release product, or as a safety interlock (e.g. to shut a plant down automatically if the level of a critical parameter goes out of preset limits or if the analyzer itself malfunctions).

The two major benefits of NIR are *speed* and *precision*. NIR analyzers provide rapid results, generally in less than a minute, both because of the scan speed and high signal-to-noise performance of the analyzers themselves and because sample preparation is generally minimal. The speed advantage can be enhanced by putting the analyzer on-line or in-line. This also eliminates the need for routine sampling of the process, which can be time-consuming and involve safety risks. The speed advantage is further enhanced by the ability of NIR to monitor multiple components simultaneously. NIR analyzers provide high-precision results, due primarily to the elimination of variability arising from sampling and sample preparation, as well as to the high stability of the analyzers themselves.

However, NIR also has significant drawbacks. The major drawback is that it takes a significant investment of time, effort, and money to implement an NIR analyzer. There are a number of reasons for this. NIR is a secondary method and must be calibrated against a primary or reference method. The calibration can take considerable time, depending on how long it takes to collect and analyze a suitably varied calibration sample set; and it takes additional time on an ongoing basis to validate and maintain the calibration. NIR can monitor multiple components, but sometimes they are ones that one did not want to monitor (i.e. interferences). It takes time and chemometric skill to ensure that the calibration model accurately monitors only the components of interest. (From observation of the interferences encountered while monitoring the process, one can often increase knowledge of the process.) NIR involves a steep learning curve not only for NIR hardware, software, and sampling issues but also for chemometrics.

Because of the large investment required to implement NIR process analyzers, there has to be a very large business value to be gained through the analyzer to make the effort worthwhile. This is why it is critical to assess the business value of a potential analyzer at the very beginning of a project. Failure to honestly assess the value (both benefits and

Table 11.1 Benefits and drawbacks of NIR

Benefits	Drawbacks
Rapid (near real time)	Indirect (secondary method)
Multi-component capability	Multi-component capability (interferences)
High precision	Large investment (time, money)
Can be on-line (no sampling)	Steep learning curve
Can be remote (with fiber optics)	Not a trace level technology
Little or no sample preparation	
Non-destructive	

costs) at the beginning makes it difficult to prioritize effort, justify capital and other expenses, or engage the cooperation of already-busy people in other organizations and functions. The benefits and drawbacks of NIR are summarized in Table 11.1.

11.2.5   *General challenges in process analyzer implementation*

Challenges arise in implementing any new technology in a manufacturing process, of course, and some of them may be difficult to anticipate. However, many of the challenges that occur during process analyzer implementations are common ones that have been experienced by many process analytical chemists in the past. Though they may or may not be show-stoppers, they will at minimum cause time and effort to be wasted. They can often be avoided or minimized if recognized and addressed early. Note that some of the most serious challenges facing the process analytical chemist are *not* technical in nature. The process for successful process analyzer implementation described in Section 11.2.1 (and in Chapter 2) is designed to allow the process analytical chemist to avoid the avoidable challenges, and recognize early and minimize the unavoidable ones.

The following factors increase the likelihood of *success* for an NIR process analyzer implementation.

- The project was undertaken for business reasons: the financial stake was sufficiently large to motivate the business to plan, support, and staff for success.
- The business calculated the financial stake for the project taking into account both the installation costs and the continuing support costs of the analyzer.
- The project was coordinated by someone in the business, typically a process engineer or chemist, a control engineer, or an analytical or control lab chemist; rather than by an 'external' resource.
- The analyzer was installed as part of a larger process improvement effort, such as installation of a distributed control system (DCS).

The following factors increase the likelihood of *failure* for an NIR process analyzer implementation.

- The project was undertaken based on insufficient and/or overoptimistically interpreted feasibility studies.
- Technology that did not meet the technical requirements for the application was chosen. Sometimes this is through ignorance, but more often it is due to not having clarified the essential technical requirements before choosing the technology.
- The hardware chosen did not meet the technical requirements or the vendor was inexperienced in the application or unable to supply necessary support.

  - Choice of vendor is particularly critical when analyzer implementation is contracted out. Avoid contract firms which have little or no experience with the specific technology being used. NIR analyzers have different challenges and issues than, for example, GC analyzers.

- There was a failure to recognize the plant-site requirements for NIR calibration and validation, such as the existence of appropriate sampling valves, well-designed sampling protocols, good laboratory reference methods, and variability in the analyte concentrations of interest.
- Model development was not thorough enough to allow the analyzer to recognize and deal with interferences. This is often due to model development being carried out by someone with insufficient training in chemometrics or insufficient knowledge of the particular chemical process.
- The project was schedule-driven rather than data-driven.

  - Bad hardware decisions are sometimes irretrievable.
  - Good model development takes time. Bad models can actually be worse than no models because they demonstrate to the process operators and engineers that the NIR results are unreliable.

- The developer (the person who did the installation and model development) left too soon – voluntarily or otherwise. The developer generally needs to remain involved to some extent well into the long-term prove-out or qualification phase (Step 8). Extreme levels of frustration can be felt by plant personnel who 'inherit' process analyzer systems before qualification is complete, especially if they have no previous NIR experience and were not involved in the earlier stages of the project. Another way to state this factor is: The business failed to adequately staff the project.
- The project encountered unforeseen difficulties such as previously unsuspected impurities, unexpected fouling of the probes, or unfixable leaks.

  - Sometimes surprises occur because of factors nobody could have known about or predicted, but sometimes surprises result from gaps in communication. The process analytical chemist should seek the input of process chemists, engineers, and operators early in the project to maximize the chances of hearing about relevant *known* problems. They should spend time at the plant talking to people. They should ask a lot of questions, and note and pursue any inconsistencies or gaps in the answers.
  - Surprises can also result from inexperience on the part of the process analytical chemist – such as not anticipating common difficulties such as fouling of probes or condensation in sampling lines. Less experienced people would be well-advised to

talk over the proposed project with more experienced process analytical chemists at an early stage.

- There was a lack of communication or cooperation among project team members. Process analyzer work always involves a diverse team. Poor communication among team members can result if a single person is not assigned to be the project manager, if the project manager is not given enough time to do the necessary work, or if the project manager is inexperienced or has poor communication skills. Poor cooperation from team members can often be attributed to lack of clarity about the business value of the work, or to failure of management to adequately prioritize and staff the current projects.
- The project objectives, roles, and responsibilities were unclear.
- Insufficient attention was paid to the general plant environment (dirt, noise, heat, cold, vibration, being hosed down with water, being cooked with a blowtorch, being run into or climbed on by mistake, etc.). These kinds of problems are relatively easy to anticipate and prevent by visiting the plant frequently (if you do not already work there) and watching and listening to the people there.

11.2.6 Approaches to calibrating an NIR analyzer on-line

Ideally, an on-line analyzer will be calibrated before it is installed in the process. It may be possible to accomplish this by calibrating it off-line with process grab samples and/or synthetic samples. It may be possible to install the analyzer in a lab-scale reactor, or in a semi-works or pilot plant. It may be possible to transfer to the on-line analyzer a method developed on an off-line analyzer or on another on-line analyzer (e.g. at a different plant site). However, sometimes none of these are possible and the analyzer will have to be calibrated on-line. The challenges of on-line model development (calibration) and validation, as well as approaches to dealing with them, are discussed below.

(1) *Limited variation in chemical composition*: This is often an issue at the point in the process where the analyzer is installed. A manufacturing plant's goal is to make only good product, with as little variability as possible. In contrast, the analytical chemist's goal (especially with chemometric methods) is to have as much variation in composition as possible represented in the calibration sample set. A commercial-scale plant is rarely willing to produce 'bad' or off-specification product just to calibrate an analyzer, especially if the plant is sold out. However, most plants make some 'bad' product at startup, at shutdown, or during product transitions. And some plants, though they are often reluctant to admit it, make bad product occasionally due to process upsets or 'excursions.' The amount of variation obtainable over a given period of time will depend on the plant: on whether it operates continuously or in batch mode, whether it produces a single product or multiple products, how long the product campaigns are (i.e. the time between product transitions), how frequently the plant starts up and shuts down, and how often they have process upsets. So one approach to increasing the composition range of the

calibration model is to include as much startup, shutdown, transition, and process upset data as possible. Another approach that can work, if the analyzer is monitoring multiple points in the process and the chemical composition is different at the different points, is to model several different points in the process together.

(2) *Covariance among composition variables (analyte concentrations)*: The covariance may be coincidental, or it may be fixed by the chemistry and therefore unavoidable. An example of a fixed covariance is in the manufacture of poly(ethylene terephthalate) or PET, where the degree of polymerization (DP) and the concentrations of both carboxyl and hydroxyl end groups all vary together: for a given DP, an increase in hydroxyl ends necessarily means a decrease in carboxyl ends. Coincidental covariance is sometimes less severe when the process is changing, such as during startups, transitions, or excursions. One approach is to intentionally break the covariance in the calibration sample set by adding odd samples, or by adding synthetic samples. Another approach is to pay attention to the regression coefficients to be sure they make spectroscopic sense: the coefficients for a given analyte should be similar to the NIR spectrum of that analyte.

Note: Monitoring an analyte indirectly, by relying on its coincidental covariance with another more easily detected analyte, is risky and should be avoided unless the covariance is 100% ironclad (which is rare).

(3) *Infrequent samples*: The plant may not sample the process very often at the point where the analyzer is installed. This could be because the lab turnaround time is so long that the results are too late to be useful, or because the operators do not have time for sampling (they are understaffed), or because they are reluctant to don the personal protective equipment necessary for that particular sample port. Reluctance to wear the 'hot suit' needed to sample a polymer melt process is a common problem, especially in hot climates and during hot seasons of the year. In any case, solving this problem requires gaining the cooperation of the process operators and engineers (and their management). One approach is to negotiate a temporary increase in sampling, with the incentive that it will be possible to reduce or nearly eliminate sampling once the analyzer is calibrated.

(4) *Non-existent samples or reference values*: Samples may not exist because the plant does not exist yet (it is under construction and has not started up yet), or because the plant does not do any sampling at the process point of interest. Reference values may not exist because the plant lab is not set up to do that particular method, or (worse) there is no established reference method for the analyte of interest. In any of these cases, one is left with a calibration set that contains no samples. There are a number of ways to approach this challenge. If there are samples but no reference values, the plant samples can be sent off-site to be analyzed. The analyte concentration of interest can sometimes be estimated based on process conditions and/or the concentrations of other analytes. (This is one place where fixed covariance can come in handy.) If there is no plant yet, it may be possible to calibrate the analyzer elsewhere (different plant, semi-works, etc.). It may also be possible (or even necessary) to attempt 'lab value-less calibration,' in which one assumes that the concentration of the analyte varies linearly with the height of an absorbance peak characteristic of that analyte (trend analysis). This works only if the spectroscopy of the system is well-characterized and if there are no significant overlaps between peaks; and in any case it will only provide qualitative data.

(5) *Unstable samples*: The chemical composition of a sample removed from an industrial process often changes after it is taken due to the sample losing or gaining moisture and/or continuing to react (i.e. not being effectively quenched). This is especially common with condensation polymers. Unstable samples result in an offset between the composition measured in the lab and the 'true' composition in the process, and the offset usually varies. Worse, the offset is usually not definable, since there is usually no way of knowing what the composition *would* have been before the sample was removed from the process. The offset appears as an increase in the variability of the lab method. One approach is to simply ignore the offset. Another approach is to minimize the variation in the offset by ensuring consistency (especially in timing) in sampling and lab measurement procedures.

(6) *Imprecise reference values*: Imprecise reference values can be due to method variability or sampling variability. Sampling is quite often the larger source of variability. Sampling variability arises from numerous causes, including errors in writing down the exact sampling time, sample inhomogeneity, and sample instability (as discussed above). Sampling variability will be largest when the process composition is changing the most, such as during startups and transitions. One approach is to use huge amounts of data during model development, in order to minimize the effect of the imprecision of any one data point. Another approach is to carefully (usually iteratively) filter out model outliers such as samples with clearly incorrect lab values (e.g. clerical errors) or samples taken during transitions where the time stamp is inaccurate.

A brief word is in order here about accuracy and precision. The process analytical chemist needs to be aware that the other members of his/her team may not be clear on the difference between analyzer accuracy and analyzer precision, especially in the case of on-line analyzers, and s/he should be prepared to explain the difference in language appropriate to their backgrounds. *Analyzer accuracy* refers to the error (difference) between the lab results and the on-line analyzer predictions for a given set of samples. It is often expressed in terms such as standard error of calibration (SEC). *Analyzer precision* refers to the variability in the on-line analyzer predictions for the same sample over and over. Since most on-line analyzers are installed in flowing streams or other systems which change continuously with time, the analyzer never really sees the same sample over and over. Consequently, analyzer precision is usually estimated from the variability of the predictions over a short period of time while the process is lined out and stable. This gives a high estimate, since it includes both process and analyzer variability. Analyzer precision is often expressed in terms such as standard deviation (SD) or relative standard deviation (RSD) of the predicted value. Analyzer accuracy and precision are independent of each other: an analyzer can easily be accurate without being precise, or precise without being accurate. Analyzer accuracy and analyzer precision are also affected by different factors, as listed below.

The factors which limit on-line NIR analyzer *accuracy* include:

- Error due to unmodeled conditions (bad model)
- Sampling error
    - Time assignment error
    - Sample inhomogeneity error
    - Sample instability error

- 'Pure' lab error (i.e. the inaccuracy of the reference method itself)
- Real process fluctuations
- Instrument factors (instrument noise, non-linearity, temperature instability, drift, etc.).

The factors which limit on-line NIR analyzer *precision* include:

- Real process fluctuations
- Instrument factors (instrument noise, non-linearity, temperature instability, drift, etc.).

Note that there are more sources of variability for accuracy than for precision. This is the basis of the common rule of thumb which states that the precision of an on-line analyzer will typically be about ten times better than its accuracy.

11.2.7 Special challenges in NIR monitoring of polymer melts

Polymers (both for plastics and for fibers) are a major part of the chemical industry. The composition of polymer melts can be of interest in several kinds of systems, including polymerization reactors, polymer melt transfer lines, and extruders. However, there are additional challenges involved when using NIR analyzers in such systems due to a number of unique factors: sustained high temperatures, sudden changes in temperature, pressure extremes (high or low), polymer flow issues, and fouling.

Sustained high temperatures can cause physical deterioration of NIR optical probes if their components are not rated for the process temperatures. This can manifest itself, for example, as either a sudden or a slow drop in the optical efficiency of the probe: the latter is difficult to distinguish from slow probe fouling unless it is possible to remove the probe(s) from the process for testing. High temperatures (especially above 200°C) also cause significant black body emission from the polymer and the process pipe itself. This is a huge source of 'stray' light. Spectra obtained with pre-dispersive NIR analyzers need to be corrected for black-body emission. High temperatures can also soften probe materials sufficiently so that the probes can be bent while being removed from the process. (With many polymer systems, the probe can be removed from the process only while the polymer is melted.) The probe body should be designed to withstand this treatment.

Sudden large changes in temperature (either heating or cooling) can crack or break probe windows. In some polymer processes, for example, it is routine practice to deal with flow blockages by heating the section of pipe or piece of equipment with a blowtorch. Careful consideration of the relative thermal expansion coefficients of the window material and the probe materials during probe design can minimize this risk. Attention paid to thermal issues when writing or modifying SOPs when an NIR analyzer is implemented can also pay dividends.

High pressures (hundreds or thousands of psig) can break probe windows, or break or loosen the seal between the window and the probe body. This is especially true when high pressures are combined with high temperatures. High pressures can also, in a worst case, push the entire probe out of the process, potentially with dangerous force. The probe itself should of course be designed and tested for the highest pressures expected to occur in the process. In addition, the mechanism by which the probe is secured to the process needs to be carefully designed and thoroughly reviewed for safety. Low pressures can also cause problems. Suction (negative pressures) can occur during a line shutdown or process upset if the polymer freezes. This can pull the window out of a probe, which not only destroys the probe and creates a process leak but also puts a piece of glass into the process stream which can damage equipment downstream. The possibility of low or even negative pressures needs to be taken into account during probe design.

Polymer flow issues are concerns on the part of plant operations personnel that can arise when one proposes to put an in-line NIR probe (or pair of probes) into a polymer reactor or transfer line. These concerns tend to be plant- or process-specific. Plant personnel are likely to be concerned if the probe will change the pressure drop in the line, if it will create a cold spot in the reactor or line, if it will protrude into the flow stream, or if it will create dead spots (e.g. recessed probes or the downstream side of protruding probes). There may also be plant- or process-specific restrictions on where probes (or analyzers) can be located, on pipe sizes, on the use of welds, and on materials of construction. It is critical to involve plant operations personnel (including process operators) as early as possible in discussions about probe design and location.

Probe fouling is common in polymer systems. The symptoms are a gradual increase in the baseline absorbance over time (although care must be taken to ascertain that it is not actual haziness in the process itself, or optical degradation of the probes, which is causing the baseline increase). There are a large number of ways to approach this problem, depending on the process. Changes in probe location, orientation (relative to the flow direction), or materials of construction can sometimes reduce fouling. Removal and cleaning of the probes at a set frequency is sometimes possible. Small baseline shifts can be eliminated by baseline-correction of the spectra. However, probe fouling often causes changes in baseline tilt and curvature as well; these effects can sometimes be dealt with ('modeled out') in the calibration modeling until the fouling levels get too high (see Section 11.3.6). Note that both the baseline-correction approach and the modeling-out approach require high instrument *y*-axis linearity, so that the peak intensities and shapes are not distorted.

11.3   Example applications

The applications described in this second part of the chapter are intended to illustrate the wide range of uses for NIR in the chemical industry. The selection of examples was intentionally limited to work done within industry and published in the open literature in order to keep the focus on work that has demonstrated business value. However, it has been the author's experience that for every industrial NIR application published in the open literature there are at least two others practiced as trade secrets for business reasons. This is especially true for on-line applications that have progressed beyond the feasibility stage, since the NIR results can reveal a great deal of information about the chemical process itself – information which industry managers are often reluctant to share. Published industrial applications should therefore be considered merely the tip of an iceberg.

The examples also serve to illustrate the points made in the first part of the chapter, with a few exceptions. First, although all of the authors discussed the business value of the project, none of them quantified it (e.g. the analyzer saved US$1.5 million annually by shortening product transitions), presumably because the exact monetary values were considered proprietary business information. Second, because these were technical papers published in technical journals, the focus was on the technical challenges involved in the work, rather than the non-technical ones. The non-technical challenges (such as inconsistent management support or lack of communication among team members) are rarely discussed in the open literature, but they can be as much of a barrier to success as the technical ones, if not more; this is why they are emphasized strongly in Section 11.2. Finally, the passive voice used in technical writing makes it difficult to appreciate the large number of people that are generally involved in process analyzer projects.

11.3.1 Monitoring monomer conversion during emulsion polymerization

Chabot *et al.*[4] at Atofina Chemicals (King of Prussia, PA, USA) used in-line NIR to monitor monomer conversion in real time in a batch emulsion polymerization process. The **business value** of this monitoring is twofold. First, emulsion polymerizations are complex processes and while much progress has been made in understanding them there are still many unknowns. Real-time monitoring allows the development of improved understanding of emulsion polymerizations, such as quantifying the effects of temperature, initiator type, initiator concentration, and other factors on the reaction dynamics. Second, accumulation of un-reacted monomer in the reactor (due to factors such as fluctuations in polymerization rate, insufficient mixing, or poor control of feed rate or reactor temperature) can lead to runaway reactions. Real-time monitoring allows control of the process to be improved, reducing the safety risk.

Chabot's work was done in a laboratory-scale batch reactor. Experiments at lab scale are a common first step in process analyzer work, since they allow technical feasibility to be demonstrated relatively quickly and inexpensively without having to interfere with production targets in a commercial-scale plant. In this work, moreover, one of the two business goals (improving process understanding) can be largely accomplished without going beyond lab-scale experimentation.

Several technical **challenges** were faced in this work. The monitoring had to be done in a complex, heterogeneous, unstable matrix, in the presence of high solid levels ($\sim$40% by weight), with large monomer droplets at the beginning of the reaction, very fine particles, and possible foaming. The matrix consisted of an aqueous mixture of seed latex particles, surfactant(s), chain transfer agent(s), one or more acrylic monomers (such as methylmethacrylate, MMA), and initiator(s), under nitrogen or argon purge, at elevated temperatures of 65–75°C. Furthermore, it was reasonably expected that an optical probe inserted into the reactor might become coated or plugged with polymer.

The emulsion polymerization process was monitored using an NIR analyzer (Foss NIRSystems Model 5000) equipped with a fiber-optic transflectance probe inserted directly into the reactor. The analyzer was calibrated for residual monomer level by withdrawing samples periodically during a run and analyzing them by the reference method, headspace GC. A high correlation ($R^2 = 0.987$) was found between the second-derivative absorbance at 1618 nm, attributed to the vinyl C—H group, and the residual monomer concentration as measured by GC. A single-wavelength multiple linear regression (MLR) model using the 1618 nm band predicted the residual monomer level with an accuracy (SEP) of 0.2% by weight, matching the accuracy of the primary GC method.

The technical challenges were dealt with as follows:

(1) The use of a transflectance probe, rather than a transmission probe, allowed good-quality spectra to be obtained despite the low light transmission due to the high solid levels.
(2) Probe fouling was minimized if the probe was placed properly in the reactor relative to the liquid surface and to the mechanical stirrer. The ideal spot, although not explicitly stated, would presumably be deep enough into the emulsion to see a representative bulk sample and to avoid any 'dead' unstirred spots, but far enough from the stirrer blade to avoid impact.
(3) The use of second-derivative spectra eliminated the baseline shifts observed during the reaction (which could be due to changes in the amounts or size distribution of particles, or small amounts of probe fouling).
(4) It was observed that the NIR-predicted monomer concentration varied greatly during the first 45 minutes after monomer addition but then stabilized; this was attributed to slow solubilization of large monomer droplets into the water phase, and slow partition of monomer between the water phase and the latex phase, leading to large local variations in monomer concentration. This was avoided by allowing a 60-minute 'soak' period after monomer addition, before the temperature was increased to start the polymerization.
(5) Finally, although temperature had a large effect on both the position (wavelength) and the intensity of the water absorption bands in the emulsion NIR spectra, careful experimentation demonstrated that the 1618 nm vinyl C—H band used in the calibration model did not shift in either position or intensity with temperature, in the temperature range used in these studies (25–75°C). Therefore, it was not necessary to correct the calibration model for temperature effects, either by the use of internal or external standards, or by including temperature variations in the calibration set.

Once calibrated, the NIR analyzer was used to investigate a number of factors expected to affect the polymerization kinetics, including reaction temperature, initiator type, and initiator concentration (relative to monomer concentration). These experiments, in addition to improving process understanding, also mimicked the effects of inadequate process control during a reaction. Figure 11.1 shows the effect of reaction temperature. The reaction rate nearly doubles when the temperature is raised from 65 to 75°C, and the concentration of un-reacted monomer after 85 minutes is reduced from 1.1 to 0.5%. In-line NIR monitoring allows unusual behavior in either reaction rates or residual monomer levels to be detected and corrected immediately.

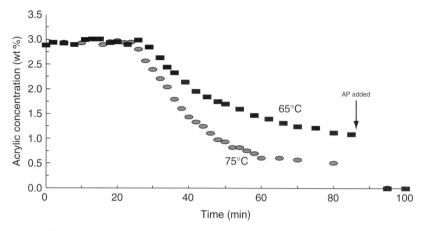

Figure 11.1   Effect of temperature on the rate of emulsion polymerization as monitored by in-line NIR.
Reprinted with permission from Chabot *et al.* (2000).[4]

11.3.2   *Monitoring a diethylbenzene isomer separation process*

Chung *et al.*[5] at SK Corporation (Ulsan, Korea) and Hanyang University (Ansan, Korea)
used off-line NIR to monitor the *p*-diethylbenzene (PDEB) separation process. The
process consists of isolating PDEB from a stream of mixed diethylbenzene isomers and
other aromatic hydrocarbons by selective adsorption on a solid adsorbent, followed by
extraction with *p*-xylene (PX). Optimal control of the process requires accurate measurement of the concentrations of the *o*-, *m*-, and *p*-diethylbenzene isomers as well as the PX
extractant. The **business value** of NIR monitoring lay primarily in reducing the analysis
time from 40 minutes (by GC) to less than 1 minute, allowing the process to be
monitored and controlled in (effectively) real time. There was some secondary benefit
to replacing the GC method, which required constant attention to achieve reliable results,
with the more rugged NIR method.

The challenges in this work concerned the fundamental limits of NIR spectroscopy.
First, would NIR, with its typically broad and highly overlapped bands, have enough
spectral resolution to distinguish the ortho, meta, and para isomers from each other and
from the extractant? Second, would NIR, with its typically weak overtone and combination bands, have enough sensitivity to quantify the minor components of the stream,
especially the ortho isomer, typically present at only ~1%?

The NIR monitoring was done off-line using a Foss NIRSystems Model 6500 analyzer
in transmission mode and a quartz cuvette with 0.5-mm pathlength. The spectra had
10 nm wavelength resolution with data points taken every 2 nm. The analyzer was calibrated
using 152 samples taken from a pilot plant simulated moving bed (SMB) adsorption unit.
The pilot plant was designed to operate under a wide range of conditions, which allowed
the compositions of the 152 samples to be intentionally varied over a wide range as well.

Table 11.2 PDEB separation calibration results

Component	Concentration		MSECV (f)	SEP
	Min	Max		
o-Diethylbenzene	0.00	4.24	0.11 (6)	0.12
m-Diethylbenzene	0.04	56.27	0.15 (8)	0.16
p-Diethylbenzene	0.15	76.15	0.16 (7)	0.15
p-Xylene	0.47	99.38	0.28 (7)	0.27

Note: All concentrations are in volume %.

The composition of the 152 samples in the calibration sample set is shown in the first three columns of Table 11.2; the second and third columns of the table indicate the maximum and minimum concentrations (in volume %) found in the calibration sample set for the components listed in the first column.

The NIR spectra of the pure isomers and the extractant are shown in Figure 11.2. Although the spectra of the three isomers and the extractant are quite similar overall, there are distinct, reproducible spectral differences, especially in the 2100–2500 nm region. The authors did extensive studies involving mid-IR spectra, spectral simulations, and principal components analysis (PCA) in order to understand the origins of these differences to ensure that they were related to the isomers themselves and not to coincidental impurities.

Calibration models were developed using 106 spectra (70% of the total) chosen at random. The remaining 46 spectra were used to validate the models. The effects of number of factors, wavelength range, and absorbance vs. second-derivative spectra were investigated. The most accurate models were obtained when the 2100–2500 nm region was included in the model, regardless of whether the rest of the spectrum was included – this makes sense since this region has the sharpest, most intense peaks. The last two columns of Table 11.2 show the results for PLS models developed using absorbance spectra in the 2100–2500 nm region only. Column 4 shows the accuracy of the model on the 106 calibration samples (MSECV: mean square error of cross-validation), with the number of PLS factors for the model in parentheses; column 5 shows the accuracy of the model on the 46 validation samples (SEP: standard error of prediction). The results are shown graphically in Figure 11.3.

The results indicate that NIR does indeed have both the resolution (or specificity) and the sensitivity needed for this application. The NIR method was successfully implemented in the plant, replacing the conventional GC method and allowing real-time control and optimization of the PDEB separation process.

11.3.3 *Monitoring the composition of copolymers and polymer blends in an extruder*

McPeters and Williams[6] at Rohm and Haas (Spring House, PA, USA) used in-line transmission NIR to monitor the compositions of heterogeneous polymer blends and

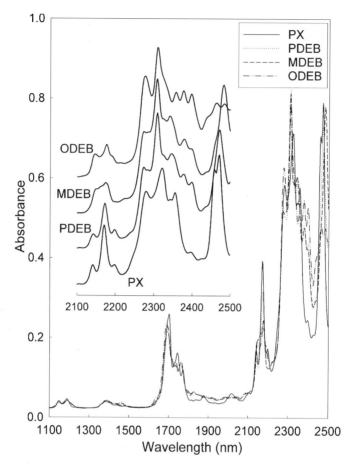

Figure 11.2 NIR spectra of *o*-diethylbenzene (ODEB), *m*-diethylbenzene (MDEB), PDEB and PX. The 2100–2500 nm range is expanded in the upper left of the plot. Reprinted with permission from Chung *et al.* (2000).[5]

terpolymers in real time at the exit of an extruder. The **business value** of this monitoring comes from four sources. First, real-time monitoring minimizes waste production at startup, since it is no longer necessary to wait several hours to get enough off-line lab data to demonstrate that the process has stabilized ('lined out'). Second, real-time monitoring allows ready determination of the extruder residence time, a key item of process information which is very tedious to determine otherwise. Third, real-time monitoring allows the detection and elimination of process 'cycling' or oscillation due to poorly tuned control loops; this greatly reduces process and product variability. Fourth, real-time monitoring rapidly increases process understanding by enabling the effects (both the magnitude and the response time) of changes in process parameters such as zone temperatures, feed rates, vent rates, or screw speed to be observed immediately.

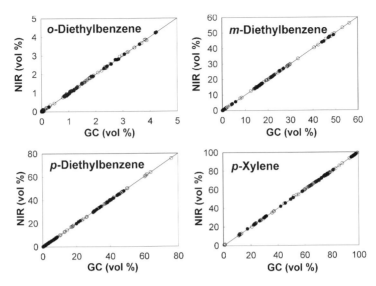

Figure 11.3 Scatter plots showing the correlation between NIR (using the 2100–2500 nm region) and GC analyses for each component. Reprinted with permission from Chung *et al.* (2000).[5]

The technical **challenges** in this work included high temperatures ($\sim$280–325°C), high pressures (several hundred psig), and NIR baseline fluctuations due to bubbles, particulates, impurities, or degraded polymer in the melt stream. The first two challenges were addressed using custom-made probes designed to withstand the simultaneous high temperatures and pressures. (Note that in addition to suitable probes, care must also be taken of how the probes are interfaced with the process, to ensure good optical alignment and reproducible pathlength, while avoiding leaks and other safety issues.) The baseline fluctuations were removed by linear baseline correction of the spectra, at 1290 nm and 1530 nm for the polymer blends, and at 1290 nm alone for the terpolymers.

The extruders were monitored using a pair of custom-made transmission probes inserted into the extruder die just downstream from the screws. Each probe consisted of a sapphire window brazed into a metal body; a quartz rod behind the window piped the light to the end of the probe; low-OH silica fiber-optic bundles connected the probe to the NIR analyzer, an LT Quantum 1200I. Optical pathlengths were typically between 0.3 and 2.0 cm.

The polymer blends were a heterogeneous mixture of an acrylate/styrene copolymer dispersed in a methacrylate/acrylate copolymer. The level of acrylate/styrene copolymer present, in wt%, is termed the 'blend level.' The NIR analyzer was calibrated over two days for blend level by obtaining spectra of seven different blends with blend levels ranging from 0 to 45%. The extruder temperature was 260°C and the optical pathlength was 2.0 cm. The reference method in this case was the known weights used to prepare the blends. All spectra were used in calibration except the transition spectra. Calibration explored both MLR and PLS methods, on absorbance, first- and second-derivative spectra. Only the spectral region from 1200 to 1650 nm (in which the absorbances ranged between 0 and 1 AU) was used in modeling. The best model was an MLR model using two wavelengths, 1211 and 1595 nm, in the absorbance spectra; it had a correlation R^2 of 0.9986 and a standard error of 0.79%. Figure 11.4 shows the transition between the 10% blend and the 25% blend as predicted by the NIR model. Both the extruder dead time, which is about 5 minutes, and the extruder transition (or mixing) time, which is about 10 minutes, are clearly evident.

The terpolymers contained three components, one major and two minor. Component 1 (a methacrylate-related repeat unit) is the major component, with levels ranging between 60 and 90%. Component 2 (a methacrylate ester repeat unit) and Component 3 (a hydroxyl-containing repeat unit) are the two minor components; Component 3 levels are typically less than 1%. The NIR analyzer was calibrated for component levels using spectra corresponding to 18 pellet grab samples. The grab samples were analyzed for Component 1 using elemental analysis, and for Component 3 using titration. Again, only the spectral region from 1200 to 1650 nm (in which the absorbances ranged between 0 and 1 AU) was used in modeling. The best model was a three-factor PLS model using second-derivative spectra; it had a standard error of 0.98% for Component 1 and 0.032% for Component 3. Because of the strong sensitivity of NIR to hydroxyl functional groups, the spectral changes due to the small variations in the very low levels of Component 3 are actually larger than those due to the much larger variations in Component 1 and 2 levels. Thus the NIR model is more accurate in absolute terms for Component 3 than for Component 1. Figure 11.5 shows the ability of NIR to catch process cycling that occurs on a time scale too fast to be caught by grab sampling once or twice an hour (at best). In this case, extra grab samples were taken, presumably to verify that the cycling was a real variation in polymer composition and not an NIR temperature or density artifact. The process cycling was traced to excessive cycling of the temperature controllers in one zone of the extruder.

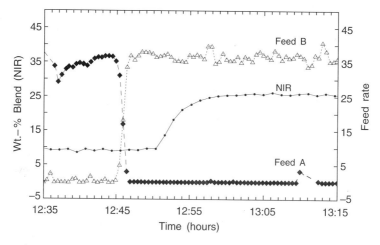

Figure 11.4 Plot of the NIR predicted blend levels for the 10–25 wt% transition region overlaid with feed rates for the two blends. Reprinted with permission from McPeters *et al.* (1992).[6]

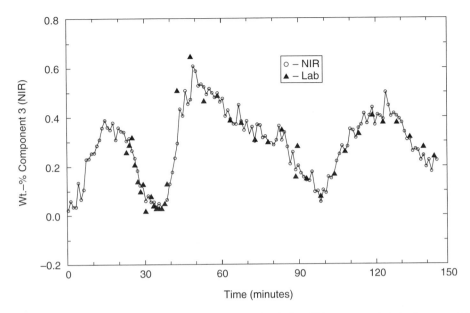

Figure 11.5 In-line NIR predicted levels of Component 3 (circles) and laboratory analysis for Component 3 in pellets taken from the extruder process (triangles). Reprinted with permission from McPeters *et al.* (1992).[6]

11.3.4 *Rapid identification of carpet face fiber*

Rodgers[7] at Solutia (Gonzalez, FL, USA) used off-line reflectance NIR to rapidly and accurately identify the face fiber on carpets returned for recycling. The **business value** of this was twofold. It allowed rapid sorting of incoming carpets, an essential aspect of any recycling program. However, there were already several commercial NIR-based instruments on the market for carpet sorting. The author developed an internal (to Solutia) system rather than using a commercially available system because of the limitations of many of the available systems from a fiber-producer's viewpoint, including the inability to include proprietary or unique samples in the calibration models, the inability to quickly update models to include new or modified products without returning the instrument to the vendor, the lack of qualitative information (to confirm identification, or to assess similarity between the evaluated sample and those used to develop the model), and the inability of many commercial systems to correctly identify carpets with widely varying color or carpet construction. Although it was not stated, the value of overcoming these limitations must have exceeded the cost of bringing method development, validation, and long-term maintenance in-house.

There were two classes of technical **challenges** faced in this work. The first challenge was how to prevent variability in carpet color (shade and darkness), carpet coloring method (dyes vs. pigments), and yarn and carpet construction (e.g. cut vs. loop pile) from interfering with accurate identification of the carpet face fiber. The second challenge was how to implement the NIR method at-line (using a fiber-optic probe that an operator can simply hold against the incoming carpet roll and minimizing the scan time) without affecting the identification accuracy. Both challenges were addressed by systematically exploring the effects of carpet and instrument factors.

The work was done using a large collection of carpet and fiber samples, containing nylon-6,6 (N66), nylon-6 (N6), polypropylene (PP), PET and acrylic (AC) polymer types. The samples in the collection varied widely in color, yarn and carpet construction, heatsetting type, and dyeing/coloring methods. NIR diffuse reflectance spectra of the samples were obtained using a Foss NIRSystems Model 6500, operating in the 1100–2500 nm range only, with either the sample transport attachment ('static' mode, for laboratory analysis of carpet or fiber pieces) or a fiber-bundle interactance probe ('remote' mode, for at-line analysis of intact carpets). In the latter case, the probe was held directly against the fiber or carpet sample during scanning. Four spectra of each sample were obtained, repacking or rotating the sample, or going to a different location on the sample, between spectra. Foss NIRSystems software, either the DOS-based NSAS/IQ2 (Identify, Qualify, Quantify) program or the Windows-based VISION^{TM} program, was used to create spectral libraries, develop calibrations, and test calibration performance. Wavelength correlation was used for identification of unknown samples, while wavelength distance was used for qualification. The results are briefly summarized here.

Polymer type: All of the polymers have very distinct, easily distinguished NIR spectra except N66 and N6 which are very similar. The use of second-derivative spectra is necessary to reliably distinguish the small differences between N66 and N6.

Color effects: The effects of color were found to be specific to the fiber type and to the coloring method (dyes vs. pigments). Color pigments caused large changes in absorbance across the entire spectrum, as shown in Figure 11.6. The use of second-derivative spectra removed the baseline shifts, but did not remove the variation in peak intensities. However, the positions of the characteristic absorption peaks for a given polymer were found not to vary significantly despite the color variations. This allowed accurate calibration models to be developed using second-derivative spectra as long as sufficient variation in carpet color was included in the calibration spectral library.

Construction effects: The effects of construction were less pronounced than the effects of color. In general, cut pile carpets had larger absorbances across the entire spectrum than did loop pile carpets, but the positions of the characteristic absorption peaks were not affected. As for color, accurate calibration models could be developed on second-derivative spectra as long as sufficient variation in carpet construction was included in the library.

Measurement speed effects: The effect of reducing the number of scans per spectrum, in order to increase measurement speed, was found to be minimal over the range from 20 scans down to 5 scans.

Sampling mode effects: The goal was to be able to measure at-line, with no sample preparation at all, using a fiber-optic probe or a remote sampling head. However, the area sampled by the fiber-optic probe is much smaller than for the sample transport

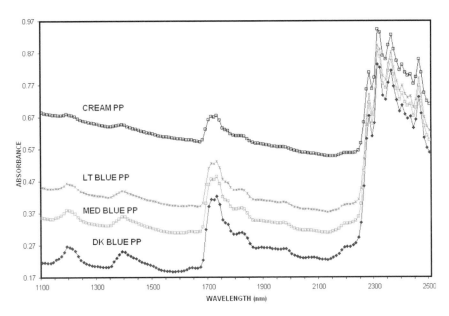

Figure 11.6 Influence of solid pigments on the NIR absorbance spectra of colored polypropylene carpets. Note that as the carpets change from dark blue (bottom trace) to cream color (top trace) the baseline shift increases, the baseline tilt increases, and the intensity of the absorbance peaks decreases. Reprinted with permission from Rodgers (2002).[7]

module. It was found that the 'remote' (probe) spectra were very similar to the 'static' (sample transport) spectra, but the baselines were shifted significantly higher and the absorbance peaks consequently reduced in intensity; as before, the characteristic peak positions were not affected. Calibration models developed using spectra obtained with the fiber-optic probe performed equivalently to those developed with the sample transport module.

The performance of the NIR analyzer in 'static' mode is shown in Table 11.3. The use of second-derivative spectra in developing the identification method, along with the inclusion of sufficient diversity in color and construction, allowed for 100% accurate identification of unknown samples.

11.3.5 Monitoring the composition of spinning solution

Sollinger and Voges[8] at Akzo Nobel (Obernburg, Germany) used off-line transmission NIR to monitor the key quality parameters in cellulose fiber (viscose) spinning solutions over time. The **business value** lay both in reducing the time and cost of analysis by consolidating four or

Table 11.3 NIR identification of polymer type

Sample type	Polymer type	Sources	Samples	% correct (absorbance model)	% correct (second-derivative model)
Fiber	N66	20	146	78	100
Fiber	N6	3	15	73	100
Carpet	PP	2	11	82	100
Carpet	N6	1	8	0	100
Carpet	N66	1	8	0	100

five different lab methods into one NIR method, and in providing more timely information to the process, enabling tighter control of product quality, higher yields, and reduced waste.

The spinning solution consists of cellulose solubilized as cellulose xanthogenate in an aqueous solution of sodium hydroxide (NaOH) and carbon disulfide (CS_2). The spinning solution is extruded through nozzles into an acidic bath, regenerating cellulose as filaments. The key quality parameters for the spinning solution are total cellulose, xanthogenate content or 'ripeness' (number of CS_2 residues per 100 glucose units of the cellulose molecule), NaOH concentration, and trithiocarbonic acid (TTC) concentration.

There were many **challenges** involved in this work.

(1) The spinning solutions are not stable over time, since CS_2 continues to react with the cellulose, increasing the xanthogenate content or 'ripeness' (which is why monitoring is needed in the first place). This was dealt with by carrying out the NIR and reference methods simultaneously, where possible, or by storing the samples at $-30°C$ (where they are stable) until the complementary analysis could be performed.

(2) There are inherent intercorrelations among many of the quality parameters due to the chemistry. This was minimized in two ways. First, the calibration sample set was created by systematic variation of the total cellulose content (6–11 wt%), the total NaOH content (4–7 wt%), and the total CS_2 content (16–32 wt%) using a central composite design, along with repeated measurement of these solutions after different ripening periods. However, this did not completely remove all intercorrelations (such as between TTC and Na_2CO_3 levels). The remaining intercorrelations were minimized by careful choice of spectral regions to include in the calibration models, to ensure that only the parameter of interest affected the model.

(3) The calibration models needed to be insensitive to variations in the levels of minor additives (<1 wt%) or process impurities. This was achieved by including samples from different production lines, containing varying concentrations (including zero) of different spinning additives, in the designed experiment sample set.

The spinning solution composition was measured off-line using a Foss NIRSystems Model 6500 VIS-NIR analyzer (wavelength range 400–2500 nm), equipped with a

Table 11.4 Multivariate calibration results for viscose spinning solutions

Parameter	Units	Concentration range	Wavelength range (nm)	Number of PLS1 factors	Regression coefficient	RMSEC	RMSEP
Cellulose	wt%	6.0–11.0	2200–2400	5	0.993	0.119	0.131
γ number	mg/100 g	25–54	1120–1250, 1600–1800, 2200–2350	5	0.983	2.0	1.5
TTC	mg/100 g	240–1350	510–590	1	0.998	17.5	22.7
NaOH	mg/100 g	2400–5500	1350–1500, 1800–2100, 2300–2450	1	0.979	131	109

thermostatted cuvette holder (30°C) and a cuvette of 1-mm pathlength. The analyzer was calibrated using the designed calibration set described above, containing 54 samples, using Camo's Unscrambler software. First-derivative spectra were used to eliminate effects due to baseline shifts. The resulting PLS1 models were validated using a separate set of samples (which also included production line samples). The results are summarized in Table 11.4 (RMSEC: root mean square error of calibration; RMSEP: root mean square error of prediction [validation]). The accuracy of the NIR predictions, expressed relative to the means of the respective calibration ranges, were between 2.3 and 3.3% for all parameters, which was considered reasonable.

11.3.6 Monitoring end groups and viscosity in polyester melts

Brearley *et al.*[9] at DuPont (Wilmington, DE, USA) used in-line transmission NIR to monitor carboxyl end groups and DP in PET oligomer and pre-polymer melt streams in a new polyester process. The **business value** was derived from several sources.

First and most important, real-time NIR monitoring enabled real-time control of the process. For a given product, the molecular weight and end-group balance in the pre-polymer exiting the 'front end' or melt part of the process must be controlled at specified levels in order for the 'back end' or solid-phase part of the process to successfully produce the intended polymer composition. In addition, the variability in pre-polymer composition must be controlled with very tight tolerances to keep the variation in final product composition within specification limits. Since the process dynamics in the front end were more rapid than those in conventional PET processes, the conventional analytical approach involving off-line analysis of samples obtained every 2–4 hours was not sufficient to achieve the desired product quality.

Second, real-time monitoring enabled particularly rapid development of process understanding, by providing otherwise-unattainable information on process dynamics and by drastically reducing the time needed to carry out designed experiments (since it was no longer necessary to remain at a given process 'state' for several hours until several lab results indicated that the process was lined out).

Finally, real-time NIR monitoring, once validated to the satisfaction of the process engineers and operators, significantly reduced the need for hot process sampling (with its attendant safety concerns) and lab analysis by allowing the sampling frequency to be greatly reduced (to near zero for some process points).

The PET melt composition was monitored using a Guided Wave Model 300P NIR analyzer, equipped with fiber-optic cable, transmission probes of 0.25-inch diameter, and DOS-based Scanner 300 software. The analyzer was located in an air-conditioned control room, with fiber optics running out to the process points. The transmission probes were inserted into the polymer transfer lines using custom-designed sapphire-windowed stainless-steel 'sleeves' to provide protection from process pressures. Custom software macros were written to scan the appropriate probes with the desired timing, apply the relevant models, and communicate the results to the plant DCS through a MODBUS interface. Calibration model development was done using Camo's Unscrambler software. There were a number of technical **challenges** involved in this work:

(1) An initial challenge was the desire to use NIR to control and optimize the process immediately upon plant startup, which required the analyzer to be calibrated ahead of time in the absence of any process samples. The approach used in this case was to implement 'provisional' models, which were simply the absorbance at a characteristic wavelength multiplied by 1000. The provisional models were expected to be neither accurate nor precise, but merely to track the level of the functional group of interest. The characteristic wavelengths were determined based on where the functional group of interest absorbed in off-line NIR spectra of PET flake samples. Peak assignments were based on the NIR literature on polyesters, extrapolation from vibrational fundamentals in the mid-IR literature on polyesters, spectra of model compounds, and observed variations in the spectra with known changes in sample composition. The provisional model wavelengths, after baseline-correction at 1268 nm, were 1590 nm for carboxyl ends, 1416 nm for hydroxyl ends, and 1902 nm for moisture. The provisional models were used for the first several weeks after plant startup, until 'real' models could be developed, and worked remarkably well. They tracked the process behavior and the lab results, revealed process oscillations due to imperfectly tuned control loops, and allowed designed experiments to be completed in minimal time.

(2) A second challenge was to rapidly develop 'real' models as soon as possible after plant startup, in spite of the relatively small variation in composition expected at a given process point. This challenge was solved by modeling all three process points (esterifier exit, pipeline reactor exit, and pre-polymerizer exit) together. Inclusion of the process variability due to the startup itself also helped. The calibration space for the first calibration set is shown in Figure 11.7.

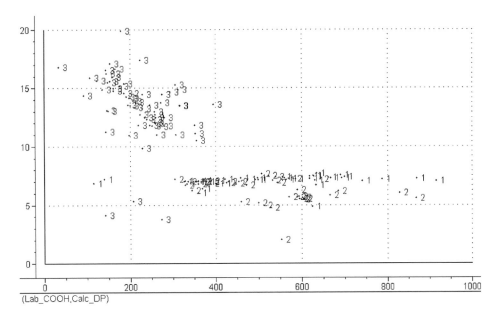

Figure 11.7 Calibration space covered by the first calibration set. The *x*-axis is lab carboxyl ends in meq/kg and the *y*-axis is DP in repeat units. The samples are labeled by process point: 1 – esterifier, 2 – pipeline reactor and 3 – pre-polymerizer. Reprinted with permission from Brearley and Hernandez (2000).[9]

(3) A third challenge was the imprecision of the reference lab values for carboxyl ends and for DP. This was due both to lab method variability (with an occasional outright mistake) and to sampling variability. Sampling variability was due to time assignment error (labeling the sample to the nearest quarter-hour introduces significant errors during transitions), sample inhomogeneity (NIR sees a large volume of sample; the lab measurements are conducted on a very small sample), and continued reaction and/or moisture pickup/loss by hot samples as they are transported to the lab. This was dealt with in two ways: by using large numbers of samples (hundreds) in the calibration sample sets in order to average out the effects of any given lab number and by careful iterative exclusion of outliers during modeling.

(4) A fourth challenge was fouling of the probes. The effect of fouling on the NIR spectra is shown in Figure 11.8. Low levels of fouling increase the baseline absorbance; this was easily remedied by baseline-correction of the spectra. At higher fouling levels, however, the baseline begins to tilt and curve as well. These effects are very difficult to correct for or model out. It was found through experimentation that the steps listed below allowed development of NIR models that predicted polymer composition accurately despite fouling levels up to 2.3 AU at 1274 nm.

Step 1 Include fouled spectra in the calibration data set.
Step 2 Use raw spectra. (Do not use baseline-corrected, derivatized or pretreated spectra.)

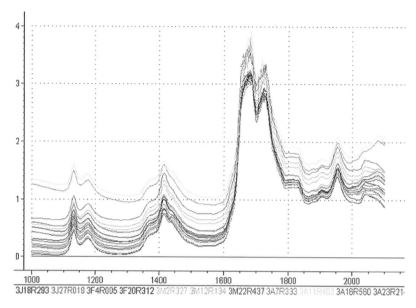

3J18R293 3J27R019 3F4R005 3F20R312 3M2R327 3M12R134 3M22R437 3A7R333 3A11R409 3A18R560 3A23R21

Figure 11.8   NIR transmission spectra of PET pre-polymer melt. The *x*-axis is wavelength in nanometers and the *y*-axis is absorbance. Reprinted with permission from Brearley and Hernandez (2000).[9]

Step 3 Exclude the short wavelength edge of the spectra (e.g. wavelengths less than 1300 nm) since they are most affected by tilt and curvature.

Step 4 Exclude the long wavelength edge of the spectra (e.g. wavelengths above 2000 nm) since they become very noisy at high fouling levels.

Calibration models were developed using process grab samples. Each sample set was split in half randomly to give independent calibration and validation sample sets. The results for the best models are shown in Table 11.5 (R_{val} is the correlation coefficient and SEP is

Table 11.5   In-line monitoring results for polyester process

Parameter	Model name	Range	Number of PLS factors	R_{val}	Accuracy (SEP)	Precision*
Esterifier: COOH ends	Ca4C3	100–700 meq/kg	3	0.9771	34 meq/kg	2.8 meq/kg
Prepolymerizer: COOH ends	Car5F5	80–460 meq/kg	5	0.9317	13.5 meq/kg	0.8 meq/kg
Prepolymerizer: DP	DP3B3	10–26 units	3	0.9935	0.54 units	0.03 units

Notes:
R_{val} – correlation coefficient
*Precision is expressed as the SD of the measurement over 30–60 minutes during a period when the process is lined out.

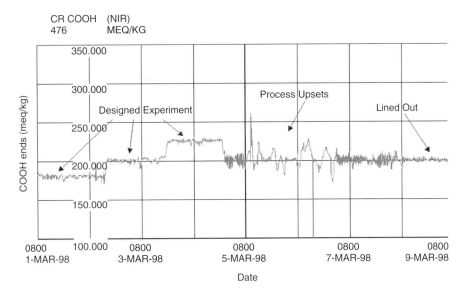

CR COOH (NIR)
476 MEQ/KG

Figure 11.9 Predictions of the NIR model for pre-polymer carboxyl ends over an eight-day period. Three states of a designed experiment, as well as a period of process upset and the return to lined-out operation, are indicated. Reprinted with permission from Brearley and Hernandez (2000).[9]

the standard error of prediction, both for the validation sample set.). The real-time monitoring results as seen by the process engineers and operators are shown in Figure 11.9.

References

1. Williams, Phil & Norris, Karl (eds); *Near-Infrared Technology in the Agricultural and Food Industries*, 2nd Edition; American Associates Cereal Chemists; St Paul MN, 2001.
2. Siesler, H.W., Ozaki, Y., Kawata, S. & Heise, H.M. (eds); *Near-Infrared Spectroscopy*; Wiley-VCH; Weinheim, 2002.
3. Burns, Donald A. & Ciurczak, Emil W. (eds); *Handbook of Near-Infrared Analysis*, 2nd Edition; Marcel Dekker; NY, 2001.
4. Chabot, Paul, Hedhli, Lotfi & Olmstead, Christyn; On-line NIR Monitoring of Emulsion Polymerization; *AT-Process* 2000, 5(1, 2), 1–6.
5. Chung, Hoeil, Ku, Min-Sik, Lee, Jaebum & Choo, Jaebum; Near-Infrared Spectroscopy for Monitoring the *p*-Diethylbenzene Separation Process; *Appl. Spectrosc.* 2000, 54(5), 715–720.
6. McPeters, H. Lee & Williams, Stewart O.; In-line Monitoring of Polymer Processes by Near-Infrared Spectroscopy; *Process Contr. Qual.* 1992, 3, 75–83.
7. Rodgers, James E.; Influences of Carpet and Instrumental Parameters on the Identification of Carpet Face Fiber by NIR; *AATCC Review* 2002, 2(6), 27–32.
8. Sollinger, S. & Voges, M.; Simultaneous Determination of the Main Constituents of Viscose Spinning Solutions by Visible Near Infrared Spectroscopy; *J. Near Infrared Spectrosc.* 1997, 5, 135–148.
9. Brearley, Ann M. & Hernandez, Randall T.; In-line Monitoring of Polymer Melt Composition; oral presentation at the Eastern Analytical Symposium; Somerset, NJ, 2 November 2000.

Chapter 12
Future Trends in Process Analytical Chemistry

Katherine A. Bakeev

12.1   Introduction

The use of process analytical chemistry (PAC) tools is expanding as the previous chapters and a growing body of literature have testified. The spread of PAC to the pharmaceutical industries is coming at a time when other industries have already implemented various technologies, and are advancing beyond the mere collection of data to predictive models of process behaviors based on data and information that have been gathered, analyzed and modeled over many years. The benefits of PAC have been discussed in all the preceding chapters. The terminology has seen some changes as well, with the historical PAC often being referred to as process analytical technology (PAT). When a process is understood, appropriate control strategies can be employed, which will allow a reduction in process variability to a point that can be improved only with more precise measurements. In order to reach the point of predictive control, relevant process measurements must be made and their values must be used to make decisions on a process. Full use of the information from various measurement devices can then be applied to modeling the process and controlling it with reduced variability. Introducing process analytical tools during product and process development can reduce the time to market by enhancing process understanding and generating knowledge critical to proper scale-up for manufacture.

Advancement in the use of process analytical tools prior to process scale-up affords opportunities to reduce product development cycles by having an earlier understanding of the parameters that influence a product's processability, as well as properties that impact the final quality. Implementation in the development phase of a product will reduce the dependence on empirical approaches to product formulation. Thorough understanding of quality attributes in a formulation or process allow one to tailor-make a product through its processing and formulation by relating properties that are important to the final product quality to the chemical and physico-chemical characteristics of the components.

Development of new process analytical tools – including sensors and data analysis – continues. Developments tend to grow out of necessity to solve a particular problem and arise from collaborative innovation between analytical chemists (in industry and academia) and industrial scientists and process engineers. Industrial scientists and engineers can drive developments because they have an understanding of their processes and their industry. Automation and control of manufacturing processes from material management to packaging, inventory control and product distribution are all part of the chain of processes that are impacted by PAC. Other areas where advances are being made in analysis drive some of the developments that are occurring in the field: in security, environmental monitoring, medical diagnostics and biotechnology.

The multidisciplinary nature of PAC requires that a team-based approach be taken to get the full benefits, as has been discussed in Chapters 2 and 11. Due to the many areas encompassed by PAC, information on this topic is published and presented in many different fields including instrumentation, analytical chemistry, chemical engineering, chemometrics, pharmaceutical science, control engineering and biotechnology, making it a large task for one to be fully versed in the area. A new American Society for Testing of Materials (ASTM) committee addressing the application of process analytical tools in the pharmaceutical industry was initiated in December 2003 and will help to bring together people from different disciplines within the pharmaceutical industry to define standards and test methods for PAT applicable to this industry. Working within ASTM allows the pharmaceutical industry to benefit from the knowledge of PAC that has been established in other industries.[1]

Several spectroscopic techniques and their applications have been discussed and reviewed in the previous chapters of this book. These tools have matured and evolved with time, and continue to undergo changes to meet the growing demands of industry. Many of the changes sought by users are not of a purely scientific nature, but pragmatic: Can it be smaller and faster? How does one get the data in the format needed? Are calibrations necessary? And does the instrument require a computer at all? The techniques covered may not be applicable to all the types of measurements that one may want to perform, though they do cover many areas where chemical composition and physical properties are important. Because the use of process analytical tools has the potential to generate large amounts of data, one must address the opportunity to interface analyzers to controllers, analyze and store the data, archive data, and communicate it to a central system. One can then use the data to gain the desired manufacturing improvements that justify the costs of the investment in the tools.

An insightful look into the future of PAT was presented by Gunnell and Dubois in 2002[2] and has served as a road map for some areas of development in terms of integrated, miniaturized systems. Their vision anticipates the need for more difficult measurements using existing technologies such as spectroscopy and chromatography, as well as new technologies such as nanotechnologies and lab-on-a-chip. They also envision computing developments including wireless connectivity, and the use of web browsers so that technicians and engineers can simultaneously access process information globally instantaneously. The use of modular sampling and sensing systems will reduce the cost of initial installation and cost of ownership over the lifetime of the process analytical tools. Another aspect of their vision is that of smart applications where key process parameters are monitored with an analyzer system that is capable of self-diagnostics and fault detection, and communicates the diagnostic information in a user-friendly way. The operator can then determine if there is a process fault or an analyzer problem. Ideally, rugged, low power-consumption, miniature devices will allow for the use of multiple sensors throughout a process. The data from these can then be combined and processed in such a manner that only a single read-out is needed for real-time monitoring of a process. The future vision of PAC given by Gunell and Dubois also entails cooperative effort by users and suppliers to ensure the continued reliability of the process analytical system throughout its lifetime.

The New Sampling/Sensor Initiative (NeSSI) consortium,[3] in existence since 1999, is now in its second generation of integrated system design and continues to pool resources from instrument manufacturers, end-users and academia to jointly define and design modular, multi-functional sampling systems and standardize key elements of the hardware that provides intelligent control of a system. The objective of NeSSI is to simplify and standardize sample system and design, for the purpose of reducing the cost to build process analyzers and lowering their cost of ownership. By designing modular systems, components, rather than analyzers, can be replaced. If components are matched exactly, the analyzer can continue operation without adjustments to calibrations following component replacement. The second generation of NeSSI is addressing questions of connectivity and communications with a push toward wireless communication and intelligent control. The advances in miniaturization and modularity will make it feasible to use multiple sensors on a single process, and throughout a plant to measure every stage of the manufacturing.

12.2   Sensor development

Many advances are being made in the area of sensors with a push toward miniaturization, micro-analytical systems and biosensors. These developments are often driven by the demand for lower-cost solutions that can provide rapid analysis on small samples. There is also a growing interest in the portability of systems, as analytical information is desired at various points in a process, and often throughout the cycle of product development. This desire for more information is leaning toward interest in multiple sensors, preferably with a single interface and single data output (perhaps a 'quality index' or quality attribute). Some of the advancements in instrumentation have come from work done for the telecommunications industry, as was the case with fiber optics that are now commonly used for process spectroscopy. There are several areas of sensor and analytical instrument research that are now reaching the point of providing commercial instrumentation worthy of being evaluated for process analytical applications. A cursory discussion on some of these areas is given below.

12.2.1   *Micro-analytical systems*

The terms 'micro-total analysis systems' and 'lab-on-a-chip' have come from the concept of having an integrated analysis system which performs complete analyses from sampling to data handling on one micro-device.[4,5] The fabrication of such devices was originally developed in the micro-electronics and telecommunications industry for micro-electrical mechanical systems (MEMS). Thus far, these devices have been used predominantly for analytical separations. Micro-total analysis systems have been applied in pharmaceutical and bioanalytical research for such applications as assays, clinical diagnostics and drug discovery. They have the advantage of being multi-functional and hence offer speed and response on a millisecond timescale. They are portable, require very small volumes of sample and reagent and may also be disposable. The on-chip detection techniques thus far have mostly utilized optical and fluorescent detection. Light (or laser)-induced fluorescence (LIF) has enabled detection at the single molecule level. Other detection techniques that have been used with lab-on-a-chip systems include electrochemical, mass spectrometry, chemiluminescence, refractive index and Raman spectroscopy.

The applications of micro-total analysis systems have been reviewed.[4–6] They cover diverse work done in the life sciences, and bioanalysis, as well as in drug enforcement, drug quality control, and detection of residual explosives. They have been used as micromixers in chemical synthesis, in DNA separation and sequencing, and in fabrication of micro-needles for drug delivery. The adaptation of microchips with various functions and detection systems provides benefits in parallel analysis at rapid speeds (milliseconds) with very small sample volumes. The integration of multiple sampling and detection devices on a single chip opens the possibility of rapid screening of many compounds as candidate drugs in development, as well as verification of authenticity of products.

12.2.2 Micro-instrumentation

Though the previously described 'lab-on-a-chip' definitely qualifies as micro-instrumentation, there are other systems in this arena that are not based predominantly on a total analysis system from sample handling to data analysis. The availability of miniaturized instrumentation increases the flexibility of having multiple process analytical devices without increasing cost for hardware and interfacing to a process. Miniaturization also has the added benefit of lower power consumption, allowing for battery operation in remote sensing applications. A micro-optical technology platform that was originally developed and proven in the telecommunications industry has been used for the development of spectroscopic instrumentation. A micro-spectrometer operating in the near-infrared (NIR) region has been designed[7] using a tunable Fabry-Perot filter interferometer, with a single-element Indium-Gallium-Arsenide (InGaAs) detector. The system can be tailored to the optical parameters needed for a specific application (i.e. bandwidth, spectral range). The micro-spectrometer is about the size of a credit card with free-space micro-optics used for all the optical couplings on a 14-mm aluminum nitride optical bench. The spectrometer is rugged, compact and reliable. These micro-spectrometers can be built to cover the entire NIR wavelength region of 1000–2500 nm, or a smaller portion of the spectrum that is applicable to specific measurements such as for water or a particular analyte that has absorbance in the defined 200-nm range of the detector. They have been developed to allow for very rapid data acquisition in reflectance or transmittance.

12.2.3 Biosensors

Biosensors are devices that use biological molecules to sense an analyte based on biological recognition. The molecules are typically enzymes, antibodies or oligonucleotides in whole cells, and are selective for particular compounds. They generate measurable signals that can be categorized as colorimetric, chemiluminescent, fluorescent, luminescent or electrochemical. The biosensors also have excellent sensitivity (μg/l to mg/l). The field of fiber-optic chemical sensors and biosensors has been reviewed[8] with 200 references, and shows that there is growing development in more specific types of biosensors such as immunosensors and gene sensors. Biosensors have been developed for analysis of glucose, lactate and cholesterol. Applications of biosensors in the areas of environmental monitoring and explosives detection in soil extracts are also being developed. A miniaturized biosensor for remote environmental monitoring based on integrated circuit technology has recently been developed, leading to the marriage of biosensors with the lab-on-a-chip.[9] As these sensors are already being used for remote monitoring, application to monitor biotechnology processes is a logical extension.

12.3   New types of PAC hardware tools

Development work is advancing technologies that can be used for rapid measurements of chemical and physical properties. New instruments suitable for in situ measurement are being developed in many different research areas. They may be relevant to the chemical and pharmaceutical industries, and if so then will surely be adopted by practitioners. Some techniques are currently being used in niche areas but may have broader application. The utility of a tool is often determined only when one has a problem to solve. The intent here is to introduce various tools, some with specific applications already defined. From this information one may find that the tool may be suited to other needs, or adaptable to them with some minor modifications. It is difficult to address all the tools that can be used, as the field is ever-growing. The reader is encouraged to take advantage of reviews published biannually in *Analytical Chemistry*, such as that on PAC.[10]

12.3.1   Thermal effusivity

Thermal effusivity is the measure of the rate at which a material can absorb heat. It is a function of the material (composition, moisture level), the physical properties of the material including particle size and morphology, and its interconnectivity or packing. Because of the various factors that affect the effusivity, variations in the measured thermal effusivity can be related to process changes in drying, granulation and blending of materials.[11] The non-destructive measurements are made with a sensor in direct contact with a material either in a laboratory or in a process setting. Effusivity is particularly effective for blending operations when the blended materials have a range of effusivity readings such that the differences between materials are detectable. One can then determine blend uniformity by the use of multiple sensors positioned at various points on a blender indicating blend homogeneity. A blending end-point can then be ascertained by the similarity of readings at multiple points. Conversely, with the use of a single sensor, consecutive measurements within a small deviation of each other are indicative of a uniform blend. As this method does not carry any chemical information, it does not provide data on the presence of contaminants, or direct information on the chemical composition of the materials. It can be used to provide a process signature and consistency of the process. Measurement of thermal effusivity of raw materials may also offer information on the variability of such material, though without any chemical specificity.

12.3.2 Acoustic spectroscopy

The use of acoustic measurements is not new, but is finding renewed interest as a process analytical tool due to its potential application to many kinds of measurements. Depending on the frequency range of the measurements, it may be known as acoustic or ultrasonic spectroscopy. Acoustic spectroscopy can be either passive or active. In passive acoustic spectroscopy the process is the source of the acoustic wave. This is often termed 'acoustic emission spectroscopy.' Acoustic emission may arise from particles impacting a reactor wall or pipe, crystal phase changes, or cavitations in flowing streams. It has been applied to the determination of end-point in high-shear granulation processes in concert with pattern recognition techniques.[12] A process signature has been determined and used to train the system to recognize the end-point in subsequent granulations.

In active acoustic spectroscopy (also called ultrasonic spectroscopy depending on the frequency of wave used) an acoustic wave is applied to a process, and its attenuation and velocity are measured following its interaction with the sample. Since one is monitoring the attenuation of the waves, this technique is also known as acoustic attenuation spectroscopy. In this technique, ultrasonic waves over a range of frequencies are propagated through a sample and the attenuation at each frequency is accurately measured.[13] Ultrasonic spectroscopy is often used for measurements on highly viscous or solid materials such as polymer melts. Because this method is based on the use of acoustics and not optical spectroscopy, it can be used for non-transparent samples including emulsions, slurries and semi-solids (i.e. fermentation broth). The ability to make measurements on concentrated solutions without the necessity to dilute samples enables accurate measurements to be made in situ on colloidal and crystallizing systems that may be perturbed by dilution. Acoustic attenuation spectroscopy allows the measurement of particle size and dispersed phase concentration in a system. It is capable of measuring particle sizes over the range of 0.01–1000 μm in solutions of 0.5–50% concentration by volume. Particle size measurement by acoustic attenuation spectroscopy requires deconvolution of the raw data by using mathematical models of the polymer-dispersant system. This technique provides a means of measuring the particle size distribution non-invasively in systems without the necessity of sample dilution, nor requiring sample opacity.

On-line particle sizing by ultrasonic (acoustic attenuation) spectroscopy was developed for use during batch crystallization processes.[14] Crystallization of the alpha polymorph of (L)-glutamic acid from aqueous solution was monitored by continuously pumping the crystallizing solution through an on-line ultrasonic spectrometer. The method enabled measurement of the crystal size distribution and solid concentration throughout the course of the crystallization process. Experimental results also provided real-time data on the nucleation, growth and breakage of crystals. From these measurements those kinetic parameters essential to process design could be estimated.

Ultrasonic spectroscopy has been utilized for real-time measurements of polymerization reactions and polymer melt extrusion.[15,16] In these applications the time required for the ultrasonic waves to propagate through the sample to a transducer was measured. The velocity of the sound wave in the medium is related to the modulus and density of the sample matrix.

Ultrasonic measurements have been made in polymer melt extrusions for process monitoring of high-density polyethylene/polypropylene (HDPE/PP) blend composition in an extruder.[15] Changes in the ultrasonic velocity, inferred from the transit time of the ultrasonic signal, arise from differences in bulk modulus and density of samples. The ultrasound technique was shown to be very sensitive to the change in polymer-blend composition in the melt, and could resolve a change of 1-wt% PP in HDPE. A disadvantage of ultrasonic measurements is that they do not have any molecular specificity, and therefore cannot be used to identify unequivocally the species causing observed changes.

The ring-opening metathesis polymerization of dicyclopentadiene was monitored by ultrasonic spectroscopy.[16] The thermoset poly(dicyclopentadiene) is formed by ring-opening and cross-linking in a reaction injection molding system. A reaction cell with a plastic window was constructed for use with pulse echo ultrasonic spectroscopy. Real-time measurements of density, longitudinal velocity, acoustic modulus and attenuation were monitored. Reaction kinetics were successfully determined and monitored using this technique.

12.3.3 Light-induced fluorescence (LIF)

Another non-invasive technique that can be applied to the characterization of processes is LIF. LIF is based on the excitation of molecules by a light source, and measurement of the light intensity emitted by these molecules. When a laser is used as the light source the technique may be called laser-induced fluorescence. Being based on molecular excitation, LIF requires that the molecule of interest be a fluorescent species, or that a tracer molecule (a fluorescent dye such as fluorescein or rhodamine B) be introduced into the system and detected. Due to the rapidity of the fluorescence emission, LIF provides real-time quantitation of a system limited only by the pulse rate of the laser (on the order of ms). Care must be taken in the use of LIF to avoid photo degradation. The production steps of mixing and blending are critical to such processes as reaction injection molding, crystallizations, precipitations and powder blending for pharmaceutical preparations. LIF has been applied to the quantitation of the mixing of miscible fluids[17] and to the monitoring of powder blend homogeneity.[18] Because many active pharmaceutical ingredients (APIs) fluoresce when excited at a particular wavelength, LIF can be used to monitor the blending of an API with excipients. As is the case with frequency-domain photon migration (FDPM, described in the next section) and thermal effusivity, blend homogeneity is defined as a measure of the residual standard deviation of consecutive measurements, indicating that there is no longer a change in the distribution of the powder blend.

12.3.4   *Frequency-domain photon migration*

One technique that is being developed for measurements in powders and colloids is FDPM. FDPM measurements are based on the time-dependent propagation character-istics of intensity-modulated light as it passes through a volume of material with multiple scattering. Specifically, with FDPM, intensity-modulated light is launched into a sample using multiple wavelengths of light in the visible or NIR wavelength regions. One then detects the average intensity attenuation, the amplitude attenuation and the phase delay at a detector point between the detected and the incident signals. The FDPM technique measures the time-dependent propagation characteristics of light and is therefore self-calibrating. With FDPM one can determine both scattering and absorp-tion of a material. The concentration of components can be determined from the absorption coefficient, while scattering measurements provide information on particle size. Blend homogeneity is determined from the residual standard deviation of consecutive FDPM measurements in a blend. Work has been done to predict the volume of powder that is probed by this technique to be 1.5–1.8 cm^3, and is thus expected to satisfy FDA regulations for sampling of powders in the pharmaceutical industry.[19] An advantage of FDPM is that it can be applied to measurements on concentrated, multiple-scattering suspensions without sample dilution. Because the absorption and scattering can be determined separately by FDPM, multivariate chemo-metric models are not necessary for the quantitation of particle size and material concen-tration, as may be necessary in reflectance spectroscopy where the physical and chemical information cannot be separated from each other.[19–22]

12.3.5   *Focused beam reflectance measurement (FBRM)*

The principle of FBRM is that a laser beam is focused into a flowing medium and the laser light is backscattered from the particles. The time period of the backscatter is then used to calculate a chord length, which gives information on the proximity of particles to each other, and correlates nicely to particle size distribution. Measurements are typically made by immersing a probe into a flowing stream where the laser beam scans through the sample media, and backscattered light is detected after the beam has traversed a particle. The measured chord length is then inverted with the aid of model-based calculations to determine particle size distribution. FBRM is sensitive to particle shape, size and number, and therefore allows for in situ measurements of the particle size and morphology of suspended particles during crystallization.[23] An advantage of FBRM is the elimination of sampling variance contribution to the measurement, and its ability to measure high solid content slurries. It has also proved to be an effective tool for in situ monitoring of polymorphic transitions.[24]

12.3.6 *Multidimensional analytical technology*

One approach that is being undertaken in the product and formulation development stage is the use of *n*-th order analyzers to aid in the R&D of a reaction process. Utilization of multiple analyzers at an early stage of development facilitates better process under-standing by collecting different types of data simultaneously in an integrated manner. It can decrease development time and help prevent processing problems during scale-up because materials and processes can be better defined. Taking the approach of using multiple analyzers in the characterization of new chemical entities can be an invaluable tool in making a smooth technology transfer from the R&D to formulation and manu-facturing. An example of this is the use of *n*-th order analyzers for the understanding of the conversion of vinyl sulfone to vinyl tributylstannane, resulting in the complete knowledge of material and energy balance with a single experiment.[25] The simultaneous application of reaction calorimetry, mass spectrometry and IR spectroscopy to a reaction is an efficient means to achieve better process understanding from the chemical and thermal information they provide.

There are many examples of second-order analyzers that are used in analytical chem-istry including many hyphenated spectroscopic tools such as FTIR-TGA, IR-microscopy, as well as GC-MS, or even two-dimensional spectroscopic techniques. Another hyphe-nated technique that is being developed for the study of solid-state transitions in crystal-line materials is dynamic vapor sorption coupled with NIR spectroscopy (DVS-NIR).[26] DVS is a water sorption balance by which the weight of a sample is carefully monitored during exposure to defined temperature and humidity. It can be used to study the stability of materials, and in this case has been used to induce solid-state transitions in anhydrous theophylline. By interfacing an NIR spectrometer with a fiber-optic probe to the DVS, the transitions of the theophylline can be monitored spectroscopically. The DVS-NIR has proven to be a useful tool in the study of the solid-state transitions of theophylline. It has been used to identify a transition that exists in the conversion of the anhydrous form to the hydrate during the course of water sorption.

A bioprocess system has been monitored using a multi-analyzer system with the multivariate data used to model the process.[27] The fed-batch *E. coli* bioprocess was monitored using an electronic nose, NIR, HPLC and quadrupole mass spectrometer in addition to the standard univariate probes such as a pH, temperature and dissolved oxygen electrode. The output of the various analyzers was used to develop a multivariate statistical process control (SPC) model for use on-line. The robustness and suitability of multivariate SPC were demonstrated with a tryptophan fermentation.

12.4   Data communication

12.4.1   *Data management*

Data overload is a reality when one begins to make real-time measurements. Add to this the fact that numerous sensors, each in a different format, are generating data and the complexity of data management is evident. The question becomes how to get the data translated into process knowledge. Before there is a clear understanding of the interrelationship of process parameters to final product quality, it is difficult to know which data are necessary. The tendency is to collect large volumes of data, and to glean from that the select few pieces of necessary data. When confronted with large amounts of data one must be assured that they are both reliable and relevant. It is from such data that valuable information can be extracted and knowledge gained.

Data relevant to a process may be discrete, intermittent or continuous. Those generated by a process analyzer are most often continuous, with measurements made at regular intervals throughout the process. Intermittent data are collected at less regular intervals, which should be more frequent than changes that occur in the process. When samples are taken to the lab for off-line measurement, discrete data are generated and these may not always be part of the electronic record of a batch. Indeed, data generated by a process analyzer may be treated as discrete when an output to the analyzer controller is recorded at a specific time on a batch record. Sometimes the data are archived and never fully analyzed. Other times the data are stored in so many different formats and areas within a plant that there is little chance of combining it in a timely manner to get a full picture of their interrelationship and importance in process control and product quality. The data may be univariate, such as a pH, temperature or pressure reading, or multivariate as provided by most spectroscopic tools. All of the data that are stored over long periods of time require durable storage media, and a means to read them in the future.

Whatever dimensionality the data have, they must be easily accessible and put to use. The data are imperative for product release, setting of product specifications that are based on true process variability, investigation of out-of-specification product, validation of the process, and trend analysis of production.[28] A cost-estimate of the inefficient use of data for decision-making in a typical pharmaceutical plant shows the impact of such behavior on the company's costs. For a single product the annual costs are estimated as:

Capacity underutilization	US$25–50M
Lost batches	US$24–48M
Delayed market entry	US$10–20M
Supply chain overburden	US$4–8M
Regulatory actions	>US$500M

With the greater availability of data about process capacity, utilization can be optimized, fewer batches will be lost due to quality issues, product can reach the market faster, and the supply chain will be less burdened due to more efficient material usage.

Often the raw data from a sensor are used to calculate another process parameter (such as measuring voltage and from that calculating temperature), or predict a process variable, or combined with data from other sources to give a broader and more-informed picture of a process. Having the right information at the right time is critical in decision-making which in turn is part of risk management. There are large potential savings in having the right information at the right place at the right time. Being able to make decisions rapidly can reduce the risk of failed materials, and increase the ability to adjust and optimize a process before it reaches completion. Data management systems that are suited to chemical and pharmaceutical manufacturing environments are evolving. These tools allow information to be made available across different areas of a company, tying R&D to production. Global availability of plant data between manufacturing and business systems aid in the decision-making and overall better control of a company's processes (from purchasing of raw materials to shipping of final product).

The first point to consider in data communication is how the data will be extracted from a device. This has implications for data acquisition, computation, communication to control systems and data archival. When each device collects and reports data in a separate format, even the work of developing chemometric models becomes more challenging. Much time can be lost in reformatting data from various sources into a format that is universal to the data analysis package that is being used. Technologies exist for the conversion of paper records into an electronic form that can be combined with data from multiple sources. Enterprise manufacturing intelligence systems can be used to convert data into a format that can be used to model manufacturing processes and can be viewed on terminals plant-wide. The assembled and filtered data can then be analyzed and aid in the release of product, investigations of out-of-specification material, and production trending. Having the data readily accessible in a useful format enables manufacturing to predict process outcomes based on available data.[28]

Before reaching the point of complete data integration as given above, there are intermediary levels of data integration that are beneficial to better analysis of data from process analyzers. The best case would be to have all the data in a human readable form that is independent of the application data format. Over the years several attempts have been made to have a universal format for spectroscopic data, including JCAMP-DX and extensible markup language (XML). Because many instrument vendors use proprietary databases, and there is not a universal standard, the problem of multiple data formats persists. This has led to an entire business of data integration by third parties who aid in the transfer of data from one source to another, such as between instruments and the plant's distributed control system (DCS).

In 2002, IUPAC initiated work in the development of terminology of a standard for analytical data. The standard format, XML, is intended to be universal for all types of analytical instrumentation, without permutations for different techniques. The XML format is designed to have information content of data defined in several layers. The most generic information is in the first layer, or core. More specific information about the instrumentation, sample details and experimental settings are stored in subsequent layers. The layers are defined as: core, sample, technique, vendor, enterprise and user.[29] The existence of a universal format will aid in the analysis of data from multiple sources, as well as in the archival and retrieval of data from historical processes.

12.4.2   Regulatory issues in data management

In March 1997, the US Food and Drug Administration (US FDA) issued a regulation on electronic record and signatures: 21 CFR Part 11.[30] The scope of 21 CFR Part 11 is to provide criteria for acceptance of electronic records, electronic signatures and hand-written signatures to electronic records. The concern is to promote the use of new technology, while maintaining the safety and security of data related to promotion of public health. The regulations are intended to ensure the integrity of electronic records, with clear audit trails in place to track any modifications that may have been made. As part of the 21 CFR Part 11, the FDA encourages the use of archival of data in a format that will allow an investigator to easily access records during an inspection. The use of common formats that can be searched, sorted or tended will not only aid the data owner in using data at later time, but can aid in easing the inspection process. Guidance on the thinking of the FDA on this regulation was published in August 2003 to aid in the interpretation of the regulation, and describes how the agency intends to enforce it while it is being re-examined.[31]

12.4.3 Digital communication

In addition to a common format for analytical data, there is also a need to have communication between devices and computers to provide real-time information in the manufacturing plant. Traditionally, analyzers provided a 4–20-mA analog signal to the process control system, which had a limited reporting range and was not instantaneous. Often this type of communication required a specific software protocol for the sending and receiving of data between the analyzer and the controller, with the software often being customized for each particular analyzer. In addition to the software requirements, there were also hardware requirements of cables and conduits linking analyzer, computer and controller. Again these items may be specific to each analyzer used in a plant. More modern digital communication protocols exist, and these are customized yet proprietary. They have reduced wiring requirements, which reduces installation cost. Standard communication protocols are needed as industry moves to broader use of process analyzers.

Object linking and embedding for process control (OPC) is an industry standard created through a collaboration of global automation, hardware and software suppliers for the purpose of moving data between different devices without the need for custom interfaces or data conversions.[32] By 1995, the amount of data being generated had grown to the point that it could not be relayed fast enough to be used for real-time control. Data are simultaneously being generated by multiple devices, and need to be available in a common format in a timely manner. All automation hardware and software that use OPC can communicate with one another, giving connectivity throughout a manufacturing plant. The use of OPC provides inter-operability so that information can be moved through an enterprise without the necessity of delays and use of multiple proprietary drivers. Some of the key features included within OPC are alarms, historical data access and security. OPC evolved from an industry need met by a consortium of providers who formed the OPC foundation. The technology is promoted by the OPC foundation[33] which has a charter to develop a standard for multi-vendor inter-operability in the manufacturing and process industries.

12.4.4 Wireless communication

Currently the connection between instrument and controllers is through some type of hardware connection. With the proliferation of wireless devices such as PDAs, the transition to wireless is also being made in analytical instrumentation and manufacturing environments. Much of the cost of implementing a process analytical system is in the wiring and communications tools. The migration to wireless communication can reduce cost, and also enable the direct interface of a sensor with a process, without the need to have the installation dictated by the necessity to have a computer or wiring in close proximity in what may be a controlled area. Wireless networking in a process control environment also provides a user-friendly interface, allowing operators to access analyzer information such as calibration and fault detection at remote locations, giving immediate access to critical information.

The US Department of Energy has a wireless technology initiative and at the time of writing has funded a project for the development of an open wireless network technology.[34] The project is expected to deliver an industrial strength wireless networking standard. This development, along with chip-based sensors, is expected to drastically reduce the cost of analyzer systems and to reduce energy consumption in manufacturing by up to US$1 billion annually. This initiative is expected to leverage technology available in the communications industry into the realm of process analysis and control. The development of a standard for wireless networking will facilitate a wider acceptance of this technology in process analysis applications.

The introduction of standardized data formats, communication tools and formats will lead to a more immediate availability of data. Again, if the data are transformed into knowledge to allow for control of a process, their availability greatly reduces manufacturing costs, and reduces risks of poor-quality material production. Data need to be available in a useful format in a relevant time – a time which enables action to be taken.

12.5 Data handling

In addition to having data in a common format transmitted to a central location, another aspect of process analysis is the application of statistics to the measurements made on a process in order to make data-driven decisions. This is termed 'SPC' and is central to measurement and control schemes. The objective of SPC is to produce quality product, rather than just measuring incoming raw materials and final products. The measurement and verification of raw materials is important in production of quality products but does not ensure that these materials will be processed in a manner that guarantees final product quality. Measuring final product is a verification of a process proceeding as expected, but if it has not, it is too late. Such a measurement has not avoided the production of off-spec product. SPC is an inseparable part of PAC, taking a data point beyond being just a number to being a significant piece of information that can be used to control a process. SPC covers the entire process from the manufacturing itself and how it runs to other contributing factors, including analyzer performance and maintenance, process control

James Hardiman Library
Self Issue Receipt

Customer name: Hennigan, Michelle
Customer ID: 04356292
Circulation system messages:
Patron in file

Title: Spectroscopy in process analysis /
ID: 31111401809239
Due: 27/08/2009 22:00
Circulation system messages:
Loan performed

Title: Process analytical technology :
ID: 31111401800212
Due: 27/08/2009 22:00
Circulation system messages:
Loan performed

Total items: 2
04/06/2009 12:10

Please take your receipt for details of return
date.

devices and the people who operate the systems. SPC includes a series of tools for reducing variability and continually improving processes by understanding the causes of variation. SPC is applicable to PAC, which is also used for reducing variability. Part of SPC may entail chemometric tools as discussed in Chapter 8 of this book, but it is more comprehensive. The topic of SPC is beyond the scope of this book, and the reader is referred to the text dedicated to SPC[35] and the application of it to process analyzer systems.[36] Here, some concepts in data handling for process analysis and control will be discussed. These concepts are now becoming recognized in industries that do not have an established history of closed-loop control as is seen in the petrochemical, food, and paper and pulp industries.

12.5.1 Chemometric tools

There are those who believe that advancements in chemometrics may have a greater impact on the future of sensor performance and process control, than will miniaturization of sensors.[37] Fast computing combined with chemometrics can provide diagnostic information about the sensor or the process. Chemometric tools such as multivariate statistical methods, as described in Chapter 8, or neural networks are used in process modeling, control and prediction. As the use of PAC increases, the application of chemometric tools will also increase. This will expand beyond the correlation of a measured value to a property as is done in developing calibration models. It will now entail the application of multivariate tools to the analysis of data from various sources. Even before calibration models may be developed (or perhaps in place of such calibrations), data are analyzed using chemometric tools to get an understanding of a process and what parameters are measurable, changing, relevant or controllable.

12.5.1.1 Multivariate data exploration

A key step in understanding the data collected on a process is to explore the data: looking closely at the information to determine how it can best be analyzed. One can make simple observations, such as patterns that may be time-dependent. One can detect outliers, or see interesting trends that may illustrate the relationship between various process parameters. Exploratory data analysis must be performed systematically as it can provide insights into the data and therefore the processes. The application of chemometrics to the data, including data pretreatments, can give diagnostic information on the process, and aid in understanding the correlations of large amounts of data that need to be compressed to a manageable size. A thorough and systematic investigation of data is important in determining the best way to use those data to make measurements. Before developing calibrations of any kind, segregating the data into categories that can be dealt with can aid in development of a fuller process understanding and more robust calibration models. Some of the multivariate tools used in exploratory data analysis are the same as those used in developing qualitative or quantitative calibrations, for example principal component analysis (PCA) and partial least squares (PLS) regression. Additionally, one may use quantile and box plots to extract features from the data.[38]

12.5.1.2 Feature extraction

Often, relationships between measured process parameters and desired product attributes are not directly measurable, but must rather be inferred from measurements that are made. This is the case with several spectroscopic measurements including that of octane number or polymer viscosity by NIR. When this is the case, these latent properties can be related to the spectroscopic measurement by using chemometric tools such as PLS and PCA. The property of interest can be inferred through a defined mathematical relation.[39] Latent variables allow a multidimensional data set to be reduced to a data set of fewer variables which describe the majority of the variance related to the property of interest. This data compression using the most relevant data also removes the irrelevant or 'noisy' data from the model used to measure properties. Latent variables are used to extract features from data, and can result in better accuracy of measurement and a reduced measurement time.[40]

12.5.1.3 Process trending and calibration-free measurements

The development of a multivariate calibration model may require significant effort, and is often limited due to the difficulty of obtaining samples with variability sufficient to model a system fully. Such calibration models also require attention to ensure their continued applicability, especially under changing process conditions. Changes that may impact calibration models include change in raw material supply, variation in process temperatures or other aspects of the process environment. In some cases progress of a process such as blending is monitored by using the residual standard deviation of successive measurements. This method has been used in the application of NIR spectroscopy,[41] thermal effusivity[11] and fluorescence.[17]

Sometimes just having a means to measure the progress of a process without having a quantitative result is sufficiently informative to improve process consistency. One can use a process signature, which indicates that a reaction is proceeding in the same manner as it was in previous cases, to show that quality product was produced. Using an in-line Raman spectrometer to follow a multi-product process, scientists at Kodak were able to improve process consistency without calibrating the Raman. The spectral data that were collected verified the complexity of the process, and enabled the use of existing sensors of temperature and flow to control the process.[42] Gemperline et al.[43] have used a fiber-optic NIR spectrometer for calibration-free monitoring and modeling of the reaction of acetic acid with butanol. In their approach, one must derive first-principle models of the reaction, which can then be applied to the spectral data to give calibration-free estimates of the reaction concentration profiles, as well as of the pure component spectra. There is not a need to collect calibration spectra of off-line reference values, but rather a reaction mechanism is postulated giving a system of ordinary differential equations that can be solved to estimate concentration profiles. These profiles are then fitted to the reaction spectra and can be used to determine reaction mechanisms, time-dependent concentration profiles, reaction yields and reaction end-points.

12.5.2 Soft sensors and predictive modeling

Soft sensors use available, readily measurable variables to determine product properties critical to prediction of product quality. Ideally the soft sensors are continuously monitored and controlled, or monitored on a relevant timescale. They need to make predictions quickly enough to be used for feedback control to keep process variability to a minimum.

Property prediction may be done using such routinely measured parameters as temperatures, pressures and flow rates when there is sufficient process knowledge to correlate these values with product quality. Sometimes process analyzers such as spectrometers are used to understand the process chemistry and kinetics, thus providing the ability to use soft sensors if the tool that helped elucidate the critical process variables is unavailable.

Model predictive control is concerned with continuous feedback of information with the object of reducing the variability of product quality by changing the set points (narrowing the range) in a plant control loop.[44] By using model predictive control, projections on batch quality can be made and mid-stream corrections made to keep a batch within the target limits of the process.

Due to the complexity of bioprocesses, and the lack of direct in-process measurements of critical process variables, much work is being done on development of soft sensors and model predictive control of such systems. Soft sensors are being used as an indirect measure of the biomass concentration in fed-batch fermentation.[45] The soft sensors can be integrated into automated control structures to control the biomass growth in the fermentation. PLS models are used to provide soft sensing. A PLS model with three latent variables was developed with the substrate feed rate being the manipulated variable. The input variables used to develop the PLS model were substrate feed rate, aeration rate, agitator power, substrate feed temperature, substrate concentration, dissolved oxygen concentration, culture volume, pH, fermentor temperature and generated heat. This PLS model can then be used in a predictive mode to control biomass growth during the fermentation. The objective is to be able to use the model of the system to predict the process outputs, and hence determine process adjustments needed to optimize the biomass growth.

The PLS model for fed-batch fermentation performs well when the fermentation is operating within conditions that are represented in the 20 training batches used in developing the model. As with many model-based systems, model performance is poor when extrapolating outside of the operating conditions of the training set.

In fermentations, as in other processes, early fault detection is needed to minimize the impact of such faults on the process and thus product quality. For fermentations, a drift in pH measurement that is part of the feedback control can be catastrophic. PLS scores can be used to detect and isolate fault sensors. Outlier conditions can be flagged, indicating that the process is operating outside of the conditions used to develop the

PLS model. In intelligent process control, the PLS model detects the fault in a measurement such as the pH. Then rather than relying on this measurement, the PLS model infers the missing variable. In this way, the performance of the controller is not grossly affected, as it would be if the faulty value were used.

Several statistics that can be used to monitor the performance of the controller have been defined.[44,45] Square prediction error (SPE) gives an indication of the quality of the PLS model. If the correlation of all variables remains the same, the SPE value should be low, and indicate that the model is operating within the limits for which it was developed. Hotelling's T^2 provides an indication of where the process is operating relative to the conditions used to develop the PLS model. Thus the use of a PLS model within a control system can provide information on the status of the control system.

An example of the use of soft sensors is given by the automation of a penicillin production dependent on strict adherence to certain limits in the fermentation process since such biological systems are sensitive to changes in operational conditions.[46] Many of the variables of the penicillin fermentation process are currently measured in the laboratory, and on-line measurement is made only of physico-chemical parameters including acidity, concentration of dissolved oxygen and carbon dioxide production rate. Control of the fermentation process is difficult with dependence on the measurement of these variables on-line, and slower off-line measurement of biomass concentration and penicillin concentration. The process has been controlled by a combination of these values, with cellular growth identified as the main control variable, and with the empirical knowledge of experts in the penicillin fermentation process. For development of soft sensors one must determine which variables can be used to predict other critical parameters. In this study, models were developed to monitor the important process variables of biomass, penicillin production and viscosity. Neural networks were trained to estimate values for the variables that cannot be directly measured on-line including biomass, penicillin and viscosity. The use of FasArt as the fuzzy logic system enabled the soft sensors to improve their knowledge during the training phase. The performance of these soft sensors was found to be accurate. The knowledge gained aids in understanding the fermentation process, and in the implementation of fault detection tools.

Soft sensors can be used for closed-loop control, but caution must be used to ensure that the soft-sensor model is applicable under all operating conditions. Presumably one would need to test any potential process condition to validate a soft-sensor model in the pharmaceutical industry, making their use in closed-loop control impracticable due to the lengthy validation requirements. An important issue in the use of soft sensors is what to do if one or more of the input variables are not available due, for example, to sensor failure or maintenance needs. Under such circumstances, one must rely on multivariate models to reconstruct or infer the missing sensor variable.[45] A discussion of validating soft sensors for closed-loop control is beyond the scope of this book.

12.6   New application areas of PAC

Process analytical chemistry is being expanded for use in areas beyond that of monitoring chemical reactions, monitoring and controlling driers, or other processes that may not be carefully monitored with on-line tools. PAC tools are being used not just in production, but also in the development of products and formulations to increase the understanding of the process, and to reduce uncertainty in technology transfer. PAC can be applied to any process that can be measured and controlled in a time frame that can improve final product quality by reducing variability. In cases where a process must currently be interrupted to make a measurement, such as blending, there is a great incentive to use PAC tools. In the heavily regulated pharmaceutical industry where so many final products are based on blended components, being able to make such measurements rapidly during the blending operation is critical. Much work is being done to determine the most efficacious way to interface instrumentation with a blender and make reliable measurements of blend homogeneity.

Bioprocesses can oftentimes be very lengthy, with some fermentations lasting several weeks. Detecting the progress of the bioprocess can lead to optimized processes if measurements can be made on-line in a timely manner. The batch-to-batch variability in bioprocesses is caused by changes in raw materials as well as the process environment. Because bioprocesses are very complex with many components in the process matrix, it is critical to define and control the key process parameters. Studies using multi-analysis systems have been performed[27] as have spectroscopic analyses of the composition.[47] Having a clear definition of critical parameters for optimal bioprocesses must begin in the product development phase. Application of PAC in such processes is increasing and will result in more optimized process performance.

Additional processes that can be monitored using spectroscopic tools of PAC are crystallization and distillation. Crystallization is an important step in manufacture of many products including APIs. Tracking the process and production of material is more valuable than testing a final product to verify that the correct crystal structure has been attained. The use of acoustic spectroscopy[14] and NIR spectroscopy[48] in industrial crystallization processes has been demonstrated and will be implemented more widely. Monitoring distillation processes, such as for solvent recovery, is another growing area of use of PAC.

An area that is undergoing much development and scrutiny currently in PAC is the validation of processes, analyzers and softwares used with them. An analyzer is often used in conjunction with a control system operating with specialized software. The analyzer itself has its own software for both data acquisition and processing. The analyzer is then interfaced with the controller. One can see that it is a large task to validate all components to achieve a closed-loop control system.

The field of PAC is continually changing, leading to innovative use of existing technologies and development of new technologies. As the field expands, it is important to have a well-trained staff to work in the process analytical applications. Since it involves such multidisciplinary skills, it is best to have a diverse team that includes engineers, chemists and statisticians. To date, most training in this field has come from experience in industry. Some large corporations with process analytical departments have served as training grounds for many current practitioners. There is a newly developed academic course in PAC at the University of Texas which will serve to support the training of process analytical chemists in the future.[49]

Whatever new tools or applications are developed for PAC, some of the same elements are needed to ensure success in PAC: clearly defined project goals, teamwork and training.

References

1. Watts, D.C.; Afnan, A.M. & Hussain, A.S., Process analytical technology and ASTM committee E55; *ASTM Standardization News* 2004, 32(5), 25–27.
2. Dubois, R. & Gunnell, J., A Review of Process Analytics in the Year 2012, presented at IFPAC, San Diego, CA, Jan. 23–25, 2002.
3. www.cpac.washington.edu/NeSSI/nessi.htm.
4. Huikko, K.; Kostiainen, R. & Kotiaho, T., Introduction to micro-analytical systems: Bioanalytical and pharmaceutical applications; *Eur. J. Pharm. Sci.* 2003, 20, 149–171.
5. Auroux, P.-A.; Iossifidis, D.; Reyes, D.R. & Manz, A., Micro total analysis systems, 2. Analytical standard operations and applications; *Anal. Chem.* 2002, 74, 2637–2652.
6. Verpoorte, E., Microfluidic chips for clinical and forensic analysis; *Electrophoresis* 2002, 23, 677–712.
7. Kotidis, P.; Atia, W.; Kuznetsov, M.; Fawcett, S.; Nislick, D.; Crocombe, R. & Flanders, D.C., Optical, tunable filter-based micro-instrumentation for industrial applications, ISA Volume 439, presented at ISA Expo 2003, Houston, TX, Oct. 2003.
8. Wolfbeis, O.S., Fiber-optic chemical sensors and biosensors; *Anal. Chem.* 2002, 74, 2663–2678.
9. Nivens, D.E.; McKnight, T.E.; Moser, S.A.; Osbourn, S.J.; Simpson, M.L. & Sayler, G.S., Bioluminescent bioreporter integrated circuits: Potentially small, rugged and inexpensive whole-cell biosensors for remote environmental monitoring; *J. Appl. Microbiol.* 2004, 96, 33–46.
10. Workman, J., Jr; Koch, M. & Veltkamp, D., Process analytical chemistry; *Anal. Chem.* 2003, 75, 2859–2876.
11. Mathis, N.E., Measurement of Thermal Conductivity Anisotropy in Polymer Materials, PhD thesis, Chemical Engineering Department, University of New Brunswick, Fredericton, N.B., Canada, 1996.
12. Belchamber, R., Acoustics: A process analytical tool; *Spectrosc. Eur.* 2003, 15(6), 26–27.
13. Alba, F.; Crawley, G.M.; Fatkin, J.; Higgs, D.M.J. & Kippax, P.G., Acoustic spectroscopy as a technique for the particle sizing of high concentration colloids, emulsions and suspensions; Colloids and Surfaces A; *Physiochem. Eng. Asp.* 1999, 15, 495–502.
14. Mougin, P.; Thomas, A.; Wilkinson, D.; White, G.; Roberts, K.J.; Herrmann, N.; Jack, R. & Tweedie, R., On-line monitoring of a crystallization process; *AIChE J.* 2003, 49(2), 373–378.
15. Coates, P.D.; Barnes, S.E.; Sibley, M.G.; Brown, E.C.; Edwards, H.G.M. & Scowen, I.J., In-process vibrational spectroscopy and ultrasound measurements in polymer melt extrusion; *Polymer* 2003, 44, 5937–5949.
16. Constable, G.S.; Lesser, A.J. & Coughlin, E.B., Ultrasonic spectroscopic evaluation of the ring-opening metathesis polymerization of dicyclopentadiene; *J. Polym. Sci.*, Part B: *Poly Phys.* 2003, 41, 1323–1333.
17. Unger, D.R. & Muzzio, F.J., Laser-induced fluorescence technique for the quantification of mixing in impinging jets; *AIChE J.* 1999, 45(12), 2477–2486.
18. Lai, C.-K.; Holt, D.; Leung, J.C.; Cooney, C.L.; Raju, G.K. & Hansen, P., Real time and noninvasive monitoring of dry powder blend homogeneity; *AIChE J.* 2001, 47(11), 2618–2622.
19. Pan, T. & Sevick-Muraca, E.M., Volume of pharmaceutical powders probed by frequency-domain photon migration measurements of multiply scattered light; *Anal. Chem.* 2002, 74, 4228–4234.
20. Pan, T.; Barber, D.; Coffin-Beach, D.; Sun, Z. & Sevick-Muraca, E.M., Measurement of low-dose active pharmaceutical ingredient in a pharmaceutical blend using frequency-domain photon migration; *J. Pharm. Sci.* 2004, 93, 635–645.
21. Sun, Z.; Huang, Y. & Sevick-Muraca, E.M., Precise analysis of frequency domain photon migration measurement for characterization of concentrated colloidal suspensions; *Rev. Sci. Instrum.* 2002, 73, 383–393.

22. Sun, Z.; Torrance, S.; McNeil-Watson, F.K. & Sevick-Muraca, E.M., Application of frequency domain photon migration to particle size analysis and monitoring of pharmaceutical powders; *Anal. Chem.* 2003, 75, 1720–1725.

23. Worlitschek, J. & Mazzotti, M., Model-based optimization of particle size distribution in batch-cooling crystallization of paracetamol; *Cryst. Growth Des.* 2004; ASAP Web Release Date: 17.2.2004; http://dx.doi.org; DOI: 10.1021/cg034179b.

24. O'Sullivan, B.; Barrett, P.; Hsiao, G.; Carr, A. & Glennon, B., In situ monitoring of polymorphic transitions; *Org. Process Res. Dev.* 2003, 7, 977–982.

25. McConnell, J.R.; Barton, K.P.; LaPack, M.A. & DesJardin, M.A., Streamlining process R&D using multidimensional analytical technology; *Org. Progress Res. Dev.* 2002, 6, 700–705.

26. Vora, K.L.; Buckton, G. & Clapham, D., The use of dynamic vapour sorption and near infrared spectroscopy (DVS-NIR) to study the crystal transitions of theophylline and the report of a new solid-state transition; *Eur. J. Pharm. Sci.* 2004, 22, 97–105.

27. Cimander, C. & Mandenius, C.-F., Online monitoring of a bioprocess based on a multi-analyser system and multivariate statistical process modelling; *J. Chem. Technol. Biotechnol.* 2002, 77, 1157–1168.

28. Neway, J.O., How new technologies can improve compliance in pharmaceutical and biotech manufacturing; presented at World Batch Forum North American Conference, Woodcliff Lake, NJ, Apr. 13–16, 2003.

29. Davies, T.; Lampen, P.; Fiege, M.; Richter, T. & Frohlich, T., AnIMLs in the spectroscopic laboratory?; *Spectrosc. Eur.* 2003, 15(5), 25–28.

30. 21 CFR Part 11 Electronic Records; Electronic Signatures; Final Rule; Federal Register 1997, 62, 13429–13466.

31. www.fda.gov/cder/guidance/5667fnl.pdf; Guidance for industry Part 11, Electronic Records; *Electronic Signatures-Scope and Application*, Aug. 2003.

32. Burke, T.J., Interoperability in automation; *InTech with Industrial Computing* 2003, 50, 43–45.

33. www.opcfoundation.org.

34. Iversen, W., Single-chip chromatography, and more; www.automationworld.com, Dec. 11, 2003.

35. Oakland, J.S., *Statistical Process Control* Third Edition; Butterworth-Heinemann; Oxford, 1996.

36. Nichols, G.D., SPC/SQC for Process Analyzer Systems. In Sherman, R.E. & Rhodes, L. (eds); *Analytical Instrumentation*; ISA; Research Triangle Park, NC; 1996; pp. 115–159.

37. Lavine, B. & Workman, J., Jr, Chemometrics; *Anal. Chem.* 2004, 76, 3365–3372.

38. Pearson, R.K., Exploring process data; *J. Proc. Cont.* 2001, 11, 179–194.

39. Brown, S.D., Chemical systems under indirect observation: Latent properties and chemometrics; *Appl. Spectrosc.* 1995, 49, 14A–31A.

40. Martinelli, E.; Falconi, C.; D'Amico, A. & Di Natale, C., Feature extraction of chemical sensors in phase space; *Sens. Actuators B* 2003, 95, 132–139.

41. Sekulic, S.S.; Ward, H.W. III; Brannegan, D.R.; Stanley, E.D.; Evans, C.L.; Sciavolino, S.T.; Hailey, P.A. & Aldridge, P.K., On-line monitoring of powder blend homogeneity by near-infrared spectroscopy; *Anal. Chem.* 1996, 68, 509–513.

42. Lippert, J.L. & Switalski, S.C., Uncalibrated in-line Raman data as a lower cost process-consistency tool for a multi-product process control; presented at IFPAC 2004, Arlington, VA, Jan. 12–15, 2004.

43. Gemperline, P.; Puxty, G.; Maeder, M.; Walker, D.; Tarczynski, F. & Bosserman, M., Calibration-free estimates of process upsets using in situ spectroscopic measurements and nonisothermal kinetic models: 4-(dimethylamino)pyridine-catalyzed esterification of butanol; *Anal. Chem.* 2004, 76, 2575–2582.

44. McAvoy, T., Intelligent 'control' applications in the process industries; *Ann. Rev. Control* 2002, 26, 75–86.

45. Zhang, H. & Lennox, B., Integrated condition monitoring and control of fed-batch fermentation processes; *J. Proc. Control* 2004, 14, 41–50.

46. Araúzo-Bravo, M.J.; Cano-Izquierdo, J.M.; Gómez-Sánchez, E.; López-Nieto, M.J.; Dimitraidis, Y.A. & López-Coronado, J., Automatization of a penicillin production process with soft sensors and an adaptive controller based on neuro fuzzy systems; *Control Eng. Pract.* 2004, 12, 1073–1090.

47. Macaloney, G.; Hall, J.W.; Rollins, M.J.; Draper, I.; Anderson, K.B.; Preston, J.; Thompson, B.G. & McNeil, B., The utility and performance of near-infrared spectroscopy in simultaneous monitoring of multiple components in a high cell density recombinant *Escherichia coli* production process; *Bioprocess Eng.* 1997, 17, 157–167.

48. Févotte, G.; Calas, J.; Puel, F. & Hoff, C., Applications of NIR spectroscopy to monitoring and analyzing the solid state during industrial crystallization processes; *Int. J. Pharm.* 2004, 273, 159–169.

49. Chauvel, J.P.; Henslee, W.W. & Melton, L.A., Teaching process analytical chemistry; *Anal. Chem.* 2002, 74, 381A–384A.

Index

Note: Page numbers in italics indicate figures.